Cryogenic Stability of Cement-Based Materials at Micro/Nano Scale

Zhengwu Jiang · Xinping Zhu

Cryogenic Stability of Cement-Based Materials at Micro/Nano Scale

Springer

Zhengwu Jiang
Tongji University
Shanghai, China

Xinping Zhu
Laboratoire Navier, École Nationale des Ponts et Chaussées
Université Gustave Eiffel, CNRS
Champs sur Marne, France

ISBN 978-981-96-1835-4 ISBN 978-981-96-1836-1 (eBook)
https://doi.org/10.1007/978-981-96-1836-1

This Springer imprint is published by the registered company Springer Nature Singapore Pte Ltd.
The registered company address is: 152 Beach Road, #21-01/04 Gateway East, Singapore 189721, Singapore

Preface

Materials subjected to ultra-low temperature often exhibit unique and unpredictable behaviors. Cryogenic concrete, designed to withstand prolonged exposure to temperatures as low as -165 °C and endures repeated ultra-low temperature freezing-thawing cycles under such extremes faces formidable challenges in maintaining structural integrity and durability. As global demand for liquefied natural gas (LNG) surges, driving the rapid expansion of LNG storage infrastructure, advancing the science of cryogenic concrete—its microstructure, performance, and tailored design—has become imperative. This research is not only critical for ensuring the safety and longevity of storage facilities but also for optimizing construction costs in an era of escalating energy needs.

For decades, the frost resistance of cement-based materials has been attributed primarily to the behavior of pore-confined freezing liquids. The dominant narrative, centered on ice crystallization, has overshadowed other critical factors—not least the intrinsic response of the cementitious matrix itself to cryogenic conditions. At the heart of this matrix lies calcium-silicate-hydrate (C-S-H), the primary binding phase that governs macroscopic performance. Subtle variations in the nanostructure of C-S-H can cascade into significant material degradation, yet its behavior under ultra-low temperatures remains poorly understood. Unraveling the cryogenic stability of C-S-H is essential for understanding the evolution of cement-based materials in such extreme environments. Moreover, insights into its micro- and nanostructural resilience offer a scientific foundation for engineering advanced cryogenic concrete, a goal of both theoretical and practical urgency.

This book delves into the micro- and nanostructures of cement-based materials in cryogenic environments across nine meticulously organized chapters. We embark on a journey that thoroughly reviews cryogenic concrete research in Chap. 1, where the application of cryogenic concrete, mechanic properties, thermal properties, and water/ice-related freezing-thawing damage theories are soundly reviewed. To present an overview of classical methodologies for characterizing cement-based materials, we outline the most-used experimental techniques for characterizing chemical composition, morphology, pore structure, nanostructure, and nanomechanics in Chap. 2. The nature of C-S-H (formation, atomistic structure, microstructures) is reviewed in Chap. 3. Starting from Chap. 4, we show how the microstructures, nanostructures, and nanomechanics of C-S-H evolve under cryogenic attack and how we can enhance the cryogenic stability of usual C-S-H by editing its molecular structure. The pore structure stability of C-S-H under cryogenic attack is described in Chap. 4. We extract fruitful pore structure information from gas sorption measurements. In Chap. 5, the cryogenic stability of C-S-H nanostructure and nanomechanics is comprehensively analyzed. The upscaling degradation pathway (from atomistic scale to mesoscale) of C-S-H under cryogenic attack is also treated. Subsequently, we introduce two molecular editing strategies to enhance the cryogenic stability of usual C-S-H in Chaps. 6 and 7. In Chap. 6, we demonstrate the possibility of aluminum incorporation in boosting the C-S-H cryogenic stability. In particular, the atomistic structure of calcium-aluminosilicate-hydrate is treated. In Chap. 7, we exhibit an innovative organic-inorganic assembly methodology for creating a C-S-H superstructure with remarkable cryogenic stability. In Chaps. 8 and 9, we discuss the mechanics and physics of C-S-H and ice at cryogenic temperatures. The elastic and failure properties of C-S-H are discussed in Chap. 8, where we show a universal scaling law to describe the nanomechanical properties of C-S-H. In Chap. 9, we present an atomistic view of ice crystallization and cryo-suction in porous media.

In fact, a key idea we attempt to convey in this book is that the cryogenic damage of cement-based materials arises not only from frost damage (ice crystallization within confined pores) but also from cryogenic attack (nanoscale degradation of cement hydrates). By reframing the problem to include both mechanisms, this book challenges conventional paradigms and opens pathways for material innovation. We embrace the axiom "less is more"—not only in our scientific approach but in our acknowledgment of this work's limitations. Scholarly progress thrives on dialogue, and we warmly invite readers to engage critically with the ideas presented here. Your feedback will enrich future explorations in this vital field.

Shanghai, China — Zhengwu Jiang
Paris, France — Xinping Zhu

Acknowledgements

We acknowledge the financial supports provided by the following sponsors.

- National Natural Science Foundation of China (U22B2076, 51878480, 52078369)
- The Fundamental Research Funds for the Central Universities.
- Program of Shanghai Academic Research Leader 22XD1403300.

We sincerely appreciate the assistance of the graduate students and collaborators who contributed to the research and the preparation of this book. Our gratitude extends to Qing Chen, Wenting Li, Zilong Deng, Xiongying Li, Nan Zhang, Chen Qian, Cong Zhang, Qiang Ren, Chen Li, Jun Li, Yi Zhang, Gan Zhou, Bin Li, Mingjun Xie, and Qiaomu Zheng from Tongji University, as well as Chao Zhang, Fan Yang, Li Xiao, Mingrui Zhao, and Bochao Zhang from CNOOC Gas and Power Group, and Matthieu Vandamme and Laurent Brochard from École Nationale des Ponts et Chaussées.

We would like to give special thanks to Bei He and Hongen Zhang from Tongji University, for their invaluable support and collaboration in ongoing research.

We extend our heartfelt thanks to Production Editors, Ms. Maggie Guo and Ms. Taeko Sato, for their significant contributions to the compilation of this book.

November 2024

Contents

About the Authors

Dr. Zhengwu Jiang is a distinguished professor at the School of Materials Science and Engineering, Tongji University, China. He is the director of the Key Laboratory of Advanced Civil Engineering Materials, Ministry of Education, China. He is also the editor-in-chief of *Journal of Building Materials*, the vice chairman of the Cement Committee and Solid Waste Committee (the Chinese Ceramic Society), the director of the Fundamental Theory and Application of Concrete Committee (the Architectural Society of China), ACI and RILEM members.

Dr. Jiang is engaged in academic research in the field of sustainable cement-based materials, properties of cement-based materials under extreme environments, self-healing cement-based materials, and high-performance concrete with manufactured sand. He is the author of more than 200 technical papers and 7 monographs and received more than 80 patents. He is the editor of more than 10 Chinese standards related to high-performance concrete with manufactured sand. Due to his contribution to the sustainable development of high-performance concrete in China, Dr. Jiang has been awarded more than 10 national, provincial, and ministerial awards, including the 2nd prize of the National Technology Invention Award and the 2nd prize of the Ministry of Education Technology Invention Award. His research outputs have been applied to more than 60 key projects, including the world's highest bridge, the Beipanjiang Bridge.

Dr. Xinping Zhu has been a CNRS researcher at Laboratoire Navier (École Nationale des Ponts et Chaussées (Institut Polytechnique de Paris), Université Gustave Eiffel, CNRS) since 2024. He specializes in multiscale modeling (molecular simulation, coarse-grained simulation) and multiscale experimental characterization of cement-based materials. His research topics include frost damage of porous media, chemistry and physics underlying the degredation of geomaterials induced by chemical attacks, gelation physics, and cement decarbonization. He received his bachelor's degree in Civil Engineering at China University of Geosciences (Wuhan, China) in 2017. He obtained his Ph.D. in Material Science and Engineering at Tongji University (Shanghai, China) in 2022, for studying the cryogenic stability of cement-based materials. During his Ph.D. study, he also worked on confined crystallization using molecular simulation at École Nationale des Ponts et Chaussées (Marne-la-Vallée, France). Before joining CNRS, he was a postdoctoral researcher at Laboratoire de Mécanique et Génie Civil (LMGC), Université de Montpellier (Montpellier, France).

Chapter 1
Cryogenic Concrete

1.1 Introduction

Cryogenic concrete is designed for use in ultra-low-temperature environments, typically below −100 °C. Such extreme conditions can be found in applications like liquefied natural gas (LNG) storage tanks, ultra-low temperature refrigerated warehouses, and lunar concrete. LNG storage tanks, in particular, must withstand prolonged exposure to temperatures as low as −165 °C, along with repeated cycles of freezing and thawing, which pose significant challenges to the durability and safety of concrete materials. As the global demand for LNG continues to rise with economic growth, the need for reliable cryogenic concrete solutions is becoming increasingly important. Ensuring a sustainable supply of bulk energy commodities is crucial for stable socio-economic operations. Therefore, establishing secure energy reserve stations is essential, highlighting the importance of research on LNG storage tanks. Currently, most LNG storage tanks follow similar construction methods: the inner wall material of the tank is made up of nickel steel and prestressed concrete, while the outer protective structure is made of cement concrete materials. Due to the high cost of the nickel steel used in the inner walls of LNG tanks, research into cryogenic concrete for LNG tanks began in the 1970s as a cost-effective alternative that could maintain long-term durability. By 2004, the idea of an all-concrete LNG tank was introduced. To ensure the safe operation of these tanks and further reduce construction costs, comprehensive research on the performance optimization and material design of cryogenic concrete remains essential.

Moreover, the emerging concept of lunar construction, gaining practical significance with the continuous progress of various spacefaring nations in the twenty-first century, has led to the development of lunar concrete. An essential characteristic of the lunar environment is its surface temperatures, which can drop as low as − 200 °C, imposing stringent requirements on the performance of lunar concrete. Exploring the microstructures, properties, and performances of concrete under

Z. Jiang and X. Zhu, *Cryogenic Stability of Cement-Based Materials at Micro/Nano Scale*, https://doi.org/10.1007/978-981-96-1836-1_1

extremely low-temperature conditions is highly needed to ensure the operational safety of structures in cryogenic engineering.

1.2 Application of Cryogenic Concrete

1.2.1 LNG Tank

The concept of an LNG concrete tank involves using specialized concrete materials and construction techniques to build storage tanks that can safely and effectively contain LNG at extremely low temperatures. These tanks are crucial components in the LNG supply chain, as LNG needs to be stored and transported at cryogenic temperatures (around − 165 °C). Figure 1.1 shows the construction of the LNG concrete tank. It is a full containment tank with a base slab, outer shell, insulation layer, inner shell, and outer concrete roof. The outer shell, often made of pre-stressed reinforced concrete, is constructed to provide mechanical protection and serve as secondary containment in case of a leak. Between the inner and outer shells, highly effective insulation, such as perlite or foam glass, is installed to prevent heat ingress and maintain the low internal temperatures essential for LNG storage. The inner shell, designed to directly hold the LNG, is constructed using cryogenic-resistant materials like 9% nickel steel, which can withstand extreme cold without becoming brittle.

Once the tanks are constructed, specialized cryogenic piping systems are installed to manage the inflow and outflow of LNG. These pipes are insulated and fitted with safety valves to prevent overpressure and facilitate emergency venting. The roof, which could be dome-shaped or suspended, is added to the structure to seal the

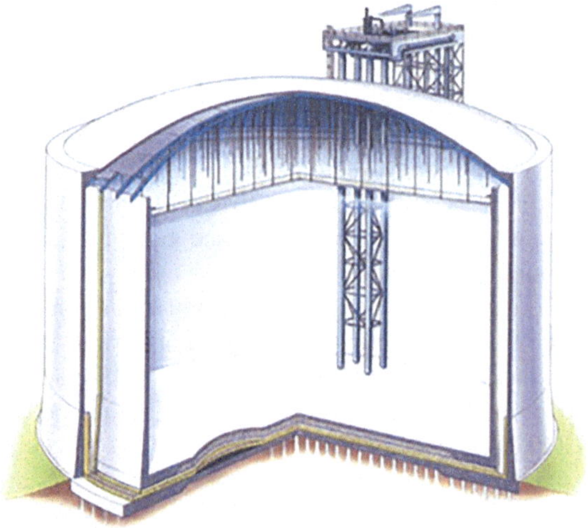

Fig. 1.1 Construction structure of LNG concrete tank [1]

tank and support further insulation. After construction, the tank undergoes a series of rigorous tests, including pressure and leak tests, to confirm its structural integrity and operational safety. A pre-chilling phase using nitrogen or small amounts of LNG is employed to gradually bring the tank down to cryogenic temperatures, preventing thermal shock that could damage the structure. Once cooled and verified, the tank is ready for initial LNG filling, equipped with continuous monitoring systems for temperature, pressure, and potential gas leaks to ensure long-term safe operation.

Given LNG's extremely low boiling point, the concrete in these tanks must retain structural integrity and functionality at cryogenic temperatures to prevent leakage or damage. Safety is paramount, as any failure to contain LNG could lead to hazardous situations, necessitating construction, and materials that meet stringent safety standards and ensure long-term serviceability. Cost-effectiveness in constructing LNG tanks is crucial, with efforts focused on designing durable and safe tanks at a reasonable cost, leveraging advancements in construction techniques and materials to enhance efficiency and reduce expenses. Its design and construction require a deep understanding of civil engineering principles and the properties of materials.

The conventional full-containment LNG tank is costly, mainly due to the expensive inner layer made of nickel steel. To lower the cost, an alternative is the all-concrete LNG tank, which breaks away from the conventional double-walled construction of LNG tanks and eliminates the inner metal lining. Instead, it directly utilizes prestressed concrete for the entire tank structure [1]. This construction method significantly reduces the construction cost of LNG tanks. However, under this structural form, concrete materials will directly contact liquefied natural gas, exposing the concrete to an extremely low-temperature environment. This poses significant challenges to concrete performance requirements [2]. Apart from the prolonged exposure to ultra-low temperatures, each cycle of gas charging and discharging in LNG tanks leads to drastic temperature fluctuations inside, which can be considered as undergoing severe cryogenic freezing–thawing cycles. Here, we refer to this type of ultra-low-temperature cycling as a "cryogenic attack".

1.2.2 Lunar Concrete

In space exploration and celestial habitation, humankind has set its sights on the next frontier: the Moon. China, the U.S. and other countries are engaged in ambitious lunar landing programs (see China's Yutu-2 rover in Fig. 1.2). As humans venture beyond Earth, establishing habitats and infrastructure on the lunar surface becomes imperative. Among the crucial facets of lunar construction, the development of lunar concrete emerges as a pivotal innovation holding promising potential for extraterrestrial colonization. The Moon, with its barren, desolate landscape and extreme environmental conditions, presents a challenging terrain for human settlement. The realization of sustainable lunar habitats involves meticulous planning and innovative construction techniques. Traditional materials and methodologies often fall short of meeting the stringent demands of lunar conditions.

Fig. 1.2 Chang'e-4 lander from Yutu-2 A week after landing, Yutu-2 snapped this picture of the Chang'e-4 lander before driving away to the northwest. Image: CNSA

With the further development of human exploration in space, the concept of establishing permanent bases on planets beyond Earth has become a priority. Lunar construction is considered a potential application for cryogenic concrete, and research into lunar concrete for construction has progressed [3–5]. However, the lunar surface temperature changes drastically between 120 °C and − 180 °C during the transition between lunar day and night [6, 7]. In such an environment, the impact of the cryogenic attack on the safe serviceability of lunar concrete is highly likely to be significant.

Lunar concrete is a pioneering technology, a revolutionary construction material poised to redefine extraterrestrial architecture. Unlike conventional concrete, lunar concrete harnesses locally available resources, primarily the regolith found on the Moon's surface, to form a composite material suitable for construction. Lunar concrete utilizes lunar regolith, a loose and powdery material abundant on the lunar surface, as its primary ingredient. Through sintering or heating, lunar regolith is transformed into a hardened substance analogous to concrete used on Earth. The resultant lunar concrete exhibits remarkable durability, structural stability, and thermal insulation properties, which are essential for sustaining lunar habitats. The utilization of lunar concrete offers several advantages. Its reliance on indigenous resources significantly reduces the need to transport construction materials from Earth, thereby cutting costs and energy consumption. Moreover, the material's adaptability to the lunar environment provides a pathway for establishing sustainable habitats.

However, the development of lunar concrete faces many challenges. Optimizing the manufacturing process, ensuring adequate mechanical strength, and addressing potential radiation exposure are among the hurdles that researchers and engineers continue to confront. As interest and investment in lunar exploration intensify, ongoing research and development efforts seek to further refine the properties of lunar concrete. Collaborative endeavors involving scientists, engineers, and space agencies worldwide aim to unlock the full potential of this innovative material for constructing lunar infrastructure.

Lunar concrete stands as a beacon of innovation, offering a promising avenue for humanity's venture into space. As advancements in materials science and construction technologies progress, the realization of sustainable lunar habitats through the application of lunar concrete draws closer, marking an exciting chapter in the story of human exploration beyond Earth.

1.3 Cryogenic Properties

1.3.1 Compressive Strength

The compressive strength of concrete at ultra-low temperature is obviously different from that at ambient temperature or low temperature, which mainly depends on the moisture content [8]. As the temperature decreases, the moisture in the internal pore of concrete gradually freezes, resulting in a significant increase in the additional resistance to external loads [9, 10]. According to different moisture content, scholars divided the concrete specimens into saturated concrete, partially dried or air-dried concrete, and dry concrete [11]. The results of different researchers on the compressive strength of concrete at ultra-low temperatures are summarized in Fig. 1.3 [12]. The compressive strength of saturated concrete at − 120 °C is 2–3 times higher than that of ambient temperature, while the compressive strength of dry concrete is relatively lower. Monfore et al. [13] found that the compressive strength of concrete would reach a maximum near − 125 °C, and then decreased with the further decrease of temperature. Tognon [14] and Miura et al. [15] thought that the strength of concrete would not increase when the temperature was cooled down to − 120 °C. In the range of 0 to −120 °C, the strength of concrete depends on temperature and moisture content. When the temperature is lower than − 120 °C, the strength is only related to the initial moisture content, not the temperature. For ultra-high-performance fiber-reinforced concrete, its compressive strength at cryogenic temperature is higher than that at ambient temperature [16]. Considering various factors, including temperature, moisture content, cooling rate, and strength grade, the different strength prediction models of concrete at ultra-low temperatures are proposed in Table 1.1. Generally, there are two methods to calculate the compressive strength of concrete at ultra-low temperatures. The first is the strength of concrete at 20 °C plus the increased value of compressive strength of concrete at cryogenic temperature [15]. The second is that the compressive strength of concrete at 20 °C multiplied by the relative increase coefficient of concrete strength at cryogenic temperature [17].

Concrete is a complex system; its water-to-cement ratio [18], relative humidity [19], pore distribution [20], the relationship between pore size and ice point [9, 21], and the properties of ice at cryogenic temperature [10] all play essential roles in the development of concrete strength at cryogenic temperature. Therefore, the accuracy

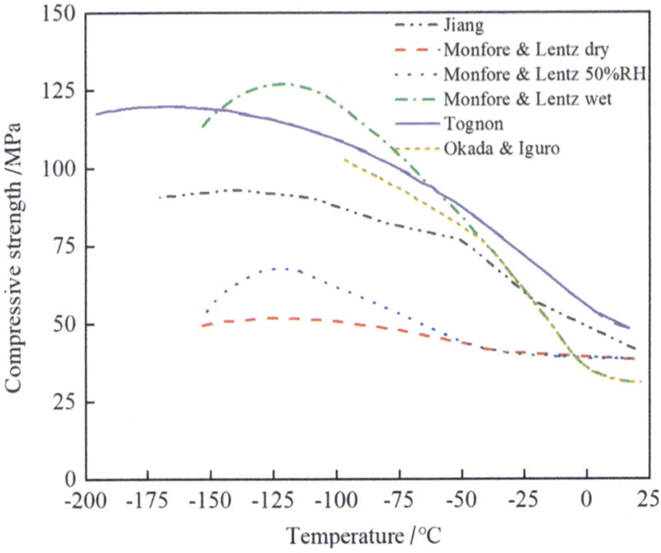

Fig. 1.3 Compressive strength of concrete with temperature [12, 25]

of the strength trend derived from the above formulas deserves further discussion. Previous studies [22] have found that the compressive strength of concrete is closely related to pore fluid at cryogenic temperature. According to Wiedemann's void model [19], the interaction between ice and pore wall, the potential energy flow of water in the pore, and the sudden change of strength and volume caused by the change of ice at cryogenic state all have effects on $\Delta\sigma_c$. Consequently, the incremental model of compressive strength of concrete at cryogenic temperature was proposed by Liu et al. [23], and the calculation formula is shown as follows:

$$\Delta\sigma_c = \frac{\int_{r_0}^{r_1} f_c^{ice}(P,T) \times V_{ice}(r, w/c)/H}{V_{concrete}/H} \tag{1.1}$$

where r_0 and r_1 are the minimum and maximum pore radii in concrete, $\int_c^{ice}(P,T)$ is the compressive strength of ice in voids with different temperatures and apertures, $V_{ice}(r, w/c)$ is the volume of pores with different diameters, $V_{concrete}$ is concrete volume, H is the height of the specimen, P is stress, T is temperature,

For the dynamic compressive strength of ice at cryogenic temperature, Zhang et al. [24] introduced stress rate into the calculation formula. The formula can quantitatively describe the relationship between compressive strength, strain rate, and temperature of ice.

$$f_c^{ice}(\mathcal{E}, T) = a\mathcal{E}^b|T^c| \tag{1.2}$$

Table 1.1 Different prediction model of the additional concrete strength

References	Calculation formula	Equation derived	Temperature range	Comments	Conditions	Research objects
Okada and Iguro [27]	$f_c = f_{co} + \Delta\sigma_c$	$\Delta\sigma_c = 54 - 8.6T - 0.028T^2$	$-10\ °C \geq T \geq -100\ °C$	Independent of the moisture content (w) and only depend on temperature (T)	Specimen was immersed in water	Saturated concrete
Goto and Miura [15, 28, 29]		$\Delta\sigma_c = \left\{ \begin{array}{l} \left[120 - \frac{(T+180)}{270}^2\right]w \\ 107w \end{array} \right.$	$\left\{ \begin{array}{l} 0 \geq T \geq -120°C \\ -120°C \geq T \geq -196°C \end{array} \right.$	Depend on both the temperature and the moisture content	Oven-dried 15 h, moisture content = 2.88%, w/c = 0.55	Concrete
Rostasy [30]		$\Delta\sigma_c = 12w\left[1 - \left(\frac{T+170}{170}\right)^2\right]$	$T \geq -170\,°C$	Depend on both the temperature and the moisture content	Oven-dried 15 h, moisture content = 2.88%, w/c = 0.55	Concrete
Browne and Bamforth [22]		$\Delta\sigma_c = \left\{ \begin{array}{l} -\frac{wT}{12} \\ -10w \end{array} \right.$	$\left\{ \begin{array}{l} 0 \geq T \geq -120°C \\ T - 120°C \end{array} \right.$	A function of temperature and the moisture content	Moisture content of 10%, the cooling rates = 2 °C/min	Non-air-entrained concrete
Chao Liu and Xudong Shi [23]		$\Delta\sigma_c = \left\{ \begin{array}{l} (0.008772T - 41.3330)w \\ (-0.1467T - 3.3565)w \\ (0.01572T + 12.8857)w \end{array} \right.$	$\left\{ \begin{array}{l} 20°C \geq T \geq -20°C \\ -20°CT \geq -100°C \\ -100°CT \geq -196°C \end{array} \right.$	The cooling rate is 1 °C/min; constant temperature for 48 h; loading rate is 0.25 MPa/s	Cubic concrete with the size of 100 × 100 × 100 mm	C60 Concrete
Zhengwu Jiang [25]		$\Delta\sigma_c = 14.0948 + 7.4721w - 0.2101w^2$	$T = -110\,°C$	The relationship between compressive strength and moisture content	The size is 40 mm × 40 mm × 160 mm	Mortar

(continued)

Table 1.1 (continued)

References	Calculation formula	Equation derived	Temperature range	Comments	Conditions	Research objects
Jian Xie [31]	$f_c = f_{co} \cdot \gamma_T$	$\gamma_T = -0.0065T + 1.1$	$20\,°C \geq T \geq -60\,°C$	The compressive strength of the specimens was measured at 20°C, 0, – 20°C, – 40°C and – 60 °C	The size is 150 mm × 150 mm × 150 mm	C50 concrete
Jian Xie [17]		$\gamma_T = \begin{cases} -0.004T + 1.08 \\ 1.56 \end{cases}$	$\begin{cases} 20°C \geq T \geq -120°C \\ -120°C T - 160°C \end{cases}.$	$\gamma_T > 1$characterizes the degree of improvement of the concrete strength at cryogenic temperatures	The size is 150 mm × 150 mm × 550 mm, the temperature of the specimen center fluctuates within 2 °C	C60 reinforced concrete
Jian Xie [32]		$\gamma_T = -0.0027T + 1.036$	$20\,°C \geq T \geq -160\,°C$	This empirical formula is obtained based on limited data. If it is used to guide design, its limitations need to be considered carefully	The size is 100 mm × 100 mm × 300 mm	C40 normal weight concrete

Notes f_c (MPa) = the compressive strength of concrete at cryogenic temperature; f_{co}(MPa) = compressive strength of concrete at 20 °C; $\Delta\sigma_c$(kg/m^2) =the increase value of compressive strength of concrete at cryogenic temperature; T (°C) = the concrete temperature; w (%) = the concrete moisture content; γ_T = the relative increase coefficient for concrete strength at cryogenic temperature T °C where ε is strain rate, a indicates a parameter associated with strain rate, b and c are comprehensive parameters.

1.3.2 Tensile Strength

Similar to compressive strength, the tensile strength of concrete at cryogenic temperature is also affected by moisture content and temperature. With the cooling of temperature, the tensile strength of concrete increases correspondingly (see Fig. 1.4). However, it is generally thought that a maximum value exists at a certain temperature. In contrast, there are different opinions about this critical temperature. Bamforth et al. [22] pointed out that the critical value is − 70 °C, while Hideo et al. [26] thought that the maximum tensile strength appears between − 30 and − 50 °C. The main reasons for the two conclusions may be the mixture ratio, curing time, water moisture, and porosity of the specimens, etc. For the increment of tensile strength, dry concrete is less affected by temperature, while the maximum value of saturated concrete is 5.2 MPa higher than that of normal temperature. In addition, for ultra-high-performance fiber-reinforced concrete (UHPFRC), its tensile properties (including strength and fracture energy) at cryogenic temperature are significantly improved by adding steel fiber [16]. Various calculations of the tensile strength of concrete at ultra-low temperatures from compressive strength are shown in Table 1.2. Given that the discreteness of the tensile strength of concrete at ultra-low temperature is larger, the predicted values of tensile strength obtained by the formula in Table 1.2 may have larger deviations.

1.3.3 Elastic Modulus

At cryogenic temperature, the increase of elastic modulus of concrete is closely related to moisture content, cooling rate, and aggregate content [34]. With the decrease in temperature, the elastic modulus of concrete increases first and then decreases, and the maximum value appears at around − 120 °C [8]. The elastic modulus increases linearly in the temperature range of 0 to − 120°C [27]. The higher the moisture content, the higher the growth rate. However, the growth rate of the elastic modulus is lower than that of the compressive strength. For the saturated concrete, the increase rate of the elastic modulus reaches 30, 50, and 70% at temperatures of − 30, − 70, and − 160 °C, respectively. The air-dried concrete is obviously different. The elastic modulus barely changes after − 30 °C [26]. At − 190 °C, the elastic modulus of saturated concrete is 1.75 times that of ambient temperature, and that of air-dry concrete is 1.65 times that of ambient temperature[20]. Marshall [12] calculated the data of several researchers and concluded that the ratio of the elastic modulus to the square root of the compressive strength could be regarded as a constant in the range of − 10 to − 100 °C, which can be estimated by the equation

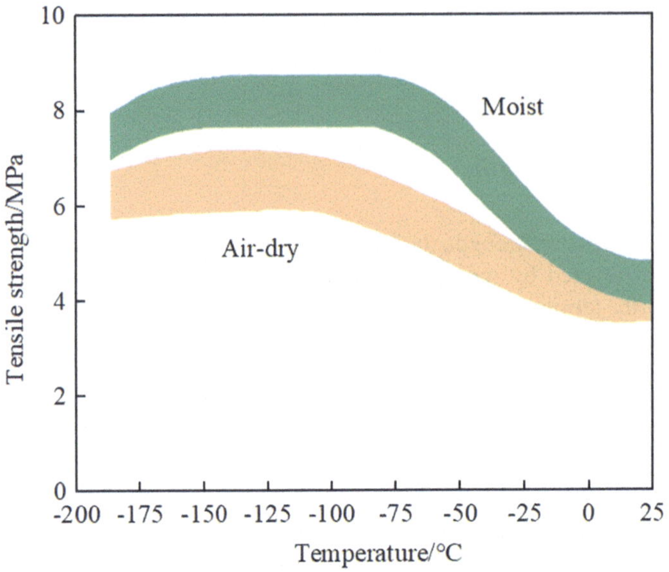

Fig. 1.4 Variation of tensile strength with temperature [20]

$E = 8\sqrt{\sigma_c}$. However, when the temperature is below − 100 °C, the ratio of the elastic modulus to the square root of the compressive strength fluctuates greatly.

Federation Internationale de la Prcontrainte (FIP) [35] proposed that the elastic modulus of concrete with relative humidity between 86% and 100% can be calculated by Eq. 1.3 when $T < 20$ °C:

$$E_c(T) = E_c(T = 20\ ^\circ\text{C}) \times \left(1 + 0.5 \times \frac{20\ ^\circ\text{C} - T}{185\ ^\circ\text{C}}\right) \quad (1.3)$$

Rostasy [30] also discussed the stress–strain relationship of concrete at cryogenic temperature and proposed Eq. 1.4 and Eq. 1.5 for calculating the peak strain of concrete at low temperature:

$$\varepsilon_{f_c}(T) = \varepsilon_{f_c}(T = 20\ ^\circ\text{C}) + \kappa \times \Delta\varepsilon_{f_c}^{max}(T = -60\ ^\circ\text{C}) \quad (1.4)$$

$$\kappa = \begin{cases} 1 - \left(\frac{T - 60\ ^\circ\text{C}}{60;^\circ\text{C}}\right)^2, T > -60\ ^\circ\text{C} \\ \frac{T - 170\ ^\circ\text{C}}{110\ ^\circ\text{C}}, -60\ ^\circ\text{C} > T > -170\ ^\circ\text{C} \end{cases} \quad (1.5)$$

where $\varepsilon_{f_c}(T = 20^\circ\text{C})$ is the strain corresponding to the peak stress of concrete at 20 °C, $\Delta\varepsilon_{f_c}{}^{max}(T = -60^\circ\text{C})$ is the maximum increment of strain corresponding to the peak stress of concrete at 60 °C. Besides, Rostasy gave $\varepsilon_{f_c}(T = 20^\circ\text{C}) = 0.2\%$, $\Delta\varepsilon_{f_c}{}^{max}(T = -60^\circ\text{C}) = 0.1\%$ for Portland cement concrete, and $\Delta\varepsilon_{f_c}{}^{max}(T =$

Table 1.2 Different prediction model of the tensile strength

References	Equation derived	Temperature range	Comments	Research objects
Tognon [27]	$\sigma_t = 0.06\ \sigma_c + 24$	$-10°\text{C} \geq T \geq -100\ °\text{C}$	–	Concrete
Rostasy et al. [30]	$\sigma_t = \begin{cases} 0.56\ \sigma_c^{0.67} \\ 0.30\sigma_c^{0.67} \end{cases}$	$20°\text{C} \geq T \geq -200\ °\text{C}$	–	{ Saturated concrete air -dried concrete
Lee et al. [28]	$\sigma_t = 5.6\ \sqrt{\sigma_c}$	$20°\text{C} \geq T \geq -70\ °\text{C}$	The regression fitting of splitting tensile strength and compressive strength of concrete at 20, – 10, – 30, – 50 and – 70 °C	C60 concrete
Miura [15]	$\sigma_t = 0.38\ \sigma_c^{0.75}$	$0 \geq T \geq -196\ °\text{C}$	Regression analysis of tensile strength and compressive strength of concrete at room temperature, – 50, – 100 °C and – 160 °C was carried out	Concrete
Jian Xie [31]	$\sigma_t = 5.5 - e^{3.85-0.06\ \sigma_c}$	$20 \geq T \geq -60\ °\text{C}$	Splitting tensile strength of cubic specimens was measured at 20, 0, – 20, – 40 and – 60 °C	C50 concrete

(continued)

Table 1.2 (continued)

References	Equation derived	Temperature range	Comments	Research objects
Jian Xie [33]	$\sigma_t = \left(1.45 - 1.02^{T-60}\right)\sigma_{to}$	$20 \geq T \geq -160$ °C	The tensile strengths of the prestressed concrete were measured at 20, − 40, − 70 and − 100 °C	Prestressed concrete
Zhengwu Jiang [25]	$\begin{cases} \sigma_t = \sigma_{to} + \Delta\sigma_t \\ \Delta\sigma_t = 3.6676 + 1.6519w - 0.0796w^2 \end{cases}$	$T = -110$ °C	The relationship between compressive strength and moisture content	Mortar

σ_t (MPa) = tensile strength of concrete at cryogenic temperature; σ_c (MPa) = the compressive strength of concrete at cryogenic temperature; σ_{to} (MPa) = the tensile strength of concrete at ambient temperature; $\Delta\sigma_t$ (kg/m^2) = the increment of tensile strength of concrete at cryogenic temperature; w (%) = the concrete moisture content

$-60°C) = 0.15\%$ for Portland slag cement concrete. Meanwhile, Rostasy developed a formula for calculating the stress–strain relationship of concrete at cryogenic temperatures in Eq. 1.6 and Eq. 1.7:

$$\sigma_c(\mathrm{T}) = f_c(T, w) \times \left[1 - \left(\frac{\varepsilon_c(T)}{\varepsilon_{f_c(T)}}\right)^n\right] \tag{1.6}$$

$$\mathrm{n} = \begin{cases} 2, T = 20\,°\mathrm{C} \\ 1 + \frac{T+170\,°\mathrm{C}}{170\,°\mathrm{C}}, \quad 20\,°\mathrm{C} > T > -170\,°\mathrm{C} \\ 1, T = -170\,°\mathrm{C} \end{cases} \tag{1.7}$$

where $\sigma_c(T)$ and $\varepsilon_c(T)$ are the stress and strain values of the designated points on the stress–strain curve of concrete, $\varepsilon_{f_c(T)}$ is the strain value corresponding to the peak stress of concrete, which is the peak strain.

For normal concrete, Xie et al. [32] proposed a computational model to predict its elastic modulus at cryogenic temperature, as shown in Eq. 1.8. However, the empirical formulas are obtained based on limited data. If they are used in concrete structure design, their limitations need to be further considered.

$$E_c(T) = (0.0012T + 0.88) \times E_c(T = 20\,°\mathrm{C}), 20\,°\mathrm{C} > T > -160\,°\mathrm{C} \tag{1.8}$$

1.3.4 Bond Strength of Reinforced Concrete

As one of the most widely used building structures, reinforced concrete structure has many advantages in practical engineering [36]. In recent years, the demand for reinforcing bars to be used in LNG tanks and cryogenic engineering structures has been increasing, requiring steel to still work in extremely cryogenic environments [37]. The bond strength between steel bar and concrete is greatly affected by the performance of steel bar. It was found that the yield strength and tensile strength of reinforcing bars increase with the decrease in temperature, but the ductility of the steel decreases [38]. When the temperature drops to − 196 °C, the yield strength and tensile strength are 1.8 times and 1.36 times that at ambient temperature, respectively. Once there is a defect on the surface of the steel bar, the lower the temperature, the more significant the reduction of the ultimate strain, while the elastic modulus has not changed significantly. Meanwhile, the deterioration mode changes from ductile mode to brittle mode [37] (see Fig. 1.5). These laws have been recognized by many scholars [20, 39].

At cryogenic temperature, the development of the bond strength between steel and concrete is similar to the compressive strength, but its growth rate is greater than that of compressive strength [20]. For reinforced concrete, due to the different thermal expansion coefficients of concrete and steel, their interaction forces are accordingly

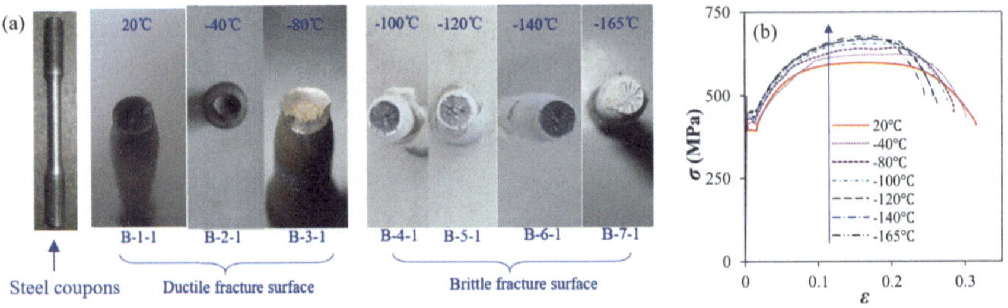

Fig. 1.5 **a** Fracture surface of the steel coupons at different cryogenic temperatures, **b** Stress–strain behavior of reinforcing steel HRB400 at cryogenic temperatures [37]

different, especially for saturated concrete [40]. Compared with the bond strength at ambient temperature, the strength of saturated concrete increased by 410% and 580% at − 70 °C and − 165 °C, respectively. The corresponding strengths of air-dry concrete were 110% and 140%, respectively [15]. In addition, the bond strengths of the two types of concrete below 0 °C increase first and then decrease with the decrease of temperature and increase with the rise of humidity [41]. At ambient temperature, humidity does not affect their bond strengths [42]. Therefore, it is reflected that the bond strength is also mainly related to cryogenic temperature and moisture content.

1.3.5 Freezing–Thawing Resistance

Under freezing–thawing conditions, the mechanical properties of cement-based materials deteriorate much faster than those of ordinary freezing–thawing conditions. The tensile and flexural strengths of cement-based materials are more sensitive to freezing–thawing cycles at cryogenic temperatures, resulting in faster losses [43, 44]. For saturated concrete, the compressive strength decreases by 50% after only three cycles when it is subjected to a freezing–thawing cycle of 20 to − 196 °C [45]. Moreover, when the relative humidity is 85%, the impact of freezing–thawing on concrete stiffness, flexural strength, and compressive strength is small [14]. When the relative humidity is higher than 85%, the influence of concrete strength increases obviously [43]. With the increase in freezing–thawing cycles and the decrease in temperature, the residual strain of concrete shows a growing trend, while the influence of temperature is not significant when the temperature is below − 50 °C [15]. The compressive strength and Young's modulus of concrete decrease by about 5% after 10 and 30 freezing–thawing cycles [28].

Many factors affect the freezing–thawing damage of cement-based materials, including the temperature difference, cooling rate, and water saturation degree [46]. Larger temperature differences in the freezing–thawing cycles are prone to faster freezing–thawing deterioration. It is noted that the degradation rate of mechanical properties of cement-based materials at freezing–thawing cycles from 20 to − 170

°C is comparable to that from 20 to − 70 °C. This is because the volume of cement-based materials after cooling down − 70 °C is not in temperature. The decrease and the expansion. This is because the volume of the cement-based material no longer expands with a decrease in temperature below − 70 °C [43]. If the cooling rate is too fast, different concrete parts will produce severe shrinkage or expansion, accelerating the destruction of materials [47]. Rostary et al. [43] frozen concrete by directly immersing it in liquid nitrogen. After a freezing–thawing cycle, the compressive and tensile strength decreased by 30 and 50%. The degradation rate was much faster than that due to freezing–thawing cycles at 20 to − 170 °C with a cooling rate of 1 °C/min. The more saturated the concrete is, the more severe the freezing–thawing damage will be. After freezing and thawing cycles, only saturated or nearly saturated concrete will cause internal damage. In the freezing–thawing cycles, the mortar gradually absorbs water from the outside. After 21 cycles, its water absorption is even greater than that in the vacuum water saturator [44] (see Fig. 1.6). Moreover, the freezing–thawing resistance of cement-based materials can be enhanced by adding fine aggregate, reducing water-cement ratio and moisture content [48].

For freezing–thawing damage, the above researchers have given qualitative explanations from a macro perspective, but there is no systematic theory. It is concluded that there are two main reasons for the cryogenic temperature damage of cement-based materials. Firstly, the volume expansion is caused by the freezing of pore water, which leads to pore damage. After freezing–thawing cycles, the proportion of macropores in hardened cement matrix increases significantly [43]. By comparing the residual strain and relative dynamic modulus of elasticity of concrete, it is considered that the deterioration mainly occurs in the stage of volume expansion of concrete at − 20 to − 50 °C [15]. Secondly, due to the different thermal expansion coefficients of each component inside the concrete, the interface of each component will produce greater internal stress due to high-temperature variation, which will lead to

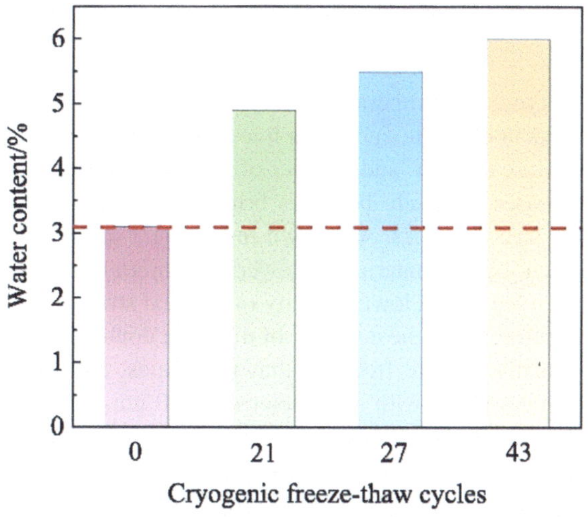

Fig. 1.6 Water content of mortar after cryogenic freezing–thawing cycles [44]

damage [5]. However, there are few studies or discussions on the microchanges and deterioration mechanism of materials during the freezing–thawing process.

1.3.6 Permeabilities

The impermeability of cryogenic concrete is the primary consideration in the design of an all-concrete LNG tank. Due to the direct contact between concrete and liquefied natural gas, the impermeability of concrete is particularly important for the effectiveness of the storage tank structure. However, few studies have focused on the permeability of cryogenic concrete, and only a few results have been different [5]. Jackson et al.[49] The impermeability of concrete should not be taken into account in the mix proportion design of cryogenic concrete. Instead, aggregates should have a thermal expansion coefficient similar to cement paste, and the water-to-cement ratio should be less than 0.45, while silica fume was used to reduce permeability. ACI Standard 376–10 also supports the addition of mineral admixtures to reduce permeability. Meanwhile, influencing factors of permeability of cryogenic concrete are fundamentally different from those at ambient temperature [50]. Ice can block some of the pores at cryogenic temperature and reduce permeability. Moreover, aggregate type is the most important factor affecting the permeability of cryogenic concrete [51]. Due to the difference in thermal expansion coefficients between aggregate and cement paste, a large number of micro-cracks are produced between aggregate and cement paste with the decrease of temperature, which provides a new channel for moisture passage. Therefore, aggregate for cryogenic concrete should be strictly selected.

1.3.7 Cracking Behaviors

At cryogenic temperature, many factors affect the formation and development of cracks in concrete, including fine aggregate, specimen shape, water-to-cement ratio, moisture content, and number of freezing–thawing cycles [48]. As the temperature decreases, concrete becomes brittle, and cracks and failure occur under the load [32] (see Fig. 1.7). As shown in Figs. 1.8 and 1.9, when the moisture content of cement-based materials is lower, the internal hydration reaction is not sufficient, which leads to a lower density of internal structure in the hardened cement matrix. Therefore, with the decrease of moisture content of cement paste, porosity increases gradually. Before freezing–thawing cycles, the internal pore in paste is basically harmless pores with a diameter of 2–10 nm. After freezing–thawing cycles, more and wider pores or cracks appear in the matrix, and the internal micro-crack width is up to 0.282 nm [48]. As shown in Fig. 1.10, the UHPFRC has a higher resistance to microcrack formation than ordinary concrete. Moreover, pre-cracked UHPFRC has excellent strength, deformation capacity, and toughness due to the formation of calcium carbonate around the cracks and the filling of microcracks at cryogenic temperature [52].

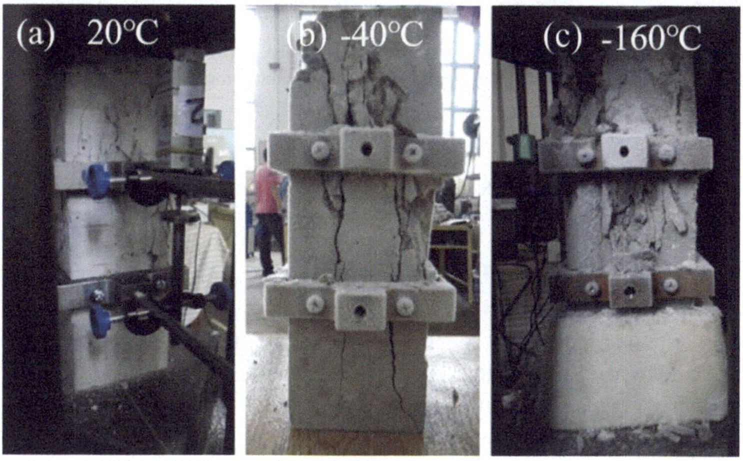

Fig. 1.7 Deterioration modes of the concrete prisms under cryogenic temperature [32].

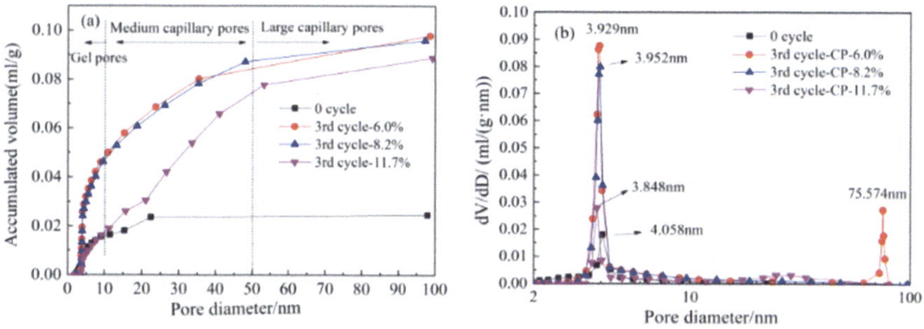

Fig. 1.8 Internal pores in pastes with different water contents: **a** Accumulated volume; **b** Pore size distribution [48]

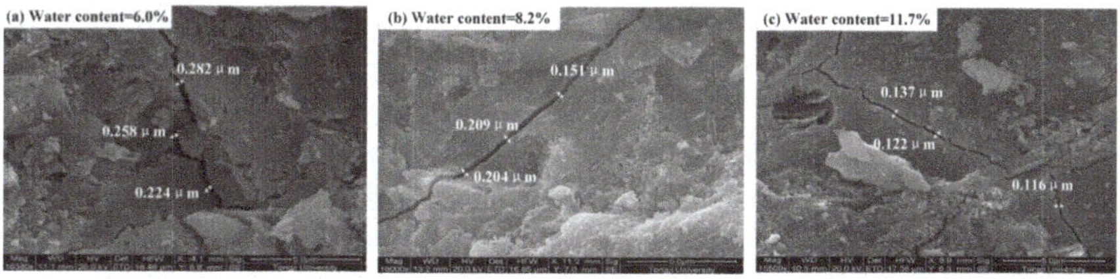

Fig. 1.9 Crack propagation of paste after freezing–thawing cycles [48]

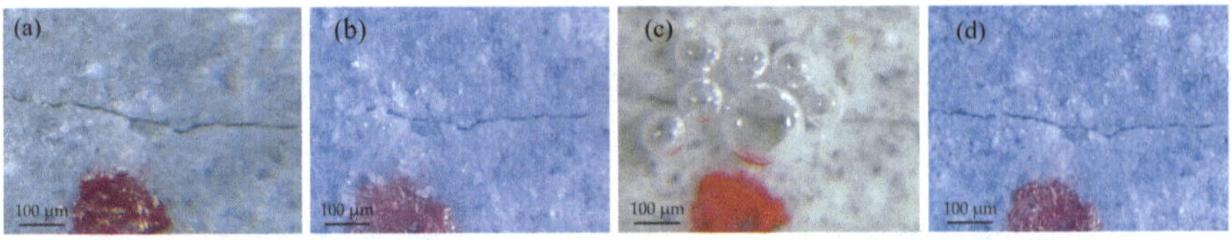

Fig. 1.10 Crack filling process in the pre-cracked UHPFRC plate after exposure to cryogenic temperature: **a** microcrack, **b** just after exposure to cryogenic temperature, **c** melting ices at the surface, **d** 3 days after [52]

1.4 Thermal Properties

1.4.1 Specific Heat Capacity

At cryogenic temperatures, the specific heat capacity of concrete is mainly affected by moisture content. However, the specific heat capacity relationship between ice and concrete at cryogenic temperature is different from that between water and concrete at ambient temperature. Generally speaking, the specific heat of water is four times that of concrete at ambient temperature [22]. In low and ultra-low temperatures, water freezes into ice, and the specific heat of ice decreases by a certain proportion, which is about twice that of concrete. However, owing to the complexity of the concrete system, the specific heat capacity of concrete will also change at cryogenic temperature. The specific heat capacity of concrete linearly decreases with decreasing temperature. However, according to the relevant theory, it is noted that the main reason for the change in the specific heat capacity of concrete is that the specific heat capacity of water is reduced by half after freezing, and the specific heat capacity of ice changes with the variation of temperature. The analysis shows that pore water freezing is a slow process; the temperature range of pore water freezing is mainly between 0 and − 50 °C, so the specific heat capacity change will not appear in the obvious step between − 20 and 0 °C in Fig. 1.11. Due to the large span of ultra-low temperature, ultra-low temperature has a significant destruction. The specific heat change of water and ice is an important factor for studying thermal performance and pore water transformation of concrete under cryogenic temperature.

1.4.2 Thermal Conductivity

At ambient and cryogenic temperatures, the thermal conductivity of cement-based materials is affected by many factors, the most important being moisture content [54]. The thermal conductivity of cement-based materials is related to the freezing point of water. The thermal conductivity of concrete with high water content increases by 60% at − 165 °C. However, the thermal conductivity of cement-based materials decreases with cooling when the temperature is higher than its freezing point [22].

The main reason for this phenomenon is pore water. The pore water will freeze to ice at 0 °C and change from liquid phase to solid phase, which will increase the thermal conductivity of the cement-based materials. As the temperature decreases, it shows a decreasing trend [26].

1.4.3 Thermal Diffusivity

Thermal diffusivity is a function of thermal conductivity, specific heat, and density, which can reflect the speed of temperature homogenization in an object with uneven temperature. Thermal diffusivity increases with temperature decreases and is greatly influenced by water content. Generally speaking, the higher the water content, the greater the increased rate of thermal diffusivity [55]. When the temperature changes within – 196–20 °C, the functional relationship between the thermal diffusivity and temperature of concrete under different moisture content is shown in Eqs. 1.9–1.11 [22].

$$\alpha = \begin{cases} \left[0.47837 \times \exp\left(\frac{-T}{100.042}\right) + 0.6263\right] \times \alpha_0 & \text{(The moisture content is 3\%)} \\ \left[0.63027 \times \exp\left(\frac{-T}{102.687}\right) + 0.4756\right] \times \alpha_0 & \text{(The moisture content is 5\%)} \\ \left[0.7238 \times \exp\left(\frac{-T}{100.89}\right) + 0.4215\right] \times \alpha_0 & \text{(The moisture content is 7\%)} \end{cases} \tag{1.9–1.11}$$

where α is the thermal diffusivity of concrete at T °C, α_0 is the thermal diffusivity of concrete at 20 °C, $\alpha_0 = 1.17201$ m^2/s.

1.4.4 Thermal Deformation

Under cryogenic temperature and freezing–thawing conditions, the thermal deformation of cement-based materials is the macroscopic manifestation of phase transformation, migration, and freezing of pore water [56], and also one of the essential properties of cement-based materials under ultra-low temperatures different from normal temperature. As the temperature decreases, cement-based materials exhibit a trend of first shrinkage, then expansion, and then shrinkage [5, 12, 15]. The expansion stage is mainly between – 20 to – 70 °C, which is caused by the expansion of pore water ice. Some residual strains remain in the hardened cement matrix after six freezing–thawing cycles (See Fig. 1.12). With the increase of freezing–thawing cycles, the residual strains gradually increase and tend to be stable [56]. Expansion mainly depends on moisture content, and absolute dry concrete will not expand. The higher the water-to-cement ratio (higher moisture content) is, the more pronounced the thermal expansion is [43]. In addition, the fine aggregate can obviously increase

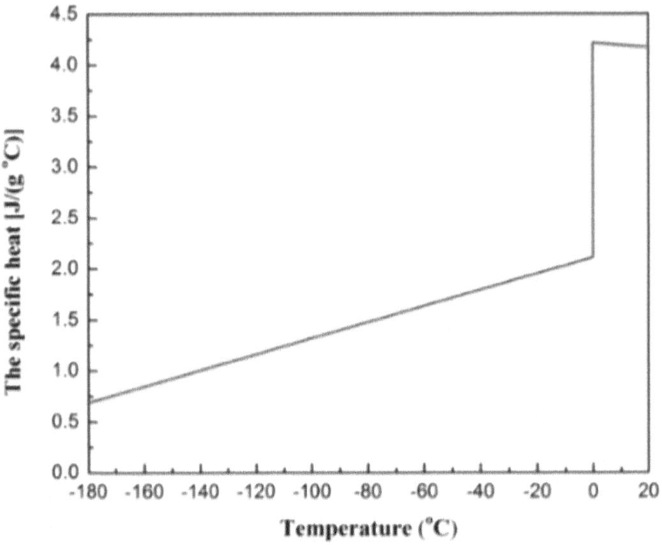

Fig. 1.11 Relationship between specific heat capacity and temperature of ice [53]

its thermal strain [56]. The main reason for the low-temperature thermal strain of cement-based materials is pore migration and frozen water.

After ultra-low temperature freezing–thawing cycles, there will be strain residues in concrete. The residual strain increases with the freezing–thawing cycles and the enlargement of the temperature difference. The residual strain will no longer increase when the temperature is below − 50 °C [15]. Meanwhile, the residual strain is linearly related to the loss of relative dynamic modulus of elasticity after freezing–thawing cycles, and the deterioration of freezing–thawing mainly occurs between − 20 to − 50 °C. Rostasy et al. [43] found that the residual strain of saturated mortar is 0.27% after 12 freezing–thawing cycles at − 170 °C.

1.4.5 Temperature Field Distribution

In the process of freezing–thawing cycles, the thermal parameters of cement-based materials will change under the environment of unsteady instantaneous heat conduction, resulting in the uneven distribution of the internal temperature field and the generation of temperature stress. Furthermore, the temperature stress acts on the internal structure of cement-based materials, which causes damage. It has been proved that many factors affect the temperature field of cement-based materials in the process of freezing–thawing cycles, including fine aggregate, specimen types, water-to-cement ratio, and moisture content. For specimens with different shapes, the thermal conductivity of the cube is better than that of the cylinder due to its smaller

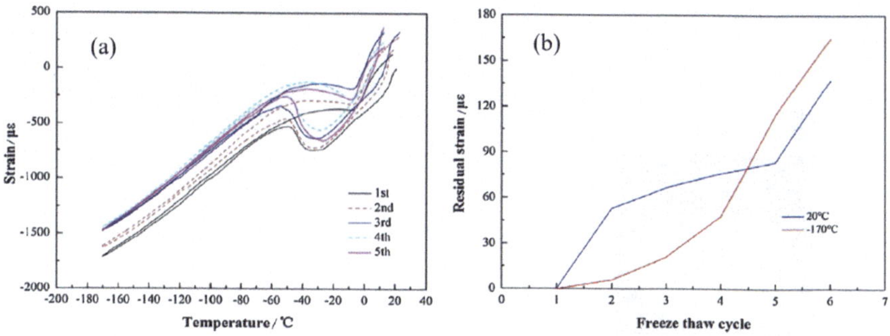

Fig. 1.12 **a** Thermal strain and **b** residual strain of cement paste after freezing–thawing cycles [56]

internal and external temperature difference (see Fig. 1.13). When the external temperature was cooled to − 170 °C, the central temperatures of cube and cylinder specimens reached − 166.4 °C and − 160.4 °C respectively [48].

1.5 Freezing Liquids in Porous Media

1.5.1 Crystallization

The theory of crystallization nucleation and ice crystal growth of pore water mainly includes nucleation crystallization theory, evaporation and condensation crystallization theory, and ice crystal infiltration growth theory.

(1) Nucleation and crystallization theory

Nucleation crystals are classified into homogeneous and heterogeneous nucleation. In a uniform single-parent phase, the embryo forms spontaneously. Only if the size of the embryo is larger than the critical size can the crystal continue to grow. The critical dimension of ice crystals decreases with decreasing temperature. In porous materials, the crystal embryo of ice crystals is limited by the size of the pores, and its size does not necessarily exceed the critical size. For this part of the small pores, the internal pore water will freeze only when the temperature continues to decrease, and its critical dimension is smaller than the size of the embryo. Heterogeneous nucleation mainly forms crystal embryos at contact with the pore walls. Nucleation crystallization requires that the embryos cross a large energy barrier in the growth process. In pure water, the in-phase nucleation temperature is approximately − 39 °C, while the heterogeneous nucleation temperature is between − 30 °C and − 5 °C [57]. Generally, the nucleation crystals we observed are heterogeneous nucleation.

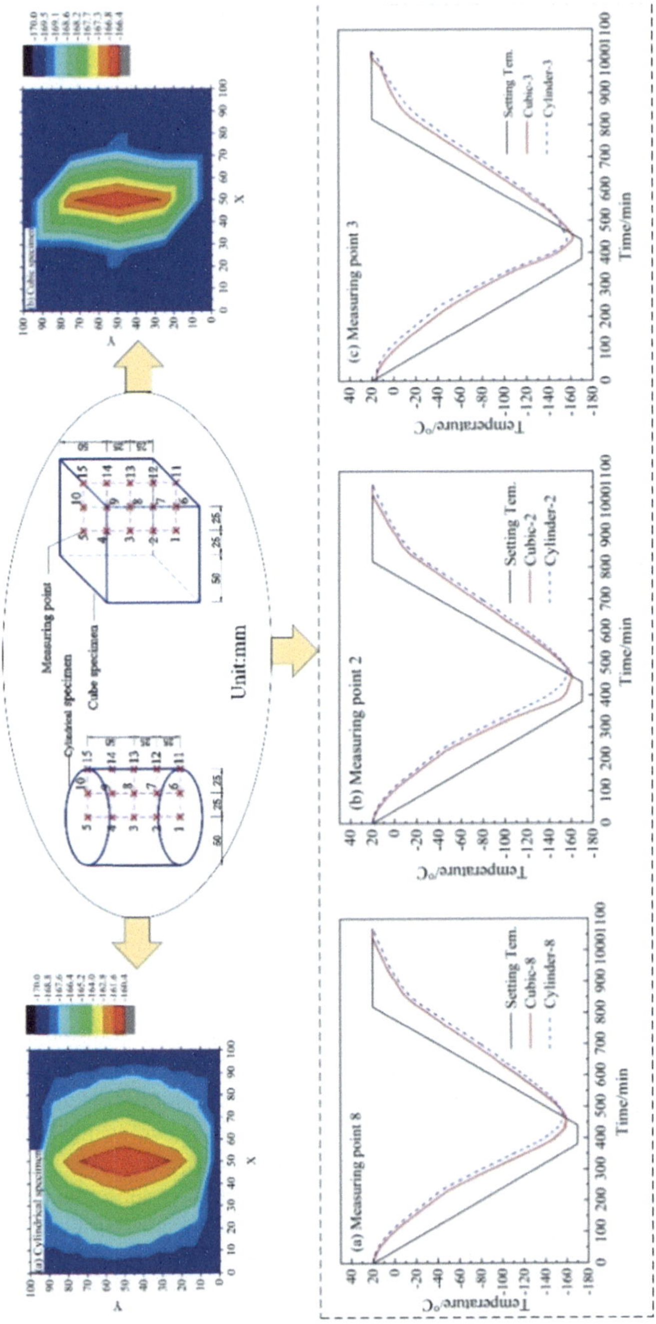

Fig. 1.13 Temperature field distribution in different specimens [48]

(2) Evaporation and condensation crystallization theory

The crystallization process is mainly the evaporation of the liquid phase into the gas phase, followed by condensation crystallization on the solid surface [58–61]. This process mainly occurs in the case of large temperature gradient. In gas, liquid, and solid equilibrium, evaporation condensation crystallization is unlikely. This crystallization process is unlikely for fully saturated-water porous materials due to the absence of a gas phase.

(3) Ice crystal infiltration growth theory

Since the energy required for pore water to continue crystallizing on the surface of ice is less than that required for nucleating the crystallization, ice crystal growth theory holds that water outside porous materials freezes, and then ice gradually infiltrates the pores [62]. However, the theory cannot explain that only small pores freeze in unsaturated porous materials, nor can it be applied to the pore structure system containing ink bottle pores.

Cement-based materials are porous materials with complex pore structures. The first is the wide pore size distribution, including the gel pores with 1–5 nm aperture, the capillary pores with 10–1000 nm aperture, and the bubble holes with 50 μm to 5 mm aperture [63]. The second is the different pass structure. The pores introduced by the air-entraining agent are primarily spherical, while the capillary and gel pores are of cylinder type. Cylinder holes are usually used to calculate the pore size of cement-based materials. The third is the complex connectivity of the pore structure, including the closed pore, the open pore, and the ink bottle structure pore. Therefore, the complex pore structure of cement-based materials makes it challenging to study the freezing of pore water.

For fully saturated cement-based materials, the freezing process of pore water is a combination of nucleation crystallization and ice crystal infiltration growth [64]. Sun and Scherer [65, 66] thought that the pore water would not achieve high supercooling, and the ice crystal growth is the main freezing form of pore water. Nucleation and crystallization occur only in very small nanopores. For cement-based materials with incomplete water saturation, all three theoretical processes mentioned above may occur during the freezing process of pore water.

1.5.2 Freezing Point

Three types of water in the hardened cement paste are chemically bound water, water absorbed on gel surfaces, and water in capillary pores [20]. Their pore size and freezing temperature are shown in Table 1.3.

Due to the freezing of pore water in concrete, the performance of concrete at cryogenic temperature is very different from that at ambient temperature. The transition temperatures of the solid, liquid, and gas phases are highly related to the interface curvature. In porous materials, the ice-water interface curvature of pore water depends

Table 1.3 Pore size and water freezing time in concrete

Types	Pore size	Freezing temperature
Bulk water	Less than 0.1 mm (4 × 10^{-3} in)	0 to – 4 °C
Water in capillary pores	Smaller than 0.01 mm (4 × 10–4 in)	– 20 to – 30 °C
Structural water in gel pores	3–10 nm (1 × 10^{-7} to 4 × 10^{-4} in)	– 30 to – 80 °C
Adsorbed water in gel pores	Smaller than 3 nm	– 160 °C

on the pore size. The equation between the three-phase transition temperature and the curvature of the two interfaces can be derived from the mathematical relationship of the Gibbs equation of state, Laplace's equation, and the chemical potential of each phase at the three-phase equilibrium point. For water-saturated porous material, the presence of gas is not considered; the equation can be simplified to [53]:

$$\Delta S_f dT + v_l d\left(\gamma_{\mathrm{ls}} \times \frac{dA_{sl}}{dV_l}\right) = 0 \tag{1.12}$$

where ΔS_f is the solidification entropy of unit mass water at temperature T, v_l is the unit mass volume of water, γ_{ls} is the surface tension of the ice-water interface, and $\frac{dA_{sl}}{dV_l}$ is the curvature of the ice-water interface in the pores.

For water-saturated porous materials, the existence of the gas phase is not considered. Assuming that the ice water interface is spherical cap shape, the interfacial curvature is:

$$\frac{dA_{sl}}{dV_l} = -\frac{2}{r} \tag{1.13}$$

Then, Eq. 1.12 can be expressed as follows.

$$\frac{1}{r} = \frac{1}{2\gamma_{ls}} \int\limits_{T_0}^{T} \frac{\Delta S_f}{v_1} dT \tag{1.14}$$

In fact, there is a layer of non-freezing water film with several molecules thick between the pore wall and ice. Brun [53] compared the results of different pore structure test methods of several porous materials and thought that the thickness of water film in porous glass was 0.8 nm. After considering the non-freezing water film, Eq. 1.14 was simplified to Eq. 1.15 [56]:

$$R_{\mathrm{P}}(\mathrm{nm}) = -\frac{64.67}{\Delta T} + 0.57 \quad (\text{Freezing, } 0 > \Delta T > -40) \tag{1.15}$$

It can be seen from Eq. 1.15 that the smaller the pore size is, the lower the pore water phase transition temperature is. In the pores with a radius of 20 nm, the water phase transition temperature is – 3.33 °C. In the pores with a radius of 10 nm,

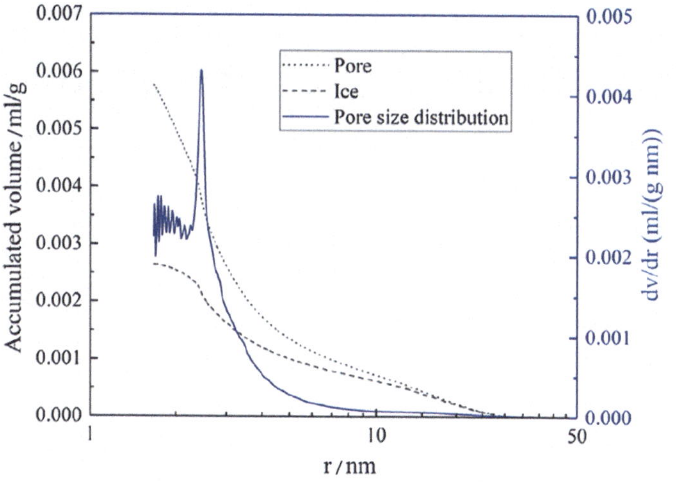

Fig. 1.14 The accumulated volume of ice, pores and pore size distribution of saturated cement paste [25]

the water phase transition temperature is − 6.86 °C. At − 40 °C, the water in the pores with a radius of 2.1 nm begins to freeze. As shown in Figs. 1.14 and 1.15, the pore volume with a radius less than 10 nm and less than 2.5 nm accounts for 85% and 61% of the total pore volume, respectively. However, the amount of ice in these pores only accounts for 75% and 21% of the total ice, respectively. Unfrozen water exists between ice and the pore wall, which accounts for 50% of the total pore volume (radius less than 40 nm). With the decrease in temperature, the water in the pores begins to freeze, and the strength of saturated concrete increases gradually. The increase in ice mass and concrete strength rapidly increases in the temperature range of 0 to − 50 °C, and then slowly grows below − 50 °C. The increase of compressive strength in the range of 0 to − 50 °C accounts for about 65% of the total increased strength [25].

The above calculation assumes that the pore water is pure water. In fact, the pore water inside cement-based materials is mainly a solution of Ca $(OH)_2$, and its phase transition temperature is lower than that of pure water.

1.5.3 Non-freezing Water Film

When pore water freezes, a layer of water molecules in contact with the pore wall will not be affected by phase transformation. A layer of non-freezing water film exists between the ice and the pore wall, as shown in Fig. 1.16. There is a strong repulsion between ice and pore walls. The sum of the energy between the ice-water interface and the pore wall is less than that between the ice and the pore wall. Therefore, when

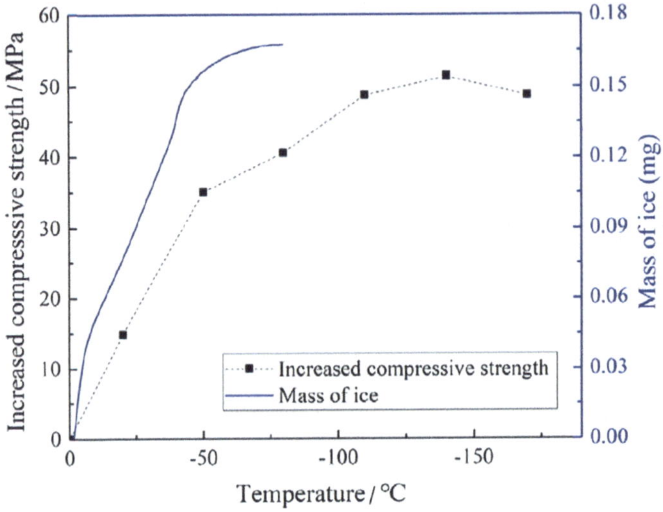

Fig. 1.15 The relationship between increased compressive strength and ice formation in pores [25]

pore water freezes, a layer of water film will be inhaled between the ice and pore wall to reduce system energy. The repulsion force between ice and pore wall mainly comes from the Van Der Waals force, hydrogen bonds, and electrostatic repulsion force in water films. Van Der Waals force accounts for about 25% of the total repulsion force. The calculation results of George [66] showed that only the Van Der Waals force between the ice and the pore wall could maintain a water film with a thickness of 4 nm. By comparing the pore structure of several porous materials with different pore structure test methods, Brun [64] deduced that the water film thickness was 0.8 nm. Sun et al. [65] thought that the pore water would be difficult to freeze when the temperature reached − 40 °C and then continued to decrease. Therefore, it is suggested that the thickness of non-freezing water film in cement-based materials should be 1.0–1.2 nm. However, Tognon [14] found that pore water still freezes until the temperature drops to − 70 °C. As long as the temperature is low enough, the previously considered "non-freezable water" will still freeze, and the thickness of the non-freezable water film depends on the temperature [67].

1.5.4 Crystal Structure of Ice

Cement-based materials can be regarded as composites of ice and cement-based materials when pore water freezes at low temperatures. The properties of ice greatly influence the properties of cement-based materials. However, under the low temperatures, many physical and mechanical properties of ice have not been fully considered in the mechanism of freezing and thawing damage of pore water in concrete. At the

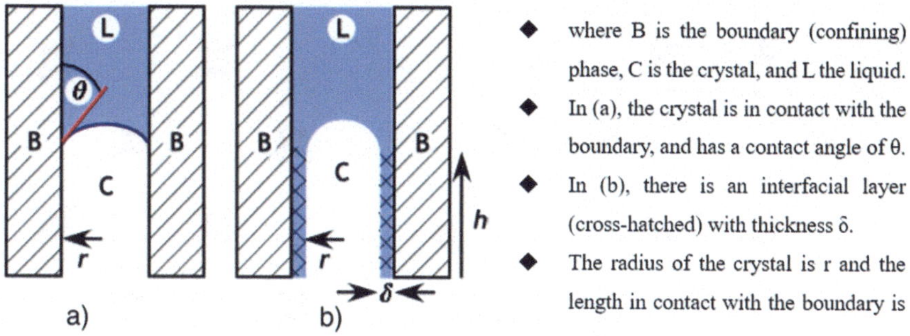

Fig. 1.16 Schematic of the layer of unfrozen water in pore [65, 68]

cryogenic temperature, ice has a more significant impact on the overall performance of materials, requiring a more thorough understanding of the physical and mechanical properties of ice. In addition, the cryogenic temperature span is large. The crystal structure and mechanical properties of ice also change obviously with the decrease of temperature.

Table 1.4 exhibits ice crystal structure at different temperatures [69]. Water can generally be divided into free water and non-freezing water. It is believed that ice crystals with a cubic structure can be formed by freezing porous free water rather than hexagonal ice crystals in a steady state (see Fig. 1.17) [70]. The main reason may be the effect of Van Der Waals force between the pore surface and water. It has also been found that cubic ice crystals and hexagonal ice crystals exist in porous ice. With the decrease in temperature, the amount of hexagonal ice crystals increases rapidly. In addition, it is believed that the crystalline structure is related to the freezing mode of pore water. For nucleation crystallization, the cubic structure is more stable than the hexagonal structure.

Morisine et al. [71] and Schreiber found that the cylindrical nanopore water was frozen and formed hexagonal pore ice by measuring the freezing process of pore water in a single cylindrical nanopore with a diameter of 2.9–3.7 nm. Schulson et al. [72] proved that pore water mainly formed ice with hexagonal crystal structure in cement-based materials. Marshall [12] considered that the hexagonal structure was a

Table 1.4 Crystal structure of bulk ice at different temperatures [69]

Temperature range	Crystal structure of ice
0 to − 3 °C	Thin hexagonal plates
− 3 to 5 °C	Needles
− 5 to − 8 °C	Hollow prisms
− 8 to − 12 °C	Hexagonal plates
− 12 to − 16 °C	Dendritic crystals
− 16 to − 25 °C	Plates
− 25 to − 50 °C	Hollow prisms

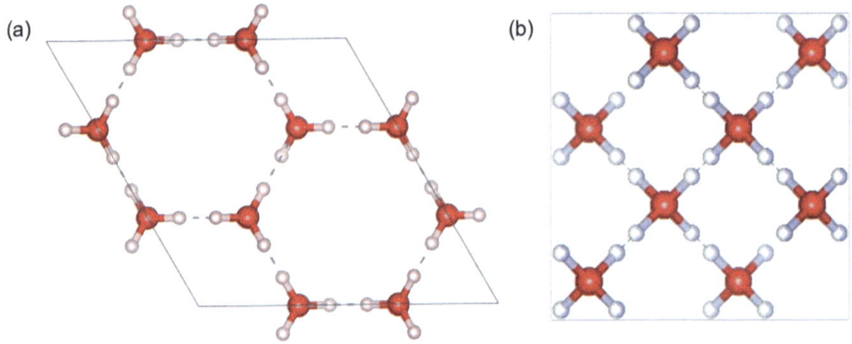

Fig. 1.17 Schematic diagram of crystal structure: **a** hexagonal ice crystal **b** cubic ice crystal [70]

common crystal type, and the tetragonal structure was not obtained by the hexagonal structure cooling but by condensation of water vapor at − 80 °C. The cubic ice crystals are irreversibly transformed into hexagonal ice crystals during the heating process. There is no sudden volume change in the transformation of the two crystal forms since their densities are similar. The transformation temperature is between − 63 °C and − 113 °C.

1.5.5 Strengthening Mechanism at Cryogenic Temperature

It is believed that the freezing of pore water is greatly responsible for the increased strength and microstructure damage of cement-based materials. Figure 1.18 reveals the mechanical enhancement mechanism of cement-based materials and the corresponding temperature range. A strong correlation can be observed between ice content and compressive strength. Strength enforcement under cryogenic temperatures is derived from four effects of frozen pore water: filling-binding effect, self-enforcement of ice body, prestress effect, and phase transformation. These effects are generally summarized as follows.

- Ice filling-binding effect, mainly at 0 to − 80°C.

Within this temperature range, pore water continues to freeze. Macroscopic pores (greater than 50nm), micropores (less than 50nm), and gel pores (less than 4nm) are gradually filled with ice crystals, and the internal microstructure of the material becomes tighter. The high bond strength of ice helps to increase the internal force of the material, making the material denser.

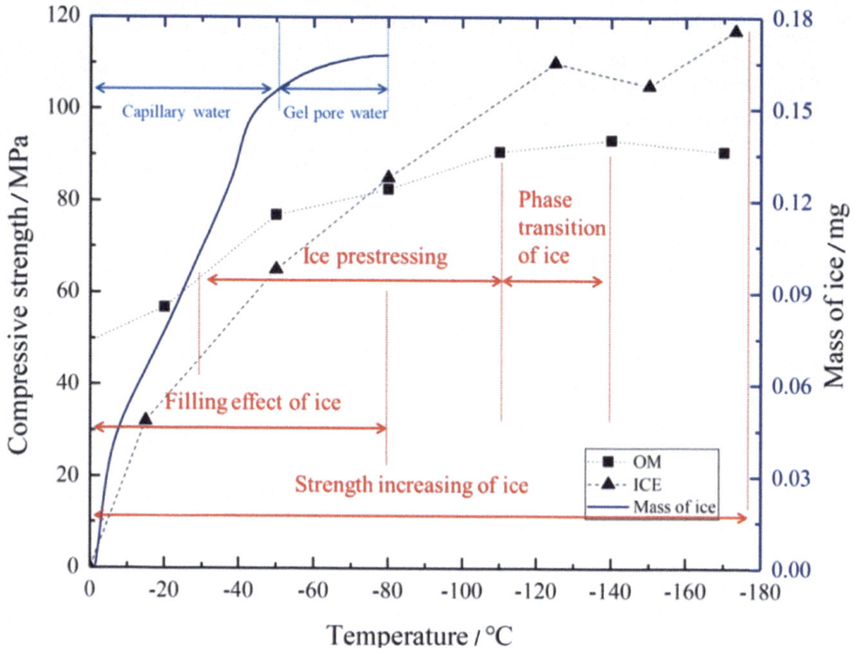

Fig. 1.18 Strength enhancement mechanism of cement-based materials at cryogenic temperature

- Self-enforcement of ice body, mainly at 0 to − 120 °C.

The compressive strength of ice increased continuously between 0 and − 120 °C, and stabilized between − 120 °C and − 170 °C. The bond strength of ice has a similar rule, and the bond strength under tensile stress is much higher than that under compressive stress.

- The prestress effect of ice, mainly at − 20 °C to − 120 °C.

At − 20 °C, most pore water freezes. As the temperature continues to drop, the prestressing effect of ice further increases the strength of the material.

- Phase transformation of ice, lower than − 120 °C.

When the temperature is − 120 °C, the crystal transformation of ice is from a hexagonal to an oblique system, reducing the volume by 20%, and reducing the compactness of the material. As a result, the compressive strength of concrete ceases to increase and even starts to decrease.

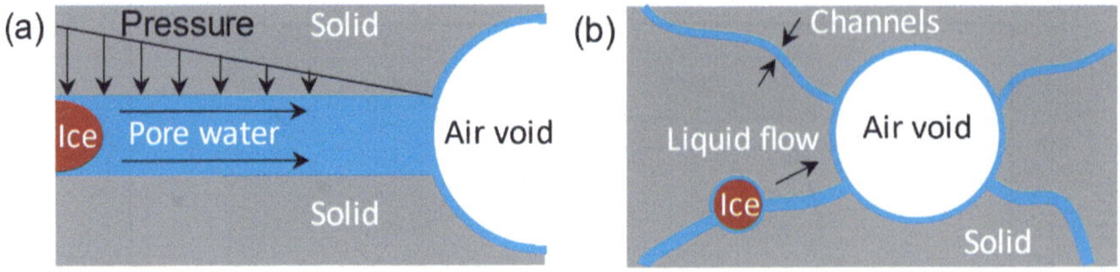

Fig. 1.19 **a** Schematic diagram of hydraulic pressure model; **b** Schematic diagram of ice forming hydraulic pressure in pores around bubbles

1.6 Freezing–Thawing Damage Theories

The freezing–thawing damage of cement-based materials at cryogenic temperature has always been one of the key directions of concrete durability. Based on the volume expansion of pore water ice and its thermodynamic process, many freezing–thawing damage models have been proposed by different scholars. However, no model or theory can fully explain the freezing–thawing damage process and phenomenon. In this section, several main freezing–thawing damage theories are introduced and compared in detail. More reviews on theoretical models of freezing–thawing damage can be found in references [63].

1.6.1 Hydraulic Pressure Theory

Powers [73, 74] hydraulic pressure theory (see Fig. 1.19) holds that when concrete is frozen, negative temperature invades from all sides of the concrete, freezes on the surface of the concrete first, and closes them up. The frozen surface pore water causes volume expansion and squeezes the unfrozen water through the capillary channels into the interior with less saturation. The ice volume increases with the decrease in temperature and continues to compress the unfrozen water. If the unfrozen water cannot flow freely, it will generate more and more pressure in the capillary pore, producing tensile stress on the wall of the capillary pore in the cement stone. When the tensile stress reaches the ultimate tensile strength, the capillary pore ruptures and produces micro-cracks in the concrete. However, the hydraulic pressure model can only explain air-entrained hardened cement pastes, not non-air-entrained specimens.

1.6.2 Osmotic Pressure Theory

When air bubbles are properly introduced into cement-based materials, the hardened cement matrix shrinks continuously with the decrease in temperature [75]. Powers et al. [76] proposed that this is due to the partial melting of the ice crystals in

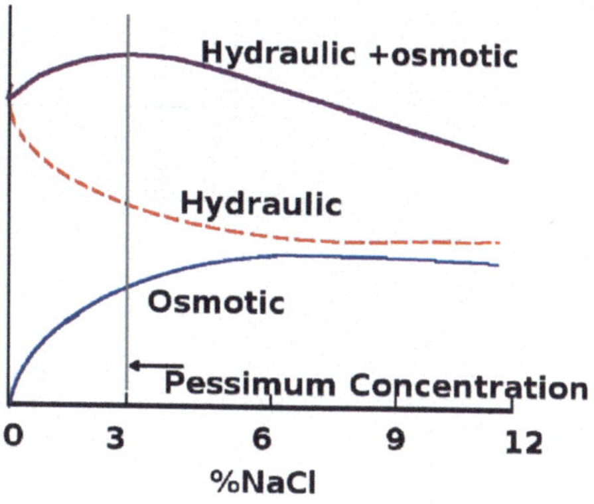

Fig. 1.20 Schematic diagram of maximum damage caused by 3% salt solution in osmotic pressure theory [76]

the capillary pores and water migration from the gel pores to the nearby frozen pores. Then, the osmotic pressure theory was developed to explain the maximum damage caused by the 3% salt solution mentioned. Under cryogenic temperatures, some of the macro-pores and capillary pores' solutions are frozen into ice. These frozen solutions make the solution's concentration larger, resulting in a concentration difference between the capillary and gel pores, causing the supercooled water to diffuse from the gel pores to the capillary pores to form the osmotic pressure. That is, the migration of supercooled water yields a significant osmotic pressure, thus causing severe damage to the internal pore structure of concrete. The higher the initial concentration of pore solution is, the greater the osmotic pressure is. Moreover, the higher the initial pore solution concentration, the lower the ice content and the lower the hydraulic pressure. Therefore, in theory, the interaction of the two pressures may lead to the formation of the maximum pore pressure at a certain concentration, as shown in Fig. 1.20. However, the osmotic pressure hypothesis has no mathematical model for quantitative calculation. It is difficult to describe its effect on freezing–thawing failure of cement-based materials.

1.6.3 Crystallization Pressure Theory

When pore water freezes, the curvature of the interface between the ice-water hemisphere and the cylindrical pore is different, and the pressure difference exists. In order to balance the pressure difference between the two interfaces, the pore wall imposes additional crystallization pressure on the ice crystal. Otherwise, the imbalance of the pressure in all directions of the ice crystal will lead to the melting of the ice crystal in the high-pressure region and the migration to the low-pressure region (see Figs. 1.21 and 1.22). If the crystallization pressure is large enough to exceed the strength limit of porous materials, the materials will be destroyed [66, 77]. In addition, in order to balance the crystallization pressure, ice crystals tend to grow in the direction of

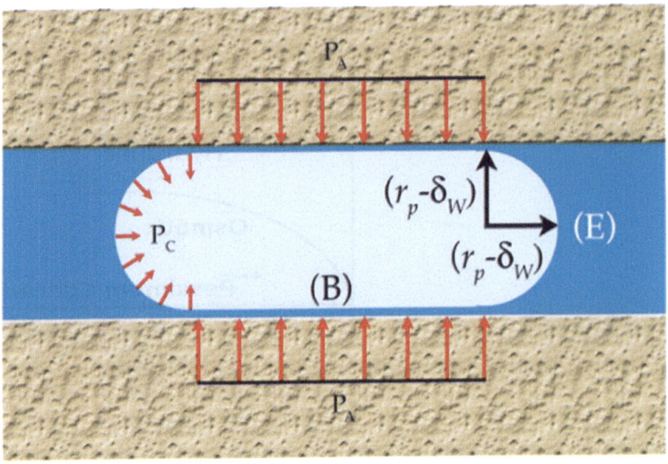

Fig. 1.21 Schematic of a cylindrical crystal in a pore of radius r_p. Both the hemispherical end (E) and the cylindrical body (B) adopt the radius, $r_p - \delta_w$. P_A is crystallization pressure [77]

low pressure and push away the pore wall, which is also the reason for the common phenomenon of freezing–thawing in nature.

1.6.4 Micro-Ice Lens Theory

According to the micro-ice lens theory [78–80], it is impossible for cement-based materials to be fully water-saturated. The pore always has three phases: liquid water, water vapor, and ice. The migration path of pore water can be deduced from the three-phase equilibrium condition. Micro-ice lens formed in the pores are similar to the cryopumps effect, which makes the internal environment of the pores reach supersaturation conditions. Micro-ice lens theory mainly focuses on heat and water transport by microcrystals. As shown in Fig. 1.23, during the cooling process, the chemical potential of pore water is higher than that of ice because of the thermodynamic action. Therefore, there is a tendency to migrate towards the freezing point of pore water or gel water near the freezing point. During the heating process, water migrates from large capillary pores occupied by ice to small capillary and gel pores, and this migration can also be achieved either by the evaporation–condensation process or by direct melting migration of ice. If there is water outside the concrete at this time, the water will migrate to the interior, resulting in the increase of water saturation in the pore with the rise of freezing–thawing cycles [78, 79].

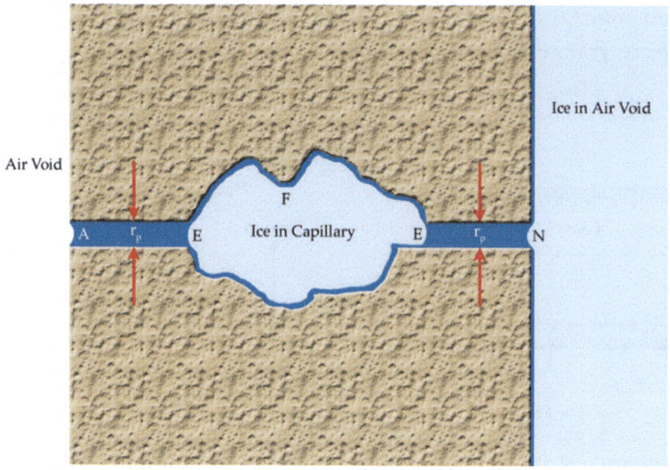

Fig. 1.22 Schematic of ice forming in a capillary cavity and in an air void. A is curved liquid/vapor meniscus, N is connection pore between equivalent liquid and crystal interface. F point represent the curvature of the crystal is negative [77]

1.6.5 Glue Spall Theory

Valenza and Scheller [77, 81–83] explained the surface erosion of cement-based materials with the glue spall mechanism (see Fig. 1.24). When there is a layer of salt solution on the surface of cement-based materials, its thermal expansion coefficient is much higher than that of cement-based materials after freezing (At − 10°C, the linear thermal expansion coefficient of ice is about $50 \times 10^{-6}K^{-1}$ and that of cement-based materials is about $10 \times 10^{-6}K^{-1}$). Therefore, the freezing salt solution layer on the surface of the substrate produces stress, resulting in the surface cracking and spalling. When the temperature decreases by − 20 °C, the surface ice layer can produce a stress of 2.6 MPa, which is sufficient to cause damage to the cement-based material [77, 81].

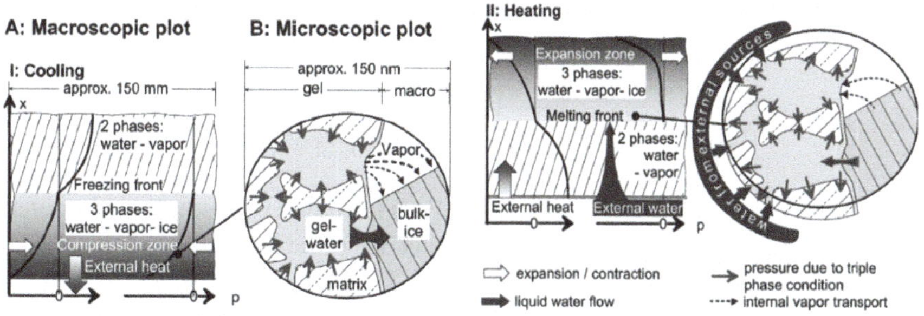

Fig. 1.23 Schematic of micro-ice-lens model [78–80]

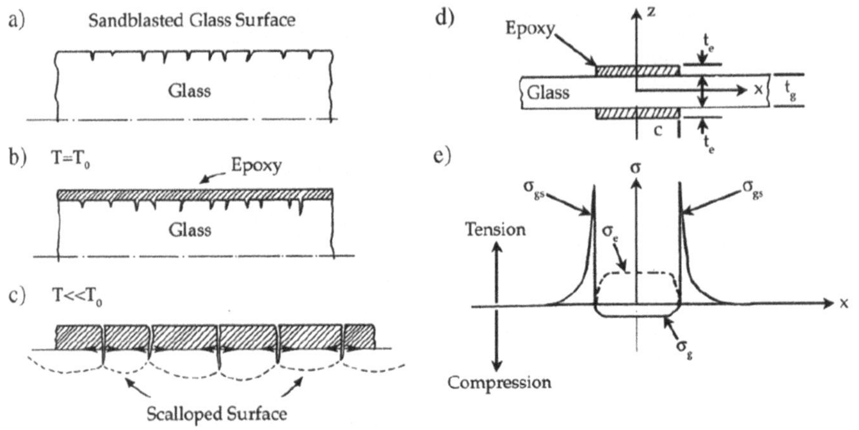

Fig. 1.24 Schematic of the glue spall mechanism [77]. **a** Rough glass surface, **b** Epoxy/glass composite at initial temperature T_0, **c** Interface of composite, illustrating the islands of epoxy and the thin scallops of glass removed when $T \ll T_0$. (**d** and **e**) Schematic representation of an epoxy/glass/epoxy sandwich seal and the stress that arises in the composite. **d** Sandwich seal, dimensions, and orientation. **e** Schematic of stress that arises in the glass surface under the epoxy, σ_g, in the epoxy, σ_e, and the glue spall stress around the boundary of the epoxy, σ_{gs}

1.6.6 Poromechanics Theory

Unsaturated poroelasticity theory has been widely used in the field of geotechnical, biological, and other porous materials. The macro-stress of porous materials is decomposed into thermal stress, the internal stress of porous media, and matrix stress [84]:

$$\sigma = K\varepsilon - \mathrm{b}P - 3K\alpha_s\Delta T \tag{1.16}$$

where ε is the macro stress of the material, b is the Biot coefficient, P is the effective pore pressure, K is the bulk modulus of the solid material, and α_s is the linear thermal expansion coefficient of the solid.

Coussy and Monteiro [85, 86] gave Eq. 1.17 for pore pressure and Eq. 1.18 for macro strain of porous materials.

$$P_L - P_{atm} \approx \left(\frac{V_C - V_L}{V_C}\right)\frac{\text{Ø}_C}{\text{Ø}_C/K_C + (1 - \text{Ø}_C)/K_L} \tag{1.17}$$

where P_{atm} is atmospheric pressure, V_{C} and V_{L} is the molar volume of ice and water, K_{C} and K_{L} is the bulk modulus of ice and water, Ø_{C} is the content of ice.

$$\varepsilon_x \approx \left(\frac{1}{3K_p} - \frac{1}{3K_S}\right)\left[\text{Ø}_C p_A + (1 - \text{Ø}_C)p_L\right] + \alpha\Delta T \tag{1.18}$$

where K_p and K_S are bulk modulus of matrix and solid phase, α is the coefficient of thermal expansion, and p_A is pressure of ice crystal on pore wall.

From the above three equations, it can be seen that the unsaturated poroelasticity theory can better calculate the pressure of porous materials and predict their macroscopic deformation.

The above classical freezing–thawing damage models are derived from pore water ice expansion and its thermodynamic process. Each model can help us understand the process and phenomenon of freezing–thawing damage more intuitively, but it cannot fully explain the mechanism of freezing–thawing damage. Hydraulic pressure, osmotic pressure, and crystallization pressure can help us understand the water migration and local stress state during the freezing–thawing process of porous materials and then understand the freezing–thawing damage process and mechanism. From the macro thermodynamic equilibrium point of view, micro-ice lens theory gives the possible path of water transfer during the freezing–thawing process, which helps understand the process of water transfer during the freezing–thawing process. The glue spall theory focuses on the salt-frost damage of concrete surfaces. Unsaturated poroelasticity theory unifies the relationship between micro-deformation and macro-strain of materials. It has a better mechanical and deformation calculation model, which helps understand the internal and external stress state of porous materials during the freezing–thawing process. Each freezing–thawing damage model has its applicable scope and conditions, which should be fully considered when applying the above models to cryogenic freezing–thawing damage.

References

1. G. Jackson., J. Powell., K. Vucinic., D. Harwood., Delivering LNG tanks more quickly using unlined concrete for primary containment, the 14th International Conference on Liquefied Natural Gas, Doha, 2004, pp. 797–809.
2. J. Xie, X. Li, J.Y. Zhu, A feasibility study of China-made allconcrete LNG storage tanks. Nat. Gas Ind. **34**(09), 113–117 (2014). ((In Chinese))
3. L. Dahmani, Thermomecanical response of LNG concrete tank to cryogenic temperatures. Strength Mater. **43**, 526–531 (2011)
4. B. Johannesson, Dimensional and ice content changes of hardened concrete at different freezing and thawing temperatures. Cem. Concr. Compos. **32**(1), 73–83 (2010)
5. R.B. Kogbara, S.R. Iyengar, Z.C. Grasley, E.A. Masad, D.G. Zollinger, A review of concrete properties at cryogenic temperatures: towards direct LNG containment. Constr. Build. Mater. **47**, 760–770 (2013)
6. G.P. Hu, Y.C. Zheng, A.A. Xu, Z.S. Tang, Lunar surface temperature of global moon: preparation of database with topographic and albedo effects. IEEE Geosci. Remote Sens. Lett. **13**(1), 110–114 (2016)
7. A.R. Vasavada, J.L. Bandfield, B.T. Greenhagen, P.O. Hayne, M.A. Siegler, J.P. Williams, D.A. Paige, Lunar equatorial surface temperatures and regolith properties from the diviner lunar radiometer experiment. J. Geophys. Res-Planet. **117** (2012) E00H18.
8. A.E.L. G. E. Monfore, Physical properties of concrete at very low temperatures, Portland Cement Assoc R & D Lab Bull (1962)

9. A.I.R.P.J. Monteiro, J. Bastacky, T.L. Hayes, Ice in cement paste as analyzed in the low-temperature scanning electron microscope. Cem. Concr. Res. **19**(2), 306–314 (1989)
10. P.J.M.K. Wang, B. Rubinsky, A. Arav, Microscopic study of ice propagation in concrete. ACI Mater. J. **93**(4), 370–377 (1996)
11. V. Van de Veen, *Properties of Concrete at Very Low Temperatures: A Survey of the Literature.* Delft University of Technology Faculty Civil Engineering & Geosciences (1987)
12. A. Marshall, Cryogenic concrete. Cryogenics **22**(11), 555–565 (1982)
13. A.L. GE Monfore, Physical properties of concrete at very low temperatures (1955)
14. G. Tognon, Behavior of mortars and concretes in the temperature range from +20 °C to -196 °C, Supplementory paper III-24 (1969) pp. 229–249.
15. T. Miura, The properties of concrete at very low temperatures. Mater. Struct. **22**, 243–254 (1989)
16. M.-J. Kim, D.-Y. Yoo, S. Kim, M. Shin, N. Banthia, Effects of fiber geometry and cryogenic condition on mechanical properties of ultra-high-performance fiber-reinforced concrete. Cem. Concr. Res. **107**, 30–40 (2018)
17. J. Xie, X. Li, H. Wu, Experimental study on the axial-compression performance of concrete at cryogenic temperatures. Constr. Build. Mater. **72**, 380–388 (2014)
18. B.C.G.J. D.E. Berner, *Static and Cyclic Behavior of Structural Lightweight Concrete at Cryogenic Temperatures*, Ocean Space Utilization '85. Springer Japan (1985)
19. G. Wiedemann, *Zum einfluss tiefer temperaturen auf festigkeit und verformung von beton Germany: Technical University of Braunschweig* (1982)
20. L. Dahmani, A. Khenane, S. Kaci, Behavior of the reinforced concrete at cryogenic temperatures. Cryogenics **47**(9–10), 517–525 (2007)
21. O.A.Z.A.V. Usherov-Marshak, Relationship between the structure of cement stone and the parameters of ice formation during stone freezing. Colloid J. **64**(2), 217–223 (2002)
22. P.B.B. R.D. Browne, The use of concrete for cryogenic storage: a summery of research, past and present. In: *Proceedings of the 1st International Conference Cryogenic Concrete*, Newcastle upon Tyne (1981), pp. 135–162.
23. X.S. Chao Liu, *Experimental investigation on mechanical property of concrete exposed to low temperatures* (Tsinghua University, 2011)
24. L. Zhang, Z. Li, Q. Jia, W. Huang, Experimental study on uniaxial compressive strength of reservoir ice. Trans. Tianjin Univ. **18**(2), 112–116 (2012)
25. Z. Jiang, Z. Deng, X. Zhu, W. Li, Increased strength and related mechanisms for mortars at cryogenic temperatures. Cryogenics **94**, 5–13 (2018)
26. H.K. Sho Yamane, and Toru Okuno, Properties of concrete at very low temperatures, in *Douglas McHenry International Symposium on Concrete and Concrete Structures*, vol. 55. (ACI Special publication, 1978), pp. 207–222.
27. O.T.I. M., Bending behaviour of prestressed concrete beams under low temperatures. J. Jap. Prestressed Concrete Eng. Ass. **20**(8),15–17 (1978)
28. T.S.S.G.C. Lee, K.C. Chang, Mechanical properties of concrete at low temperature. J. Cold Reg. Eng. **2**, 13–24 (1988)
29. Y. Goto, T. Miura, Experimental studies on properties of concrete cooled to about minus 160 °C, Technology Reports. Tohu University **44**(2), 357–385 (1979)
30. F.S. Rostasy, *Verfestigung und versprodung von beton durch tiefe temperaturen, Fortschritte im Konstruktiven Ingenieurbau* (Ernst & Sohn, Berlin, 1984), pp.229–239
31. J.X. C.X. Wang, H.J. Li,, Experimental research on the properties of concrete under low-temperature. Eng. Mech. **28** (S2), 182–186 (2011)
32. J.-B. Yan, J. Xie, Behaviours of reinforced concrete beams under low temperatures. Constr. Build. Mater. **141**, 410–425 (2017)
33. J. Xie, X. Chen, J.-B. Yan, G. Lei, L. Zhu, Ultimate strength behavior of prestressed concrete beams at cryogenic temperatures. Mater. Struct. **50**(1), 81–94 (2016)
34. B.D. E., *Behavior of Prestressed Concrete Subjected to Low Temperatures and Cyclic Loading* (California Univ., Berkeley, USA, 1984)

35. F.I.d.L. Precontrainte, Cryogenic behaviour of materials for prestressed Concrete, State of the Art Report, FIP (1982)
36. S.S. Zhang, T. Yu, G.M. Chen, Reinforced concrete beams strengthened in flexure with near-surface mounted (NSM) CFRP strips: current status and research needs. Compos. B Eng. **131**, 30–42 (2017)
37. J.-B. Yan, J. Xie, Experimental studies on mechanical properties of steel reinforcements under cryogenic temperatures. Constr. Build. Mater. **151**, 661–672 (2017)
38. H.C.M. Elices, J. Planas, Behavior at cryogenic temperatures of steel for concrete. ACI J. Proc. **83**(3), 405–411 (1986)
39. U.C.M. Bruneau M, Sabelli S E R. *Ductile Design of Steel Structures* (McGraw Hill Professional, 2011)
40. C.H. Elices M, Planas J., *Behaviour at Cryogenic Temperatures of Steel for Concrete Reinforcement*. ACI. Technical Paper (1986), pp. 405–411
41. J. Xie, E.S. Simiyu, G. Lei, Z. Nie, Experimental study on bonding properties between steel strand and concrete at cryogenic temperatures. Trans. Tianjin Univ. **22**(4), 308–316 (2016)
42. L. Vandewalle, Bond between a reinforcement bar and concrete at normal and cryogenic temperatures. J. Mater. Sci. Lett. **8**, 147–149 (1989)
43. S.U.F.S. Rostásy, Wiedemann G, Behavior of mortar and concrete at extremely low temperatures. Cem. Concr. Res. **9**, 365–376 (1979)
44. D.Z. Jiang Zhengwu, L. Wenting, Z. Lei., Freeze-thaw effect at ultra-low temperature on properties of mortars. J. Chi. Ceram. Soc. **42**(5), 596–600 (2014)
45. M.P. Kamal Henry Khayat, Cryogenic frost resistance of lightweight concrete containing silica fume, vol. 114 (ACI Spec. Publ., 1989), pp. 915–928
46. Z.-W. Jiang, W.-T. Li, Z.-L. Deng, Z.-G. Yan, Experimental investigation of the factors affecting accuracy and resolution of the pore structure of cement-based materials by thermoporometry. J. Zhejiang Univ-Sci. A **14**(10), 720–730 (2013)
47. B.D. Eric, *Behavior of Prestressed Concrete Subjected to Low Temperatures and Cyclic Loading*. University of California, Berkeley ProQuest Dissertations & Theses (1984)
48. B. He, M. Xie, Z. Jiang, C. Zhang, X. Zhu, Temperature field distribution and microstructure of cement-based materials under cryogenic freeze-thaw cycles. Constr. Build. Mater. **243** (2020)
49. J.P. Gordon Jackson, Vucinic, K., D. Harwood, Delivering LNG tanks more quickly using unlined concrete for primary containmen, in *The 14th International Conference on Liquefied Natural Gas*, Doha, Qatar. vol. 14 (2004) PO-09.1-PO-09.13
50. P.J.S. Ariel Hanaor, Factors affecting concrete permeability to cryogenic fluids. Mag. Concr. Res. **35**(124), 142–150 (1983)
51. R.B. Kogbara, S.R. Iyengar, Z.C. Grasley, S. Rahman, E.A. Masad, D.G. Zollinger, Relating damage evolution of concrete cooled to cryogenic temperatures to permeability. Cryogenics **64**, 21–28 (2014)
52. S. Kim, M.-J. Kim, H. Yoon, D.-Y. Yoo, Effect of cryogenic temperature on the flexural and cracking behaviors of ultra-high-performance fiber-reinforced concrete. Cryogenics **93**, 75–85 (2018)
53. L.A. Brun M, Quinson J F, et al., A new method for the simultaneous determination of the size and shape of pores: the thermoporometry, Thermochim. Acta **21**(1), 59–88 (1977)
54. S.Z.-W. X. Jian-zhuang, Z. Feng, An experimental study on thermal conductivity of concrete. J. Build. Mater. **13**(1) (2010) 17–21
55. W.M. Chirdon, W. Aquino, K.C. Hover, A method for measuring transient thermal diffusivity in hydrating Portland cement mortars using an oscillating boundary temperature. Cem. Concr. Res. **37**(5), 680–690 (2007)
56. J. Zhengwu, Z. Cong, D. Zilong, Z. Xinping, L. Wenting, Thermal strain of cement-based materials under cryogenic temperatures and its freeze-thaw cycles using fibre Bragg grating sensor. Cryogenics **100**, 1–10 (2019)
57. P.W. Wilson, A.F. Heneghan, A.D.J. Haymet, Ice nucleation in nature: supercooling point (SCP) measurements and the role of heterogeneous nucleation. Cryobiology **46**(1), 88–98 (2003)

58. G.G. Litvan, The mechanism of frost action in concrete: theory and practical implications, National Research Council Canada, Institute for Research in Construction (1988)
59. G.G. Litvan, Phase transitions of adsorbates: VI, effect of deicing agents on the freezing of cement paste. J. Am. Ceram. Soc. **58**(1–2), 26–30 (1975)
60. G.G. Litvan, Phase transitions of adsorbates: IV, mechanism of frost action in hardened cement paste. J. Am. Ceram. Soc. **55**(1), 38–42 (1972)
61. G.G. Litvan, Phase transitions of adsorbates. III. Heat effects and dimensional changes in nonequilibrium temperature cycles. J. Colloid Interface Sci. **38**(1) (1972) 75–83
62. D. Everett, The thermodynamics of frost damage to porous solids. Trans. Faraday Soc. **57**, 1541–1551 (1961)
63. P.K.M.P.J.M. Monteiro, Concrete: Microstructure, Properties and Materials. McGraw-Hill Professional **13**(4), 499–499 (2006)
64. A.L. Maurice Brun, J.-F. Quinson, C. Eyraud, A new method for the simultaneous determination of the size and shape of pores: the thermoporometry. Thermochimica Acta **21**(1), 59–88 (1977)
65. G.W.S. Zhenhua Sun, Pore size and shape in mortar by thermoporometry. Cem. Concr. Res. **40**(5) (2010) 740–751
66. G.W. Scherer, Crystallization in pores. Cem. Concr. Res. **29**(8), 1347–1358 (1999)
67. M.S. Eddy Walther Hansen, Ralf Schmidt, Low-temperature phase transition of water confined in mesopores probed by NMR. influence on pore size distribution. J. Phys. Chem. **100**, 2195–2200 (1996)
68. A.L.M. Brun, J.F. Quinson, C. Eyraud, Changement d'état liquid–solide dans les milieux poreux. II. Étude théorique de la solidification d'un condensat capillaire, (Liquid–solid change of state in porous media. II. Theoretical study of the solidification of a capillary condensate). J. Chim. Phys. **6**, 979–989 (1973)
69. A.Z. Sahin, An analytical study of frost nucleation and growth during the crystal growth period. Heat Mass Transf. **30**(5), 321–330 (1995)
70. F.N.H, Structural aspects of the ice-water system. Rep. Progress Phys. **34**, 913–994 (1971)
71. K.K. Kunimitsu Morishige, Freezing and melting of water in a single cylindrical pore: the pore-size dependence of freezing and melting behavior. J. Chem. Phys. **110**(10), 4867–4872 (1999)
72. I.P.S. Erland M. Schulson, Thomas M. Holden, Charles J. Korhonen. , Hexagonal ice in hardened cement. Cem. Concr. Res. **30**(2), 191–196 (2000)
73. T.C. Powers, A working hypothesis for further studies of frost resistance of concrete. J. Proc. 245–272 (1945)
74. T.C.W. Powers, TF, The air requirement of frost resistant concrete, Highway Research Board Proceedings (1950), pp. 184–211
75. Z. Sun, G.W. Scherer, Effect of air voids on salt scaling and internal freezing. Cem. Concr. Res. **40**(2), 260–270 (2010)
76. T.C. Powers, R. Helmuth, Theory of volume changes in hardened portland-cement paste during freezing, Highway research board proceedings (1953), pp. 285–297
77. G.W.S. John J Valenza II, A review of salt scaling: II. Mechanisms. Cem. Concr. Res. **37**(7) (2007) 1022–1034
78. M.J. Setzer, *Development of the micro-ice-lens model* (RILEM Publications SARL, International RILEM Workshop on Frost Resistance of Concrete, 2002), pp.133–145
79. M.J. Setzer, Micro-ice-lens formation in porous solid. J. Colloid Interface Sci. **243**(1), 193–201 (2001)
80. M.J. Setzer, Micro ice lens formation, artificial saturation and damage during freeze thaw attack. Materials for Buildings and Structures **6**, 175–182 (2005)
81. G.W.S. John J Valenza II, Mechanisms of salt scaling. Mater. Struct. **38**(4), 479–488 (2005)
82. G.W.S. John J Valenza II, A review of salt scaling: I. Phenomenology. Cem. Concr. Res. **37**(7) (2007) 1007–1021
83. G.W.S. John J Valenza, Mechanism for salt scaling, J. Am. Ceram. Soc. **89**(4) (2006) 1161–1179
84. O. Coussy, *Mechanics and Physics of Porous Solids*. John Wiley & Sons (2011)

85. P.J.M. Olivier Coussy, Poroelastic model for concrete exposed to freezing temperatures. Cem. Concr. Res. **38**(1), 40–48 (2008)
86. O. Coussy, P.J.M. Monteiro, Poroelastic model for concrete exposed to freezing temperatures. Cem. Concr. Res. **38**(1), 40–48 (2008)

Chapter 2
Micro/Nano-characterization Methodologies for Cement-Based Materials

2.1 Introduction

Understanding the micro and nanostructure of cement-based materials stands as a cornerstone in modern construction research and development. The performance, durability, and sustainability of these materials rely heavily on their intricate microstructural and nanoscale properties. Advances in characterization techniques have played a pivotal role in unraveling the complexity of these materials, allowing for precise analysis of their composition, morphology, and structural features.

This chapter provides a comprehensive overview of various characterization methods employed to explore the micro and nanostructure of cement-based materials. From traditional techniques to cutting-edge advancements, a multitude of analytical methods are utilized to delve deep into the intricate world of these materials, shedding light on their properties at different length scales.

The methods covered in this chapter encompass a broad spectrum of analytical tools. Techniques such as X-ray diffraction (XRD) offer insights into the crystalline phases and structural arrangement within cement-based materials. Thermogravimetric analysis (TGA) aids in understanding thermal behavior and decomposition processes. Energy-dispersive X-ray spectroscopy (EDS) integrated with Scanning electron microscopy (SEM) allows for elemental analysis and imaging at the microscale.

Moving into the micro/nanoscale level, transmission electron microscopy (TEM) and atomic force microscopy (AFM) provide high-resolution imaging and characterization of material surfaces and structures. Nitrogen sorption and helium pycnometry techniques offer information on pore size distribution and total porosity. Mercury intrusion porosimetry (MIP) evaluates pore structure and connectivity within these materials. Advancements in imaging techniques like X-ray computed tomography (X-ray CT) have facilitated three-dimensional visualization and analysis of internal structures.

Z. Jiang and X. Zhu, *Cryogenic Stability of Cement-Based Materials at Micro/Nano Scale*, https://doi.org/10.1007/978-981-96-1836-1_2

Spectroscopic methods such as fourier transform infrared spectroscopy (FTIR), raman spectroscopy, nuclear magnetic resonance (NMR), and X-ray photoelectron spectroscopy (XPS) offer insights into molecular compositions, chemical bonding, and phase identification.

Regarding the mechanical properties, specialized indentation techniques, including AFM-based indentation and micro/nano-indentation, provide valuable data on micro and nanoscale mechanical properties. Additionally, high-pressure X-ray diffraction techniques offer unique insights into the structural behavior of these materials under extreme conditions.

The integration and application of these diverse characterization methods provide a holistic understanding of the micro and nanostructure of cement-based materials. Such insights enable us to tailor material compositions, optimize properties, predict behavior under various conditions, and ultimately enhance the performance and durability of concrete structures in diverse applications.

2.2 Composition Characterization

2.2.1 X-ray Power Diffraction

XRD provides detailed insights into the crystallographic structure, composition, and properties of materials. It determines the atomic arrangement within a material by measuring the diffraction pattern generated when X-rays interact with crystalline solids. Over the years, XRD has evolved into an indispensable tool for characterizing and understanding various substances at the atomic and molecular levels.

The fundamental principle of XRD relies on the interaction of X-rays with a crystalline sample, resulting in the scattering of X-rays at specific angles. When X-rays strike a crystalline solid, they interfere with the regularly spaced atomic planes within the crystal lattice, causing constructive and destructive interference patterns. The diffraction angles and intensities of the scattered X-rays are recorded, forming a unique diffraction pattern that provides crucial information about the crystal structure and phases present in the sample.

XRD is extensively used for phase identification, determining crystalline structure, analyzing crystal defects, measuring lattice parameters, and quantifying phase composition in complex materials. X-ray diffractometers are the primary instruments used for conducting XRD experiments. These instruments consist of an X-ray source, a sample holder, and a detector. Modern diffractometers employ high-intensity X-ray sources, such as rotating anode generators or synchrotron radiation, to enhance data collection accuracy and speed. The sample is prepared in powdered form, ensuring a random orientation of crystallites, which is essential for obtaining reliable diffraction patterns. The incident X-rays interact with the sample, and the resulting diffraction pattern is recorded as a series of peaks corresponding to the diffraction angles and intensities. Advanced data analysis software assists in interpreting these patterns to extract valuable structural information.

XRD has emerged as a cornerstone technique in the analysis of cementitious materials, providing valuable insights into their crystallographic composition, phase identification, and structural changes. Understanding the phase assemblage, hydration products, and the evolution of cementitious compounds in construction materials is fundamental to optimizing their performance, durability, and long-term behavior. It also enables quantification of crystalline phases in cementitious systems, offering information on the formation of hydration products and the evolution of phases during hydration and aging. Advanced techniques, such as Rietveld refinement and quantitative phase analysis, allow for the precise determination of phase compositions, crystallite sizes, and amorphous content within cementitious materials, enhancing the accuracy of phase quantification and structural analysis. The technique aids in determining the degree of hydration, polymorphism, and the presence of unreacted clinker phases, providing critical data for optimizing cement formulations and understanding material behavior under various environmental conditions.

2.2.2 Thermogravimetric Analysis

TGA stands as a vital technique in investigating the thermal behavior, composition, and stability of cement-based materials. In construction materials, understanding the thermal properties and degradation mechanisms of cementitious compounds is crucial for enhancing their performance, durability, and sustainability. TGA provides insights into the thermal stability, decomposition kinetics, and compositional changes of cement-based materials, aiding in optimizing their engineering applications.

In the study of cementitious systems, TGA involves subjecting a sample to a controlled temperature ramping in an inert or oxidative atmosphere. This method tracks the weight change of the sample as a function of temperature, offering valuable insights into phase transitions, dehydration, decarbonation, and organic decomposition processes within the material. The analysis of weight loss or gain curves enables the determination of characteristic temperatures, including dehydration onset and peak temperatures, as well as the estimation of remaining mass fractions. It helps identify the evolution of bound water content, phase transformations, and the decomposition of organic and inorganic additives or admixtures in cementitious matrices. TGA coupled with evolved gas analysis (EGA) techniques, such as mass spectrometry or fourier-transform infrared spectroscopy, allows for simultaneous assessment of released gases, enabling a better understanding of degradation mechanisms.

2.2.3 Energy Dispersive X-ray Spectroscopy

EDS utilizes an electron beam to excite the sample, causing the emission of characteristic X-rays that are detected and analyzed to determine elemental composition. The technique's capability to identify elements from carbon to uranium allows for identifying major, minor, and trace elements in cement-based materials. Advanced techniques, such as elemental mapping and line scan analysis, enable the visualization of elemental distribution and quantification within the microstructure.

EDS has emerged as a powerful analytical technique for the elemental characterization and mapping of cement-based materials. It enables the qualitative and quantitative analysis of elements present within these materials, offering valuable insights into their composition, distribution, and microstructural features. In particular, the elemental mapping provides spatial information on element distribution and their association with specific phases or microstructures, which can be used to characterize various phases, identify impurities, and monitor alterations in elemental content due to environmental exposure or chemical changes during hydration.

2.3 Morphology Characterization

2.3.1 Scanning Electron Microscopy

SEM operates on the principle of using an electron beam to scan the surface of a sample. As the electron beam interacts with the material, signals such as secondary electrons, backscattered electrons, and X-rays are generated, providing information about the surface topography and elemental composition, of the sample.

SEM has become a cornerstone technique in the analysis of cement-based materials. It enables the visualization of hydration products, pore structure, cementitious phases, and interfacial transition zones (ITZs). This visualization aids in understanding the microstructural evolution during hydration, durability, mechanical behavior, and performance of cement-based materials. Effective sample preparation is crucial for SEM analysis. In particular, when performing backscattered electrons SEM, it involves careful handling, mounting, polishing, and coating (if necessary) to ensure a clean, flat, and conductive surface for imaging and analysis.

Recent advancements in SEM technology have led to improved resolution, faster imaging, and enhanced analytical capabilities. Furthermore, the integration of SEM with other microscopy techniques, such as focused ion beam (FIB) milling and correlative microscopy, has expanded the scope of cement-based materials characterization. Future directions include the development of in-situ SEM techniques, automation in data analysis, and the application of machine learning for image analysis and interpretation.

2.3.2 Transmission Electron Microscopy

TEM is a high-resolution imaging technique that has widely used for studying the nanostructure and properties of cement-based materials. TEM operates by transmitting electrons through a thin specimen, generating a high-resolution image based on the interaction of electrons with the sample. The technique utilizes electromagnetic lenses to focus the electron beam, allowing nanoscale resolution imaging and analysis. It provides information about crystallographic structure, phase identification, elemental distribution, defects, and interfaces within materials.

TEM offers invaluable insights into the nanostructure and properties of cementitious materials. It facilitates the visualization and analysis of various components, including C–S–H phases, ettringite, portlandite, clinker phases etc. Understanding these nanostructures is crucial for comprehending material behavior, performance, and durability.

TEM sample preparation is critical for obtaining accurate and meaningful results. Techniques such as thin sectioning, mechanical polishing, ion milling, and FIB milling are employed to create ultra-thin sections (typically < 100 nm) for TEM analysis. These techniques preserve the sample's integrity and allow for high-quality imaging and analysis.

TEM operation involves setting appropriate imaging conditions, including accelerating voltage, beam current, and aperture settings, to optimize resolution and contrast. Imaging modes such as bright-field imaging, dark-field imaging, selected area electron diffraction (SAED), and high-resolution TEM (HRTEM) enable various analyses, including crystallography, phase identification, and lattice imaging.

Recent advancements in TEM technology have led to improved resolution, aberration correction, and analytical capabilities. In-situ TEM techniques, combined with environmental TEM (ETEM), offer real-time observations of dynamic processes in cement-based materials. Furthermore, the integration of TEM with other analytical techniques, such as electron energy-loss spectroscopy (EELS) and electron tomography, expands the scope of nanoscale characterization.

2.3.3 Atomic Force Microscopy

AFM is a powerful imaging and probing technique capable of examining surfaces at the nanometer scale. AFM uses a sharp tip mounted on a cantilever to scan the sample surface. The tip and surface interaction generates force measurements, enabling topographical imaging and surface property mapping with exceptional resolution. AFM offers insights into surface roughness, morphology, adhesion, and mechanical properties of materials. It enables the visualization and analysis of cement hydrates, C–S–H phases, ettringite, as well as surface defects and microstructural features.

Sample preparation for AFM analysis involves obtaining suitable surfaces for scanning. Techniques such as polishing, cleaving, and ultrasonic treatment are employed to prepare flat, clean, and defect-free surfaces. Careful sample preparation ensures accurate and high-quality AFM imaging and measurements. We will show the typical sample preparation procedures in Sect. 2.6.

AFM operational modes include contact mode, tapping mode, and dynamic mode, each offering specific advantages for surface imaging and measurement. Imaging parameters like scan rate, setpoint, and tip-sample interaction force are adjusted to optimize resolution and minimize sample damage during scanning.

Recent advancements in AFM technology have led to improved resolution, multi-functional capabilities, and specialized modes for specific applications. Advanced AFM modes like force spectroscopy, nanomechanical mapping, and nanomanipulation enhance the characterization and manipulation of cement-based materials at the nanoscale.

2.4 Pore Structure Characterization

2.4.1 Nitrogen Sorption

Nitrogen sorption measurement is a fundamental technique used to analyze the pore structure, surface area, and porosity characteristics of cement-based materials. It is based on gas adsorption and desorption principles using nitrogen as the probe molecule. The Brunauer–Emmett–Teller (BET) theory is widely applied to determine the specific surface area, while pore size distribution is often assessed through the Barrett-Joyner-Halenda (BJH) method, providing insights into pore characteristics within materials.

The preparation of samples for nitrogen sorption analysis requires careful handling to ensure accurate measurements. Typically, samples are degassed to remove adsorbed moisture or impurities before analysis. Preparation techniques involve grinding, sieving, and drying to obtain representative and homogenous samples for analysis.

Nitrogen sorption measurements are conducted using specialized instruments such as gas sorption analyzers, employing the principle of adsorption and desorption isotherms at various pressures and temperatures. The analysis involves the determination of adsorbed gas quantity, from which surface area and pore size distribution are calculated using appropriate models. We will introduce this technique with more details in Chap. 4.

2.4.2 Helium Pycnometry

Helium pycnometry is widely used to assess the pore structure, bulk density, and porosity characteristics of cementitious materials. It aids in understanding the void content, total pore volume, and interconnected porosity. It relies on Archimedes' principle, where the volume of a solid material is determined by the amount of helium gas displaced. It measures the apparent density of a sample by comparing the difference in volume between the empty and helium-filled sample chamber.

Preparing samples for helium pycnometry requires careful consideration to obtain accurate measurements. Representative and homogenized samples must be obtained through grinding, sieving, and drying processes to ensure consistency in results. Additionally, samples should be free from moisture and entrapped air to prevent errors during measurement. Helium pycnometry measurements are conducted using specialized pycnometers, where the sample is filled with helium gas at a controlled pressure and temperature. The volume of the gas displaced by the sample determines its apparent volume, allowing for the calculation of bulk density and porosity. Due to the small diameter of helium gas, which is only 0.260 nm [1], it can be considered capable of passing through almost all types of pores within cement-based materials.

2.4.3 Mercury Intrusion Porosimetry

MIP is a technique used to determine the pore structure of materials by measuring mercury intrusion into their pore network under controlled pressure. It operates based on the principle that mercury infiltrates pores of various sizes when subjected to increasing pressure. MIP serves as a versatile method for investigating several critical aspects of cement-based materials. It provides precise quantification of total porosity by measuring the volume of mercury intruded into the pore space, helping to understand the overall void space that influences mechanical properties and durability. MIP enables the characterization of pore size distribution, distinguishing between micro, meso, and macropores, which affects the material's permeability and susceptibility to deleterious agents. It also evaluates pore connectivity and tortuosity, offering insights into fluid transport properties by assessing the pathways for liquid and gas movement, which is crucial for predicting durability and resistance to environmental effects.

MIP offers several advantages in cement research, providing precise quantitative data on porosity and pore size distribution, which facilitates accurate comparisons between different materials, mix designs, or curing conditions. It spans a broad range of pore sizes, from nanometers to micrometers, capturing the full spectrum of pores present in cement-based materials. Additionally, MIP is a non-destructive method, allowing repeated measurements on the same sample to study changes in pore structure over time without compromising sample integrity. However, challenges in MIP analysis include the need for representative sample preparation, potential artifacts from pore collapse, and limitations in accurately characterizing very small or very large pores. Ongoing research aims to address these issues by refining sample preparation techniques, developing correction algorithms, and integrating MIP with other complementary techniques for a more comprehensive analysis of pore structures.

2.4.4 X-ray Tomography

X-ray Tomography provides detailed three-dimensional (3D) insights into the internal microstructure and porosity of cement-based materials. It operates on the principle of X-ray attenuation through materials, which generates contrast variations. It involves the acquisition of a series of 2D X-ray projections from multiple angles around the sample, which are reconstructed into a 3D volume dataset. This technique allows visualization and quantification of internal structures with high resolution. This technique is crucial in assessing the microstructure, porosity, and connectivity of pores within cementitious materials. It provides valuable information about pore size distribution, tortuosity, crack propagation, aggregate distribution, and the interfacial transition zone.

Preparation of samples for X-ray Tomography involves careful handling and specific procedures to ensure accurate imaging. Techniques such as embedding, sectioning, and polishing are employed to create flat surfaces and minimize artifacts that may affect the tomographic reconstruction process.

X-ray Tomography utilizes synchrotron or laboratory-based X-ray sources to obtain 2D radiographic projections from various angles around the sample. Reconstruction algorithms, such as filtered back projection or iterative methods, are used to generate 3D volumetric images. Advanced techniques like phase contrast and high-resolution imaging enhance the quality and resolution of tomographic datasets. Recent advancements in X-ray Tomography include improved spatial resolution, faster acquisition times, and advanced software for quantitative analysis and visualization. Furthermore, the integration of X-ray Tomography with other techniques, such as X-ray diffraction or electron microscopy, enhances the comprehensive characterization of cement-based materials.

2.5 Nanostructure Characterization

2.5.1 Fourier Transform Infrared Spectroscopy

FTIR is a powerful analytical technique extensively employed in characterizing the chemical composition, structure, and molecular interactions within various materials, including cement-based materials. It plays a pivotal role in unraveling the complex chemistry and properties of cementitious compounds. FTIR operates based on the principle of infrared spectroscopy, utilizing the interaction between infrared radiation and the molecular vibrations within a material. The absorption of specific wavelengths corresponds to the vibrational frequencies of different chemical bonds present in the sample. In cementitious systems, FTIR spectroscopy helps identify and characterize key functional groups, including silicates, aluminates, aluminosilicates, and hydration products like C–S–H. It enables real-time monitoring of the hydration process by identifying changes in functional groups, providing insights into the cement hydration mechanism. FTIR, in combination with chemometric methods, facilitates quantitative analysis to determine the concentration of phases or constituents present in cementitious materials.

Sample preparation is crucial for accurate FTIR analysis of cement-based materials. Samples need to be finely ground and homogenized to ensure representative spectra. Pressed KBr pellets or attenuated total reflectance (ATR) sampling methods are commonly employed for solid samples. Solutions or suspensions of powdered samples can also be analyzed to extract soluble components.

2.5.2 Raman Spectroscopy

Raman spectroscopy operates on the principle of inelastic scattering of photons, where incident light interacts with a material, resulting in energy changes that reveal information about molecular vibrations and rotational states. By analyzing the scattered light, Raman spectroscopy provides detailed insights into chemical bonds, crystal structures, and chemical compositions. This technique is versatile in investigating various aspects of cement-based materials. It aids in identifying and characterizing mineral phases such as calcium silicates, aluminates, and sulfates, assessing their distribution, and monitoring transformations during hydration and aging processes. Raman spectroscopy enables real-time monitoring of hydration kinetics by detecting changes in molecular structures and the formation of hydration products, offering insights into the progress and mechanisms of cement hydration. It also facilitates the detection of chemical composition alterations due to environmental exposure, additives, or degradation processes, such as carbonation or sulfate attack, helping assess material stability and degradation mechanisms. Additionally, Raman spectroscopy provides spatially resolved information essential for microstructural characterization, including particle morphology, crystallinity, and pore structure, thus aiding in understanding material properties and heterogeneity.

Raman spectroscopy offers several advantages in cement research, including high chemical specificity that enables the identification and differentiation of various chemical species and phases within cement-based materials. Its non-destructive nature allows for the analysis of samples without altering their integrity, permitting multiple measurements and longitudinal studies on the same material. Additionally, Raman spectroscopy can achieve high spatial resolution with appropriate instrumentation, providing detailed insights into localized variations and heterogeneity within cementitious materials.

2.5.3 Solid-State NMR

NMR spectroscopy is a sophisticated analytical technique offering profound insights into the atomic-level structure, molecular dynamics, and chemical composition of materials, including cementitious systems. Solid-state NMR operates by analyzing the interactions between atomic nuclei with inherent magnetic properties and an external magnetic field. In cement-based materials, nuclei like ^{1}H, ^{27}Al, ^{29}Si, and ^{23}Na are commonly studied using NMR spectroscopy. Through radiofrequency pulses and subsequent detection of emitted signals during relaxation, solid-state NMR provides information on local atomic environments, crystal structures, and molecular dynamics. This technique aids in identifying different phases and crystalline structures present in cementitious materials, such as C–S–H, aluminates, and silicates. It enables monitoring the hydration process by examining changes in chemical shifts, revealing insights into the evolution of hydrated phases over time.

2.5.4 X-ray Photoelectron Spectroscopy

XPS operates on the principle of analyzing the elemental composition and chemical environment of a material's surface. It involves irradiating the sample surface with X-rays, which results in the ejection of photoelectrons from the material. By measuring the kinetic energy and intensity of these emitted electrons, XPS provides detailed information about the elemental composition, chemical bonding, and oxidation states present on the material's surface.

XPS enables precise identification and quantification of elemental composition on their surfaces, including elements such as calcium, silicon, aluminum, iron, and sulfur. This capability provides insights into the distribution and concentration of these elements. Understanding the chemical bonding and states of these elements is crucial for comprehending the reactivity and stability of cement-based materials. XPS can delineate oxidation states, detecting changes in chemical environments due to aging, hydration, or exposure to external factors. Surface modifications such as carbonation, hydration, or the presence of contaminants profoundly impact material properties. XPS can identify and quantify these alterations, aiding in assessing material degradation, understanding environmental impacts, and guiding preservation strategies. Furthermore, XPS facilitates interface analysis between various components within cement composites, helping evaluate adhesion, interfacial bonding, and potential weaknesses in the composite structure. With exceptional sensitivity, XPS detects elements and compounds even at low concentrations, offering specificity in identifying chemical states for precise characterization. Its non-destructive analysis preserves sample integrity, enabling multiple measurements in the same area and facilitating longitudinal studies to track changes over time. Moreover, XPS provides quantitative elemental analysis, delivering detailed information about elemental concentrations and their surface distribution in cement-based materials.

Despite its numerous advantages, XPS also faces challenges in analyzing cement-based materials. Issues such as sample heterogeneity, surface contamination, and charging effects can influence the accuracy of results. Improvements in sample preparation techniques, data interpretation methods, and instrument sensitivity are areas of ongoing research to enhance the reliability and applicability of XPS in cement studies.

2.6 Nanomechanical Properties Characterization

2.6.1 AFM-Based Indentation

AFM-based indentation is a powerful technique for probing mechanical properties at the nanoscale. It is a technique used to measure mechanical properties such as stiffness, elasticity, and hardness at the nanoscale level. Herein, we show how this technique is applied to measure the nanomechanical properties of cement-based materials.

2.6.1.1 Sample Preparation

The sample for AFM-based indentation requires an ultra-flat surface. Here, we take the synthetic C–(A)-S–H product as an example. Sample preparation procedures for this measurement are depicted in Fig. 2.1. The synthetic C-S–H should be ground into fine particles. Then, 4.5 g powders are used for each group to compact into the disc at a maximum loading pressure of 95 MPa. Before sample compaction, the dried synthetic C–S–H solids are ground and filtered through a 75 μm sieve. The compaction regime is generally similar to that used for nanoindentation measurements in Refs. [2, 3]. Those discs are set with epoxy, and then they are polished to obtain flat surfaces.

Before AFM measurements, all discs are vacuum-impregnated with epoxy resin and cured in a dictator at ambient temperature for 24 h. The impregnated specimens are sequentially ground with SiC paper in 400, 600, 800, 1000, 1200, 1500, 2000, and 2500 grit at 15 N load for about 2 min for each. To be clean, the specimens are ultra-sonic cleaned in ethanol for each grinding level. This procedure is to obtain a flat surface. After that, the flat surface is progressively polished with 3 μm diamond wheels at 15 N for 10 min and 250 nm diamond wheels for another 15 min.

2.6.1.2 AFM-Based Indentation Measurement

(1) Surface measurement

The topographic features and nanomechanical properties can be obtained from atomic force microscopy. These measurements are all performed on C–S–H discs prepared before (see Sect. 2.6.1.1). AFM observations are performed on those polished C–S–H discs to evaluate their topographic features. In this measurement, a Si-doped probe of Bruker scansyst-air model is selected and operated in ScanAsyst mode to

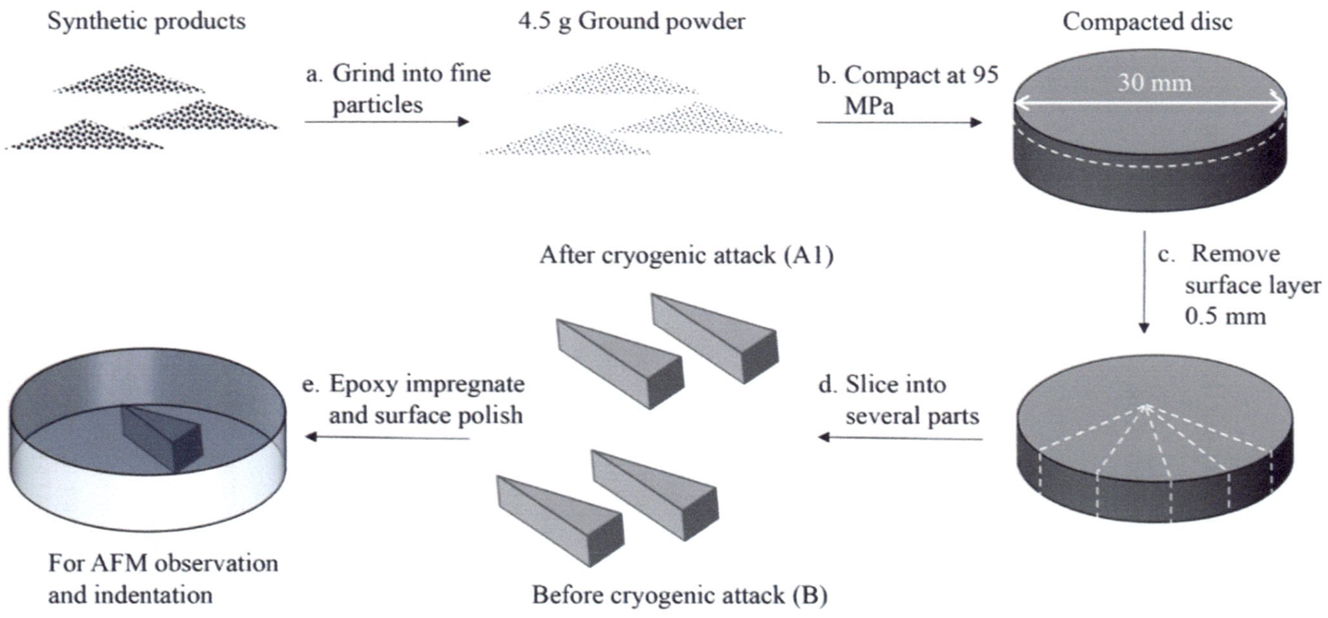

Fig. 2.1 Sample preparation for AFM-based indentation and AFM observation

scan the sample surface. The topographic maps of the C–S–H surface are recorded in the scanning scales of 2 μm, 500 nm, and 200 nm. Whilst, the surface roughness of the polished sample can be assessed from three scanning areas of 2 μm × 2 μm by using arithmetic mean roughness (R_a) and root mean square roughness (R_q). R_a and R_q can be calculated by Eq. 2.1 and Eq. 2.2, respectively:

$$R_q = \sqrt{\frac{1}{N}\sum_{i=1}^{N}\sum_{j=1}^{N}\left(z_{ij} - z_{av}\right)^2} \tag{2.1}$$

$$R_a = \frac{1}{N}\sum_{i=1}^{N}\sum_{j=1}^{N}\left|z_{ij} - z_{cp}\right| \tag{2.2}$$

where z_{ij}, z_{av}, z_{cp} denote the absolute height, average height, and height of the center plane, respectively. N represents the number of scanning points of the vision, which was 256 in this study. In order to ensure better statistics, different areas can be selected and scanned, obtaining an average roughness value for every sample.

(2) Indentation test

For the indentation test, contact mode is chosen for the tip operation. A stiff probe with a rectangular cantilever, e.g., Bruker RTESP-300 model, is used, which has a resonance frequency of 300 kHz and 42 N/m spring constant. Before formal indentation measurements, the deflection sensitivity and spring constant of the cantilever shall be calibrated first. The deflection sensitivity is calibrated by performing indentations on standard silica glass. The cantilever's spring constant can be calibrated by the thermal tune method on the standard stiff silica glass surface. The detailed calibration information can be referenced in Refs. [4, 5]. In formal measurements of the samples, the indentations are performed on 10 points in an area of 500 nm × 500 nm, and 20 locations are randomly selected. Overall, more than 200 indentations are performed for every specimen.

We can also apply AFM-based indentation mapping. In this mode, the sample is observed using an optical microscope to select the measured area. For each sample, indentations are performed on 20 × 20 grids within 2 × 2 μm^2. Hence, 400 indentations are obtained for each, as illustrated in Fig. 2.2b. The sample has a large exposed area over 5 × 5 mm^2; we can select optically the region in which we perform 2 × 2 μm^2 indentation grids to ensure that the grids are located in the measured sample and not in epoxy. Then, statistical analysis is performed for these indentation data using the Kernel Density Estimation (KDE) method [5]. From the statistical analysis, the probability density distribution of Young's moduli can be obtained.

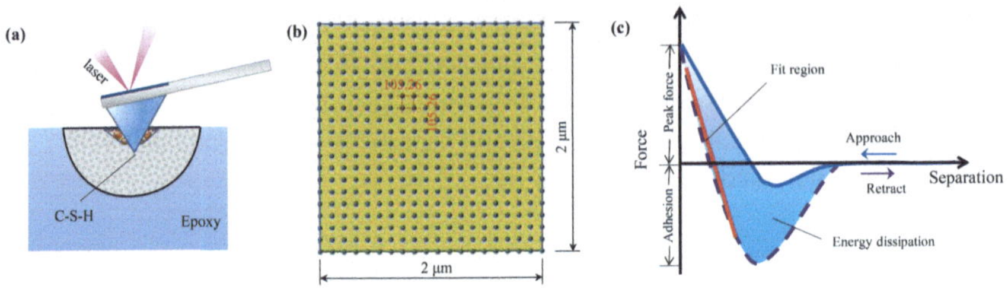

Fig. 2.2 Schemes of AFM-based indentation mapping technique. **a** Indentation of AFM tip on the sample (not to scale, since the sample region is much larger than the indentation depth); **b** indentation mapping 20 × 20 points within 2 × 2 μm^2 area located in C–S–H, the distance between two neighboring points is fixed at ~ 105.26 nm; **c** force-separation curve fit

2.6.1.3 Young's Modulus

The Hertz and Sneddon analytical theories are the most classical and commonly used for the AFM-based indentation. Here, the Hertz model is applied, which assumes the indenter shape is a sphere with a radius R. According to the Hertz theory, when the indenter is subjected to an applied load F, the correlation between the elastomer deformation δ and the load can be presented as Eq. 2.3:

$$F=\frac{4}{3}E^{*}R^{1/2}\delta^{3/2} \tag{2.3}$$

where E^{*} is the reduced Young's modulus of the elastomer, which can be determined from Eq. 2.4:

$$\frac{1}{E^{*}}=\frac{1-\nu_1^2}{E_1}+\frac{1-\nu_2^2}{E_2} \tag{2.4}$$

where υ_1, E_1 are the Poisson's ratio and the elastic Young's modulus of the elastomer, respectively, and υ_2, E_2 are that of the indenter. In our system, the elastic modulus of the indenter is high enough to leave out its contribution to the E^{*} [6]. Thus, the Eq. 2.4 can be expressed as:

$$\frac{1}{E^{*}}\approx\frac{1-\nu_1^2}{E_1} \tag{2.5}$$

The raw data obtained from the indentation is Photodiode deflection vs. Z displacement curves, which should be further processed into force-distance curves. As we have already calibrated the cantilever deflection sensitivity S, the cantilever deflection D can be obtained from:

$$D=D_{pd}\times S \tag{2.6}$$

where D_{ph} is the photodiode deflection, and S is the cantilever deflection sensitivity. Using the Hooke's law, the load F applied to the indenter can be computed from:

$$F = kD \tag{2.7}$$

Then, the indentation depth δ can be calculated from:

$$\delta = Z - D \tag{2.8}$$

where Z is the piezo displacement. Actually, because of the thermal drift in the detection system and also the stresses in the cantilever, the free cantilever has a deflection offset. Similarly, an offset exists in the piezo displacement Z [4]. Therefore, the actual cantilever deflection and piezo displacement should subtract the related offset. Thus, the load F and the indentation depth δ are modified as follows:

$$F = k(D - D_0) \tag{2.9}$$

$$\delta = (Z - Z_0) - (D - D_0) \tag{2.10}$$

where D_0 and Z_0 are the offset of cantilever deflection and that of piezo displacement, respectively.

The typical force-distance curve is presented in Fig. 2.2. The elastic Young's modulus is determined by fitting the retraction line (the green line in Fig. 2.2), with the fitting range from 10 to 90% [7]. The adhesion force between the tip and the contact surface is considered. Therefore, for the model fitting, the adhesion force F_{adh} is added into the applied force F, yielding the effective loading force F_{eff}, as presented in Eq. 2.11:

$$F_{eff} = F + F_{adh} \tag{2.11}$$

2.6.1.4 Probability Distribution Analysis

The calculated elastic moduli of each sample can be further statistically analyzed using KDE. By using this technique, the experimental probability density function (PDF) of those elastic moduli data can be obtained, which is presented as:

$$\hat{f}(x) = \frac{1}{hN} \sum_i K\left(\frac{|x - x_i|}{h}\right) \tag{2.12}$$

where $\hat{f}(x)$ is the probability density at the elastic modulus value of x, h denotes the window width, N is the amount of data, $K(\bullet)$ is the kernel function.

Given the heterogeneous nature of C-S–H gels, it is meaningful to identify individual phases and determine their properties and statistical spatial distribution. This requires a deconvolution of those calculated elastic moduli data. The deconvolution is done by fitting several probability density functions to the experimental PDF of elastic moduli, and the Gaussian function fits these PDFs. The deconvolution aims to identify each phase and solve its unknown statistical parameters of μ^E, σ^E, and f_j. They are calculated from:

$$f_j = \frac{N_j}{N} \tag{2.13}$$

$$\sum_{J=1}^{m} f_j = 1 \tag{2.14}$$

$$\mu_j^E = \frac{1}{N_j} \sum_{n=1}^{N_j} \chi_i^E \tag{2.15}$$

$$\left(\sigma_j^E\right)^2 = \frac{1}{N_j - 1} \sum_{n=1}^{N_j} \left(\chi_i^E - \mu_j^E\right)^2 \tag{2.16}$$

where N_j, f_j are the amount of *j*-th phase data and its surface fraction, respectively, N denotes the total amount of data, μ_j^E and σ_j^E stand for the mean deviation and standard deviation, respectively, χ_i^E represents the *i*-th calculated elastic modulus.

2.6.2 Nano/Micro-Indentation

Nano/Micro-Indentation is a technique that involves applying controlled, localized forces with a sharp indenter tip onto the surface of a material and measuring the resulting indentation to extract mechanical properties. This method operates on scales ranging from nanometers to micrometers, offering detailed information about the material's mechanical response at the microstructural level. This technique involves using a sharp indenter, typically made of diamond or sapphire, to apply a controlled force to the surface of a material. By measuring the depth of penetration as a function of applied force, one can obtain valuable information about the material's hardness, elastic modulus, and other mechanical properties. Compared with AFM-based indentation, the indenter used in micro/nano-indentation is much larger (> 200 nm), which allows it to measure the average local mechanical properties of cement-based materials.

In the case of cement-based materials, nanoindentation allows for the characterization of the complex microstructure and mechanical behavior of these materials at the nanoscale. Measurements of indentation hardness and elastic modulus offer crucial insights into the material's resistance to localized deformation and its ability to recover its original shape after deformation, which is essential for evaluating strength and stiffness at the microscale. Assessing fracture toughness through Nano/Micro-Indentation is vital for predicting durability, as it reveals the material's resistance to crack propagation and failure. By performing indentations at multiple locations, these techniques can also reveal variations in porosity, phase composition, and microstructure. Additionally, mapping mechanical properties across the surface of cement-based materials provides spatially resolved data, highlighting variations and identifying potential weak points or areas prone to deterioration.

2.6.3 High-Pressure XRD

The high-pressure XRD technique has emerged as a powerful tool in material science, enabling researchers to study the mechanical properties of materials under extreme conditions. This technique, which combines X-ray diffraction with high-pressure chambers, allows for the investigation of the behavior of materials at pressures up to several gigapascals.

One of the key advantages of the high-pressure XRD technique is its ability to provide detailed information about the crystal structure of materials under high pressure. By analyzing the diffraction patterns produced by X-rays interacting with the crystal lattice of a material, researchers can determine the changes in interatomic distances, bond angles, and unit cell parameters that occur as the material is subjected to increasing pressure. This information is crucial for understanding the mechanical response of materials to external forces and can provide valuable insights into the mechanisms governing their deformation and failure.

In addition to studying the crystal structure of materials, the high-pressure XRD technique can also be used to investigate their mechanical properties, such as their elastic moduli, strength, and ductility. By subjecting materials to controlled pressure environments and measuring their diffraction patterns in real-time, researchers can observe how their mechanical properties change as a function of pressure. This allows for the determination of important parameters such as the bulk modulus, shear modulus, and Poisson's ratio of materials, which are essential for predicting their behavior under different loading conditions.

One of the key challenges in using the high-pressure XRD technique to study the mechanical properties of materials is the need to design and construct high-pressure chambers that can withstand the extreme pressures required for these experiments. These chambers typically consist of a diamond anvil cell or a multi-anvil press, which can generate pressures in excess of several gigapascals. In addition, specialized X-ray diffraction equipment is required to generate high-intensity X-ray beams and collect diffraction data from materials under high pressure.

Despite these challenges, the high-pressure XRD technique has been successfully used to study a wide range of materials, including metals, ceramics, polymers, and minerals. By combining the unique capabilities of X-ray diffraction with the ability to apply controlled pressure to materials, researchers have been able to gain valuable insights into the mechanical behavior of materials at the atomic level. This knowledge is essential for developing new materials with tailored mechanical properties for applications in fields such as aerospace, automotive, and energy storage.

References

1. X. Yang, Z.P. Sun, L.L. Shui, Y.L. Ji, Characterization of the absolute volume change of cement pastes in early-age hydration process based on helium pycnometry. Constr. Build. Mater. **142**, 490–498 (2017)
2. J.J. Kim, E.M. Foley, M.M. Reda Taha, Nano-mechanical characterization of synthetic calcium–silicate–hydrate (C–S–H) with varying CaO/SiO_2 mixture ratios, Cem. Concr. Compos. **36**, 65–70 (2013)
3. E.M. Foley, J.J. Kim, M.M. Reda Taha, Synthesis and nano-mechanical characterization of calcium-silicate-hydrate (C-S-H) made with 1.5 CaO/SiO_2 mixture. Cem. Concr. Res. 42(9), 1225–1232 (2012)
4. S.S. Tan, R.L. Sherman, W.T. Ford, Nanoscale compression of polymer microspheres by atomic force microscopy. Langmuir **20**(17), 7015–7020 (2004)
5. X. Zhu, Z. Jiang, B. He, C. Qian, Investigation on the physical stability of calcium-silicate-hydrate with varying CaO/SiO_2 ratios under cryogenic attack. Constr. Build. Mater. **252**, 119103 (2020)
6. C.A. Jones, Z.C. Grasley, J.A. Ohlhausen, Measurement of elastic properties of calcium silicate hydrate with atomic force microscopy. Cem. Concr. Compos. **34**(4), 468–477 (2012)
7. G.H. Zeng, K. Dirscherl, J. Garnaes, Toward accurate quantitative elasticity mapping of rigid nanomaterials by atomic force microscopy: Effect of acquisition frequency, loading force, and tip geometry. Nanomaterials **8**(8), 616 (2018)

Chapter 3
Calcium–Silicate–Hydrate

3.1 Introduction

Among the key constituents of cement-based materials, the C–S–H phase stands as a linchpin, dictating the fundamental properties and performance of these materials. The intricate nature of C–S–H, a primary hydration product resulting from cement hydration, has captivated the interest of researchers and engineers for its profound influence on the strength, durability, and microstructural characteristics of cementitious systems.

The formation of C–S–H, arising from the reaction between calcium silicates and water during cement hydration, remains a subject of intense study. Synthesis methods for generating C–S–H in controlled environments have evolved, ranging from conventional techniques to advanced methodologies involving precursors and additives. The structural-chemical models proposed for C–S–H attempt to decipher its complex atomic arrangement and composition. Various microstruture models, from the classic C–S–H gels to more sophisticated representations, endeavor to capture the intricate nature of its nanostructure and bonding configurations. Furthermore, exploring the nanomechanical properties of C–S–H unveils crucial information on its stiffness, strength, elasticity, and deformation behavior at the micro and nano scales. Such insights contribute to understanding the material's behavior and aid in predicting the mechanical performance of cementitious materials. Advancements in computational methods, particularly molecular dynamics simulations, have offered a glimpse into the molecular-scale behavior and dynamics of C–S–H. These simulations provide a platform for investigating atomic interactions, diffusion mechanisms, and structure–property relationships within C–S–H, enriching our understanding of its behavior at the atomic level.

In what follows, we endeavor to present a comprehensive overview of research and developments related to the formation, micro/nanostructure, and properties of C–S–H. Some basic concepts related to C–S–H micro/nanostructure will be introduced here. In particular, the classical chemical-structural and microstructure models of

Z. Jiang and X. Zhu, *Cryogenic Stability of Cement-Based Materials at Micro/Nano Scale*, https://doi.org/10.1007/978-981-96-1836-1_3

C–S–H are summarized. This makes it easier to discuss the cryogenic stability of C–S–H in following chapters. The molecular simulation method in cement science is also briefly introduced.

3.2 Formation of Calcium–Silicate–Hydrate

3.2.1 *Cement Hydration*

The hydration process of cement is accompanied by the formation of C–S–H. In ordinary Portland cement, C_3S accounts for 50% to 70% of its content, and the hydration of C_3S is the main reaction process for producing C–S–H. The hydration of cement is a complex process, and to date, its specific hydration mechanism has not been thoroughly clarified. Based on the early hydration heat release process of C_3S, the classical viewpoint divided it into five stages: the pre-induction period, the induction period, the acceleration period, the deceleration period, and the steady period. However, in 2011, Bullard et al. [1] proposed a review suggesting it could be broadly divided into four main stages: the initial period, the period of slow reaction, the acceleration period, and the deceleration period, as shown in Fig. 3.1. In 2019, Scrivener et al. [2] further suggested it could be divided into three main stages: the induction period (3 h), the major hydration peak period (3 to 24 h), and the hydration after the main peak period (> 24 h). The growth morphology of C–S–H varies in these stages. The specific reaction processes are as follows:

C_3S dissolution:

$$Ca_3SiO_5 + 5H_2O \rightarrow 3Ca^{2+} + 6OH^- + H_4SiO_4 \tag{3.1}$$

C_x–S–H_y precipitation:

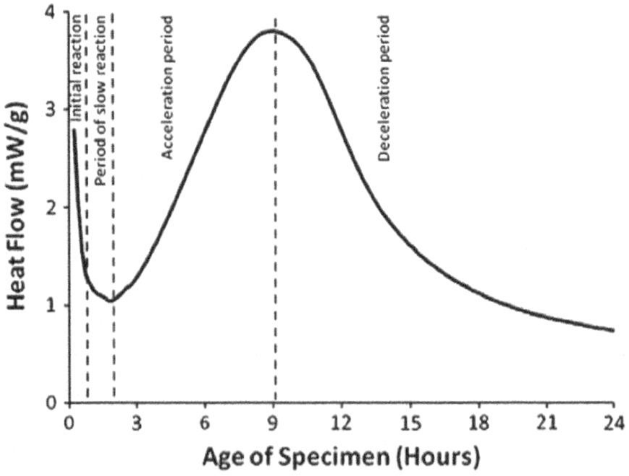

Fig. 3.1 C_3S hydration heat flow [1]

$$xCa^{2+} + 2xOH^{-} + H_4SiO_4 \rightarrow (CaO)_x - (SiO_2) - (H_2O)_b + (2 + x - b)H_2O \quad (3.2)$$

$Ca(OH)_2$precipitation:

$$yCa^{2+} + 2yOH^{-} \rightarrow yCa(OH)_2 \downarrow \quad (3.3)$$

There are two main hypotheses regarding the formation mechanism of the initial stage of C_3S hydration. The first is the metastable barrier hypothesis (also known as the protective membrane theory) [3, 4], while the second is the slow dissolution hypothesis [5, 6]. Both hypotheses explain the initial decrease in hydration heat by attributing it to the reduction in C_3S solubility to comply with the kinetic characteristics of hydration. The difference lies in the cause of the reduction in solubility between the two hypotheses. The metastable barrier hypothesis suggests that the initial slowdown in the hydration rate of C_3S is due to the rapid formation of a continuous thin layer of metastable C–S–H on the surface of C_3S. This metastable thin layer hinders the unhydrated inner C_3S from coming into contact with water in the liquid phase, preventing it from reaching equilibrium with the hydrates. On the other hand, the slow dissolution hypothesis proposes that the initial decrease in solubility of C_3S is because its surface becomes hydroxylated upon contact with water, making it difficult for C_3S to dissolve. This theory was first proposed by Barret et al. [5, 6] and was further explained by Garrault et al. [7]. Their research indicated that the actual dissolution rate of C_3S is much lower than the calculated dissolution rate. When the concentration of $Ca(OH)_2$ in the liquid phase increases, the solubility of C_3S significantly decreases. Only after the C–S–H in the liquid phase becomes supersaturated does it nucleate on the surface of C_3S particles. Due to the small surface area of the particles, the initial growth of C–S–H is relatively slow. Observations using cryo-electron microscopy by Nicoleau et al. [8] essentially rejected the first hypothesis. They continued the viewpoint of the second hypothesis and validated it by monitoring ion dissolution rates and concentrations. The hydration theory of C_3S proposed by Nicoleau et al. [9] is gradually gaining acceptance in the field.

According to existing research findings, the formation of C–S–H involves two primary processes: nucleation and growth. The exact timing of C–S–H nucleation remains undetermined. Thomas [10] and Scherer [11], using boundary nucleation growth models, successfully fitted data from isothermal calorimetry and chemical shrinkage experiments. They indicated that the initial nucleation time of C–S–H is close to the mixing time. Additionally, in studies on the microscopic structural development of C_3S hydration, Bullard et al. [12, 13] further noted that C–S–H undergoes rapid nucleation within a few minutes after the contact of C_3S with water. When considering the metastable barrier and slow dissolution hypotheses, these points in time correspond respectively to the early stage of nucleation growth and the later stage of initial hydration. The growth of C–S–H determines its structural morphology. Currently, two main models have been proposed for the structural morphology of C–S–H: the nanoscale particle aggregation model and the defective C–S–H chain-like model, which will be detailed in Sect. 3.4. According to the nanoscale particle aggregation model proposed by Allen et al. [14–16], C–S–H particles stop growing and reach a stable state after reaching several nanometers. These stable C–S–H nanoparticles facilitate the nucleation of new particles on their surface in the liquid phase or promote aggregation between nanoparticles [17]. Regarding the defective C–S–H chain-like model, Gartner [18] proposed a corresponding growth mechanism. This model suggests that tetrahedral bridging of silica forms silicate chains, which then compose two-dimensional silicate sheet layers. Calcium ions (Ca^{2+}) and hydroxide ions (OH^-) are embedded between these silicate sheet layers, forming structures similar to Tobermorite or Jennite-like C–S–H. Initial embryos nucleate and grow in an organized layered manner, forming crystal structures approximately 5 nm in size. However, after the formation of two-dimensional silicate sheets, the increased growth points of C–S–H might lead to an increase in potential defects within the interlayer silicate chains. The rise in defects within the silicate chains could cause lattice strain, leading to distortions between the silicate sheet layers and eventually resulting in the amorphous structure of C–S–H. Between these two mechanisms for C–S–H formation, the defective C–S–H chain-like model aligns better with the kinetic characteristics and nanostructural features of C–S–H.

3.2.2 Precipitation

The primary raw materials for synthesizing C–S–H gel using the chemical precipitation method are alkali silicate, calcium salt, NaOH, and deionized water. Alkali silicate is often sodium silicate ($Na_2SiO_3{\cdot}5H_2O$ or $Na_2SiO_3{\cdot}9H_2O$), while calcium salt can be either calcium nitrate ($Ca(NO_3)_2{\cdot}4H_2O$) or calcium chloride ($CaCl_2{\cdot}2H_2O$). The solution should have a pH greater than 12, and the synthesis can occur at room temperature. Additionally, to maintain the solution's ionic strength at $I = 0.245$ [19], a certain amount of NaCl may be added as needed. The process of C–S–H generation in the solution can be represented by Eq. 3.4.

$$x\mathrm{Ca}^{2+} + 2x\mathrm{OH}^- + H_4\mathrm{SiO}_4 \rightarrow (\mathrm{CaO})_x - (\mathrm{SiO}_2) - (\mathrm{H_2O})_y + (2 + x - \mathrm{y})\mathrm{H_2O} \tag{3.4}$$

During the synthesis process, the C–S–H gel is highly susceptible to carbonation and requires necessary measures to prevent the ingress of CO_2 at each stage. Before preparing the solution, it is advisable to remove small amounts of CO_2 present in deionized water by thoroughly purging it with nitrogen gas. Additionally, depending on specific research requirements, the entire synthesis process can range from several minutes to several days. Throughout this process, it is essential to maintain a nitrogen environment consistently. Once the samples are prepared, they should be stored in airtight containers to protect them from environmental exposure.

The chemical precipitation method for synthesizing C–S–H is often used to study its early crystallization formation process [19–21]. The reaction process can be relatively controlled, allowing sufficient time to observe the formation of C–S–H. It should be noted that the water-to-solid ratio in the solution used for chemical precipitation is extremely high, indicating an almost unrestricted growth space for C–S–H. This significantly differs from the hydration process of actual cementitious materials, potentially leading to differences in the growth morphology of C–S–H.

3.2.3 Pozzolanic Reaction

Pozzolanic reaction describes the chemical reaction between reactive silica, $Ca(OH)_2$, and water. The following formula can represent the basic chemical reaction equation:

$$x\mathrm{Ca(OH)_2} + y\mathrm{SiO_2} + z\mathrm{H_2O} \rightarrow (\mathrm{CaO})_x - (\mathrm{SiO_2})_y - (\mathrm{H_2O})_{x+z} \tag{3.5}$$

The pozzolanic reaction is a process that allows for reasonable control of the Ca/Si ratio in the desired reaction products. Various methods have emerged based on different synthesis conditions, including ambient temperature volcanic ash reaction synthesis, thermo-mechanochemical synthesis, and hydrothermal synthesis. Under ambient temperature conditions, the pozzolanic reaction typically requires a higher water-to-solid ratio and longer reaction times, often several weeks to months, to obtain relatively pure C–S–H. In contrast, thermo-mechanochemical synthesis significantly accelerates the reaction rate, with complete reactions usually achieved within several hours. However, this method requires continuous grinding of reactants, increasing the complexity of the synthesis process. Hydrothermal synthesis is more efficient and straightforward than the previous two methods, offering high operability.

Hydrothermal synthesis utilizes hydrothermal conditions to accelerate the dissolution rate of poorly soluble reactants, enhance the activity of reactants, and expedite the reactions between ions in solution. Controlling different temperatures and pressures makes it possible to produce hydration products with varying degrees of crystallinity. This method is widely used to synthesize crystalline minerals, such as tobermorite [22]. It is generally believed that the amorphous form of C–S–H acts as a precursor to crystalline tobermorite [23, 24]. The crystallinity of the C–S–H precursor increases with higher hydrothermal temperatures, tending to form tobermorite crystal structures.

C–S–H hydrothermal synthesis primarily involves CaO, SiO_2, and deionized water as the main raw materials. Prior to the hydrothermal reaction, the CaO undergoes high-temperature calcination at 1000 °C to remove any moisture, organic impurities, and carbides that might be present in the raw material, ensuring the production of pure CaO. The calcined CaO is then finely ground and passed through a sieve with a mesh finer than 200, resulting in highly refined reaction materials with a larger specific surface area. This increased surface area enhances the reactivity of the materials, thereby accelerating the formation of C–S–H.

Temperature, pressure, and reaction time are the main factors influencing the hydrothermal synthesis of C–S–H. High temperatures and pressures favor the synthesis of crystalline tobermorite, although typically, the synthesis of C–S–H is achieved using atmospheric pressure heating methods. Typical heating temperatures range from 60 to 95 °C, with some researchers employing room temperature and pressure conditions [25]. Higher temperatures can be anticipated to result in a faster rate of C–S–H synthesis. Hydrothermal synthesis is relatively more straightforward than the chemical precipitation method and does not require continuous stirring.

3.2.4 Sol–gel Synthesis

Sol–gel synthesis is a condensation reaction after the hydrolysis of metal alkoxides. The commonly used silicon precursor is tetraethyl orthosilicate (TEOS), and its reaction equation is as follows:

$$\mathrm{Si(OEt)_4} + n\mathrm{H_2O} \rightarrow \mathrm{Si(OEt)_{4-n}(OH)_n} + \mathrm{nEtOH} \tag{3.6}$$

$$\mathrm{Si(OEt)_{4-n}(OH)_n} \rightarrow \mathrm{(OEt)_{4-n}Si - O - Si(OEt)_{4-n})} + \mathrm{H_2O} \tag{3.7}$$

This reaction is significantly influenced by pH. When the pH is below 7 or in the presence of salts, it forms a gel network. Conversely, when the pH is above 7, it results in the formation of a sol.

Usually, calcium ethoxide ($\mathrm{Ca(OEt)_2}$) is used as the calcium precursor. During the aforementioned reaction, introduced $\mathrm{Ca^{2+}}$ ions from the calcium precursor enter the colloidal network and undergo the following reaction:

$$\mathrm{Ca^{2+}} + 2\mathrm{Si(OEt)_{4-n}(OH)_n} \rightarrow \left(\mathrm{(OEt)_{4-n}Si - O-}\right)_2\mathrm{Ca} + \mathrm{H^+} \tag{3.8}$$

The duration of this reaction varies from several minutes to several hours. After the reaction, the sample must be repeatedly washed with ethanol and water to remove impurities.

3.3 Chemical-Structural Model of C–S–H

Over the past century, concrete has continuously undergone technological innovations toward higher strength, durability, and functionality. Scientists have attempted to study the structure and properties of concrete at various scales, including macroscopic, microscopic, nanoscopic, and even molecular levels, aiming to fundamentally enhance the structural performance of concrete from the micro-nanoscale perspective. To this end, it is imperative to understand the fundamental structure of C–S–H, given it is the primary hydration product of cement. Over recent decades, efforts have persistently aimed to thoroughly understand the formation mechanism, molecular structure, and its correlation with the macroscopic properties of C–S–H.

Advancements in science and technology have greatly enriched the research means in concrete science, making this exploration achievable. Researchers such as Cong et al. [26] employed solid-state nuclear magnetic resonance spectroscopy to identify the Q^n distribution of silicon tetrahedra in C–S–H. Xu et al. [27] were the first to observe clear layered images of C–S–H gel at the nanoscale using transmission electron microscopy, revealing crystalline structural characteristics at a 5 nm scale, demonstrating both short-range order and long-range disorder. Tennis et al. [28] introduced the existence of two different packing densities phases in C–S–H, termed LD C–S–H and HD C–S–H, using neutron scattering (SANS), small-angle scattering (SAXS), and NMR techniques, a theory validated by Plassard et al. [29] through nanoindentation techniques. Bonaccorsi et al. [30] utilized XRD to elucidate the nanocrystalline structural parameters of Jennite-type C–S–H. Nonat et al. [31] and Plassard et al. [29] used atomic force microscopy to analyze C–S–H finely. Allen et al. [32] quantitatively analyzed the molecular formula of C–S–H as $(CaO)_{1.7}\cdot(SiO_2)\cdot(H_2O)_{1.8}$ using SANS and SAXS techniques. Geng et al. [33] resolved the crystal structure parameters of C-(A)-S–H and calculated its superior mechanical properties using X-ray synchrotron high-pressure powder diffraction. Gmira et al. [34] studied the creep behavior of C–S–H at the nanoscale using Monte Carlo simulations, while numerous researchers used first-principles calculations [35, 36], molecular dynamics simulations [37, 38], Monte Carlo simulations [39, 40], and other methods to conduct molecular simulation studies on C–S–H. Kulik and Kersten [41, 42] used chemical kinetics methods to simulate and investigate the early growth stages, conducting extensive research on the mutual influence of various ions in the solution on C–S–H growth. Our understanding of C–S–H has continuously deepened along with the evolution of research technologies and the development of theoretical methods. Let's review a few classical models for describing the atomistic structure of C–S–H.

3.3.1 Stade-Kurczyk-Schwiete Model

In 1962, Kurczyk et al. [43], in their research on the hydration of C_3S and β-C_2S, suggested that C–S–H possesses a tobermorite-like sandwich-layered structure. According to their speculations, the silicon-oxygen chains in C–S–H are infinitely long, and they proposed that the Ca/Si molar ratio results from the presence of Ca^{2+} and OH^- ions between layers. Their proposed molecular formula for C–S–H is $Ca[Ca_4Si_6O_{16}(OH)_2][Ca(OH)_2]_{5\to7}\cdot mH_2O$.

Stade et al. [44] further refined and expanded on the structural model proposed by Kurczyk and Schwiete. They envisioned the presence of CaO_x polyhedra between two

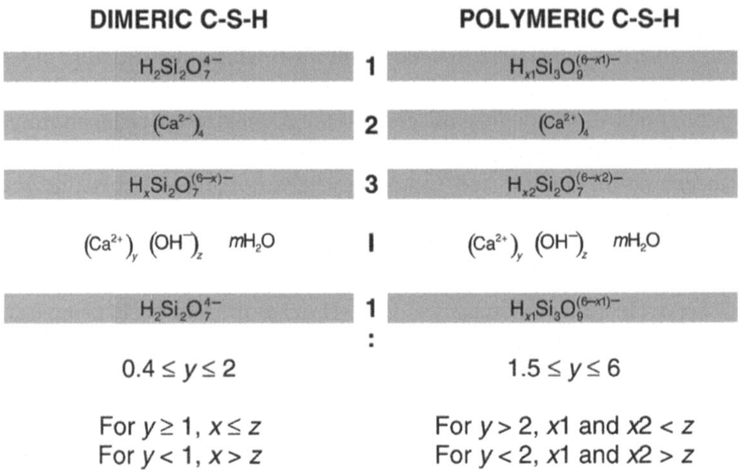

Fig. 3.2 Scheme of Stade-Kurczyk-Schwiete model [45]

silicon-oxygen chain layers, forming a sandwich structure. Between these sandwich structures, they postulated the existence of Ca^{2+}, OH^-, H_2O, and other types of ions. Unlike Kurczyk's model, in Stade's model, the silicon-oxygen chains in C–S–H are not infinitely long. According to their assumptions, the silicon-oxygen chains in C–S–H exist in dimers and oligomers. Based on this, Richardson [45] created a model of this structure, as shown in Fig. 3.2. In this model, the molecular formula of C–S–H is represented as $Ca_{4+y}(OH)_z[(H_{1+x/2}Si_2O_7)]_2 \cdot mH_2O$.

3.3.2 Taylor Model

In 1986, Taylor [46] proposed that C–S–H is composed of two phases, C–S–H (I) and C–S–H (II), which correspond to defective structures of tobermorite (Ca/Si < 1.5) at 1.4 Å and jennite (Ca/Si > 1.5) respectively. According to Taylor's theory, C–S–H possesses a sandwich structure comprising silicon-oxygen chain and calcium–oxygen layers. The silicon-oxygen chains in C–S–H consist of dimers of silicon tetrahedral units and Dreierkettes chains. Each Dreierkettes structural unit comprises one bridging silicon tetrahedron and two opposing silicon tetrahedra, with the oxygen atoms in the bridging silicon tetrahedron coordinated with a CaO unit, some of which may be replaced by hydroxyl groups, as depicted in Fig. 3.3. Taylor highlighted that the length of the silicon-oxygen chains in C–S–H depends on the Ca/Si molar ratio. The length of the silicon-oxygen chains follows the $3n$–1 rule, implying that the silicon-oxygen chains could only consist of 2, 5, 8… $3n$–1 silicon tetrahedral units. This finding was further supported by Richardson et al. [47].

The Taylor model can explain numerous experimental phenomena, such as the distribution of silicon in different chemical environments after the hydration and

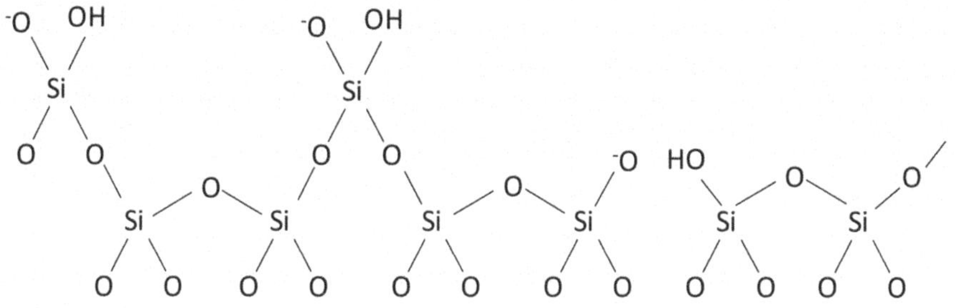

Fig. 3.3 Silicate chains of C–S–H [46]

hardening of ordinary Portland cement, the degree of disorder, chain length, and polymerization of C–S–H. However, it fails to explain the phenomenon observed in cement paste where, with the aging process, C–S–H transitions from dimers to oligomers [48].

3.3.3 Richardson-Groves Model

Richardson and Groves [47, 49, 50] suggested that the Taylor model represents just one particular case of the C–S–H structure. They proposed a generalized model for C–S–H based on the relative content of various ions within C–S–H. This model consists of the T/J (tobermorite-jennite) and T/CH (tobermorite-$Ca(OH)_2$) models. It can be represented by a generalized chemical formula (3.9), where $1 \leq y \leq 5$.

$$\left[\mathrm{Ca}_{2n}H_n\mathrm{Si}_{(3n-1)}O_{(9n-2)}\right]\cdot(\mathrm{OH})_{n(y-1)}\cdot\mathrm{Ca}_{\frac{n\cdot y}{2}}\cdot mH_2O \tag{3.9}$$

when $y = 1$ and $y = 5$, this chemical formula is represented as Eqs. 3.10 and 3.11, respectively, indicating the T/J structure and T/CH structure, which are the situations within the Taylor model.

$$\mathrm{T(3n-1)}:\left[\mathrm{Ca}_{2n}H_n\mathrm{Si}_{(3n-1)}O_{(9n-2)}\right]\cdot\frac{n}{2}\mathrm{Ca}\cdot mH_2O \tag{3.10}$$

$$\mathrm{J(3n-1)}:\left[\mathrm{Ca}_{4n}H_n\mathrm{Si}_{(3n-1)}\mathrm{OH}_{(9n-2)}(\mathrm{OH})_{4n}\right]\cdot\frac{n}{2}\mathrm{Ca}\cdot m\mathrm{H_2O} \tag{3.11}$$

Considering that in real scenarios, the OH^- ions in C–S–H vary with the water-to-cement ratio of the cementitious material and its surrounding environment, Richardson and Groves [47] modified Esq. 3.9–3.12 in order to maintain the interlayer charge balance of C–S–H.

$$\left[\mathrm{Ca}_{2n}H_w\mathrm{Si}_{(3n-1)}O_{(9n-2)}\right]\cdot(\mathrm{OH})_{w+n(y-1)}\cdot\mathrm{Ca}_{\frac{n\cdot y}{2}}\cdot m\mathrm{H_2O} \tag{3.12}$$

where $n = 1, 2, 3\ldots$, when $0 \leq y \leq 2$, $n(2 - y) \leq w \leq 2n$; when $2 \leq y \leq 4, 0 \leq w \leq 2n$; when $4 \leq y \leq 6, 0 \leq w \leq n(6 - y)$.

Additionally, considering the presence of alkali metal ions, especially Al^{3+} ions that substitute for Si^{4+} in the bridging tetrahedra of C–S–H in systems like fly ash and mineral powder, Richardson et al. [50] further modified the model to the form shown in Eq. 3.13.

$$\{\mathrm{Ca}_{2n}\mathrm{H}_w(\mathrm{Si}_{1-a}\mathrm{R}_a^{[4]})_{(3\mathrm{n}-1)}\mathrm{O}_{(9n-2)}\} \cdot I^{c+}_{\frac{a}{c}(3n-1)} \cdot (\mathrm{OH})_{w+n(y-2)} \cdot \mathrm{Ca}_{\frac{n \cdot y}{2}} \cdot m\mathrm{H_2O} \quad (3.13)$$

where R [4] represents trivalent cations, primarily Al^{3+}, located within tetrahedral coordination; I^{c+} denotes interlayer ions in C–S–H, representing monovalent alkali metal ions or Ca^{2+}, which, after substituting Si^{4+} by R^{3+} ions, maintain the charge balance of C–S–H.

3.3.4 Cong-Kirkpartric Model

Cong and Kirkpatrick [26, 51, 52] proposed the defect Tobermorite model, which can be described by two equations representing the structural composition relationship, depicted as Eqs. 3.14 and 3.15.

$$Ca/Si = r/\left(1 - \frac{m}{3}\right) \quad (3.14)$$

$$Q^1/\sum Q^i = 2/\left(\frac{m}{3} - 1\right) \quad (3.15)$$

In this model, r represents the initial Ca/Si ratio when the silicon-oxygen chain is infinitely long, m signifies the ratio of missing bridging silicon tetrahedra, and Q^i denotes the silicon tetrahedra at different sites. In this context, a higher m value correlates with a higher Ca/Si ratio, indicating a greater number of missing bridging silicon tetrahedra, resulting in thinner layers of silicon-oxygen chains. This aligns with Cong's XRD testing results. Richardson [53] conducted a comparative analysis between this model and the Richardson-Groves model, concluding that there is no fundamental difference between the two. After conversion, both models yield the same results, as detailed in Ref. [53].

In the Cong-Kirkpartric model, different Ca/Si ratios of C–S–H are considered defective structures formed by 14 Å tobermorite. According to the proportion of missing bridging silicon tetrahedra units, these defective structures have two types. One involves a minority of missing bridging silicon tetrahedra at specific sites, resulting in partial depolymerization and twisting of the silicon-oxygen chains. The other type involves a higher proportion of missing bridging silicon tetrahedra, leading to increased dimers, significant depolymerization, severe twisting, and the presence of Ca^{2+} ions, OH^- ions, and H_2O between the layers of the silicon-oxygen chains,

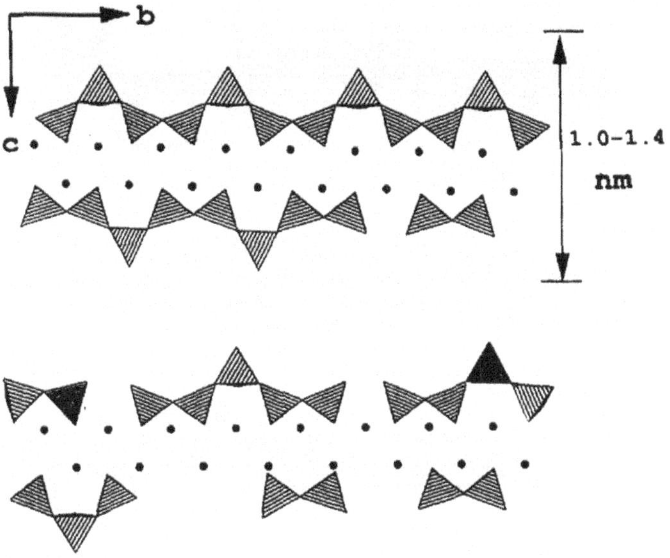

Fig. 3.4 Scheme of Cong-Kirkpartric model [26]

as depicted in Fig. 3.4. The ^{29}Si MAS NMR test results of Espinose et al. [54] suggest that the silicon tetrahedra at different sites follow a Qn distribution, with Q^0 accounting for 13%, Q^1 for 67%, and Q^2 for 20%. Subsequent molecular simulation studies [37, 55, 56] have used this conclusion to construct C–S–H structure models. Furthermore, Cong et al. [26] considered the bonding of interlayer ions from the perspective of charge balance and proposed that the bonding between Ca–OH and Si–OH depends on the Ca/Si ratio.

3.3.5 Cross-Linked Substituted Tobermorite Model

In consideration of the calcium–(sodium)–aluminosilicate-hydrate (C–(N)–A–S–H) products formed in alkali-activated slag systems, Myers et al. [57] expanded on the Richardson-Groves model and proposed a Cross-linked Substituted Tobermorite Model. This model allows for determining the Ca/Si ratio, Al/Si ratio, and average chain length (MCL) within the C–(N)–A–S–H system. As shown in Fig. 3.5, the C–(N)–A–S–H structure comprises both cross-linked and non-cross-linked forms, each with distinct chemical structures corresponding to different scenarios.

For the crosslinked tobermorite model, the chemical structural formula of the dreierketten unit can be represented as:

$$\left[(CaSiO_{3.5})^-\right]_2 \cdot \left\{\left(\left[SiO_{1.5}^+\right]\cdot\left[Si_{(1-\alpha)}Al_\alpha O_{1.5}^{(1-\alpha)+}\right]\cdot\left[Na_\alpha^{\alpha+}\right]\right)_{(1-\delta)}\cdot\right.$$

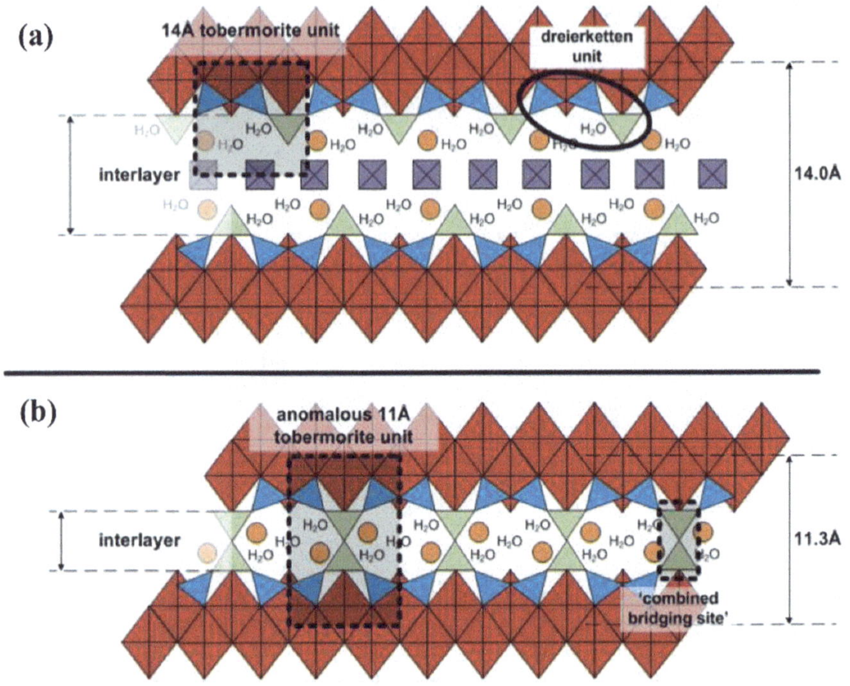

Fig. 3.5 The crosslinked (**a**) and non-crosslinked (**b**) model of calcium–(sodium)–aluminosilicate-hydrate [57]

$$\left.\left(\mathrm{Ca}_{\omega}\mathrm{H}_{2(1+\delta-\omega)}\right)^{2(1+\delta)+}\right\}_{0.5}\cdot m\mathrm{H_2O} \tag{3.16}$$

The variables are as follows: α represents the substitution rate of aluminum atoms in the cross-linked tobermorite; δ stands for the proportion of bridging vacancies per tobermorite unit, calculated as $\delta = 1/(\sigma + 1)$; σ indicates the quantity of cross-linked tobermorite per chain; ω signifies the content of calcium in the interlayer of tobermorite units, where $\omega = u/2$; u denotes the hydroxylation rate within the silicate chains.

The parameters such as MCL, Al/Si ratio, Ca/Si ratio, Ca/(Al + Si) ratio, δ, and α can be determined by the following equations:

$$\mathrm{MCL}_{[C]} = \frac{4\left[\mathrm{Q}^1 + \mathrm{Q}^2 + \mathrm{Q}^2(1\mathrm{Al}) + \mathrm{Q}^3 + 2\mathrm{Q}^3(1\mathrm{Al})\right]}{\mathrm{Q}^1} \tag{3.17}$$

$$\mathrm{Al/Si}_{[C]} = \frac{4\mathrm{Q}^3(1\mathrm{Al})}{\mathrm{Q}^1 + \mathrm{Q}^2 + \mathrm{Q}^2(1\mathrm{Al}) + \mathrm{Q}^3 + \mathrm{Q}^3(1\mathrm{Al})} \tag{3.18}$$

$$\mathrm{Ca/Si}_{[C]} = \frac{4+\omega}{4+(1-\delta)(2-\alpha)} \tag{3.19}$$

$$\mathrm{Ca/(Al + Si)}_{[C]} = \frac{4 + \omega}{2(3 - \delta)} \tag{3.20}$$

$$0 \leq \omega \leq 1 + \delta \tag{3.21}$$

$$\delta = \frac{6}{\mathrm{MCL}_{[C]} + 2} \tag{3.22}$$

$$\alpha = \frac{\mathrm{Al/Si}_{[C]}[4 + 2(1 - \delta)]}{\left[\mathrm{Al/Si}_{[C]} + 1\right](1 - \delta)} \tag{3.23}$$

where C represents the crosslinked structure.

For the non-crosslinked tobermorite model, the chemical structural formula of its dreierketten unit can be represented as:

$$\left[(\mathrm{CaSiO}_{3.5})_2^-\right] \cdot \left\{\left(\left[\mathrm{Si}_{(1-a)}\mathrm{Al}_a\mathrm{O}_2\right] \cdot \left[\mathrm{Na}_a^+\right]\right)_{(1-v)} \cdot \left(\mathrm{Ca}_\varphi \mathrm{H}_{2(1-\varphi)}\right)^{2+}\right\} \cdot m\mathrm{H_2O} \tag{3.24}$$

The parameters, Ca/Si, a, v, can be calculated as follows:

$$\mathrm{Ca/Si}_{[NC]} = \frac{2 + \varphi}{2 + (1 - v)(1 - a)} \tag{3.25}$$

$$0 \leq \varphi \leq 1 \tag{3.26}$$

$$\mathrm{Ca/(Al + Si)}_{[NC]} = \frac{2 + \omega}{2 + (1 - v)} \tag{3.27}$$

$$v = \frac{3}{\mathrm{MCL}_{[C]} + 1} \tag{3.28}$$

$$a = \frac{\mathrm{Al/Si}_{[NC]}[2 + (1 - v)]}{\left[\mathrm{Al/Si}_{[NC]} + 1\right](1 - v)} \tag{3.29}$$

3.3.6 Richardson's C–(A)–S–H(I) Model

In 2015, Richardson [58] proposed the C–(A)–S–H(I) model. For the non-crosslinked tobermorite or C–S–H (I), its structural chemical formula can be represented as Eq. 3.30.

$$\mathrm{Ca}_4\left[\mathrm{Si}_{1-f-v}\mathrm{Al}_f\square_v O_{3-2v}\right]H_{2i}\mathrm{Ca}_{2-i}(\mathrm{Ca,Na_2,K_2})_{3f} \cdot \mathrm{mH_2O} \tag{3.30}$$

$$i = 6\left[1 - \left(\frac{C}{S}\right)(1 - v)\right] \tag{3.31}$$

It considers the aluminum substitution scenario. □ represents the vacancy rate of the silicate tetrahedral site, while f and v represent the substitution rate and the vacancy rate of aluminum in the silicate chain, respectively. i indicates the degree of charge balance for H^+ or Ca^{2+}.

The alkaline cations (Ca^{2+}, Na^+, K^+) within parentheses are primarily there to maintain charge balance and exist within the silicate chain layers.

Regarding the crosslinked C-(A)-S–H, the structural chemical formula can be represented as:

$$Ca_4[Si_{1-f-v}Al_f\square_v O_{(17/6)-2v}]_6H_{2i}Ca_{2-i}(Ca, Na_2, K_2)_{3f} \cdot mH_2O \tag{3.32}$$

Combined formula (1.30), the mixed phase of cross-linked and non-cross-linked C-(A)-S–H can be described by the following formula:

$$Ca_4\left[Si_{1-f-v}Al_f\square_v O_{[(18-d)/6-2v]}\right]H_6Ca_{2-i}(Ca, Na_2, K_2)_{3f} \cdot mH_2O \tag{3.33}$$

where d is the fraction of crosslinked chains, $0 \le d \le 1$. These variables can be determined through ^{29}Si MAS NMR measurements.

3.3.7 Other Structural-Chemical Formula Model

Nonat [31] and Chen [59], and others have proposed their respective C–S–H structure models but have not deviated from the basic framework of the above-mentioned models. The model by Nonat is based on two fundamental assumptions: firstly, there are no Si–OH groups in the tobermorite structural unit, and secondly, the interlayer Ca participates in forming Si–O-Ca-OH chains instead of being the main layer Ca. In Chen's model, a higher Ca-OH content corresponds to a higher Ca/Si ratio, and conversely, the value is lower. Gartner et al. [60] introduced their latest structural model, representing the chemical formula of the basic unit of C–S–H as $[Ca_2Si_2O_7]^{2-}$ and the interlayer cations as $[Ca(H_2O)_6]^{2+}$ or $[Ca(OH)(H_2O)_5]^+$. This model explains various phenomena, such as the crystallinity of C–S–H, the irreversibility of some water loss during initial drying, and irrecoverable shrinkage phenomena.

3.4 Microstructure Model of C–S–H

The structural-chemical model of C–S–H is suitable for explaining its molecular composition and structure but falls short in explaining the formation of concrete strength, C–S–H pore distribution, specific surface area, shrinkage, creep, interparticle forces, and other related issues. Hence, many researchers have attempted to analyze its structure and properties from a colloidal scale perspective, proposing corresponding colloid models for the microstructure.

3.4.1 P-B Model

The P-B colloidal model was proposed by Powers and Brownyard [61, 62]. In this model, Powers [61] suggested that C–S–H gel consists of needle- or plate-shaped crystalline particles and gel pores, as depicted in Fig. 3.6, with the crystalline particles being $3CaO \cdot 2SiO_2 \cdot 3H_2O$. Regarding the formation of C–S–H strength, Powers [61] suggested that it is primarily provided by physical forces such as van der Waals forces between gel particle surfaces. Brunauer [63] modified this model by proposing that C–S–H gel is a layered structure, aggregated from fibrous particles, with gel particles composed of 2–3 layers of low-crystallinity gel particles.

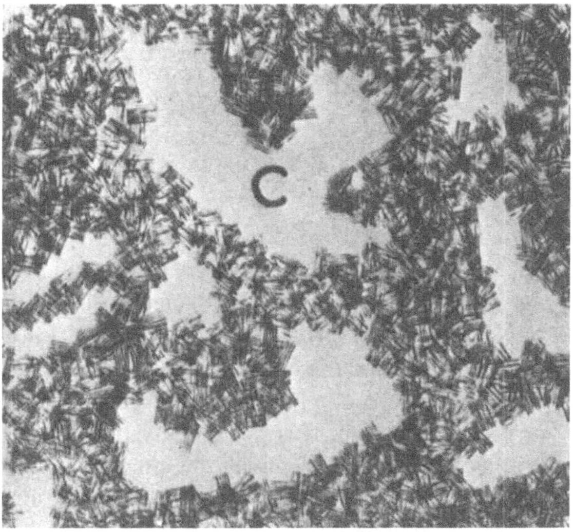

Fig. 3.6 Scheme of power model [62]

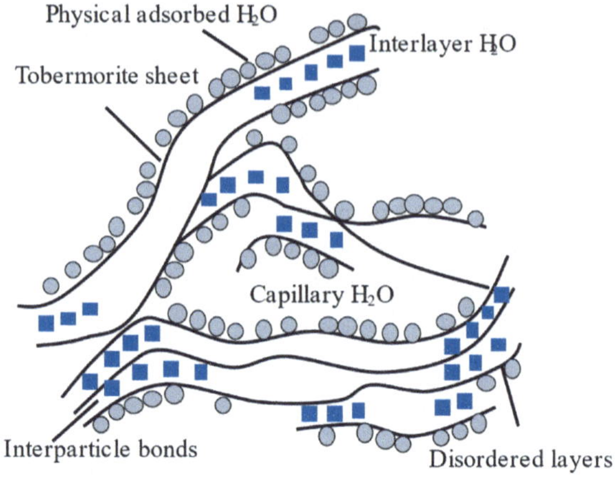

Fig. 3.7 Scheme of F-S model [48, 66]

3.4.2 *F-S Model*

Feldman and Sereda [64, 65] further refined the colloidal model proposed by Powers and Brownyard, introducing the layered colloidal model as depicted in Fig. 3.7. In this model, interlayer water is considered a part of the C–S–H gel structure, constituting chemically bound water that affects creep and shrinkage. Simultaneously, chemical bonds are formed at contact points between adjacent layers. Lowering the water-to-cement ratio densifies the structure of C–S–H, creating more binding sites for structural water.

3.4.3 *Munich Model*

The Munich model was proposed by Wittmann [67] in 1976 with the primary objective of explaining the shrinkage and creep phenomena of hardened cement paste under different humidity conditions. Wittmann developed a three-dimensional model of the C–S–H gel, considering it as a dry gel made up of colloidal particles, with water filling the spaces between these colloidal particles, exerting a certain degree of constraint on them, as illustrated in Fig. 3.8. Wittmann suggested that the shrinkage of cement paste is caused by an increase in attractive forces between the gel particles under low environmental humidity or under the influence of compressive stress, reducing the distance between them, which is manifested macroscopically as shrinkage. However, when the relative humidity of the environment is high, water molecules enter the spaces between gel particles, resulting in an expansion effect on the C–S–H gel as a whole. According to test results, relative humidity significantly affects long-range forces and surface energy, explaining why hardened cement paste exhibits greater creep under higher relative humidity. However, the Munich model fails to explain

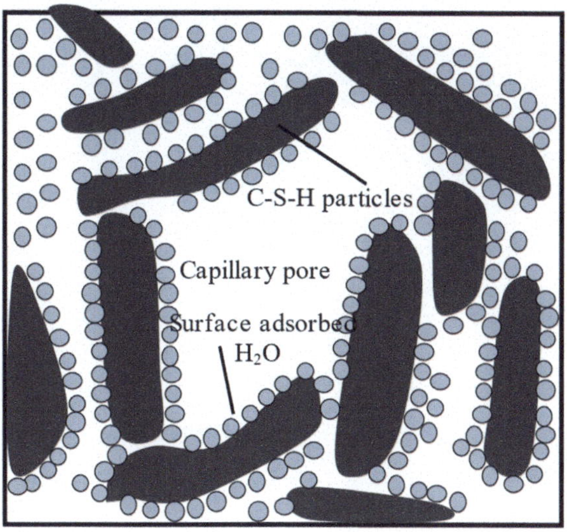

Fig. 3.8 Scheme of Munich model [66, 67]

the changes in the internal nano-scale pores of C–S–H caused by creep or drying shrinkage [48].

3.4.4 CM-I Model

Jennings [15] conducted tests using SANS and SAXS to measure the specific surface area and density of C–S–H in cement paste on a scale of 1–100 nm. He proposed the CM-I colloid model, depicted in Fig. 3.9. In this model, the fundamental unit of C–S–H is a dense spherical colloidal particle with a diameter of about 2.2 nm. These small colloidal units aggregate to form gel clusters approximately 5.6 nm in size, which further assemble to create the C–S–H gel.

According to Tennis and Jennings [28], C–S–H can be classified into high-density C–S–H (HD) and low-density C–S–H (LD) based on their packing densities. They proposed that nitrogen more easily permeates LD C–S–H compared to HD C–S–H. The ratio of LD C–S–H can be determined through nitrogen adsorption, as shown in Eq. 3.34

$$M_r = \frac{S_{N_2} M_D}{S_{LD} M_t} \tag{3.34}$$

where S_{N2} represents the specific surface area of the cured cement paste after drying, M_D denotes the mass of the cement paste, S_{LD} stands for the area per unit mass of dried LD C–S–H, and M_t represents the total mass of C–S–H

Generally, it is believed that HD C–S–H grows at the anhydrous interface, constituting an internal hydration product; whereas LD C–S–H pervades the matrix,

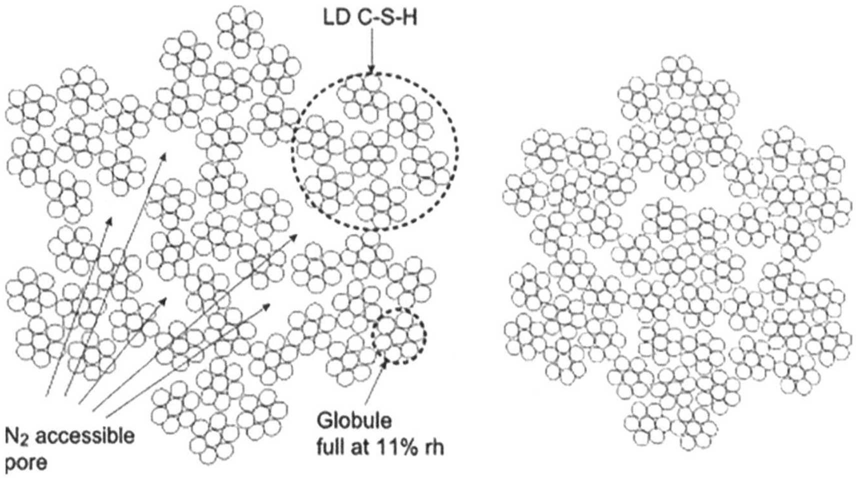

Fig. 3.9 CM-I model [15]

exhibiting a higher content during the early hydration stages and falls under external hydration products. Due to its higher density, HD C–S–H displays increased stiffness and hardness. Ideally, LD C–S–H and HD C–S–H have approximately 37% and 24% porosity, respectively, with elastic moduli around 22 GPa and 29 GPa [15, 68]. Allen and Jennings [38] determined through SAXS and hydrogen/deuterium (H/D) isotope techniques that the chemical formula of saturated C–S–H is $(CaO)_{1.7}\cdot(SiO_2)\cdot(H_2O)_{1.8}$, with a density of 2.605 g/cm^3 and particle size of 4.2 nm.

However, the CM-I model cannot explain the phenomena of thermodynamic adsorption as well as the relationship between the elastic or viscoelastic behavior of cement paste and relative humidity.

$$M_r = 3.017(w : c)\alpha - 1.374\alpha + 0.538 \tag{3.35}$$

where w:c represents the water to cement ratio, and α denotes the degree of hydration.

The volume fraction of HD C–S–H is:

$$V_{HD} = \frac{M_t - (M_r M_t)}{\rho_{HD}} \tag{3.36}$$

The volume of LD C–S–H can be expressed by:

$$V_{LD} = \frac{M_r M_t}{\rho_{LD}} \tag{3.37}$$

3.4.5 CM-II Model

To address the recoverable and non-recoverable shrinkage phenomena observed in C–S–H during drying and compression, Jennings made further modifications to his proposed CM-I model, referred to as the CM-II model [69]. Similar to CM-I, this model views C–S–H colloids as assemblies of brick-like colloidal particles. The internal water within colloidal particles and the water present in the pores between particles also constitute a part of the C–S–H gel. The most significant feature of this model integrates a layered structure similar to the F-S model, enabling the CM-I model to explain the shrinkage during isothermal adsorption and the creep of colloids under compressive stress. Jennings divided the gel pores into three levels: interlayer pores within the basic colloidal unit, small gel pores of 1–3 nm formed between these basic units, and larger gel pores of 3–12 nm resulting from the accumulation of basic colloidal units, as illustrated in Fig. 3.10. It is suggested that the irrecoverable shrinkage occurs due to the loss of water within the larger gel pores.

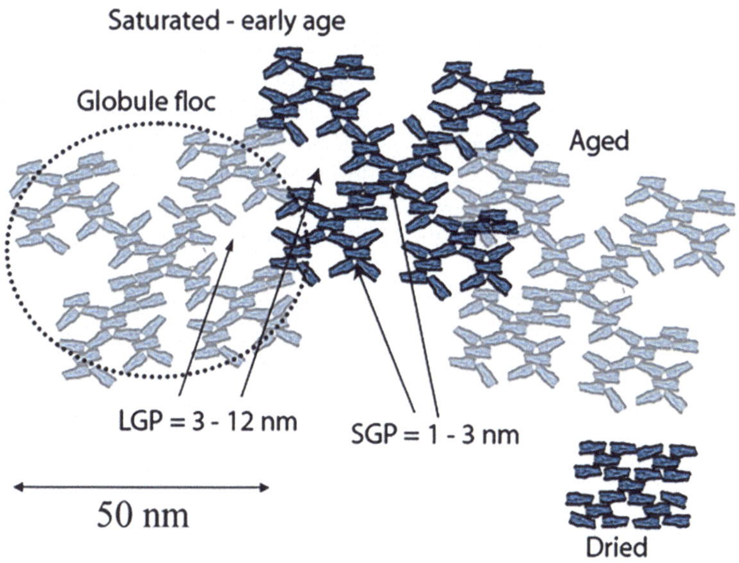

Fig. 3.10 CM-II model [69]

3.5 Microscopic Characteristics of C–S–H

3.5.1 C–S–H Gel Water

Water plays a crucial role in the structure and properties of C–S–H. The water inside the C–S–H gel can be divided into chemically bound water, physically adsorbed water, and free water. According to Jennings' model [15, 69], the water inside C–S–H is believed to be located in pores formed between clustered nanoparticles. In contrast, Feldman-Sereda's model [64] suggests that the gel water is attached to small pores between silicon layers. Both models indicate that a small portion of non-evaporable water is tightly bound to the silicon layer in C–S–H.

According to Muller et al. [70], the gel water in cement paste can be divided into gel pore water and interlayer water at the nanoscale based on the difference in proton relaxation times. The relaxation time of gel pore water is about three times that of interlayer water, indicating that gel pore water is located in larger pores. The ratio of gel pore water to interlayer water decreases with increasing hydration degrees for different hydration degrees of cement paste. At the beginning of hydration, this ratio is about 2, and as the hydration degree further increases, its value will approach 1. Gajewicz et al. [71] found that C–S–H loses most of its interlayer water and gel pore water in a dry environment at 60 °C. However, when the C–S–H that has undergone this drying process is immersed in water again, interlayer water will gradually recover first, while gel pore water recovers more slowly. When the signal of gel pore water is enhanced, the signal of interlayer water is weakened, indicating that the transport channel of gel pore water in C–S–H must pass through the interlayer pore.

According to Powers [72], pore water is defined as water that is not affected by the surface thermodynamics of C–S–H, which means that capillary water is at least two molecular layers above the surface. Jennings [73] defines gel water as pore water that can still exist stably when the relative humidity is 85% RH. Different types of water in C–S–H will be lost to varying degrees at different relative humidities. According to Ref. [74], when the relative humidity drops from 100% RH to 90% RH, some of the water in larger capillaries is lost, and the cement undergoes slight shrinkage. When the relative humidity drops to 40% RH, most of the capillary and gel water is lost, and the cement undergoes significant shrinkage accompanied by large stress. When the relative humidity drops to 20% RH, most adsorbed water or free water in small gel pores is lost, resulting in slight shrinkage. When the relative humidity drops from 20% RH to 0% RH, the interlayer water in C–S–H is lost, resulting in irreversible structural deformation.

According to Kelvin's formula, when the relative humidity drops to 85% RH, all water in pores larger than 10 nm will be lost, causing a pressure of about 10 MPa in the pores and resulting in partial shrinkage deformation [15]. When the relative humidity drops to 40% RH, LD C–S–H undergoes collapse and significant shrinkage deformation [15]. In the CM-II model [69], there is no water in the 1.5 nm gel pores. C–S–H almost does not shrink below 40% RH. At 11% RH, multi-layer adsorbed water begins to form in the interlayer pores and on the surface of gel particles, and there is no other free water except interlayer water. Jennings [69] believes that when the relative humidity drops from 100% RH to 50% RH, the shrinkage of C–S–H is irreversible, the range of 50% ~ 11% RH is recoverable, and < 11% RH is irreversible. Muller et al. [70] further pointed out through 1H NMR testing that interlayer water in C–S–H will be lost rapidly when the relative humidity is lower than 25% RH. The ^{1}H NMR test results of Gajewicz et al. [71] show that drying the gel pores for a period of time in a 40 °C environment for 2 days or 23% RH can remove the water in the gel pores, but interlayer water in C–S–H is still retained.

3.5.2 Micro/Nanomechanical Properties

As mentioned in Sect. 3.4.4, C–S–H exists in HD C–S–H and LD C–S–H forms, depending on the packing density. Constantinides et al. [75] measured two different hardnesses of C–S–H using nanoindentation testing, which were 29.4 GPa and 21.7 GPa, respectively. This test result is consistent with the model proposed by Jennings [28] and confirms the existence of HD C–S–H and LD C–S–H. The elastic modulus of C–S–H also varies depending on the characteristic water-to-cement ratio of the cement paste. Constantinides et al. [69] defined the elastic modulus range of LD C–S–H as 13–26 GPa and that of HD C–S–H as 26–39 GPa. Ulm et al. [76, 77] further conducted nanoindentation testing on cement paste with different water-to-cement ratios and found that, in addition to LD C–S–H and HD C–S–H, there is another type of C–S–H with ultra-high density, which they called UHD C–S–H, and its elastic modulus is as high as 63 GPa. However, Chen et al. [78] subsequently confirmed through nanoindentation and SEM–EDX testing that this UHD C–S–H with ultra-high elastic modulus and hardness is actually caused by nano-calcium hydroxide embedded in C–S–H. Some nanoindentation test results [79] show that the higher the Ca/Si ratio of C–S–H, the greater its elastic modulus, while the high-pressure XRD test results of Li et al. [80] are opposite to this. The reason for this difference may be due to the influence of sample orientation on the nanoindentation test results, while high-pressure XRD avoids this problem well.

3.6 Force Fields for Molecular Dynamics Simulation of Cement

Molecular dynamics simulation is an important research method for studying the structural characteristics and physical properties of molecules or molecular systems. Thanks to the rapid development of computer technology, molecular simulation has become an essential means of understanding the physicochemical properties of materials at the molecular level after experimental research and theoretical derivation. The earliest molecular dynamics simulation research can be traced back to the 1950s. Alder and Wainwright [81] simulated the motion characteristics of a rigid small ball molecular system. They further extended their research to the structure and properties of molecular systems with square well potential [82]. In the following decades, the corresponding theories and technologies of molecular dynamics simulation have made great progress. Through molecular dynamics simulation, researchers can easily obtain basic information such as the thermodynamic properties, dynamic characteristics, and structural evolution process of materials under specific conditions.

Molecular dynamics simulation has been successfully applied to the scientific research of cement concrete. With the help of molecular dynamics simulation research methods, people have begun to understand the creep [83], ion transport characteristics [84], composite structure characteristics [85, 86], and the thermodynamic and mechanical properties [87–90] of cement-based materials under various conditions - at the molecular and atomic levels, which has greatly improved the research level of cement concrete science.

The interactions between molecules in a molecular system need to be described using a molecular force field, that is, a potential function. Whether the potential function can accurately describe the atomic interactions in the molecular system is the key to the success of molecular dynamics simulation, which determines whether the simulation results are comparable to experimental results. For the empirical potential function of cementitious materials, several have been developed, including Born–Mayer-Huggins potential [91], ClayFF potential [92], CSHFF potential [93], CementFF potential [94], IFF potential [95], ReaxFF potential [96], and UFF potential [97], etc. In the following, we will introduce the basic principles of molecular dynamics simulation and several essential potential functions related to cementitious systems.

3.6.1 Theoretical Foundation of Molecular Dynamics Simulation

The basic theory of molecular dynamics simulation is developed based on classical mechanics. From the perspective of classical mechanics, complex molecular systems can be appropriately simplified. As we all know, molecular systems are composed of atoms, and the main mass of atoms is concentrated in the atomic nucleus. The mass of electrons outside the nucleus can be ignored, so a single atom can be simplified as a single particle. Therefore, the molecular system can be regarded as a particle mechanics system composed of many atoms. At the same time, the intermolecular interactions and intramolecular interactions between atoms in the molecular system are both conservative forces. Therefore, in this simplified molecular system, the total energy is conserved, and its Hamiltonian function is:

$$H = K + u \tag{3.38}$$

where K represents the total kinetic energy of the molecular system, and u represents its potential energy, which is only related to the spatial position of the atoms. In the Cartesian coordinate system, K and u can be expressed as:

$$K = \frac{1}{2}\sum_{i=1}^{N} m_i\left(\dot{x}_i^2 + \dot{y}_i^2 + \dot{z}_i^2\right) \tag{3.39}$$

$$u = u(x_1, y_1, z_1, \ldots, x_i, y_i, z_i, \ldots, x_n, y_n, z_n) \tag{3.40}$$

where N is the number of atoms in the molecular system, m_i is the mass of the i-th atom, x_i, y_i, and z_i are the position vector of the i-th atom.

Since momentum $\boldsymbol{p}$ is related to velocity as follows

$$\begin{cases} p_{i,x} = m_i\dot{x}_i \\ p_{i,y} = m_i\dot{y}_i \\ p_{i,z} = m_i\dot{z}_i \end{cases} \tag{3.41}$$

Equation 3.38 can be rewritten as follows:

$$H = \sum_{i=1}^{N} \frac{1}{2m_i}\left(p_{i,x}^2 + p_{i,y}^2 + p_{i,z}^2\right) + u(x_1, y_1, z_1, \ldots, x_i, y_i, z_i, \ldots, x_n, y_n, z_n) \tag{3.42}$$

In the case of molecular systems, using Cartesian coordinates to describe the movement of atoms within the molecular system is not very convenient. To address this apparent limitation, generalized coordinates are commonly used in actual calculations to describe atomic movements. These coordinates, denoted by q_i (where i = 1, 2, ..., f), refer to different types of coordinates such as center-of-mass coordinates, Euler angles, internal coordinates within the molecule, and others. In the generalized coordinate system, for a mechanical system with f degrees of freedom, the center-of-mass coordinates of an individual particle can be expressed as $\boldsymbol{q} = (q_1$, q_2, ..., q_i, ..., $q_f)$ using f generalized coordinates. Similarly, its motion state can also be represented by f generalized momenta $\boldsymbol{p}$, which are denoted as $\boldsymbol{p} = (p_1, p_2, \ldots, p_i, \ldots, p_f)$. In the generalized coordinates, the Hamiltonian function of the molecular system can be expressed as:

$$H = \sum_{i=1}^{f}\sum_{j=1}^{f} a_{ij}p_ip_j + u\left(q_1, q_2, \ldots, q_i, \ldots, q_f\right) \tag{3.43}$$

Based on this, the equations governing the motion of a molecular system can be expressed as the Hamiltonian equations:

$$\begin{cases} \frac{\partial H}{\partial q_i} = -\dot{p}_i \\ \frac{\partial H}{\partial p_i} = \dot{q}_i \end{cases} i = 1, 2, \ldots f \tag{3.44}$$

In practical applications, for unconstrained molecular systems, the Hamiltonian equations can be simplified to Newton's equations:

$$\frac{\partial}{\partial t}p_i = -\frac{\partial}{\partial q_i}H \tag{3.45}$$

$$m_i \frac{d^2}{dt^2}\boldsymbol{q}_i = -\frac{\partial}{\partial q_i}u(\boldsymbol{q}) \tag{3.46}$$

Equation 3.46 represents the form of Newton's second law, that is:

$$\boldsymbol{F}_i = m_i \boldsymbol{a}_i \tag{3.47}$$

Obviously, molecular dynamics is based on Newtonian mechanics, which simulates the motion of molecular systems. It samples different system states to compute thermodynamic quantities and other macroscopic properties within the system.

3.6.2 ClayFF Force Field

ClayFF force field was initially developed to describe the interactions within mineral–water interfaces of clay minerals. In the development of the actual potential functions, due to the highly complex coordination conditions of Ca^{2+} ions in structures like C–S–H and AFm, an alternative approach was proposed. This method considered most bonding interactions in the crystal as pseudo ions (i.e., not bonded formally) and required that the appropriate crystal structure and local coordination could be maintained solely by the balance between atomic electrostatics and van der Waals repulsion terms within the force field [92]. This simplified approach allows molecular simulations of complex molecular systems using a small number of force field parameters. However, it's essential to note that for most silicate minerals, the ClayFF force field overlooks the nature of their chemical bonds, which are primarily covalent with some ionic bonding [98].

The total potential energy of the simulation system in ClayFF is typically composed of the electrostatic interaction (E_{Coul}), short-range non-bonded interaction (E_{VDW}), and limited explicit bonded interactions ($E_{\mathrm{Bond\ Strength}}$ and $E_{\mathrm{Angle\ Bend}}$), represented as follows:

$$E_{\mathrm{total}} = E_{Coul} + E_{VDW} + E_{BondStrentch} + E_{AngleBend} \tag{3.48}$$

The electrostatic interaction can be represented by Coulomb's law:

$$E_{Coul} = \frac{e^2}{4\pi\varepsilon_0}\sum_{i\neq j}\frac{q_i q_j}{r_{ij}} \tag{3.49}$$

where q_i and q_j are the residual charges of the atoms, *e* represents the charge of an electron, and ε_0 is the vacuum permittivity. The short-range non-bonding interaction terms include the dominant energy increase due to short-range repulsion when two atoms approach each other and the attractive dispersion energy as the distance between atoms increases:

$$E_{VDW} = \sum_{i\neq j} D_{0,ij}\left[\left(\frac{R_{0,ij}}{r_{ij}}\right)^{12} - 2\left(\frac{R_{0,ij}}{r_{ij}}\right)^{6}\right] \tag{3.50}$$

where $D_{0,ij}$ and $R_{0,ij}$ are empirical parameters that can be computed for different atoms based on Lorentz-Berthelot rule between two types of atoms:

$$R_{0,ij} = \frac{\left(R_{0,j} + R_{0,j}\right)}{2} \tag{3.51}$$

$$D_{0,ij} = \sqrt{D_{0,i}D_{0,j}} \tag{3.52}$$

The explicit bonding interaction includes the stretching potential energy between covalent bonds and the angular bending potential energy, which can be expressed as:

$$E_{BondStrentchij} = k_i\left(r_{i,j} - r_0\right)^2 \tag{3.53}$$

$$E_{AngleBendij} = k_2\left(\theta_{i,j} - \theta_0\right)^2 \tag{3.54}$$

where k_1 and k_2 are harmonic force constants. r_0 and $\theta_{0,}$ respectively, denote the equilibrium values for bond length and bond angle.

3.6.3 CSHFF Force Field

CSHFF is a force field developed specifically for C–S–H based on its structure and stiffness constants [93]. It is an improved version derived from the ClayFF force field and is aimed at predicting structural features and higher-order properties, such as elastic constants. In comparison to other force fields like Core–Shell and ReaxFF, CSHFF has lower computational costs, making it more efficient for large-scale systems. CSHFF does not consider short-range Lennard–Jones interactions between cations and assumes that the Coulombic repulsion among cations is sufficient. Compared to ClayFF, its formulation is simpler.

3.6.4 INTERFACE Force Field

INTERFACE force field finds significant applications in describing interface structures of different molecular systems, such as inorganic–organic interfaces and inorganic-biomolecule interfaces. The atomic models established based on IFF parameters can compute various properties of systems, including lattice constants, density, surface energy, solid–liquid interfacial tension, anisotropy in interfacial energies on different crystal facets, atomic structural features, and thermodynamic properties [99].

In the INTERFACE force field, the total potential energy of a molecular system can be represented by Eqs. 3.55 or 3.56:

$$E_{pot} = \sum_{ijbonded} K_{r,ij}\left(r_{ij} - r_{0,ij}\right)^2 + \sum_{ijbonded} K_{\theta,ij}\left(\theta_{ij} - \theta_{0,ij}\right)^2 + \frac{1}{4\pi\varepsilon_0}\sum_{\substack{ijbonded\\(1,3excl)}} \frac{q_i q_j}{r_{ij}} + \sum_{\substack{ijbonded\\(1,3excl)}} \left[2\left(\frac{\sigma_{ij}}{r_{ij}}\right)^9 - 3\left(\frac{\sigma_{ij}}{r_{ij}}\right)^6\right] \tag{3.55}$$

$$E_{pot} = E_{bonds} + E_{angles} + E_{coul} + \sum_{\substack{ijbonded\\(1,3excl)}} \varepsilon_{ij}\left[\left(\frac{\sigma_{ij}}{r_{ij}}\right)^9 - 2\left(\frac{\sigma_{ij}}{r_{ij}}\right)^6\right] \tag{3.56}$$

The total potential energy represented by Eqs. 3.55 and 3.56 consists of covalent bond stretching potential, bond angle bending potential, Coulomb potential, and van der Waals potential.

Compared to other force fields, the IFF (Interface Force Field) does not rely on quantum mechanical calculations of atomic charges due to the significant uncertainty associated with atomic charges obtained from any quantum chemistry calculation [98, 100]. In the IFF, atomic charges are approximated by assuming that only one charge is precisely distributed for each dipole moment, thus avoiding the substantial uncertainty related to quantum–mechanical electron density.

3.6.5 ReaxFF Forcefield

The ReaxFF force field, initially developed by Van Duin et al. [96] to study combustion reactions, has subsequently been widely applied in molecular dynamics simulations of inorganic materials. In ReaxFF, the total potential energy E_{total} of a molecular system is composed of the bond stretching energy E_{bond}, angle bending energy E_{angle} , torsion energy $E_{torsion}$, penalty energy E_{overl}, van der Waals force E_{vdW}, Coulombic energy E_{Coul}, and specific potential energy $E_{specific}$. It is expressed as follows:

$$E_{total} = E_{bond} + E_{angle} + E_{torsion} + E_{over} + E_{vdW} + E_{coul} + E_{specific} \quad (3.57)$$

Unlike non-reactive force fields, the ReaxFF force field does not establish potential energy relationships between atomic distances or angles and bond interactions using constant functions. Instead, the potential energy is a continuous function dependent on bond order. Bond order represents the regional electron density between two atoms, thereby reflecting their bonding strength. Bond orders are computed from atomic positions at each simulation step and introduced into the potential to balance interactions. If the bond order decreases between two atoms, their interaction diminishes until it reaches zero, indicating an effective bond breakage. The last term, $E_{specific}$, incorporates specific energy terms for certain systems, such as hydrogen bonds in water molecules.

References

1. J.W. Bullard, H.M. Jennings, R.A. Livingston, A. Nonat, G.W. Scherer, J.S. Schweitzer, K.L. Scrivener, J.J. Thomas, Mechanisms of cement hydration. Cem. Concr. Res. **41**(12), 1208–1223 (2011)
2. K. Scrivener, A. Ouzia, P. Juilland, A.K. Mohamed, Advances in understanding cement hydration mechanisms, Cem. Concr. Res. 124 (2019).
3. H.N. Stein, J.M. Stevels, Influence of silica on the hydration of 3 CaO, Sio2. J. Appl. Chem. **14**(8), 338–346 (1964)
4. H.M. Jennings, P.L. Pratt, An experimental argument for the existence of a protective membrane surrounding portland cement during the induction period. Cem. Concr. Res. **9**(4), 501–506 (1979)
5. P. Barret, D. Ménétrier, D. Bertrandie, Mechanism of C3S dissolution and problem of the congruency in the very initial period and later on. Cem. Concr. Res. **13**(5), 728–738 (1983)
6. P. Barret, D. Ménétrier, Filter dissolution of C_3S as a function of the lime concentration in a limited amount of lime water. Cem. Concr. Res. **10**(4), 521–534 (1980)
7. S. Garrault, A. Nonat, Hydrated layer formation on tricalcium and dicalcium silicate surfaces: Experimental study and numerical simulations. Langmuir **17**(26), 8131–8138 (2001)
8. L. Nicoleau, A. Nonat, D. Perrey, The di- and tricalcium silicate dissolutions. Cem. Concr. Res. **47**, 14–30 (2013)
9. L. Nicoleau, A. Nonat, A new view on the kinetics of tricalcium silicate hydration. Cem. Concr. Res. **86**, 1–11 (2016)
10. J.J. Thomas, J.J. Biernacki, J.W. Bullard, S. Bishnoi, J.S. Dolado, G.W. Scherer, A. Luttge, Modeling and simulation of cement hydration kinetics and microstructure development. Cem. Concr. Res. **41**(12), 1257–1278 (2011)
11. G.W. Scherer, J. Zhang, J.J. Thomas, Nucleation and growth models for hydration of cement. Cem. Concr. Res. **42**(7), 982–993 (2012)
12. J.W. Bullard, A determination of hydration mechanisms for tricalcium silicate using a kinetic cellular automaton model. J. Am. Ceram. Soc. **91**(7), 2088–2097 (2008)
13. R.A. Livingston, J.S. Schweitzer, C. Rolfs, H.W. Becker, S. Kubsky, Characterization of the induction period in tricalcium silicate hydration by nuclear resonance reaction analysis. J. Mater. Res. **16**(3), 687–693 (2001)

14. A.J. Allen, R.C. Oberthur, D. Pearson, P. Schofield, C.R. Wilding, Development of the fine porosity and gel structure of hydrating cement systems. Philos. Mag. B **56**(3), 263–288 (1987)
15. H.M. Jennings, A model for the microstructure of calcium silicate hydrate in cement paste, Cem. Concr. Res. **30** (2000) 101–116.
16. H.M. Jennings, J.W. Bullard, J.J. Thomas, J.E. Andrade, J.J. Chen, G.W. Scherer, Characterization and modeling of pores and surfaces in cement paste: Correlations to processing and properties. J. Adv. Concr. Technol. **6**(1), 5–29 (2008)
17. J.J. Thomas, H.M. Jennings, J.J. Chen, Influence of nucleation seeding on the hydration mechanisms of tricalcium silicate and cement. J. Phys. Chem. C **113**(11), 4327–4334 (2009)
18. E.M. Gartner, A proposed mechanism for the growth of C-S-H during the hydration of tricalcium silicate. Cem. Concr. Res. **27**(5), 665–672 (1997)
19. N. Krautwurst, L. Nicoleau, M. Dietzsch, I. Lieberwirth, C. Labbez, A. Fernandez-Martinez, A.E.S. Van Driessche, B. Barton, S. Leukel, W. Tremel, Two-step nucleation process of calcium silicate hydrate, the nanobrick of cement. Chem. Mater. **30**(9), 2895–2904 (2018)
20. M. Schönlein, J. Plank, A TEM study on the very early crystallization of C-S-H in the presence of polycarboxylate superplasticizers: Transformation from initial C-S-H globules to nanofoils. Cem. Concr. Res. **106**, 33–39 (2018)
21. G. Artioli, L. Valentini, M. Voltolini, M.C. Dalconi, G. Ferrari, V. Russo, Direct imaging of nucleation mechanisms by synchrotron diffraction micro-tomography: Superplasticizer-induced change of C–S–H nucleation in cement. Cryst. Growth Des. **15**(1), 20–23 (2015)
22. M. Diez-Garcia, J.J. Gaitero, J.S. Dolado, C. Aymonier, Ultra-fast supercritical hydrothermal synthesis of tobermorite under thermodynamically metastable conditions. Angew. Chem. Int. Ed. Engl. **56**(12), 3162–3167 (2017)
23. T. Mitsuda, H.F.W. Taylor, Influence of aluminium on the conversion of calcium silicate hydrate gels into 11 Å tobermorite at 90°C and 120°C. Cem. Concr. Res. **5**(3), 203–209 (1975)
24. 童雪莉, 刘公诚, 托勃莫来石类水化硅酸钙形成过程的研究 硅酸盐学报 (03) (1963) 129–138.
25. X.L. Tong, G.C. Liu, Tuo bo mo lai shi lei shui hua gui suan gai xing cheng guo chen de yan jiu. J. Chin. Ceram. Soc. **03**, 129–138 (1963). ((In Chinese))
26. Y. He, L. Lu, L.J. Struble, J.L. Rapp, P. Mondal, S. Hu, Effect of calcium–silicon ratio on microstructure and nanostructure of calcium silicate hydrate synthesized by reaction of fumed silica and calcium oxide at room temperature. Mater. Struct. **47**(1–2), 311–322 (2014)
27. X. Cong., R.J. Kirkpatrick., ^{29}Si MAS NMR study of the structure of calcium silicate hydrate, Adv. Cem. Based Mater. **3** (1996) 144–156.
28. Z. Xu., W. Viehland., Observation of a mesostructure in calcium silicate hydrate gels of portland cement. Phys. Rev. Lett. **5**(5) (1996) 952–955.
29. P.D. Tennis., H.M. Jennings., A model for two types of calcium silicate hydrate in the microstructure of Portland cement pastes. Cem. Concr. Res. **30**, 855–863 (2000)
30. C. Plassard, E. Lesniewska, I. Pochard, A. Nonat, Investigation of the surface structure and elastic properties of calcium silicate hydrates at the nanoscale. Ultmi **100**(3–4), 331–338 (2004)
31. E. Bonaccorsi, S. Merlino, H.F.W. Taylor, The crystal structure of jennite, $Ca_9Si_6O_{18}(OH)_6{\cdot}8H_2O$. Cem. Concr. Res. **34**(9), 1481–1488 (2004)
32. A. Nonat, X. Lecoq, The structure, stoichiometry and properties of C–S–H prepared by C_3S hydration under controlled condition, in *Nuclear Magnetic Resonance Spectroscopy of Cement-Based Materials*. ed. by P. Colombet, H. Zanni, A.-R. Grimmer, P. Sozzani (Springer, Berlin Heidelberg, Berlin, Heidelberg, 1998), pp.197–207
33. A.J. Allen, J.J. Thomas, H.M. Jennings, Composition and density of nanoscale calcium–silicate–hydrate in cement. Nat. Mater. **6**(4), 311–316 (2007)
34. G. Geng, R.J. Myers, J. Li, R. Maboudian, C. Carraro, D.A. Shapiro, P.J. Monteiro, Aluminum-induced dreierketten chain cross-links increase the mechanical properties of nanocrystalline calcium aluminosilicate hydrate. Sci. Rep. **7**, 44032 (2017)

35. A. Gmira, M. Zabat, R.J.M. Pellenq, H. Van Damme, Microscopic physical basis of the poromechanical behavior of cement-based materials. Mater. Struct. **37**(265), 3–14 (2004)
36. A. Ayuela, J.S. Dolado, I. Campillo, Y.R. de Miguel, E. Erkizia, D. Sanchez-Portal, A. Rubio, A. Porro, P.M. Echenique, Silicate chain formation in the nanostructure of cement-based materials. J. Chem. Phys. **127**(16), 164710 (2007)
37. H. Manzano, A. Ayuela, J.S. Dolado, On the formation of cementitious C–S–H nanoparticles. J. Comput.-Aided Mater. Des. **14**(1), 45–51 (2007).
38. A. Kunhi Mohamed, S.C. Parker, P. Bowen, S. Galmarini, An atomistic building block description of C–S–H - Towards a realistic C-S-H model, Cem. Concr. Res. **107**, 221–235 (2018)
39. P. Faucon., J.M. Delaye., J. Virlet., Molecular dynamics simulation of the structure of calcium silicate hydrates I. $Ca_{4+x}Si_6O_{14+2x}(OH)_{4-2x}(H_2O)_2$ ($0 \leq x \leq 1$), J. Solid State Chem. **127**, 92–97 (1996)
40. L.B. Skinner, S.R. Chae, C.J. Benmore, H.R. Wenk, P.J. Monteiro, Nanostructure of calcium silicate hydrates in cements. Phys. Rev. Lett. **104**(19), 195502 (2010)
41. R. González-Teresa, V. Morales-Florez, H. Manzano, J.S. Dolado, Structural models of randomly packed Tobermorite-like spherical particles: a simple computational approach. Mater. Constr. **60**(298), 7–15 (2010)
42. D.A. Kulik, M. Kersten, Aqueous solubility diagrams for cementitious waste stabilization systems: II, end-member stoichiometries of ideal calcium silicate hydrate solid solutions. J. Am. Ceram. Soc. **84**(12), 3017–3026 (2001)
43. M. Kersten, Aqueous solubility diagrams for cementitious waste stabilization systems. 1. The C-S-H solid-solution system. Environ. Sci. Technol. **30**(7), 2286–2293 (1996)
44. H. Kurczyk, H. Schwiete., Concerning the hydration products of C_3S and β-C_2S, in *The 4th International Symposium on the Chemistry of Cement, United States* (Government Printing Office., Washington D.C., 1962), pp. 349–358.
45. H. Stade, W. Wieker, On the structure of ill-crystallized calcium hydrogen silicates. I. Formation and properties on an ill-crystallized calcium hydrogen dislicated phase. Chemischer Informationsdienst **11**(42), 1–10 (1980)
46. I.G. Richardson, The calcium silicate hydrates. Cem. Concr. Res. **38**(2), 137–158 (2008)
47. H.F.W. Taylor, Proposed structure for calcium silicate hydrate gel. J. Am. Ceram. Soc. **69**(6), 464–67 (1986)
48. I.G. Richardson., G.W. Groves., Models for the composition and structure of calcium silicate hydrate(C-S-H) gel in hardened tricalcium silicate pastes, Cem. Concr. Res. **22**, 1001–1010 (1992)
49. S. Papatzani, K. Paine, J. Calabria-Holley, A comprehensive review of the models on the nanostructure of calcium silicate hydrates. Constr. Build. Mater. **74**, 219–234 (2015)
50. I.G. Richardson., G.W. Groves., The incorporation of minor and trace elements into calcium silicate hydrate (C–S–H) gel in hardened cement pastes. Cem. Concr. Res. **23**, 131–138 (1993)
51. I.G. Richardson, A.R. Brough, G.W. Groves, C.M. Dobson, The characterization of hardened alkali-activated blast-furnace slag pastes and the nature of the calcium silicate hydrate (C-S-H) phase. Cem. Concr. Res. **24**(5), 813–829 (1994)
52. R.J. Kirkpatrick., G.E. Brown., N. Xu., X. Cong., Ca X-ray absorption spectroscopy of C–S–H and some model compounds. Adv. Cem. Res. **9**(33), 31–36 (1997)
53. R.J. Kirkpatrick, J.L. Yarger, P.F. McMillan, P. Yu, X. Cong, Raman spectroscopy of C–S–H, tobermorite, and jennite. Adv. Cem. Based Mater. **5**, 93–99 (1997)
54. I.G. Richardson, Tobermorite/jennite- and tobermorite/calcium hydroxide-based models for the structure of C-S-H: applicability to hardened pastes of tricalcium silicate, β-dicalcium silicate, Portland cement, and blends of Portland cement with blast-furnace slag, metakaolin, or silica fume. Cem. Concr. Res. **34**(9), 1733–1777 (2004)
55. J. d'Espinose de la Caillerie, N. Lequeux, Lecture on the structure of CSH, AFm and AFt phases, Physique, Chimie et Mécanique des Matériaux Cimentaire, ed Ecole ATILH-CNRS (2008)

56. R.J.M. Pellenq, A. Kushima, R. Shahsavari, K.J. Van Vliet, M.J. Buehler, S. Yip, F.J. Ulm, A realistic molecular model of cement hydrates. Proc. Natl. Acad. Sci. U.S.A. **106**(38), 16102–16107 (2009)
57. A. Zaoui, Insight into elastic behavior of calcium silicate hydrated oxide (C–S–H) under pressure and composition effect. Cem. Concr. Res. **42**(2), 306–312 (2012)
58. R.J. Myers, S.A. Bernal, R. San Nicolas, J.L. Provis, Generalized structural description of calcium–sodium aluminosilicate hydrate gels: The cross-linked substituted tobermorite model. Langmuir **29**(17), 5294–5306 (2013)
59. I.G. Richardson, Model structures for C-(A)-S-H(I), Acta Crystallogr. B: Struct. Sci. Cryst. Eng. Mater. **70**(6), 903–923 (2014)
60. J.J. Chen, J.J. Thomas, H.F.W. Taylor, H.M. Jennings, Solubility and structure of calcium silicate hydrate. Cem. Concr. Res. **34**(9), 1499–1519 (2004)
61. E. Gartner, I. Maruyama, J. Chen, A new model for the C–S–H phase formed during the hydration of Portland cements. Cem. Concr. Res. **97**, 95–106 (2017)
62. T.C. Powers, T.L. Brownyard, Studies of the physical properties of hardened cement paste. J. Am. Concrete Inst. **43**, 101–132 (1946)
63. T.C. Powers, Structure and physical properties of hardened Portland cement paste. J. Am. Ceram. Soc. **41**(6), 1–6 (1958)
64. S. Brunauer, R.S. Mikhail, E.E. Bodor, Some remarks about capillary condensation and pore structure analysis. J. Colloid Interface Sci. **25**(3), 353–358 (1967)
65. R.F. Feldman., P.J. Sereda., A model for hydrated Portland cement paste as deduced from sorption-length change and mechanical properties, Materiaux et Const. **1**(6), 509–520 (1968)
66. R.F. Feldman, Application of the helium inflow technique for measuring surface area and hydraulic radius of hydrated portland cement. Cem. Concr. Res. **10**, 657–664 (1980)
67. S. Tang, Y. Wang, Z. Geng, X. Xu, W. Yu, H. A, J. Chen, Structure, fractality, mechanics and durability of calcium silicate hydrates. Fractal Fractional **5**(2), 47 (2021)
68. F.H. Wittmann, *The structure of hardened cement paste - a Basis for a better understanding of the materials properties, Hydraulic cement paste: their structure and properties* (Tapton Hall, University of Sheffield, Cement and Concrete Association, 1976), pp.96–116
69. G. Constantinides, F.J. Ulm, The effect of two types of C-S-H on the elasticity of cement-based materials: results from nanoindentation and micromechanical modeling. Cem. Concr. Res. **34**(1), 67–80 (2004)
70. H.M. Jennings, Refinements to colloid model of C-S-H in cement: CM-II. Cem. Concr. Res. **38**(3), 275–289 (2008)
71. A.C.A. Muller, K.L. Scrivener, A.M. Gajewicz, P.J. McDonald, Densification of C-S-H measured by 1H NMR relaxometry. J. Phys. Chem. C **117**(1), 403–412 (2013)
72. A.M. Gajewicz, E. Gartner, K. Kang, P.J. McDonald, V. Yermakou, A 1H NMR relaxometry investigation of gel-pore drying shrinkage in cement pastes. Cem. Concr. Res. **86**, 12–19 (2016)
73. T.C. Powers, The thermodynamics of volume change and creep. Matériaux et Construction **1**(6), 487–507 (1968)
74. H.M. Jennings., Colloid model of C−S−H and implications to the problem of creep and shrinkage, Mater. Struct. **37**, 59–70 (2004)
75. H. Roper, Dimensional change and water sorption studies of cement paste, in *Symposium on structure of Portland Cement Paste and Concrete* (Highway Research Board, Washington, DC, 1966), pp. 74–83
76. G. Constantinides, F.-J. Ulm, The nanogranular nature of C–S–H. J. Mech. Phys. Solids **55**(1), 64–90 (2007)
77. F.-J. Ulm, Nano-engineering of concrete. Arab. J. Sci. Eng. **37**(2), 481–488 (2012)
78. M. Vandamme, F.-J. Ulm, P. Fonollosa, Nanogranular packing of C–S–H at substochiometric conditions. Cem. Concr. Res. **40**(1), 14–26 (2010)
79. J.J. Chen, L. Sorelli, M. Vandamme, F.-J. Ulm, G. Chanvillard, A coupled nanoindentation/SEM-EDS study on low water/cement ratio Portland cement paste: evidence for C-S-H/Ca(OH)2 nanocomposites. J. Am. Ceram. Soc. **93**(5), 1484–1493 (2010)

80. J.J. Kim, E.M. Foley, M.M. Reda Taha, Nano-mechanical characterization of synthetic calcium–silicate–hydrate (C–S–H) with varying CaO/SiO_2 mixture ratios. Cem. Concr. Compos. **36**, 65–70 (2013).
81. J. Li, W. Zhang, P.J.M. Monteiro, Structure and intrinsic mechanical properties of nanocrystalline calcium silicate hydrate. ACS Sustainable Chem. Eng. **8**(33), 12453–12461 (2020)
82. B.J. Alder, T.E. Wainwright, Phase transition for a hard sphere system. J. Chem. Phys. **27**(5), 1208–1209 (1957)
83. B.J. Alder, T.E. Wainwright, Studies in molecular dynamics. I. General method, J. Chem. Phys. **31**(2), 459–466 (1959)
84. A. Morshedifard, S. Masoumi, M.J. Abdolhosseini Qomi, Nanoscale origins of creep in calcium silicate hydrates. Nat. Commun. **9**(1), 1785 (2018)
85. D. Hou, Z. Li, Molecular dynamics study of water and ions transport in nano-pore of layered structure: a case study of tobermorite. Microporous Mesoporous Mater. **195**, 9–20 (2014)
86. M. Eftekhari, S. Mohammadi, Molecular dynamics simulation of the nonlinear behavior of the CNT-reinforced calcium silicate hydrate (C–S–H) composite. Compos. - A: Appl. Sci. Manuf. **82**, 78–87 (2016)
87. C. Feng, P. He, Atomistic investigation on the diffusion mechanism of Pt nanoclusters on well-aligned multi-walled carbon nanotubes. Comput. Mater. Sci. **103**, 157–164 (2015)
88. D. Hou, H. Ma, Y. Zhu, Z. Li, Calcium silicate hydrate from dry to saturated state: Structure, dynamics and mechanical properties. Acta Mater. **67**, 81–94 (2014)
89. S. Hajilar, B. Shafei, Nano-scale investigation of elastic properties of hydrated cement paste constituents using molecular dynamics simulations. Comput. Mater. Sci. **101**, 216–226 (2015)
90. D. Hou, J. Zhang, Z. Li, Y. Zhu, Uniaxial tension study of calcium silicate hydrate (C–S–H): structure, dynamics and mechanical properties. Mater. Struct. **48**(11), 3811–3824 (2014)
91. D. Tavakoli, A. Tarighat, Molecular dynamics study on the mechanical properties of Portland cement clinker phases. Comput. Mater. Sci. **119**, 65–73 (2016)
92. P. Faucon., J.M. Delaye., J. Virlet., J.F. Jacquinot., F. Adenotg., Study of the structural properties of the C–S–H(I) by molecular dynamics simulation. Cem. Concr. Res. **27**(10), 1581–1590 (1997)
93. R.T. Cygan., J.-J. Liang., A.G. Kalinichev., Molecular models of hydroxide, oxyhydroxide, and clay phases and the development of a general force field. J. Phys. Chem. B **108**, 1255–1266 (2004)
94. R. Shahsavari, R.J. Pellenq, F.J. Ulm, Empirical force fields for complex hydrated calcio-silicate layered materials. Phys. Chem. Chem. Phys. **13**(3), 1002–1011 (2011)
95. S. Galmarini, A. Aimable, N. Ruffray, P. Bowen, Changes in portlandite morphology with solvent composition: atomistic simulations and experiment. Cem. Concr. Res. **41**(12), 1330–1338 (2011)
96. H. Heinz, T.-J. Lin, R. Kishore Mishra, F.S. Emami, Thermodynamically consistent force fields for the assembly of inorganic, organic, and biological nanostructures: The INTERFACE force field, Langmuir **29**(6), 1754–1765 (2013)
97. A.C.T. van Duin, A. Strachan, S. Stewman, Q.S. Zhang, X. Xu, W.A. Goddard, ReaxFF(SiO) reactive force field for silicon and silicon oxide systems. J. Phys. Chem. A **107**(19), 3803–3811 (2003)
98. A.K. Rappe, C.J. Casewit, K.S. Colwell, W.A. Goddard, W.M. Skiff, UFF, a full periodic table force field for molecular mechanics and molecular dynamics simulations. J. Am. Chem. Soc. **114**(25), 10024–10035 (1992)
99. H. Heinz, U.W. Suter, Atomic charges for classical simulations of polar systems. J. Phys. Chem. B **108**(47), 18341–18352 (2004)
100. R.K. Mishra, R.J. Flatt, H. Heinz, Force field for tricalcium silicate and insight into nanoscale properties: Cleavage, initial hydration, and adsorption of organic molecules. J. Phys. Chem. C **117**(20), 10417–10432 (2013)
101. K.C. Gross, P.G. Seybold, C.M. Hadad, Comparison of different atomic charge schemes for predicting pKa variations in substituted anilines and phenols*. Int. J. Quantum Chem. **90**(1), 445–458 (2002)

Chapter 4
Pore Structure of C-S-H Under Cyclic Cryogenic Attack

4.1 Introduction

As the primary binding phase of cement-based materials, C-S-H microstructures have attracted much attention [1–4]. The pore structure of C-S-H plays a crucial role in contributing to the transport properties, frost resistance, and strength of cement-based materials [5–8]. As far as concrete in extremely cryogenic engineering, its pore structure is considered one of the critical factors responsible for the degradation in concrete microstructures and properties under the cryogenic attack (usually refers to an ultra-low-temperature condition of nearly − 170 °C) [9, 10]. It has been experimentally observed that the harshly cryogenic attack can lead to an apparent nanoscale degradation on C-S-H, weakening its nanomechanical stability [11], which is also thought to correlate with its nano/microstructure, i.e., pore structure of C-S-H. Hence, it is of great significance to comprehensively investigate the pore structure of C-S-H under cryogenic attack.

C-S-H is a porous amorphous phase in the cement hydration system, growing with a complicated microstructure [12, 13]. It has a wide and continuous pore size range from ~ 1 nm up to ~ 10 μm. Therefore, a defined classification for these pores is impossible. It is classified into three types, namely micropore (< 2 nm), mesopore (2 ~ 50 nm), and macropore (> 50 nm), respectively [14, 15]. A two-scale classification method is also widely adopted to describe the pores in C-S-H, i.e., gel pores (< 10 nm) and capillary pores (2 nm ~ 10 μm). In the colloid model proposed by Jennings [16], the gel pore in C-S-H was defined more specifically with two primary types, namely small gel pore (1 ~ 3 nm) and large gel pore (3 ~ 12 nm). The gel pores are viewed as part of the intrinsic C-S-H structure, while capillary pores are mainly empty spaces initially occupied by hydration water. It is noted that the total volume and distribution of gel pores and capillary pores are dynamically changing as the hydration time of cement increases [17, 18]. The total volume of the capillary pore decreases when the hydration products continuously fill in and expand, whereas the total volume of the gel pore undergoes an increase [17]. The pores with a pore size between 10 and

Z. Jiang and X. Zhu, *Cryogenic Stability of Cement-Based Materials at Micro/Nano Scale*, https://doi.org/10.1007/978-981-96-1836-1_4

100 nm are substantially occupied in a high volume fraction, especially at the late aging stage. Due to the micro-ice-lens effect, these substantially present capillary pores are more destructive in the presence of water during frost damage or cryogenic attack [19, 20].

Typical parameters for describing pore structure include cumulative pore volume, pore size distribution (PSD), specific surface area (SSA), fractal dimension, and tortuosity. Various techniques can be utilized to obtain these characterizing parameters, such as gas adsorption method (nitrogen or water vapor) [21–24], nuclear magnetic resonance [25, 26], mercury intrusion porosimetry [27, 28], thermoporometry [10, 29], small angle X-ray scattering [30], and image analysis [31]. None of these techniques can pose pore structure information overwhelmingly without additional testing methods. The adsorption method is widely adopted when it primarily concentrates on pores smaller than 100 nm. In contrast to nitrogen adsorption, the primary challenge for water vapor sorption is that C-S-H is inclined to undergo swelling and arrangement in its microstructure with the water vapor intrusion [16, 32]. Besides unlikely water molecules, nitrogen molecules are inactive to cement hydrates, which avoids any potential chemical effects [22]. Ample pore structure information can be shown using the nitrogen adsorption method and different analytical approaches to analyze the adsorption and desorption data. Among these theories, Brunauer–Emmett–Teller (BET) [33] and Barrett-Joyner-Halenda (BJH) [34] are used most in nitrogen adsorption to calculate SSA, PSD, and total pore volume of cement-based materials. Notably, the BJH approach is based on the Kelvin equation and capillary condensation, which is generally reasonable in most cases as the adsorption occurs at capillary pores [35]. While when it comes to micropores and narrow mesopores, a significant error can be produced by using this approach [36, 37], since the liquid meniscus does not exist in these pores [38]. According to the simulation study [39], the lower bound on the pore size range calculated using the classical thermodynamic approach approximates at 7.5 nm. Actually, the adsorption in micropores and narrow mesopores is mainly pore filling rather than capillary condensation. Schemes for the adsorption behaviors in different pores of C-S-H are illustrated in Fig. 4.1. Therefore, it is insufficient to reflect the pore structure of C-S-H by using the BJH approach only. To overcome this drawback, it has developed specific analytical theories to consider the mesopores in porous materials, mainly including density functional theory (DFT) [40], Dubinin-Radushkevich (DR) equation [41], and Horvath-Kawazoe (H–K) method [42]. Although these analytical theories have been widely and successfully used to characterize the PSD and pore capacity of micropores in various materials, they are barely adopted for analyzing the pore structure of cement-based materials.

The pores in cement-based materials usually vary in different sizes, and their connectivities are spatially independent as well, making the transport medium, e.g., fluid and gas, tortuously flow through the path instead of straightly moving across it. The tortuosity is invented to characterize such a tortuous feature of the pore structure, therefore correlated to porous media's diffusivity, permeability, and even electrical conductivity [43]. Regarding the methods to determine the tortuosity of cement-based material, various models and approaches, such as Kozeny's equation [44] and

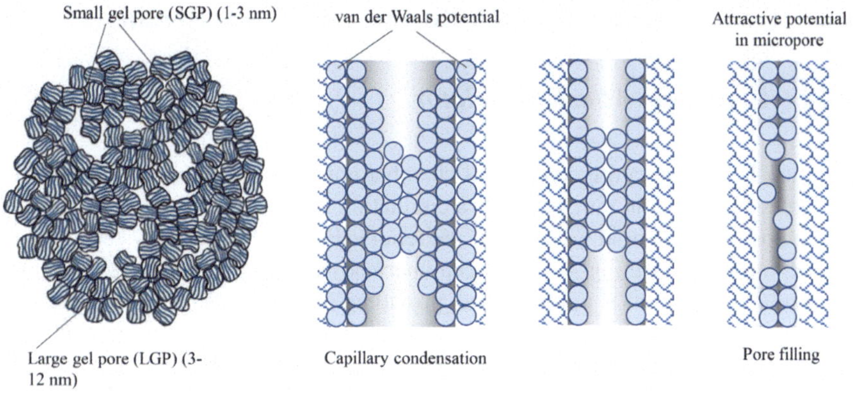

Fig. 4.1 Gas adsorption in capillary pores and micropores of C-S-H

stereological approach [45], were proposed. It is noted that most of these methods are pretty technical in obtaining a tortuosity factor for one material. Fortunately, the development of the corrugated pore structure model (CPSM) [46–48] makes it possible to calculate the tortuosity factor from the nitrogen sorption hysteresis data. One can easily calculate the tortuosity by fitting the sorption isotherms based on the CPSM model.

In this chapter, we aim to show how the cyclic cryogenic attack reshapes the pore structure of C-S-H. The cryogenic attack impacts on the pore structure of C-S-H were comprehensively evaluated by nitrogen adsorption measurement. Specifically, the gel and capillary pores smaller than 100 nm were mainly focused on. C-S-H samples with four different bulk Ca/Si molar ratios from 0.84 to 2.0 were synthesized by a hydrothermal method. Then, the C-S-H samples were treated in a cryogenic chamber where the temperature was down to − 170 °C with three cryogenic attack cycles at most. The PSD and pore volume were determined using the BJH method and the DFT approach. By using DR and DFT approaches, the micropore information was additionally obtained. SSA was calculated using the BET approach, and the segmented surface fractal dimension was determined using the Frenkel-Halsey-Hill (FHH) model. In addition, the tortuosity was also computed using the CPSM model, which provided additional evidence to estimate the cryogenic attack's influence on the pore structure of C-S-H.

4.2 Cyclic Cryogenic Attack Regime and C-S-H Synthesis

4.2.1 Cryogenic Attack Simulator

The ultra-low temperature control test chamber consists of a cryogenic chamber, a self-pressurizing liquid nitrogen tank, a circulating fan, a liquid nitrogen distributor, low-temperature solenoid valves, pressure relief valves, and a control system, among other components (see Fig. 4.2). The chamber body is made of stainless steel, while the insulation material is polyurethane, providing excellent insulation. An electromagnetic valve controls the evaporation of liquid nitrogen, automatically regulating the temperature of the chamber. The pressure relief valve expels the nitrogen gas after heat exchange. The principle behind controlling the cooling and heating rates involves using dual-set time relays to control the switching interval of the low-temperature solenoid valve adjusting the cooling rate by changing the interval time. Once the set temperature is reached, the heating principle works via a thermoresistor to raise the temperature. The chamber is equipped with a temperature sensor, enabling real-time measurement and display of the internal environmental temperature. The temperature control switch regulates the inflow of liquid nitrogen, maintaining a constant temperature once the set temperature inside the chamber is achieved. The operating temperature range is from -190 to 100 °C, with a temperature change rate between 0.2 and 10 °C/min. In the testing process, embedded thermocouples can be used to monitor the surface and core temperature of the test specimen throughout. Adjustments in the testing temperature range, heating and cooling rates, and constant temperature periods can be made in real time according to test requirements. The physical depiction of the device is shown in Fig. 4.2a, and the operational schematic is illustrated in Fig. 4.2b.

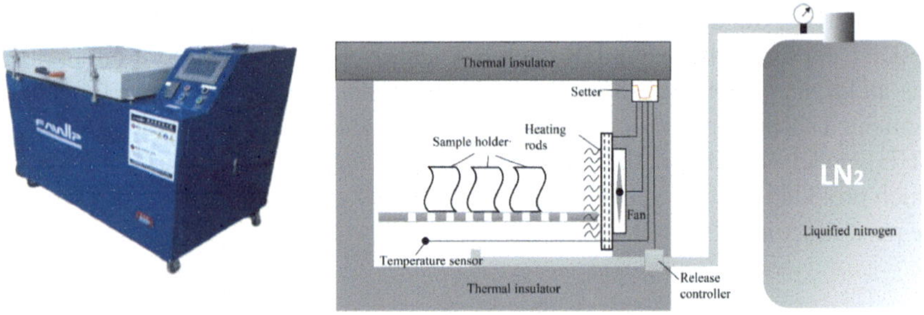

(a) Cryogenic attack simulator (b) The schematic diagram of the cryogenic attack simulator

Fig. 4.2 Cryogenic attack simulator used to perform the cryogenic attack tests

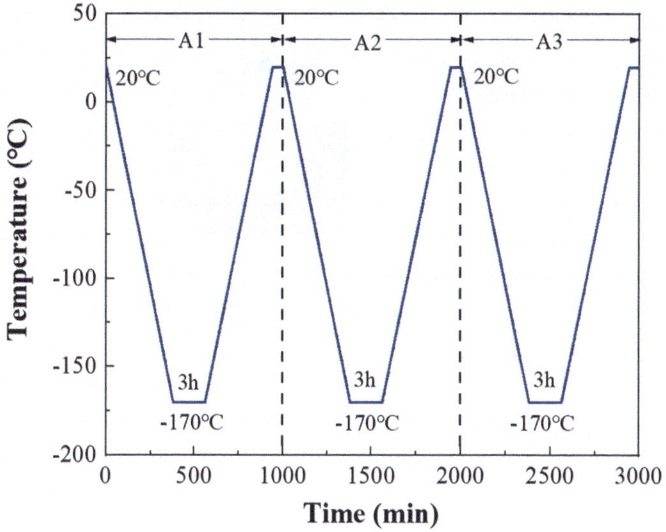

Fig. 4.3 Temperature variation for the cyclic cryogenic attack

4.2.2 Cryogenic Attack Regime

The harshly cryogenic attack treatment was realized in a cryogenic chamber using liquid nitrogen. The cryogenic attack regime is shown in Fig. 4.3. The temperature was controlled by changing from 20 °C to − 170 °C at a constant rate of 0.5 °C/min for 3 cycles. It was kept at the lowest temperature of − 170 °C for 3 h for each cryogenic attack cycle, then heated up to 20 °C. The samples subjected to 0 ~ 3 cryogenic attack cycles were labeled with B, A1, A2, A3, respectively.

4.2.3 C-S-H Synthesis

C-S-H solids were synthesized using a hydrothermal method by mixing a slurry of lime and fumed silica with 98% and 99% purity. The synthesis procedures are shown in Fig. 4.4. Lime and fumed silica were pretreated under 1000 °C and 200 °C respectively, to remove impurities and physically adsorbed water before being used for formal synthesis, and deionized water was boiled up to remove CO_2. The slurry was mixed with a Ca/Si ratio of 0.84, 1.5, and 2.0 and a constant water/solid ratio of 12. To obtain a homogenous synthetic liquor, the slurry was first stirred in a sealed three-neck flask at a rate of 400 r/min for 10 h, meanwhile heating at a constant temperature of 60 °C under the protection of N_2 atmosphere. After that, the homogenous slurry was transferred into a 500 mL thermal reactor and heated at 200 °C for another 10 h.

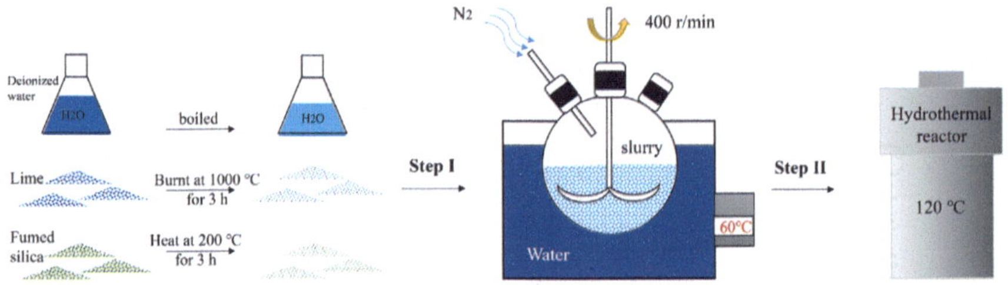

Fig. 4.4 Hydrothermal synthesis procedures. Step I: Premix and pre-react for 10 h. The mixed slurry was stirred at 400 r/min in a thermostatic water bath at a constant temperature of 60 °C with the protection of N2. Step II: hydrothermally react for 10 h. The pre-reacted slurry was transferred into a hydrothermal reactor

A > 95vol.% ethanol solution washed the synthetic solids and then vacuum oven dried at 50 °C for 3 days.

4.3 Nitrogen Sorption

4.3.1 Nitrogen Adsorption Measurement

The C-S-H samples were tested for their pore structures by utilizing nitrogen adsorption–desorption measurements at a temperature of 77.3 K. Before formal nitrogen sorption running in the adsorption station for each measurement, the test samples were degassed for 4 h at vacuum and a temperature of 40 °C in the desorption station. All the measurements were performed on 3H-2000PS2 apparatus (Beishide Instruments, China), where the degassing vacuum was set at 1×10^{-3} mbar, and the relative pressure for the adsorption–desorption running was set in a range of 0.001 ~ 0.998.

4.3.2 The BJH Approach for Mesopores

The BJH method [34] is proved to be applicable to almost all types of porous materials. Therefore, it is the most widely used analytical theory to calculate PSD and total pore volume of cement-based materials in nitrogen adsorption measurement. It is theoretically based on the Kelvin equation, which forges a link between the relative pressure and the radius of curvature of a liquid meniscus, i.e., Kelvin radius. In BJH theory, the pores are assumed to have a cylindrical shape with a radius r_i equal to the sum of Kelvin radius $r_{k,i}$ and the thickness of absorbed film t_i, as presented in Eq. 4.1:

$$r_i = r_{k,i} + t_i \tag{4.1}$$

The desorption isotherm branch obtained in the measurement is utilized for analysis. The basic equation for the calculation can be expressed as Eq. 4.2:

$$\Delta V_i = Q_i\left(\Delta v_i - C\Delta t_i \sum_{j=1}^{i=1} \Delta S_j\right) \tag{4.2}$$

where ΔV_i is desorbed pore volume, Δv_i denotes desorbed gas volume. Δt_i represents the reduced thickness of the adsorbed film, C is a correction coefficient and is assigned a value of 0.85. The thickness of adsorbed film t can be determined by using the Haley equation:

$$t = 0.345(-5/\ln(P/P_0))^{1/3}\,\text{nm} \tag{4.3}$$

where P/P_0 is relative pressure. The exposed surface area ΔS_j in each desorption can be calculated by Eq. 4.4:

$$\Delta S_j = 2\Delta V_i/\overline{r}_j \tag{4.4}$$

Q_i is a conversion coefficient, which can be expressed as:

$$Q_i = \left(\overline{r}_i/\left(\overline{r}_i - \overline{t}_i\right)\right)^2 \tag{4.5}$$

The BJH approach has modified part of the inaccuracy of the Kelvin equation by taking into account the absorbed film, which is basically applicable to mesopores.

4.3.3 The DFT Approach for Micro/mesopores

The DFT approach is grounded in statistical mechanics to calculate the sorbent property. In contrast to the classical thermodynamic approach, e.g., the BJH method, the strong interactions between absorbed films on opposing pore walls and the strong, attractive force between adsorbed films and solid pore walls are taken into account [38], which allows it to be applicable for analyzing micropores apart from mesopores. In addition, the analytical pore model in this approach is based on a slit shape. The basic equation to calculate PSD is expressed as:

$$N_{\exp}(P/P_0) = \int_{D_{\min}}^{D_{\max}} N_{DFT}(P/P_0, D) f(D) dD \tag{4.6}$$

where $N_{\exp}(P/P_0)$ is the experimental adsorption isotherm, $N_{DFT}(P/P_0, D)$ is the theoretical isotherm for the pore with a size of D, and $f(D)$ denotes the unknown PSD function. By using Eq. 4.6, the PSD function can be numerically solved.

Since it has been proposed to analyze nitrogen adsorption data, the DFT approach has been developed into two main styles: conventional non-local density functional theory (NLDFT) approach and quenched solid density functional theory (QSDFT) approach. The primary difference between the two approaches is that QSDFT is a further advanced way to describe realistic structures by considering the chemical and geometrical heterogeneity of pore walls on the molecular level [49]. Nevertheless, the conventional NLDFT is still theoretically more realistic than the BJH approach to calculate the PSD of materials containing massive micropores and narrow mesopores.

4.3.4 The DR Approach for Micropores

The DR approach was explicitly developed to characterize the adsorbents in a microporous nature, in which mesopores and macropores may exist, although it is the micropore that decisively determines their adsorptive behavior. This approach postulates that the adsorption in micropores is based on pore filling instead of layer-by-layer formation of a film [41]. The fundamental theory used in this approach is the Polanyi potential theory of adsorption. A gas adsorption potential A is defined by:

$$A = RT\ln(P_0/P) \tag{4.7}$$

The relation between the adsorption volume W of micropores at a relative pressure of P/P_0 and the limiting pore volume W_0 for pore filling is expressed as:

$$W = W_0 \exp\left[-\left(\frac{RT\ln(P_0/P)}{E_0\beta}\right)^2\right] \tag{4.8}$$

where E_0 represents the characteristic adsorption energy of reference benzene vapor, and β is the similarity constant ($\beta = 1$ for reference benzene vapor, $\beta = 0.33$ for nitrogen vapor). Then Eq. 4.8 can be transformed in a linear form:

$$\log W = \log W_0 - 2.3026(RT/(E_0\beta))^2(\log(P_0/P))^2 \tag{4.9}$$

By fitting the adsorption data in a linear plot, $\log W$ vs. $(\log P_0/P)^2$, the total micropore volume W_0 can be calculated from the intercept.

4.3.5 The H–K Approach for Micropores

The basic H–K equation for slit pore [42] is shown as:

$$RT\ln(P/P_0) = N_{Av}\frac{N_sA_s + N_AA_A}{\sigma^4(L - 2d_0)} \times \left[\frac{\sigma^4}{3(L - d_0)^3} - \frac{\sigma^{10}}{9(L - d_0)^9} - \frac{\sigma^4}{3d_0^3} + \frac{\sigma^{10}}{9d_0^9}\right] \tag{4.10}$$

where T is measurement temperature, R is the universal gas constant, L refers to the distance of two opposite planes, d_0 means the average diameter of adsorbent and adsorbate molecules, N_{Av} denotes Avogadro constant, N_s and N_A represent the number of adsorbent and adsorbate per unit area, respectively, A_s and A_A stand for the constant of Lenard-Jones of adsorbent and adsorbate, respectively. The Lenard-Jones constant of one substance is associated with its polarizability and magnetisability. Here, the parameters relating to the adsorbent, i.e., C-S-H, in this study are adopted from the zeolite.

4.3.6 *Specific Surface Area from BET Theory*

The BET approach is widely adopted to determine the SSA of solid materials with open porosity. Unlikely the Langmuir approach, it is based on a hypothesis that the adsorbates are densely packed in the form of a multimolecular layer on the surface of solid materials in the BET approach. The BET equation is expressed in a linear form as Eq. 4.11:

$$\frac{P/P_0}{V(1 - P/P_0)} = \frac{1}{V_mC} + \frac{(C - 1)P/P_0}{V_mC} \tag{4.11}$$

where P/P_0 is the relative pressure, V denotes the adsorbed volume, V_m represents the volume of the monolayer, and C is viewed as a unique constant. Then, it fits the adsorption data in a linear form of $(P/P_0)/[V(1 - P/P_0)]$ vs. P/P_0 at a low relative pressure range, typically in a range of 0.05 ~ 0.35. Thus, the value of V_m and C can be determined from the coordinated intercept and the slope of the plot. After calculating the value of V_m, the SSA S can be finally determined by:

$$S = 4.356V_m \tag{4.12}$$

4.3.7 *Surface Fractal Dimension from FHH Model*

The surface fractal dimension is an important parameter to characterize the surface roughness of pores. It has been proved that the surface fractal dimension is closely associated with the transportation and heat transfer properties of cement-based materials. This parameter is calculated from the FHH model in this study, which is expressed as:

$$\ln V = C + h\ln(-\ln(P/P_0)) \tag{4.13}$$

where V represents the adsorbed volume at a relative pressure of P/P_0, C is a constant determined from the ordinate intercept for a fitting plot of $\ln V$ vs. $\ln(-(\ln P/P_0))$, and h is the slope of the fitting plot.

The surface fractal dimension D is related to h. When the predominant bonding force between adsorbent and adsorbed films is controlled by van der Waals and surface tension [50], the surface fractal dimension D can be determined from Eqs. 4.13 and 4.14, respectively.

$$h = (D - 3)/3 \tag{4.14}$$

where $2 \le D < 3, -1/3 \le h < 0$.

$$h = D - 3 \tag{4.15}$$

where $-1 \le h < 0$.

4.3.8 Tortuosity

The CPSM-tortuosity model was developed by Salmas et al. [48]. CPSM theory simplifies the pore shape as a corrugated pore model (see Fig. 4.5). A single pore tunnel can be simplified as a combination of several pore segments with different pore diameters. The layer thickness of adsorbed film t on the pore wall at a given relative pressure P/P_0 is calculated by Eq. 4.16, which is a revised Halsey equation.

$$t = n\left[-\frac{5}{\ln(P/P_0)}\right]^{1/3}(P/P_0)^m \tag{4.16}$$

For simulating the hysteresis loop of nitrogen sorption isotherm, the Kelvin equation Eq. 4.17 is adopted to describe the capillary condensation, yielding a Kelvin diameter D_K.

$$D_K = \frac{1.906}{P/P_0}\cos\theta \tag{4.17}$$

The fundamental equations to simulate the adsorbed and desorbed nitrogen volumes in the hysteresis loop are Eqs. 4.18 and 4.19.

$$V_{ads} = \left[1/\int_{D_{\min}}^{D_{\max}} D^2F(D)dD\right]\left[4t\int_{D_h}^{D_{\max}} (D - t)F(D)dD\right.$$

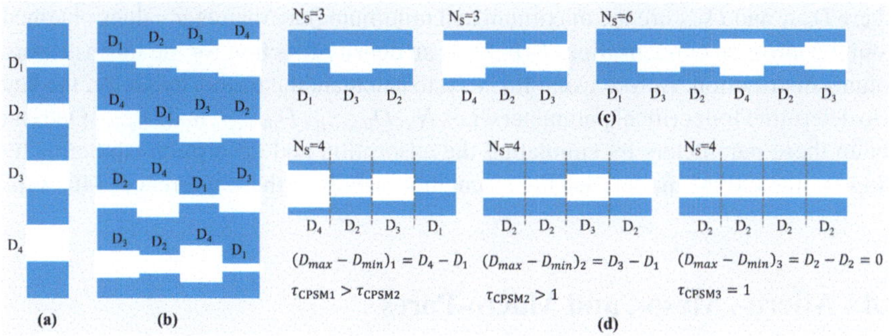

Fig. 4.5 Illustration of the corrugated pore model. **a** Single pore segment with varying pore diameters. **b** The corrugated pore model is constructed by a set of pore segments. **c** The nominal pore length NS and its effect on the tortuosity. Compared with the first two corrugated pore models (from the left to the right), the third model has a longer nominal pore length, yielding a higher tortuosity value. **d** The effect of pore size distribution on the tortuosity. From the left to the right, the pore size distribution is more and more converged and concentrated, consequently yielding a less tortuous pore structure. When the pores have the same size, the τ_{CPSM} equals to 1

$$+\int_{D_{\min}}^{D_C} D^2F(D)dD + \int_{D_C}^{D_h} D^2F(D)dD\left(\frac{1}{N_S}\sum_{j=1}^{N_S}P_j\right)$$

$$+\,4t\int_{D_C}^{D_h}(D-t)F(D)dD\left(\frac{1}{N_S}\sum_{j=1}^{N_S}(1-P_j)\right) \tag{4.18}$$

$$V_{der} = \int_{D_h}^{D_{max}}(D-2t)^2F(D)dD \times \left(\frac{2}{N_S}\frac{1-q^{N_S}}{1-q} - q^{N_S-1}\right)\Big/\int_{D_{min}}^{D_{max}} D^2F(D)dD \tag{4.19}$$

where $F(D)$ is the intrinsic pore segment number distribution function, which can be found in Ref. [48], $D_{\max}$ and $D_{\min}$ represent the maximum and minimum pore diameters determined by nitrogen sorption measurement.

Via the simulation of the nitrogen sorption hysteresis loop, the CPSM tortuosity factor τ_{CPSM} is computed by Eq. 4.20.

$$\tau_{CPSM} = 1 + 0.69\left(\frac{D_{\max,eff} - D_{\min,eff}}{D_{mean}}\right)(N_S - 2)^{0.58} \tag{4.20}$$

where τ_{CPSM} is the tortuosity factor, $D_{\max,eff}$, $D_{\min,eff}$, are determined from Eq. 4.21, D_{mean}, denotes the mean pore diameter of the intrinsic pore volume distribution, N_S represents the number of pore segments forming a corrugated pore.

$$\int_{D_{\max,eff}}^{D_{\max}} D^2F(b;D)dD = \int_{D_{mn}}^{D_{\min,eff}} D^2F(b;D)dD = 0.025 \tag{4.21}$$

where $D_{\max}$ and $D_{\min}$ are the maximum and minimum pore diameter values obtained from the nitrogen condensation, $F(b; D)$ is an density function for the intrinsic pore volume distribution. Before using Eq. 4.20 to calculate the tortuosity factor, the key is to determine four critical parameters, i.e., N_S, $D_{\max,eff}$, $D_{\min,eff}$, and D_{mean}. One can obtain these parameters by simulating the adsorption and desorption isotherms by adopting the CPSM model. See the calculation details in these references [46, 48].

4.4 Micro-, Meso-, and Macro-Pores

4.4.1 Microstructure Characterization

The SEM micromorphology and AFM topography of synthetic C-S-H solids are shown in Fig. 4.6. From previous studies [51, 52], C-S-H hydration products in hydrated cement systems have a wide range of morphological features such as needle-like, plate-like, fibrillar, round, rectangular, and spindle, different in varying physical and chemical conditions. As shown in Fig. 4.6a, c and d, the synthetic C-S-H has a shape of the fibrillar structure, particularly when the Ca/Si ratio is at 1.5, which is very close to the morphology of one-year hydrated C_3S attained in Ref. [51]. This indicates that the hydrothermal synthesis method adopted here helps accelerate the C-S-H nanocrystal assembling reaction, therefore obtaining more mature C-S-H products. The morphology of C-S-H at C/S2.0 shows somewhat granular characteristics, presenting a denser surface morphological structure than the other two C-S-H samples with lower Ca/Si ratios. Typical amorphous and porous micromorphology from SEM images can be seen for the synthetic C-S-H. AFM also captured stacking orientation order features. As shown in AFM images, the small building block constructs the C-S-H clusters. Those small building blocks have a ball-like structure with a typical diameter of around 50–80 nm, which is also very similar to the observation for hydrated cement [53]. These results reflect the high efficiency and effectiveness of the approach used here to obtain C-S-H products.

4.4.2 Sorption Isotherms

The nitrogen adsorption and desorption isotherms of C-S-H samples are illustrated in Fig. 4.7. According to the International Union of Pure and Applied Chemistry (IUPAC) classifications, there are typically 6 types of adsorption isotherms. As shown in Fig. 4.7, most of the isotherms are not associated with the typical isotherms defined in IUPAC. This is understandable since it is usually a combination of two or more different types of isotherms in the actual situation. Hysteresis loop can be observed in every sorption isotherm. The inception points of these hysteresis loops are all positioned at approximately a high relative pressure of 0.80, indicating that

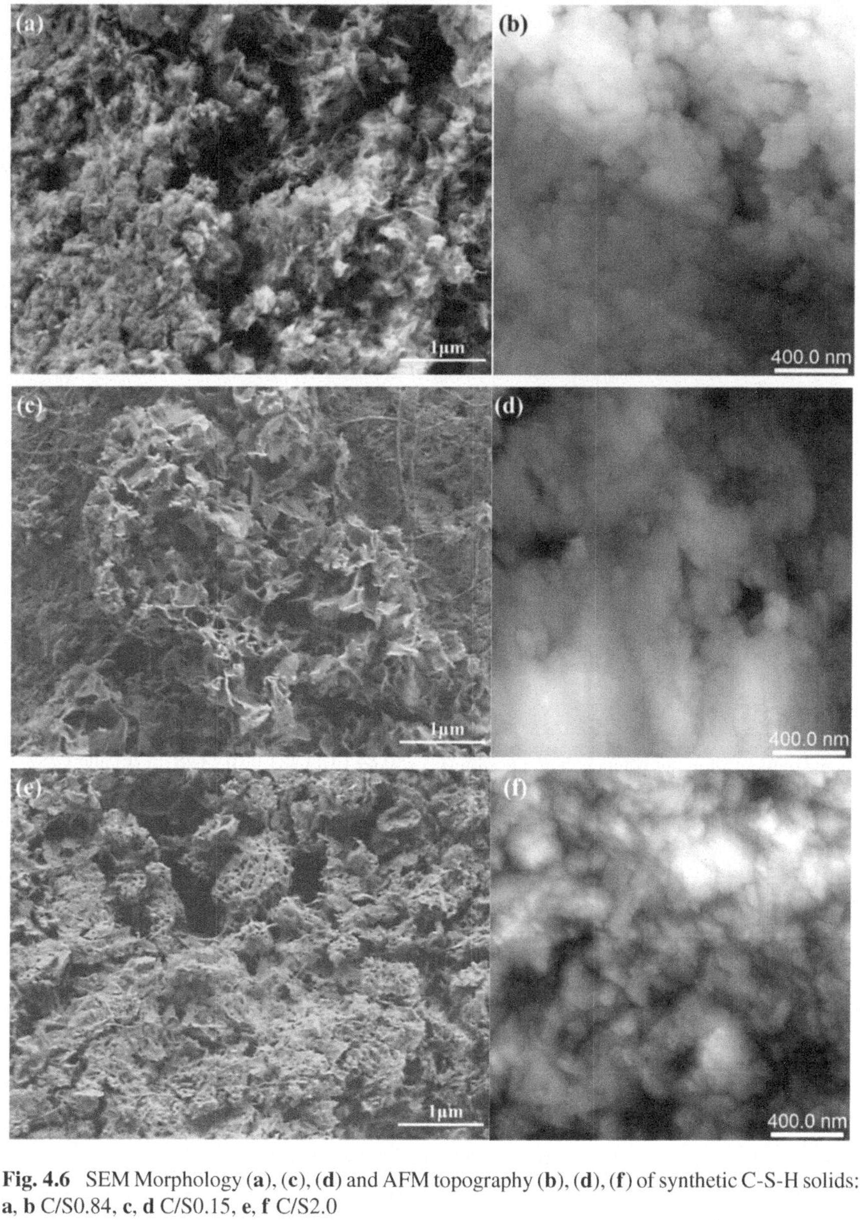

Fig. 4.6 SEM Morphology (**a**), (**c**), (**d**) and AFM topography (**b**), (**d**), (**f**) of synthetic C-S-H solids: **a**, **b** C/S0.84, **c**, **d** C/S0.15, **e**, **f** C/S2.0

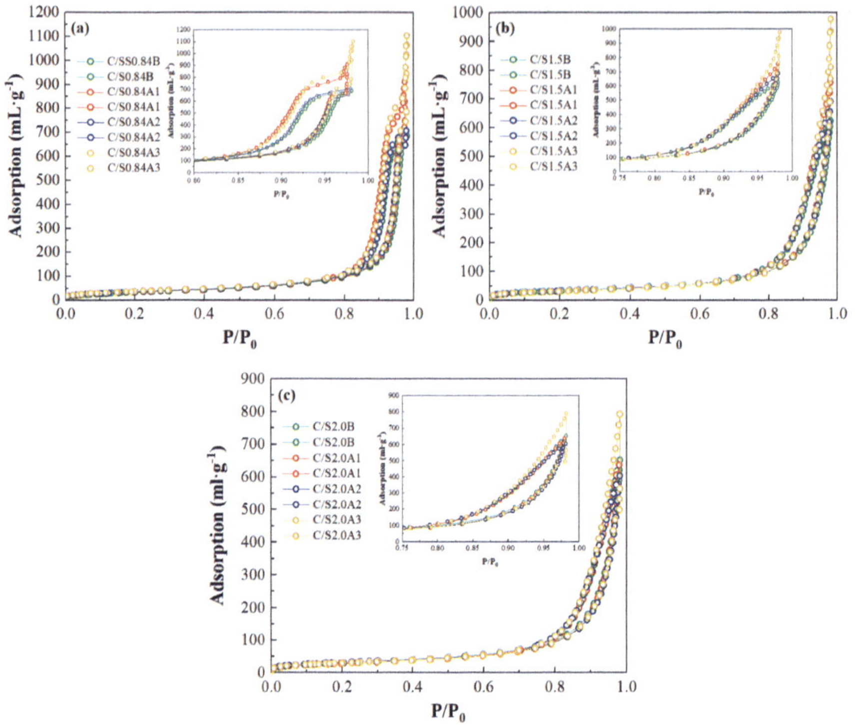

Fig. 4.7 Adsorption curves of C-S-H samples of C/S0.84 (**a**), C/S1.0 (**b**), C/S1.5 (**c**) and C/S2.0 (**d**)

capillary condensation occurs first in the mesopore with a large radius. In other words, large mesopores are predominantly present in the pore systems of the synthetic C-S-H solids. These large mesopores contributed more significantly to the amount of adsorbed gas in C-S-H samples.

Notably, the amount of adsorbed gas volumes for C-S-H samples that sustained cryogenic attack is clearly higher than that without experiencing the attack, particularly for the C-S-H samples that have undergone triple times of cryogenic attack. The increased amount of absorbed gas volume primarily originated at high relative pressure, which means the number of large mesopores or macropores was probably increased due to the attack. Among these isotherms, those for C-S-H with lower C/S ratio were changed more significantly under the influence of harshly cryogenic attack. The most significant increase in the amount of adsorbed gas volume was seen for C/S0.84, and the differentiation point was shifted to a lower relative pressure compared with isotherms for other C-S-H samples. These apparent differences indicate that the pore system in C-S-H with a lower C/S ratio is more likely to be damaged by the cryogenic attack. Instead, C-S-H with a high C/S ratio has better stability in its pore structure.

4.4.3 Specific Surface Area

Nitrogen specific surface area highly corresponds to the tested sample's pore structure. It is even hypothesized by Zhang et al. [22] that a higher surface area indicates better preservation of the microstructure of cement-based materials. This view is inherited in the current study, and our further expectation is that the more stable the specific surface area, the more stable the C-S-H pore structure is maintained under cryogenic attack. Since the C-S-H samples used were synthesized hydrothermally and were produced in high purity [11, 54], unlikely in the hydrated cement [22], the concerns for the decomposition of existing phases in tested samples are unnecessary. Hence, this hypothesis is valid without any preconditions.

The specific surface areas of C-S-H samples that experienced 0 to 3 cryogenic attack cycles are calculated by the BET method and shown in Fig. 4.8a. For the synthetic solids before the cryogenic attack, a higher surface area is seen for the C-S-H with a lower C/S ratio, which indicates that C-S-H is probably more porous in a lower C/S ratio. It is noted that C-S-H usually has a better structure ordering when it is in a lower C/S ratio. In comparison, C-S-H with a high C/S ratio has less structure ordering and typically presents amorphous features. Concerning this, a denser structure is furtherly observed for C-S-H basic building blocks with a higher C/S ratio in Ref. [11]. The trend seen for specific surface areas in this study confirms that C-S-H with a higher C/S ratio has a denser microstructure.

The cryogenic attack clearly devastated the original microstructures of all C-S-H samples. As shown in Fig. 4.8a, C/S0.84 has the most significant variation in its surface area. The measured specific surface area for C/S0.84 has an increasing trend with the cryogenic attack cycle increased. Such a trend indicates that the corresponding C-S-H solids became more porous and exposed to higher surface area under the continuous cryogenic attack. C/S1.5 has surprisingly witnessed a reversely decreasing trend, suggesting that pore collapse probably occurred on C-S-H, resulting in a denser microstructure and a lower surface area [22]. C/S2.0 shows a relatively

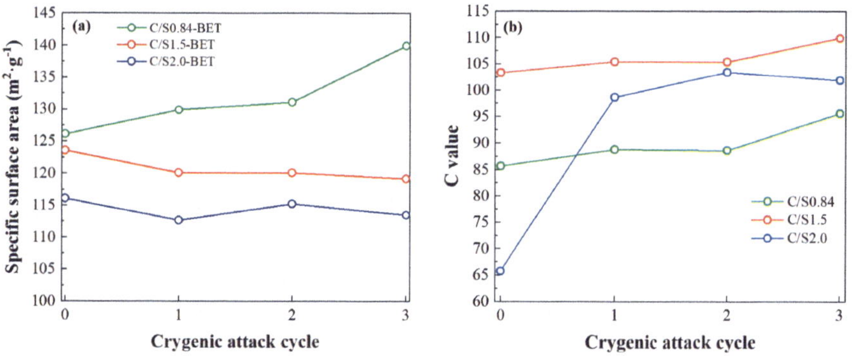

Fig. 4.8 Specific surface area (**a**) and constant C value (**b**) of C-S-H samples obtained from BET method

gentle variation in its specific surface area, just a limited fluctuation, while it still presents a declining trend overall. All these changes in the surface areas of different C-S-H samples are associated with the gel pores since they are significant contributors to the BET surface area [22]. From this perspective, it seems that the gel pore structure of C-S-H with a higher C/S ratio provides better stability under the harshly cryogenic attack, which will be provided additional evidence and discussed later.

The C value obtained from the BET approach is treated as a unique constant to characterize the adsorption energy of the first adsorbed nitrogen molecule layer, further reflecting the interaction between pore walls and nitrogen molecules [22]. Figure 4.8b depicts the variations of C values calculated for three C-S-H samples. An increasing trend of constant C value was seen for every sample under triple cryogenic attacks, which means that the adsorption energy of the first adsorption layer was increased due to the attacks. Therefore, it can be further read that the cryogenic attacks have chemically created a new surface condition of C-S-H, forming a strong interaction between the C-S-H pore walls and nitrogen adsorbates. This finding is surprisingly coherent with the observations for the hydrophilic nature of C-S-H in Ref. [54]. It was found that a stronger interaction between the C-S-H and water molecules was obtained after the cryogenic attack [54]. From these results, a consensus is that the chemical state of C-S-H surface is probably converted to have a stronger interaction with the absorbates.

4.4.4 Mesopores

Total pore volumes of different C-S-H samples were calculated by the BJH and DFT approaches (see Fig. 4.9). All the samples show fluctuant trends in both BJH and DFT calculations, which is inconsistent with that of the surface area. Such inconsistency between surface area and pore volume was also witnessed for cement pastes [22]. It is understandable to see such a trend since the connectivity in the pore systems may be reconstructed under the cryogenic attack, making some pores accessible or inaccessible to nitrogen probes. Nevertheless, it still can be seen from Fig. 4.9 that the total pore volumes of all these C-S-H samples were increased to much higher values after triple times of cryogenic attack cycles, showing more porous microstructures. It is noted that the calculated BJH total pore volume is always about 6–8% greater than the corresponding DFT total pore volume. This difference is highly associated with the essence of BJH and DFT approaches.

Figure 4.10 displays the pore size distribution and cumulative pore volume of C-S-H samples determined from BJH and DFT methods. According to Liu et al. [55], dV/dD format is more appropriate for presenting PSD data from nitrogen adsorption. Herein we adopted this suggestion to present the PSD results. The obtained cumulative pore volume curves from the two approaches are generally present in similar trends, while the corresponding pore size distributions are significantly different. Since the conventional condensation approaches, e.g., BJH [34] and DH [56], inevitably overestimate the capillary condensation pressure for the narrow

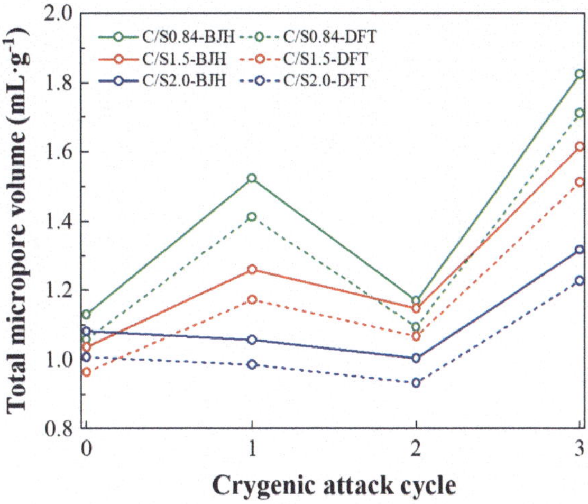

Fig. 4.9 Total pore volumes of C-S-H samples evaluated by BJH and DFT methods

mesopores and micropores due to the presence of fluid–solid and fluid–fluid attractions. Thus, their pore sizes are usually underestimated by using these approaches [57, 58], which is confirmed in this study. As shown in Fig. 4.10, the pore volumes at the range of 0.5–7 nm obtained from the BJH approach are smaller than that from the DFT approach, consequently obtaining lower cumulative pore volumes corresponding to these micropores and narrow mesopores. The BJH approach even failed to access information on these fine pores. The measured pores from this approach are primarily present in a range of 2–100 nm, while those micropores are unreached. In comparison, a much more comprehensive range of pore sizes from 0.5 nm to 120 nm is reached by the DFT approach, and abundant information is obtained for micropores existing in C-S-H. Apart from this, another apparent difference between the two approaches is that the derivative pore volumes of mesopores are underestimated in DFT calculation compared with BJH calculation, and the most probable pore size upshifts to a higher value.

Although the PSD results from BJH and DFT approaches are significantly different, typical characteristics for the variations resulting from cryogenic attacks are still identifiable. It is clearly shown in Fig. 4.10 that the C-S-H samples with varying C/S ratios overall obtained higher cumulative pore volumes after the cryogenic attack. After experiencing triple times of cryogenic attack, the increase in the cumulative pore volume was significant for every sample. In contrast, considerable variations were not observed for pores smaller than 100 nm of cement-based materials subjected to normal freezing-thawing according to the results shown in Ref. [59]. This means that the harshly cryogenic attack is more destructive to C-S-H than the normal freezing-thawing, even without the interference of freezing pore water. Combined with BJH and DFT calculations, it is indicated that the C/S ratio may control the stability of micropores and mesopores in corresponding C-S-H phases. As shown in Fig. 4.10, the micropores of C/S0.84 are better preserved under cryogenic attack [59], while significant reorientations are witnessed for C/S1.5 and C/

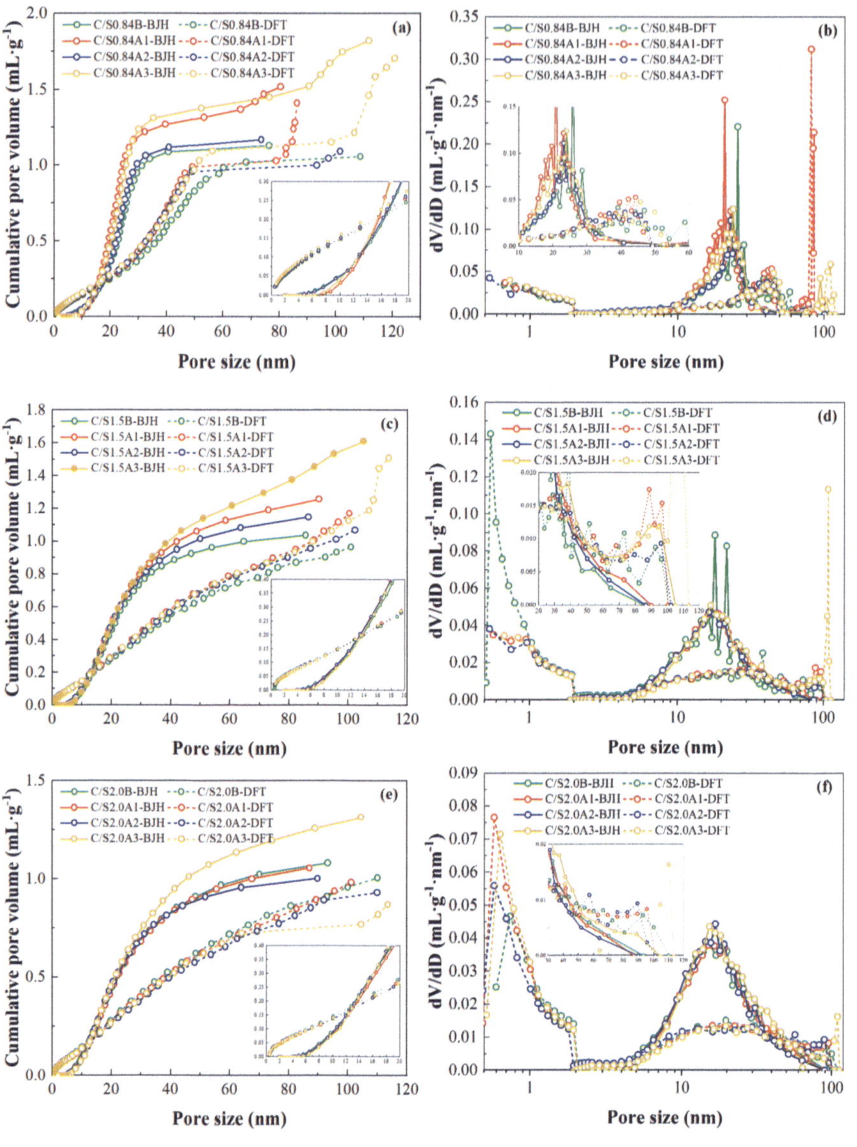

Fig. 4.10 Cumulative pore volume (**a**), (**c**), (**e**) and pore size distribution (**b**), (**d**), (**f**) of C-S-H samples evaluated by BJH and DFT methods

S2.0. It is also noteworthy that the ultramicropores (< 0.7 nm) seem more unstable. Apparent collapse is seen for ultramicropores in C/S1.5 after only a single cryogenic attack, while they are newly constructed in C/S2.0, and a higher amount of ultramicropore is formed after the cryogenic attack. However, the mesopores are better preserved in C-S-H with a higher C/S ratio, which both BJH and DFT calculations can confirm. Among these C-S-H samples, C/S2.0 has the most stable mesopore structure under cryogenic attack, even with triple cycles. A significant increase is observed for the derivative pore volume of large mesopore (> 10 nm) in C/0.84. Apart from micropores and mesopores, the condition of macropores is reconstructed as well. Both BJH and DFT results show that all the C-S-H samples are inclined to have a more considerable amount of macropores after the cryogenic attack. Additional peaks are formed at 80-110 nm for most samples, leading to a multimodal PSD. According to Ref. [11], The increased number of mesopores and macropores are assigned to be newly formed microcracks in C-S-H clusters, indicating that the harsh cryogenic attack has damaged the microstructure of C-S-H.

4.4.5 Micropores

As stated previously, the heterogeneous C-S-H has a complex pore system containing intraglobular pores (<, 1 nm) small gel pores (1~3 nm), large gel pores (3–12 nm), other large mesopores (12–50 nm), and macropores (> 50 nm). Micropores are primarily assigned to small gel pores, which is usually neglected in analyzing the pore structure using the BJH method, which is believed to be trustworthy in analyzing pores within the range of 2–50 nm [60]. While those micropores also contribute to the measured BET surface area, it is necessary to consider them when evaluating the pore volume. Although the DFT approach has been utilized in the last section and presents part of the information for micropores in C-S-H, additional methods, i.e., H-K, DR, and DFT, are adopted here to analyze micropores.

The total micropore volumes determined by using H–K, DR, and DFT methods and micropore size distributions are depicted in Fig. 4.11. From Fig. 4.11a, the general trends of total micropore volume obtained from the three methods are basically synchronized with each other, though small gaps are present in their corresponding calculated pore volumes. The micropore volumes obtained from DR and DFT approaches are exceedingly closer. Compared with them, about a percentage of 10–15% lower value was obtained for the micropore volume from the H–K approach. This conclusion applies to every set of calculations. Such an apparent difference probably originated from the adsorbate parameters used in the H–K equation. In the H–K model, the characteristics of the adsorbate, mainly including the arithmetic mean diameter, the number of oxide ions per unit area, the polarizability, and the magnetic susceptibility of an adsorbate molecule [42, 61, 62], are highly associated with the accuracy of the calculation. In the current study, since these parameters for the amorphous C-S-H are unknown, they are adopted from that for zeolite, triggering this inaccuracy.

The general trends for the PSD calculated from the H–K approach are similar to those obtained from the DFT approach (see Fig. 4.10), although the corresponding derivative micropore volumes are lower than those of other methods. As shown in Figs. 4.10b, d f, and 4.11b, c and d, significant peaks can be seen for the derivative pore volume of ultramicropores (< 0.7 nm) and small supermicropores (< 1.1 nm)

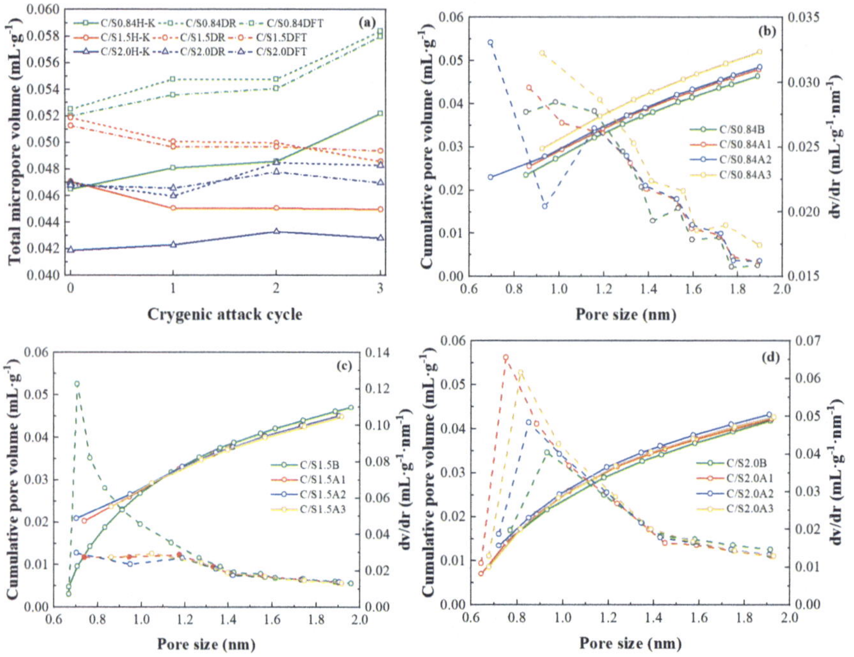

Fig. 4.11 Total micropore volume of C-S-H samples evaluated by H–K, DR and DFT methods (**a**), and micropore size distribution and cumulative micropore volume of C-S-H samples of C/S0.84 (**b**), C/S1.5 (**c**) and C/S2.0 (**d**), evaluated by H–K method

in most C-S-H samples, showing that the C-S-H solids contained abundant smaller micropores. Such small micropores are supposed to be intraglobular pores present in C-S-H. DFT and H–K calculations show that the ultramicropores and small supermicropores were most likely damaged under cryogenic attack. Nevertheless, other larger micropores have withstood the harsh attack, showing a much more stable structure even after undergoing triple attacks.

Combined with Figs. 4.8 and 4.11a, it is surprising to see the trends of total micropore volumes are positively coherent with that of SSA. Figure 4.12 further shows the relationship between the increment of total micropore volume (ΔV) and SSA (ΔS) increment. Linear regressions were performed on the results calculated from H–K, DR, and DFT approaches. It was found that the increment of total micropore volume had satisfactory linearity with the increment of SSA, and the R^2 value corresponding to the fitting of DFT calculations even reached 0.928, showing the best linear relation. Commonly, micropores are ignored when analyzing the SSA of C-S-H, and much attention is usually paid to the mesopores and macropores. However, the reconstruction of these mesopores and macropores due to the harshly cryogenic attack observed in the current study showed poorly linear relations with SSA, while the increment of micropore volume was perfectly consistent with that of SSA calculated for C-S-H. Hence, the micropores existing in C-S-H are believed to

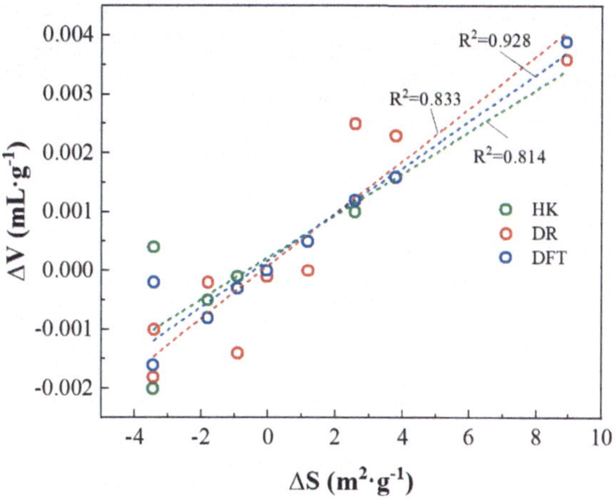

Fig. 4.12 Correlations between ΔS (increment of SSA) and ΔV (increment of total micropore volume)

significantly contribute to its SSA. By saying this, it does not mean mesopores and macropores in C-S-H do not contribute to SSA. Instead, they play predominant roles in providing BET surface area. The critical insight is that the reconstruction, e.g., collapse or expansion of micropores in the original C-S-H pore systems, is mainly responsible for the variation of SSA leading from cryogenic attack.

4.4.6 Surface Fractal Dimension

Fractal dimension *D* is an essential parameter to characterize the fractal geometry of a given solid, quantitively showing its surface and structural irregularities [63]. Two popular models, i.e., the FHH model and the Neimark model [64], are widely adopted to calculate *D* value. Since the FHH model has better sensitivity to PSD (particularly to small pores) than the Neimark model, as suggested in Ref. [50], therefore, the FHH model is used here. Figure 4.13 and Table 4.1 show the regression results by using the FHH equation. The slopes (h values) of all fitting lines are beyond the range of -1/3 ~ 0, suggesting that the predominant bonding force relating to C-S-H surface and nitrogen is surface tension. Similar observations were also found in Ref. [50]. In the current study, a higher surface fractal dimension indicates that the C-S-H solids are more heterogeneous with varying pore sizes, and a more significant variation in the value of *D* suggests the pore structure is more harshly damaged by the cryogenic attack. Therefore, Eq. 4.15 is used here to calculate the fractal dimension.

As shown in Table 4.1, for the C-S-H samples before the cryogenic attack, C/S0.84B had the smallest *D* value among these samples, while C/S1.5B and C/S2.0B obtained higher *D* values, which indicates that the C/S ratio plays a role in altering the surface irregularity of C-S-H. The influence of cryogenic attacks on the surface fractal dimension is apparent. Compared with those C-S-H samples without experiencing a cryogenic attack, higher surface fractal dimensions are obtained after the cryogenic

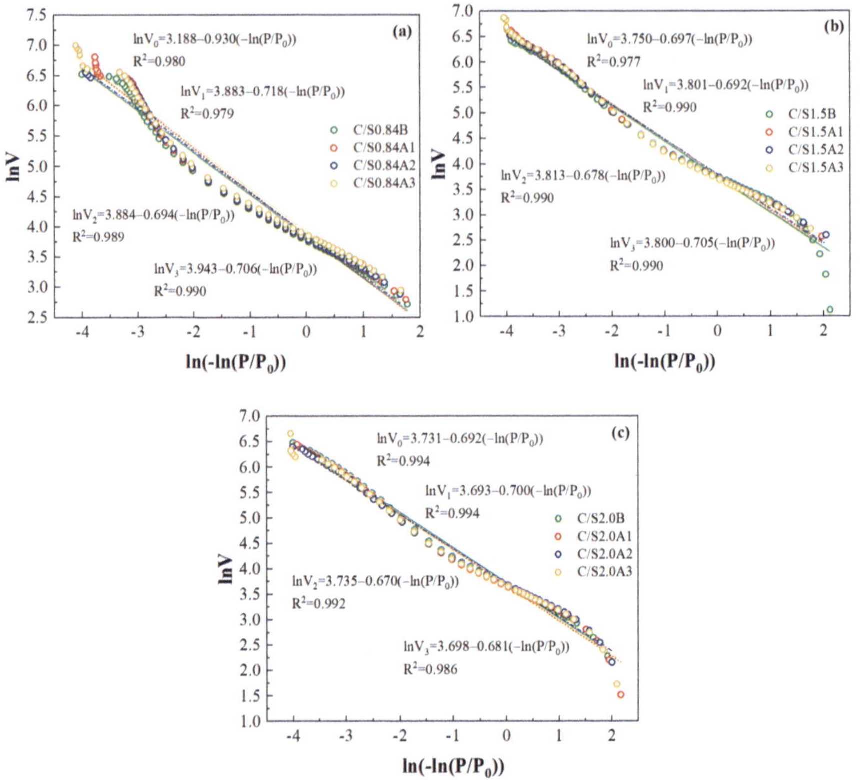

Fig. 4.13 Surface fractal dimension analysis for C-S-H samples by using FHH equation: **a** C/S0.84, **b** C/S1.5; **c** C/S2.0. The grey dash lines show the boundary of different fitting scales for C-S-H samples before the cryogenic attack

Table 4.1 Surface fractal dimension of C-S-H samples calculated from FHH equation

C/S ratio	Sample	R^2	h	$D1 = 3 + 3 \times h$	$D2 = 3 + h$
0.84	C/S0.84B	0.980	− 0.930	0.210	2.070
	C/S0.84A1	0.979	− 0.718	0.846	2.282
	C/S0.84A2	0.989	− 0.694	0.919	2.306
	C/S0.84A3	0.990	− 0.706	0.883	2.294
1.5	C/S1.5B	0.977	− 0.697	0.908	2.303
	C/S1.5A1	0.990	− 0.692	0.924	2.308
	C/S1.5A2	0.990	− 0.678	0.965	2.322
	C/S1.5A3	0.990	− 0.705	0.885	2.295
2.0	C/S2.0B	0.968	− 0.692	0.925	2.308
	C/S2.0A1	0.987	− 0.700	0.902	2.301
	C/S2.0A2	0.986	− 0.670	0.989	2.330
	C/S2.0A3	0.989	− 0.681	0.957	2.319

attack, which is seen for all C-S-H samples regardless of varying C/S ratios. This is consistent with the observed reconstruction of mesopores and macropores due to the cryogenic attack in Sect. 4.4.4. It is noted that C/S0.84 has a more significant increase in its surface fractal dimension than the other two groups of C/S1.5 and C/S2.0, which indicates that the harshly cryogenic attack leads to more severe damage on the pore geometry of C-S-H with a lower Ca/Si ratio. Such a difference agrees with the pore volume and PSD variations observed for these C-S-H samples. The increase in the surface fractal dimension can be attributed to the newly formed microcracks between C-S-H grains after the cryogenic attack [11], which consequently leads to a more heterogeneous nature of C-S-H, increasing the complexity and irregularity of mesopores.

4.4.7 Tortuosity

From the above sections, we know how the micropores and mesopores evolved under the cryogenic attacks, while the connectivity and the geometric tortuosity of the pore structure systems are still unclear. As mentioned before, the tortuosity factor can be used to characterize the pore structure network, i.e., the geometric complexity for the path of the flow medium, therefore providing additional criteria for estimating the evolution of the C-S-H pore structure. Table 4.2 displays the results for calculating the tortuosity factors from the CPSM model.

As defined in Eq. 4.20, the tortuosity factor highly depends on the pore length and shape. The N_S and D values obtained from the calculations characterize the frequency of pore size variation, i.e., nominal pore length and shape, respectively. Specifically, a higher value for the N_S means a longer pore tunnel with varying pore cross-sectional areas along the diffusion path, and the pore shape factor is further determined by $(D_{\max,eff} - D_{\min,eff})/D_{mean}$ which is related to the pore size distribution. As shown in Table 4.2, C-S-H with a lower Ca/Si ratio has a higher nominal pore length N_S value before the cryogenic attack, indicating that low-Ca/Si-ratio C-S-H hosts more cross-section pore tunnels along the path. In contrast, the pore shape factor of low-Ca/Si-ratio C-S-H is lower due to the more concentrated pore size distribution than high-C/S-ratio C-S-H as indicated by D values in Table 4.2, in line with the results shown in Sect. 4.4.4 for the BJH pore size distribution. Nonetheless, the leading tortuosity factor still reached C/S0.84. These observations indicate that the tortuosity of the C-S-H pore structure is highly dependent on its Ca/Si ratio, specifically decreasing with the increase of the Ca/Si ratio.

Table 4.2 Calculated tortuosity factors of C-S-H samples from the CPSM-tortuosity model

C/S ratio	Samples	N_S	$D_{\text{min,eff}}$ (nm)	$D_{\text{max,eff}}$ (nm)	D_{mean} (nm)	$\frac{D_{max,eff} - D_{min,eff}}{D_{mean}}$	τ_{CPSM}
0.84	C/S0.84B	7.39	14.35	39.09	26.39	0.94	2.72
	C/S0.84A1	6.64	15.54	41.35	28.93	0.89	2.50
	C/S0.84A2	6.57	13.83	40.05	25.29	1.04	2.73
	C/S0.84A3	5.24	15.72	40.93	29.47	0.86	2.17
1.5	C/S1.5B	5.54	11.93	36.07	22.63	1.07	2.53
	C/S1.5A1	4.94	14.11	42.01	27.14	1.03	2.32
	C/S1.5A2	4.82	12.85	40.93	24.71	1.14	2.43
	C/S1.5A3	5.08	17.49	39.36	33.87	0.65	1.86
2.0	C/S2.0B	4.22	12.32	40.82	23.98	1.19	2.30
	C/S2.0A1	4.61	12.67	39.83	24.59	1.10	2.33
	C/S2.0A2	5.46	11.89	36.00	23.03	1.05	2.48
	C/S2.0A3	3.58	14.28	45.16	26.79	1.15	2.04

The results in Table 4.2 provide more details about the evolution of the C-S-H pore structure under the successive cryogenic attack. Interestingly, the nominal pore lengths of C/S0.84 and C/S1.5 decreased after the cryogenic attacks. Meanwhile, the pore size distribution of these C-S-H phases was more concentrated in larger pores, leading to an increasing trend of the pore shape factor. As a result, their pore structure tortuosity factors tended to shift to lower values after cryogenic attacks. We should note that the seemingly abnormal observation for the decrease of nominal pore length did not mean that the cryogenic attacks had no damaging impact on the C-S-H pore structure. Instead, it is a reasonable consequence of forming large microcracks out of the range of capillary condensation-evaporation hysteresis. The formation of local microcracks broke the pore tunnels, giving rise to shorter pore lengths. This observation provides perfect evidence to validate the schematic deduction made in Ref. [11] that microcracks form after the harshly cryogenic attack. C/S2.0 behaved slightly differently than the other two phases under the cryogenic attacks. Although its pore size distribution was more concentrated in larger pores, its nominal pore length increased after the first and second cryogenic attacks. This observation suggests that the expansion or formation of microcracks did not initially split up the original pore tunnels. Instead, the blocked pores present in the pore networks were probably further connected, consequently increasing the nominal pore length. After further expansion under the continuous cryogenic attack, the microcracks finally broke the pore tunnels as they did for C/S0.84 and C/S1.5. Therefore, its nominal pore length significantly decreased after the third cryogenic attack. Such a difference between C/S2.0 and the other two phases was probably due to the smaller grain size. The AFM images in Fig. 4.6 show that the C-S-H grains of C/S2.0 are smaller than C/S0.84 and C/S1.5. The quantitative measurements in Ref. [11] also indicate that the C-S-H grain size decreases with the increase of Ca/Si ratio. As stated before, the small grains in C/S2.0 were more stable under the cryogenic attack. Unlike the large grains in low-Ca/Si-ratio C-S-H, they can avoid forming internal microcracks [11].

4.5 Concluding Remarks

In this chapter, we have comprehensively discussed the pore structure evolution under cyclic cryogenic attack. C-S-H is a very complex porous phase that has a wide range of pore sizes from microscale to macroscale. A comprehensive nitrogen adsorption evaluation is conducted to investigate the evolution of these pores. The surface area, pore size distribution, pore volume, surface fractal dimension, and tortuosity were analyzed using various theories, including BET, BJH, NLDFT, H–K, DR, FHH, and CPSM approaches. Conclusions are summarized as follows:

(1) The cryogenic attack can lead to the C-S-H surface forming higher interactions with the absorbates. The adsorption energy of the first nitrogen adsorbed layer is increased with the cryogenic attack cycles. This may result in the higher hydrophilicity of C-S-H under cryogenic attack.
(2) The mesopore of C-S-H with a higher Ca/Si ratio can be preserved well under the harshly cryogenic attack. BJH and NLDFT calculations show macropores are inclined to emerge in C-S-H after the cryogenic attack.
(3) The ultramicropores and small supermicropores are more sensitive to the cryogenic attack than other larger micropores. Both NLDFT and H–K calculations show that these micropores are reconstructed under the cryogenic attack. The reconstruction of micropores in C-S-H is relate to the variation of surface areas. Linear regression is well-fitted for the increment of specific surface area and the increment of micropore volume.
(4) The cryogenic attack has resulted in higher complexity and irregularity of mesopores and macropores of C-S-H. This transition is probably a consequence of the formation of microcracks in C-S-H grains.
(5) The propagation of large microcracks may decrease the nominal pore length and tortuosity under cyclic cryogenic attacks.

References

1. Q. Huang, Z. Jiang, X. Gu, W. Zhang, B. Guo, Numerical simulation of moisture transport in concrete based on a pore size distribution model. Cem. Concr. Res. **67**, 31–43 (2015)
2. H.M. Jennings, J.J. Thomas, J.S. Gevrenov, G. Constantinides, F.-J. Ulm, A multi-technique investigation of the nanoporosity of cement paste. Cem. Concr. Res. **37**(3), 329–336 (2007)
3. J. Mitchell, J. Webber, J. Strange, Nuclear magnetic resonance cryoporometry. Phys. Rep. **461**(1), 1–36 (2008)
4. J.Q. Li, G.Q. Geng, R. Myers, Y.S. Yu, D. Shapiro, C. Carraro, R. Maboudian, P.J.M. Monteiro, The chemistry and structure of calcium (alumino) silicate hydrate: a study by XANES, ptychographic imaging, and wide- and small-angle scattering. Cem. Concr. Res. **115**, 367–378 (2019)
5. O. Coussy, P.J.M. Monteiro, Poroelastic model for concrete exposed to freezing temperatures. Cem. Concr. Res. **38**(1), 40–48 (2008)
6. S. Ghabezloo, Association of macroscopic laboratory testing and micromechanics modelling for the evaluation of the poroelastic parameters of a hardened cement paste. Cem. Concr. Res. **40**(8), 1197–1210 (2010)
7. N.P. Mayercsik, M. Vandamme, K.E. Kurtis, Assessing the efficiency of entrained air voids for freeze-thaw durability through modeling. Cem. Concr. Res. **88**, 43–59 (2016)
8. C. Li, Z. Jiang, R.J. Myers, Q. Chen, M. Wu, J. Li, P.J.M. Monteiro, Understanding the sulfate attack of Portland cement–based materials exposed to applied electric fields: Mineralogical alteration and migration behavior of ionic species. Cem. Concr. Compos. **111**, 103630 (2020)
9. Z. Jiang, B. He, X. Zhu, Q. Ren, Y. Zhang, State-of-the-art review on properties evolution and deterioration mechanism of concrete at cryogenic temperature. Constr. Build. Mater. **257**, 119456 (2020)
10. J. Zhengwu, D. Zilong, Z. Xinping, L. Wenting, Increased strength and related mechanisms for mortars at cryogenic temperatures. Cryogenics **94**, 5–13 (2018)
11. X. Zhu, Z. Jiang, B. He, C. Qian, Investigation on the physical stability of calcium-silicate-hydrate with varying CaO/SiO_2 ratios under cryogenic attack. Constr. Build. Mater. **252**, 119103 (2020)
12. J. Li, W. Zhang, K. Garbev, G. Beuchle, P.J.M. Monteiro, Influences of cross-linking and Al incorporation on the intrinsic mechanical properties of tobermorite. Cem. Concr. Res. **136**, 106170 (2020)
13. C. Li, J. Li, A. Telesca, D. Marchon, K. Xu, M. Marroccoli, Z. Jiang, P.J.M. Monteiro, Effect of polycarboxylate ether on the expansion of ye'elimite hydration in the presence of anhydrite. Cem. Concr. Res. **140**, 106321 (2021)
14. S. Storck, H. Bretinger, W.F. Maier, Characterization of micro- and mesoporous solids by physisorption methods and pore-size analysis. Appl. Catal. A-Gen. **174**(1–2), 137–146 (1998)
15. N. De Belie, J. Kratky, S. Van Vlierberghe, Influence of pozzolans and slag on the microstructure of partially carbonated cement paste by means of water vapour and nitrogen sorption experiments and BET calculations. Cem. Concr. Res. **40**(12), 1723–1733 (2010)
16. H.M. Jennings, Refinements to colloid model of C-S-H in cement: CM-II. Cem. Concr. Res. **38**(3), 275–289 (2008)
17. M. Tziotziou, E. Karakosta, I. Karatasios, G. Diamantopoulos, A. Sapalidis, M. Fardis, P. Maravelaki-Kalaitzaki, G. Papavassiliou, V. Kilikoglou, Application of H-1 NMR to hydration and porosity studies of lime-pozzolan mixtures. Microporous Mesoporous Mater. **139**(1–3), 16–24 (2011)
18. Y. Zhang, Z. Jiang, Y. Zhu, J. Zhang, Q. Ren, T. Huang, Effects of redispersible polymer powders on the structural build-up of 3D printing cement paste with and without hydroxypropyl methylcellulose. Constr. Build. Mater. **267**, 120551 (2021)

19. M.J. Setzer, Micro ice lens formation, artificial saturation and damage during freeze thaw attack. Mater. Build. Struct. **6**, 175–182 (2000)
20. M.J. Setzer, Micro-ice-lens formation in porous solid. J. Colloid Interface Sci. **243**(1), 193–201 (2001)
21. C.S. Zhou, F.Z. Ren, Q. Zeng, L.Z. Xiao, W. Wang, Pore-size resolved water vapor adsorption kinetics of white cement mortars as viewed from proton NMR relaxation. Cem. Concr. Res. **105**, 31–43 (2018)
22. Z.D. Zhang, G.W. Scherer, Evaluation of drying methods by nitrogen adsorption. Cem. Concr. Res. **120**, 13–26 (2019)
23. M. Wu, B. Johannesson, M. Geiker, Application of water vapor sorption measurements for porosity characterization of hardened cement pastes. Constr. Build. Mater. **66**, 621–633 (2014)
24. Q. Ren, Z. Zeng, Z. Jiang, H. Li, Functionalization of renewable bamboo charcoal to improve indoor environment quality in a sustainable way. J. Cleaner Prod. **246**, 119028 (2020)
25. A.C.A. Muller, K.L. Scrivener, J. Skibsted, A.M. Gajewicz, P.J. McDonald, Influence of silica fume on the microstructure of cement pastes: New insights from ^{1}H NMR relaxometry. Cem. Concr. Res. **74**, 116–125 (2015)
26. A.C.A. Muller, K.L. Scrivener, A.M. Gajewicz, P.J. McDonald, Densification of C-S-H measured by 1H NMR relaxometry. J. Phys. Chem. C **117**(1), 403–412 (2013)
27. R. Tislova, A. Kozlowska, R. Kozlowski, D. Hughes, Porosity and specific surface area of Roman cement pastes. Cem. Concr. Res. **39**(10), 950–956 (2009)
28. A.M. Kaja, H.J.H. Brouwers, Q.L. Yu, NO_x degradation by photocatalytic mortars: the underlying role of the CH and C-S-H carbonation. Cem. Concr. Res. **125**, 105805 (2019)
29. Z.-W. Jiang, W.-T. Li, Z.-L. Deng, Z.-G. Yan, Experimental investigation of the factors affecting accuracy and resolution of the pore structure of cement-based materials by thermoporometry. J. Zhejiang Univ-Sci. A **14**(10), 720–730 (2013)
30. Y.Q. Shen, M. Deng, A.Q. Lu, Structural evolution of hydrated cement compacts. Mater. Struct. **44**(10), 1735–1743 (2011)
31. A. Tripathy, A. Kumar, V. Srinivasan, K.H. Singh, T.N. Singh, Fractal analysis and spatial disposition of porosity in major Indian gas shales using low-pressure nitrogen adsorption and advanced image segmentation. J. Nat. Gas Sci. Eng. **72** (2019)
32. C.S. Zhou, F.Z. Ren, Z.D. Wang, W. Chen, W. Wang, Why permeability to water is anomalously lower than that to many other fluids for cement-based material? Cem. Concr. Res. **100**, 373–384 (2017)
33. S. Brunauer, L. Deming, W. Deming, E. Teller, On a theory of the Vander Waals adsorption of gases. J. Am. Chem. Soc. **62**, 1723–1732 (1940)
34. E.P. Barrett, L.G. Joyner, P.P. Halenda, The determination of pore volume and area distributions in porous substances. I. computations from nitrogen isotherms. J. Am. Chem. Soc. **73**(1), 373–380 (1951)
35. M. Luisa Ojeda, J. Marcos Esparza, A. Campero, S. Cordero, I. Kornhauser, F. Rojas, On comparing BJH and NLDFT pore-size distributions determined from N_2 sorption on SBA-15 substrata. Phys. Chem. Chem. Phys. 5(9), 1859–1866 (2003)
36. C.M. Lastoskie, K.E. Gubbins, Characterization of porous materials using molecular theory and simulation. Adv. Chem. Eng. (Academic Press, 2001), pp. 203–250
37. M. Thommes, Physical adsorption characterization of nanoporous materials. Chem. Ing. Tech. **82**(7), 1059–1073 (2010)
38. C. Lastoskie, K.E. Gubbins, N. Quirke, Pore-size distribution analysis of microporous carbons: a density-functional theory approach. J. Phys. Chem. **97**(18), 4786–4796 (1993)
39. J.P.R.B. Walton, N. Quirke, Capillary condensation: a molecular simulation study. Mol. Simul. **2**(4–6), 361–391 (1989)

40. P.I. Ravikovitch, G.L. Haller, A.V. Neimark, Density functional theory model for calculating pore size distributions: pore structure of nanoporous catalysts. Adv. Colloid Interface Sci. **76–77**, 203–226 (1998)
41. M. Dubinin, L. Radushkevich, The equation of the characteristic curve of activated charcoal. Chem. Zentr **1**(55), 327–329 (1946)
42. G. Horvath, K. Kawazoe, Method for the calculation of effective pore-size distribution in molecular-sieve carbon. J. Chem. Eng. Jpn. **16**(6), 470–475 (1983)
43. B. Ghanbarian, A.G. Hunt, R.P. Ewing, M. Sahimi, Tortuosity in porous media: a critical review. Soil Sci. Soc. Am. J. **77**(5), 1461–1477 (2013)
44. S. Ahmad, A.K. Azad, An exploratory study on correlating the permeability of concrete with its porosity and tortuosity. Adv. Cem. Res. **25**(5), 288–294 (2013)
45. P. Stroeven, A stereological approach to roughness of fracture surfaces and tortuosity of transport paths in concrete. Cem. Concr. Compos. **22**(5), 331–341 (2000)
46. G.P. Androutsopoulos, C.E. Salmas, A new model for capillary condensation−evaporation hysteresis based on a random corrugated pore structure concept: prediction of Iintrinsic pore size distributions. 1. Model formulation. Ind. Eng. Chem. Res. 39(10), 3747–3763 (2000)
47. G.P. Androutsopoulos, C.E. Salmas, A new model for capillary condensation−evaporation hysteresis based on a random corrugated pore structure concept: Prediction of intrinsic pore size distribution. 2. Model application. Ind. Eng. Chem. Res. **39**(10), 3764–3777 (2000)
48. C.E. Salmas, G.P. Androutsopoulos, A novel pore structure tortuosity concept based on nitrogen sorption hysteresis data. Ind. Eng. Chem. Res. **40**(2), 721–730 (2001)
49. J. Landers, G.Y. Gor, A.V. Neimark, Density functional theory methods for characterization of porous materials. Colloids Surf., A **437**, 3–32 (2013)
50. P. Duan, Z.H. Shui, W. Chen, C.H. Shen, Influence of superplasticizer on composition and pore structure of C-S-H. Constr. Build. Mater. **44**, 87–91 (2013)
51. L. Zhang, K. Yamauchi, Z. Li, X. Zhang, H. Ma, S. Ge, Novel understanding of calcium silicate hydrate from dilute hydration. Cem. Concr. Res. **99**, 95–105 (2017)
52. A. Ouzia, K. Scrivener, The needle model: A new model for the main hydration peak of alite. Cem. Concr. Res. **115**, 339–360 (2019)
53. A. Nonat, The structure and stoichiometry of C-S-H. Cem. Concr. Res. **34**(9), 1521–1528 (2004)
54. X. Zhu, C. Qian, B. He, Q. Chen, Z. Jiang, Experimental study on the stability of C-S-H nanostructures with varying bulk CaO/SiO_2 ratios under cryogenic attack. Cem. Concr. Res. **135**, 106114 (2020)
55. K. Liu, M. Ostadhassan, The impact of pore size distribution data presentation format on pore structure interpretation of shales. Adv. Geo-Energy Res. **3**(2), 187–197 (2019)
56. D. Dollimore, G.R. Heal, Pore-size distribution in typical adsorbent systems. J. Colloid Interface Sci. **33**(4), 508–519 (1970)
57. C.M. Lastoskie, K.E. Gubbins, Characterization of porous materials using density functional theory and molecular simulation. Stud. Surf. Sci. Catal. **128**, 41–50 (2000)
58. J. Villarroel-Rocha, D. Barrera, K. Sapag, Introducing a self-consistent test and the corresponding modification in the Barrett, Joyner and Halenda method for pore-size determination. Microporous Mesoporous Mater. **200**, 68–78 (2014)
59. L. Liu, X.C. Wang, J. Zhou, H.Q. Chu, D.J. Shen, H.S. Chen, S.N. Qin, Investigation of pore structure and mechanical property of cement paste subjected to the coupled action of freezing/thawing and calcium leaching. Cem. Concr. Res. **109**, 133–146 (2018)
60. K. Sing, The use of nitrogen adsorption for the characterisation of porous materials, Colloids Surf., A **187**, 3–9 (2001)
61. S.U. Rege, R.T. Yang, Corrected Horvath-Kawazoe equations for pore-size distribution. AIChE J. **46**(4), 734–750 (2000)

62. N. Saeidi, M. Parvini, Accuracy of Dubinin-Astakhov and Dubinin-Radushkevich adsorption isotherm models in evaluating micropore volume of bentonite. Period. Polytech. Chem. Eng. **60**(2), 123–129 (2016)
63. F.M. Wang, S.F. Li, Determination of the surface fractal dimension for porous media by capillary condensation. Ind. Eng. Chem. Res. **36**(5), 1598–1602 (1997)
64. A. Neimark, A new approach to the determination of the surface fractal dimension of porous solids. Physica A **191**(1–4), 258–262 (1992)

Chapter 5
Cryogenic Stability of C-S-H Nanostructure and Nanomechanical Properties

5.1 Introduction

C-S-H has the most complex nanostructure among the hydration products of cement. Its (CaO)/(SiO_2) (C/S) molar ratio ranges from 1.2 to 2.3 [1, 2] with a mean value of 1.75 [2, 3] when it is hydrated from neat Portland cement. The addition of supplementary cementitious materials makes its composition much more complicated, expanding the range of C/S ratios to 0.7–2.1 [2]. C-S-H is well known for its non-crystalline phase structure, also described as phases with relatively better crystallinity, such as C-S-H(I) [3] and C-S-H(II) [4]. C-S-H(I) is considered a structurally imperfect form of tobermorite [5], also denoted as the nanocrystalline phase [6]. According to the differences in layer thickness obtained from X-ray powder diffraction, tobermorite is described into three types, namely 14 Å, 11 Å, and 9 Å tobermorite, which are further revealed in approximating to compositions $C_5S_6H_9$, $C_5S_6H_5$ and C_5H_6H, respectively [7]. The 11 Å tobermorite has two family members with two different subcell descriptions in crystallography. One is orthotobermorite, characterized by orthorhombic subcells, and the other is clinotobermorite by monoclinic subcells [7]. The 'dreierketten' [8] was proposed to describe the chemical structure of the silicate phase. The Ca-O polyhedral layer is clasped with silicate chains in a repeat of three silicate tetrahedra on both sides, composed of the disordered sandwiched structure. Between the interlayer of two Ca-O sheets, additional Ca^{2+} and H_2O molecules exist. According to the C-S-H structure described in Ref. [9–11], protons and hydroxyl groups are also in the interlayer, participating in the compensation of charge. Based on the distance between silicate tetrahedra to Ca-O sheet, the silicate tetrahedra are named with paired tetrahedra (closest) and bridging tetrahedra (far away), respectively. Specifically, the two bridging silicate tetrahedra units of silicate chains on the lateral sides of Ca-O sheets may link together and form double silicate chains, since the bridging tetrahedra of adjacent layers are close enough to share an O atom [7]. However, this bridging structure does not form in 14 Å

Z. Jiang and X. Zhu, *Cryogenic Stability of Cement-Based Materials at Micro/Nano Scale*, https://doi.org/10.1007/978-981-96-1836-1_5

tobermorite due to the long distance between two adjacent silicate layers, therefore forming a single silicate chain structure.

The first chemical structure model proposed to describe C-S-H was Taylor's model [12], which considers the C-S-H as a combination of imperfect layers of jennite ($Ca_9[H_2Si_6O_{18}](OH)_8{\cdot}6H_2O$) and a small portion of 14 Å tobermorite ($Ca_5[H_2Si_6O_{18}]{\cdot}8H_2O$). Another widely discussed model was proposed by Richardson and Groves [13]. This model assumes that C-S-H is composed of single-chain tobermorite and jennite or calcium hydroxide, representing the T/J model and T/CH model, respectively, which enables to flexibly characterize the C-S-H with Ca/Si ranging from 0.67 to 2.5. Richardson [8] proposed another more general chemical model called the C-(A)-S-H(I) model in 2014. The C-(A)-S-H model can be used to describe both single-chain and double-chain C-(A)-S-H phases. However, it cannot be used to characterize C-S-H with a high C/S ratio and containing abundant Ca-rich phases in the interlayer. Mohamed et al. [9] proposed another C-S-H Brick model, which can characterize C-S-H with varying C/S ratios and describe the complexity in the nanoparticulate nature of C-S-H. From He et al. [14], the crosslink between silicate chain layers was observed as C/S below 0.8, which was not seen as C/S higher than 0.8. At a low C/S ratio ($0.66 < C/S < 1$), the structure of C-S-H is approximated to that of tobermorite [12], which is characterized by infinite silicate chains. Silicate chains of the same or neighboring layers may also interconnect at a low C/S ratio [15]. As the C/S ratio increases, there are growing numbers of calcium ions participating in the substitution for protons in the interlayer [9], accompanied by rupture and depolymerization of silicate chains [9, 16], leading to a more disordered [16] and more unstable silicate chain structure [12]. These facts have been confirmed by solid-state ^{29}Si NMR studies [9, 17] and molecular dynamics simulation [18, 19]. As the C/S ratio ranges from 1 to 1.5, the bridging silicate tetrahedra are gradually missing from the silicate chains [9], leading to a more severe breakage to the chains [16] and exposing more vacant sites [20]. As the C/S ratio rises to more than 1.5, the infinite silicate chains are broken into dimers [21], excessive calcium hydroxide may be incorporated in the interlayer space [17].

Considering the vital role of C-S-H in cement-based materials, any changes on its nanostructures are likely to result in macro-scale degradation [22, 23]. However, what may happen to its properties and structure under cryogenic attack is still unclear and needs to be systematically clarified. Thus, it would be of great significance to study the degradation role of C-S-H under cryogenic attack to better understand its impacts on cement-based materials. Understanding the stability of different C-S-H structures is also paramount to fulfilling the sustainability of cement-based materials used in ultra-low-temperature environments.

This chapter focuses on the detrimental effects of cryogenic attack on C-S-H gel. It specifically investigates the influence of cryogenic attack on the nanostructure of C-S-H with various chemical compositions. Further evaluation involves the physical stability of C-S-H under cryogenic attack, primarily encompassing nanomechanical properties and the stability of C-S-H micro-nanostructures. We employed multiscale characterization techniques to comprehensively analyze the multiscale structural features of C-S-H. The multiscale experimental observations finally led us to find out a bottom-up degradation pathway for C-S-H under cryogenic attack.

5.2 Morphology of Synthetic C-S-H

5.2.1 Phase Assemblage

The XRD results (Fig. 5.1) show that the C-S-H phases are predominant in the final synthetic products of each sample. Meanwhile, the portlandite (CH) peaks were apparently observed in the diffraction pattern of C/S2.0. According to Cong and Kirkpatrick [16], the CH phase exists in the reacted products by the reaction of fumed silica and highly reactive β-C_2S. However, the CH phase was not identified in the reaction products at this C/S molar ratio of 1.7 in He's synthetic system [14], in which the C-S-H was synthesized by reaction of fumed silica and calcium oxide at room temperature. As the C/S molar ratio increased, the synthetic products detected the CH phase. This phenomenon was also found in our hydrothermal method and other synthetic systems [14, 24]. From this perspective, 1.7 is probably a critical bulk C/S molar ratio for the reactants to fully react into C-S-H. The basal reflection (of 2θ 5° ~ 10°) indicates the 3D ordering structure of C-S-H. The XRD patterns also show that a lower C/S molar ratio can obtain a higher ordering C-S-H phase. The C/S0.84 sample showed a much better structure ordering than the other three groups from the XRD patterns. As the C/S ratio increases, the peak near 30° is wider, and the peaks near 7° and 16° are weaker, which indicates the increasing disorder of the silicate chain structure of C-S-H. Although protective methods had been applied in the process of synthesis and storage, a small part of carbonation was still inevitable. The calcite phase was detected in C/S2.0.

The results of TGA performed on synthetic C-S-H products are illustrated in Fig. 5.2. Combined with the XRD results, the major mass losses corresponding to the compositions can be identified from the derivative mass loss traces. As depicted in Fig. 5.2, typical peaks centered at 120 °C and 830 °C were observed in all DTG curves. According to previous knowledge [25, 26], they were associated with the C-S-H bound water loss and C-S-H crystallization, respectively. Different from the other two samples of C/S0.84 and C/S1.5, the peaks at around 430 °C and 680 °C were observed in the DTG curve of C/S2.0. They were assigned to decompositions of $Ca(OH)_2$ and $CaCO_3$, respectively. These observations are consistent with the XRD results. It is noted that a small peak at around 625 °C was witnessed in the DTG curves of both C/S0.84 and C/S1.5. This was also associated with the decomposition of $CaCO_3$, indicating that the synthetic solids in the two groups also had a small amount of calcium carbonate phase.

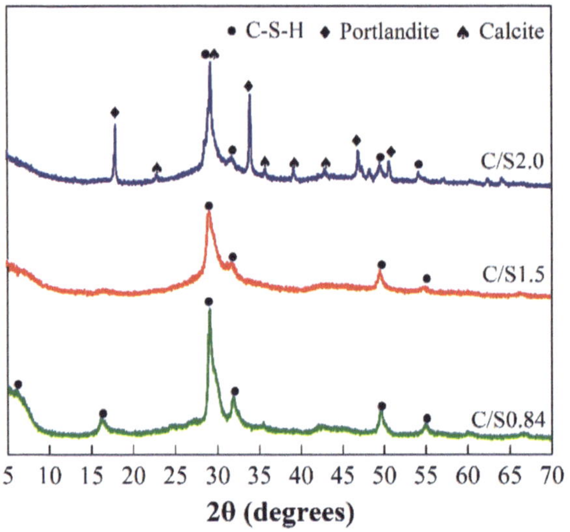

Fig. 5.1 XRD patterns of synthetic C-S-H

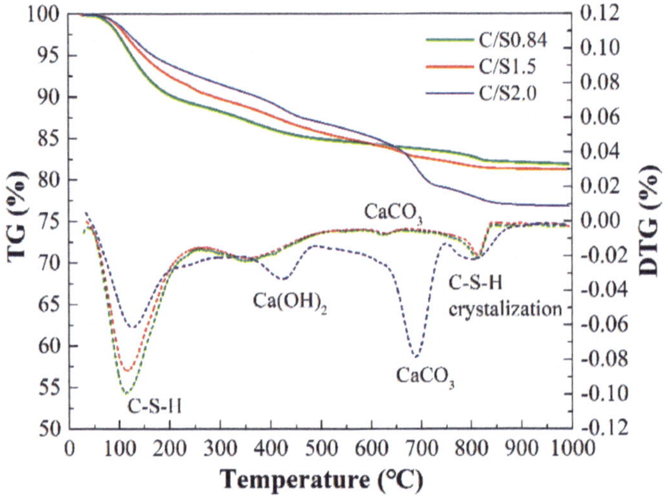

Fig. 5.2 Thermogravimetric analysis of synthetic C-S-H products

5.2.2 SEM Morphology

The SEM surface microstructure of the synthesized C-S-H is shown in Fig. 5.3 C-S-H in cement hydration systems exhibits diverse morphological characteristics such as needle-like, plate-like, fibrous, circular, rectangular, and spindle-shaped structures. These structures vary under different physical and chemical conditions. As depicted

Fig. 5.3 Scanning electron microscope surface morphology of C-S-H synthesized products: **a** C/S 0.84, **b** C/S 1.50, **c** C/S 2.0

in Fig. 5.3a, b and c, the synthesized C-S-H exhibits plate-like structural characteristics, especially when the Ca/Si ratio is 1.5. The morphology of this structure closely resembles the C-S-H formed after the hydration of C_3S at an early age. This indicates that the hydrothermal synthesis method employed in this study accelerates the aggregation reaction of C-S-H, resulting in a more mature C-S-H product.

Comparing the samples with Ca/Si ratios of 0.84 and 1.5 to that with a Ca/Si ratio of 2.0, the C-S-H morphology in the Ca/Si 2.0 sample demonstrates a networked granular characteristic. Moreover, it presents a denser surface morphology, displaying typical amorphous features consistent with the trend indicated by the XRD results.

5.2.3 AFM Morphology

The topographic maps of C-S-H recorded by AFM observation are depicted in Fig. 5.4, which unambiguously show their characteristic nanogranular nature. The surface roughness values obtained from Eqs. 4.1 and 4.2 for each C-S-H sample are listed in Table 5.1. It is illustrated that the surface conditions of these polished samples are satisfactory. Except for sample C/S2.0 A1, all other samples have similar average R_a and R_q values in the 20–27 nm range and 25–34 nm, respectively. Although the average R_a and R_q values of sample C/S2.0 A1 are relatively higher than other samples, they are still less than 59 nm, the maximum R_q value reached by Trtik et al. [27] in their AFM-based indentation measurements. These surface roughness values indicate that the polished C-S-H samples were appropriate for AFM-based indentation.

The nanogranular feature of C-S-H is challenging to graphically observe in a hydrated cement system since its compositional phases are complex and randomly distributed. The topographic features of C-S-H phases with different C/S ratios before and after cryogenic are illustrated in Fig. 5.4g–l. Typically, the C-S-H clusters are agglomerated from several spherical grains, characterized by sizes of tens of nanometers. According to Jennings's model [28], these grains are further composed of smaller globules of around 5 nm, identified as the basic building blocks of C-S-H. In the ScanAyst mode of AFM operation in this study, it was difficult to obtain a satisfactory scanning image as the scale was smaller than 100 nm. Therefore, these small

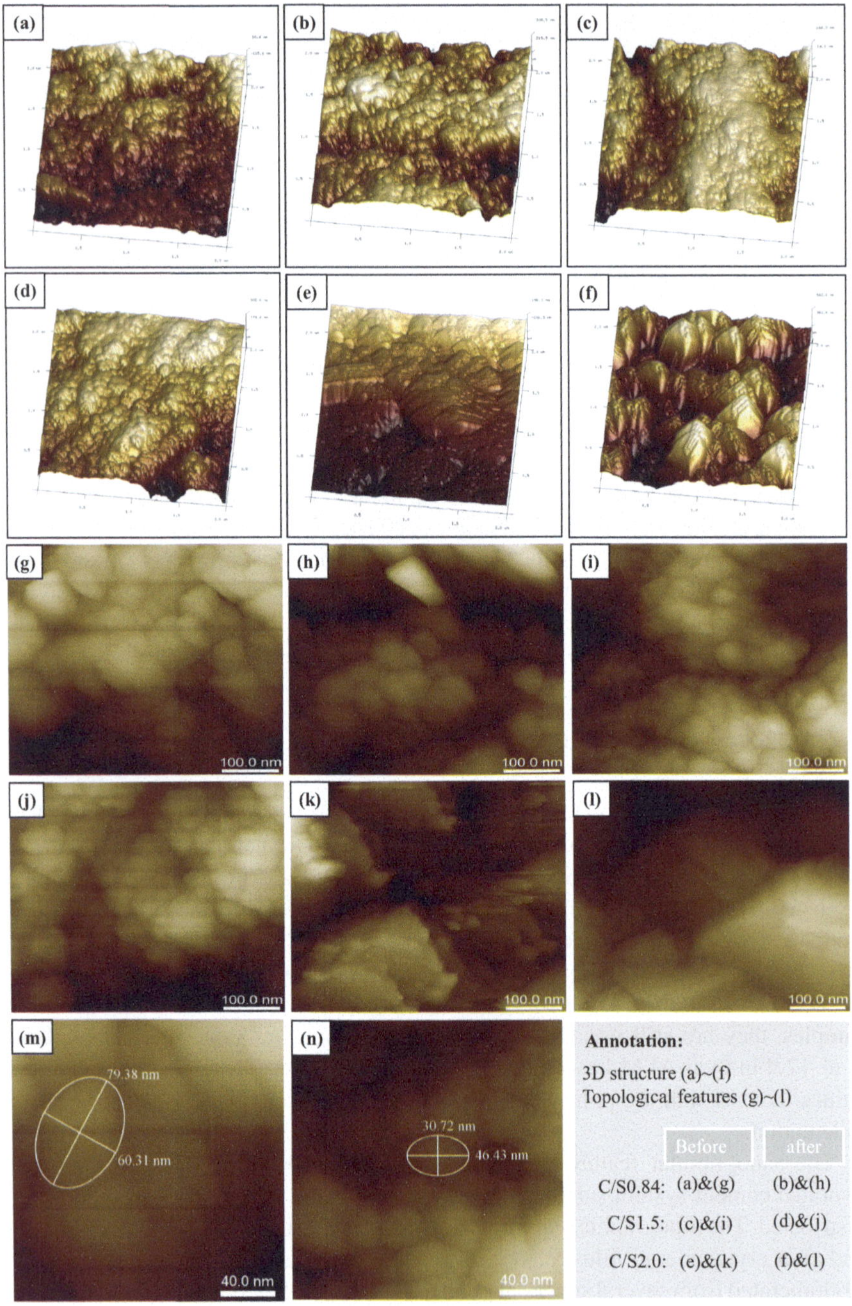

Fig. 5.4 Surface roughness evaluation (**a**)–(**f**), topographic feature (**g**)–(**l**) of C-S-H discs before and after cryogenic attack, and grain size comparison for C-S-H with lower C/S ratio C/S084 (**m**) and high C/S ratio C/S1.5 (**n**)

Table 5.1 Surface roughness of polished C-S-H samples

Sample	Average R_a (nm)	Average R_q (nm)
C/S0.84 B	21.57	26.83
C/S0.84 A1	20.40	25.53
C/S1.5 B	27.00	33.67
C/S1.5 A1	25.27	31.50
C/S2.0 B	24.23	30.83
C/S2.0 A1	46.87	58.43

globular units were still not observed. The maps obtained in this study showed that C/S ratios greatly regulate the topographic features of C-S-H. As illustrated in Fig. 5.4, the C-S-H phases of group C/S0.84 (Fig. 5.4g and h) are made up of larger packing grains that are even more homogeneous with each other, and these grains are packed in a more ordered way.

In comparison, the C-S-H phases of group C/S1.5 (Fig. 5.4i and j) are more disordered and more inhomogeneous since their packing grains are generally smaller and more randomly distributed than that of C/S0.84. From Fig. 5.4m, the typical size of C-S-H grains in C/S0.84 is at around 60–80 nm, while that of C/S1.5 is much smaller at around 30–40 nm. While in the group of C/S2.0 (Fig. 5.4k and l), large angular blocks are widely distributed in the scanning area, which is unlikely to be C-S-H due to their prominent crystalline characteristics. From the XRD results (Fig. 5.1), portlandite and calcite were the two main mineral compositions of group C/S2.0; therefore, those angular blocks observed in (Fig. 5.4k and l) were attributed to these two minerals, which obstructed to observe the accurate topographic maps of C-S-H beneath. The topographic features of these C-S-H phases seemed consistent with the structure ordering revealed by the XRD results. The more topographically ordered C-S-H of C/S0.84 had a better structure ordering among the three groups. Correspondingly, the C-S-H grains are larger in size. From these maps shown in Fig. 5.4, the shape and the size of C-S-H packing grains generally maintained in their original form, which means the topographically nanogranular features of C-S-H had no identifiable change resulting from the cryogenic attack.

5.3 Cryogenic Stability of C-S-H Nanostructure

5.3.1 Phase Stability

Figure 5.5 illustrates the comparison of the XRD patterns of the C-S-H synthesized products before and after the cryogenic attack. Comparatively, no significant differences were observed in the XRD spectra of the samples after the cryogenic attack. This indicates that the cryogenic attack does not induce the formation of new crystalline or semi-crystalline phases within the C-S-H. Notably, following the cryogenic

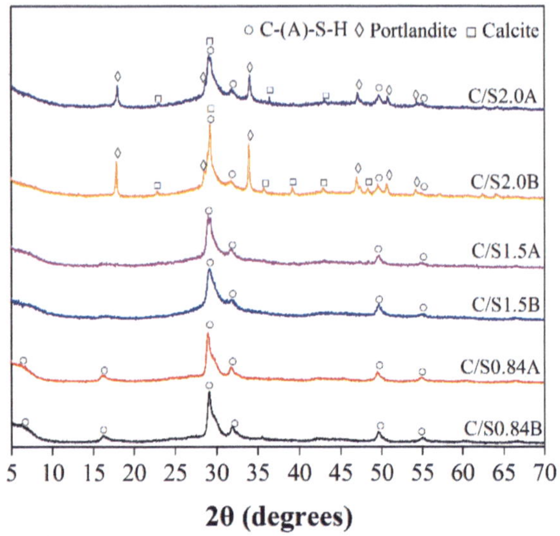

Fig. 5.5 XRD spectra of synthetic C-S-H phases. Label B represents before cryogenic attack, Label A represents after the cryogenic attack

attack, a relatively weaker intensity is observed in the characteristic peaks of C-S-H. Though not particularly pronounced, this phenomenon suggests the potential reconstruction of the semi-crystalline structure of these C-S-H products under cryogenic attack. For the C/S 2.0 sample, the diffraction peaks of calcium carbonate show a certain increase in intensity after cryogenic attack, indicating a certain degree of carbonation occurring in the sample during this process.

5.3.2 Composition Stability

5.3.2.1 Weight Loss Identification

TGA was performed on those samples before and after cryogenic attack to evaluate the weight fraction of different phases existing in C-S-H. The TG-DTG analysis results of these C-S-H samples are shown in Fig. 5.6. Four major weight loss stages can be observed in every TG curve. However, the samples showed incongruous weight loss behaviors after the cryogenic attack. Therefore, a quantitative analysis of the content of each composition is necessary. In combination with XRD results, we can explicitly identify most of the weight loss stages depicted in TGA. XRD analysis results illustrated in Fig. 5.1 have provided clear evidence for the predominant presence of C-S-H in every measurement sample. Calcite and portlandite were only detected in C/S2.0B and C/S2.0A. No new detectable crystalline or semi-crystalline phase formed in those samples after the cryogenic attack.

According to [25, 26], the weight loss centered at around 150 °C is assigned to the bound water loss of C-S-H. The peaks positioned at around 680 °C in DTG curves

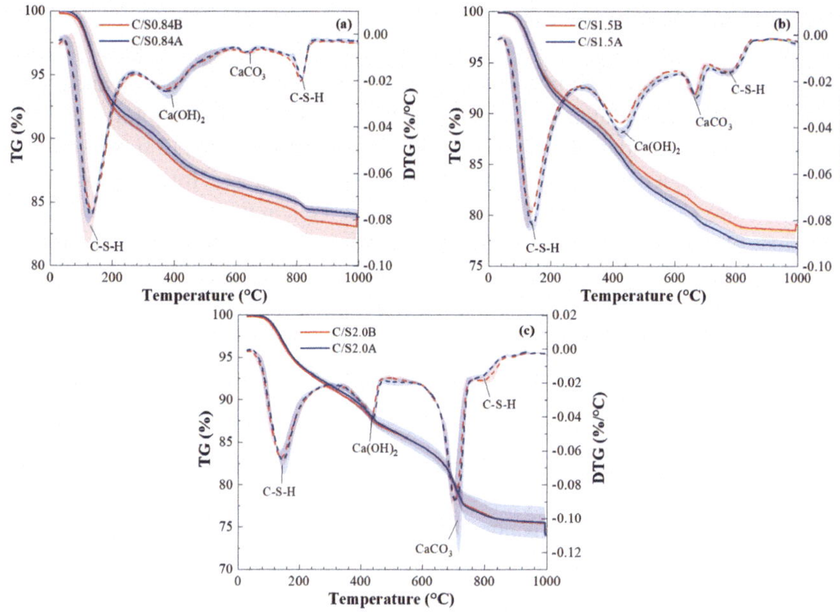

Fig. 5.6 Comparisons of weight loss of C-S-H phases before and after cryogenic attack. **a–c** are for C/S0.84, C/S1.5, and C/S2.0, respectively. Label B represents before the cryogenic attack, Label A represents after the cryogenic attack. These curves are based on three repeated TG experiments

are associated with the decomposition of calcium carbonate [25, 26]. Although X-ray diffraction peaks of calcite were only shown up in C/S2.0, a small fraction of calcite phases are still likely to exist in the test samples of C/S0.84 and C/S1.5 which can be disclosed by TGA. As temperature continuously increases, further decomposition of C-S-H will happen, resulting in wollastonite formation [5, 26]. Hence, the peaks at around 820 °C are attributed to the decomposition of C-S-H. Apart from these ascertainable weight loss stages, there is another significant peak from 350 to 500 °C in every derivative weight loss trace. Commonly, this peak is most likely to be the decomposition of calcium hydroxide (portlandite) [25, 26], which is based on the solid evidence for the presence of portlandite detected in C/S2.0 (B&A) by XRD analysis. However, portlandite was not detected in the XRD measurements of C/S0.84 (B&A) and C/S1.5 (B&A). The weight losses corresponding to this possible portlandite phase shown in Fig. 5.6a, b are significant and unignorable, which means it has a considerable amount in the samples. Most importantly, all these derivative weight loss peaks relevant to this phase are generally centered at a similar temperature of around 400 °C. Similar results are also obtained in other studies [29–32]. Kim et al. [32] ignored this contradiction and considered it the decomposition of calcium hydroxide. This view is adopted in this analysis, but our further insight is that it belongs to the decomposition of interlayer $Ca(OH)_2$ existing in the missing bridging silicate sites for charge compensation according to the C-S-H model proposed by Mohamed et al. [9] as well as the experimental data presented in Ref. [10].

5.3.2.2 Bound Water Fraction

Researchers hold different opinions on the temperature boundary of bound and physical water in thermogravimetric studies. Gallucci et al. [25] and L'Hôpital et al. [29] chose 80 °C as the boundary, but the ^{1}H NMR relaxometry detection [33] showed that the gel water was not completely evaporated at this temperature. However, the oven-dry temperature (110 °C or 105 °C) was much more violent to the C-S-H, leading to a slight loss of interlayer water, collapse in the layer structure, and polymerization in the silicate chains [33]. Due to this reason, 100 °C was chosen as the starting point of C-S-H dehydration in Ref. [34]. In this analysis, the bound water loss, ranging from 100 to 350 °C, was considered, and the up boundary was chosen according to Ref. [32] as well as our own test results. Thus, the weight fraction of bound water in C-S-H samples can be determined from TGA results by using:

$$f_{H_2O} = ml_1\% \tag{5.1}$$

where f_{H_2O} represents the bound water fraction, ml_1 is the mass loss of bound water in TGA in the range of 100–350 °C. The calculation results of these samples are shown in Fig. 5.7. Significant increase in the bound water fraction was seen in C/S1.5 after the cryogenic attack, while the bound water fractions of C/S0.84 and C/S2.0 appeared to be much more stable, only showing very mild reductions in the average values. These inconsistent variations in different C-S-H samples indicate that the stability of bound water molecules in the C-S-H interlayer may greatly depend on the C/S ratio of its composition. This will be further investigated by ^{29}Si NMR and FTIR analysis later. It is underlined that the bound water is highly related to C-S-H interlayer water since the interlayer water is usually chemically bound or structural bound according to [35, 36].

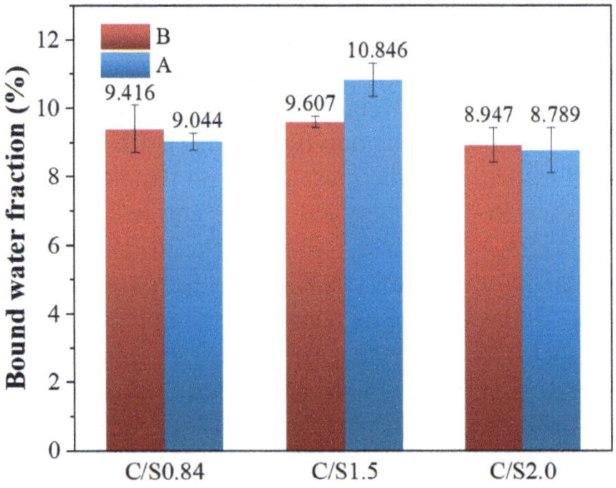

Fig. 5.7 Bound water fractions of C-S-H phases before and after cryogenic attack. Label B represents before the cryogenic attack, Label A represents after the cryogenic attack

5.3.2.3 $Ca(OH)_2$ Fraction

As supported by XRD analysis, crystalline $Ca(OH)_2$, namely portlandite, has been explicitly determined in C/S2.0B and C/S2.0A. Interlayer $Ca(OH)_2$ is believed to exist in all C-S-H samples. Combining the TGA results and previous knowledge [25, 26], the weight loss in the 350–500 °C range is associated with $Ca(OH)_2$ decomposition. Its fraction can be computed by:

$$f_{\mathrm{CH}} = \frac{M_{\mathrm{CH}} \times ml_2\%}{M_{\mathrm{H_2O}}} \tag{5.2}$$

where f_{CH} represents the fraction of $Ca(OH)_2$, ml_2 is the mass loss of $Ca(OH)_2$, and M_{CH} and $M_{\mathrm{H_2O}}$ are the molar mass of calcium hydroxide and water molecules, respectively.

Comparisons for the calcium hydroxide fraction of different groups are shown in Fig. 5.8. Interestingly, the calcium hydroxide fraction was ultimately changed due to the cryogenic attack. A sharp increase was observed for calcium hydroxide fraction in C/S1.5 after the cryogenic attack. The calcium hydroxide fractions of C/S0.84 and C/S2.0 were maintained at a more stable level. Nonetheless, a slight increase was still noticeable for the calcium hydroxide content in C/S2.0 under the attack. Unlike the other two groups, the calculated calcium hydroxide fractions of C/S0.84B and C/S0.84A were similar without significant differences. As a result, it was believed that the C-S-H in C/S0.84 could be kept more stable. Although the properties of interlayer $Ca(OH)_2$ remain mysterious, it can be deduced that the fraction variations towards calcium hydroxide are highly associated with interlayer $Ca(OH)_2$. Most importantly, these variations further indicate that the calcium ions may leach out from C-S-H, forming interlayer $Ca(OH)_2$ under the cryogenic attack. This inference will be further verified in subsequent sections.

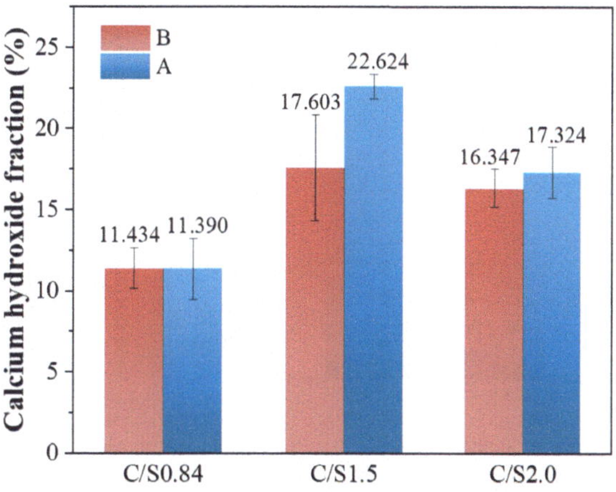

Fig. 5.8 Calcium hydroxide fractions of C-S-H phases before and after cryogenic attack. Label B represents before the cryogenic attack, Label A represents after the cryogenic attack

5.3.2.4 $CaCO_3$ Fraction

The decomposition of calcium carbonate is positioned at the range of 600–750 °C [25, 26]. From TG results, the calcium carbonate fraction can be calculated by:

$$f_{CC} = \frac{M_{CC} \times ml_3\%}{M_{CO_2}} \tag{5.3}$$

where f_{CC} represents the fraction of $CaCO_3$, ml_3 denotes the mass loss of $CaCO_3$, and M_{CC} and M_{CO_2} is the molar mass of calcium hydroxide and water molecules, respectively.

Considering the inevitable carbonation during the cryogenic treatment or other tests, it would be more precise to consider the changes of calcium carbonate fraction in the analysis of the calcium hydroxide fraction. It is assumed here that the $CaCO_3$ came from the $Ca(OH)_2$, which included the leached $Ca(OH)_2$. Therefore, the total $Ca(OH)_2$ fraction, i.e., f_{CHt} can be calculated by:

$$f_{CHt} = \frac{M_{CH}}{M_{CC}} \times f_{CC} + f_{CH} \tag{5.4}$$

Variations in the calcium carbonate f_{CC} and total calcium hydroxide f_{CHt} fractions are basically synchronized with that in the calcium hydroxide fraction f_{CH} as presented in Figs. 5.8 and 5.9. Literally, the calcium carbonate fraction of C/S1.5 significantly increased after the cryogenic attack, while the fractions in the other two groups kept in relatively constant levels. A small decrease and a slight increase can be observed in C/S0.84 and C/S2.0, respectively, though. By combining the carbonated $Ca(OH)_2$, the increment of the $Ca(OH)_2$ fraction in the C/S1.5 and C/S2.0 were even higher, as shown in Fig. 5.9b. These results further provide a confirmative support for the variations of calcium hydroxide in different groups as presented in Sect. 5.3.3.

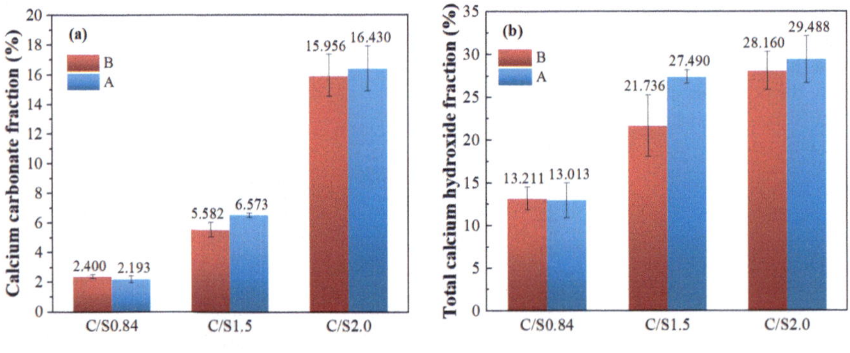

Fig. 5.9 Calcium carbonate fractions (**a**) and total calcium hydroxide fractions (**b**) of C-S-H phases before and after cryogenic attack. Label B represents before the cryogenic attack, Label A represents after the cryogenic attack

5.3.2.5 Stoichiometric Formula

The stoichiometric formula of C-S-H is presented as $(CaO)_{C/S}$•(SiO_2)•$(H_2O)_x$. The C/S and x can be determined from Eqs. 5.4 and 5.6, respectively, in which ml_1, ml_2, and ml_3 denote the mass loss of bound water, calcium hydroxide, calcium carbonate, respectively, and $\overline{C/S}$ is the bulk C/S molar ratio. This was also used by Kim et al. [32].

$$C/S = \frac{\overline{C/S}(100 - \%ml_1 - 4.113\%ml_2 - 2.7274\%ml_3) - 3.336\%ml_2}{100 - \%ml_1 - \%ml_2 - 2.274\%ml_3} \tag{5.5}$$

$$x = \frac{\%ml_1\left(3.115\overline{C/S} + 3.337\right)}{100 - \%ml_1 - \%ml_2 - 2.274\%ml_3} \tag{5.6}$$

Based on the abovementioned rules, the parameters of the C-S-H stoichiometric formula are calculated and illustrated in Table 5.2. Notably, the bound water fraction in the C/S1.5 group had the most significant increase, leading to its formula parameter x increasing from 0.96 to 1.13. In comparison, the values of x in C/S0.84 and C/S2.0 stood steadily without apparent changes, consistent with the bound water fraction variations. These results indicate that the bound water of C-S-H is still dynamically changeable under cryogenic attack, and the variation (bound water loss or increase) seems highly dependent on the C/S ratio.

From the calculations, it is underlined here that the actual C/S ratios of the synthetic products were all smaller than the bulk C/S values of the compositions. The stoichiometric formula parameter C/S is driven by calcium hydroxide fraction. The remarkable increase of calcium hydroxide fraction observed in C/S1.5 resulted in its stoichiometric formula parameter C/S significantly declining from 1.05 to 0.90. For C/S2.0, a small decrease was also calculated for its stoichiometric C/S value, which was much milder compared with that in C/S1.5, though. The C/S value of C/S0.84 was very stable, keeping at around 0.64. A basic consensus reached from these results is that the bound water and calcium ions in the C-S-H structure are unstable under the influence of cryogenic temperature. The inconsistent calcium hydroxide fraction variation trends reflect different removability of calcium ions in C-S-H at different C/S ratios.

Table 5.2 TGA results for the stoichiometric of C-S-H before and after cryogenic attack

$\overline{C/S}$	C/S					x				
	B[a]	Error	A[a]	Error	Increment	B[a]	Error	A[a]	Error	Increment
0.84	0.64	± 0.02	0.64	± 0.03	0.00	0.66	± 0.05	0.63	± 0.02	– 0.03
1.5	1.05	± 0.09	0.90	± 0.02	– 0.15	0.96	± 0.03	1.13	± 0.06	0.17
2.0	1.37	± 0.07	1.33	± 0.08	– 0.04	1.21	± 0.10	1.20	± 0.13	– 0.01

[a] Label [B] Represents before the cryogenic attack; label [A] Represents after the cryogenic attack

5.3.3 Calcium Leaching

SEM–EDS analysis was performed on those C-S-H phases that sustained a cryogenic attack. Several spots were selected to perform EDS analysis in the observed vision. Among these chosen random spots, two typical spectrums were depicted and compared with each other, as shown in Fig. 5.10. These spectrums present clear information for the element fraction of the chosen spots. Specific attention is focused on the atomic content of Ca and Si and their molar ratios. For EDS analysis of C/S0.84A1 shown in Fig. 5.10a, there is no significant fluctuation in the atomic fraction of Ca and Si. The Ca/Si ratio of spectrum 1 and spectrum 2 is 0.818 and 0.782, respectively. While for C/S1.5A (Fig. 5.10b), the Ca/Si ratio (1.720) of spectrum 2 is prominently higher than that of spectrum 1 (1.223) as well as the designed bulk Ca/Si ratio (1.5), which indicates that calcium ions are possibly enriched at spot 2. This could be supportive evidence of calcium leaching from the internal structure. Similar phenomena were also found in C/S2.0A1, as depicted in Fig. 5.10c. In this sample, the Ca/Si ratio of spectrum 1 and spectrum 2 is 1.330 and 2.524, respectively. These results present evidence for the calcium leaching in C/S1.5 and C/S2.0 at the micro-level and indicate that the leaching process may lead to the enrichment of calcium phases due to the cryogenic attack. From the TEM results (see Fig. 5.11), the enriched phases were perhaps portlandite and vaterite that had been observed in C-S-H of C/S1.5 after cryogenic attack.

Due to the pronounced leaching of Ca^{2+} ions in sample C/S1.5, our objective was to ascertain whether the C-S-H in this sample transformed into crystalline $Ca(OH)_2$ or if the leached interlayer $Ca(OH)_2$ could further reassemble to form crystalline $Ca(OH)_2$. However, the XRD testing in Fig. 5.1 did not detect the presence of crystalline $Ca(OH)_2$, thus failing to provide evidence for this hypothesis. Consequently, we conducted further analysis using TEM and SAED. Figure 5.11 illustrates the TEM images and SAED patterns of the C-S-H sample from C/S1.5 after cryogenic attack. It's crucial to emphasize that $Ca(OH)_2$ (portlandite) and $CaCO_3$ (calcite) were not observed in the C-S-H of C/S1.5B, while they were present in the C-S-H sample of C/S1.5A. Notably, they lack statistical significance due to the significant influence of sampling methods on these observations. Nonetheless, this phenomenon remains noteworthy.

5.3.4 C-S-H Atomistic Structure Variation

The silicate tetrahedra moieties in dreierketten chains of C-S-H are basically characterized into five types, namely monosilicates (Q^0), disilicates (Q^1), the chain middle groups (Q^2), layers and chain branching sites (Q^3), as well as the three-dimensional networks (Q^4) [37]. Q^1 and Q^2 are the predominant moieties in C-S-H with varying C/S ratios [16], which has been confirmed by other vibrational spectroscopy studies

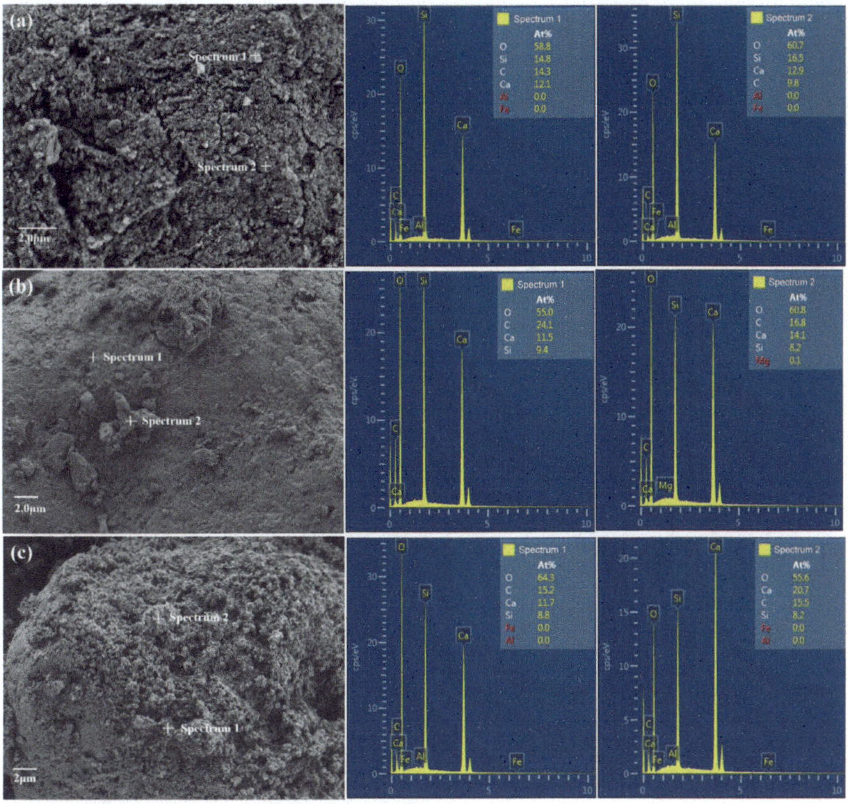

Fig. 5.10 SEM–EDS analysis on the surface of C-S-H after cryogenic attack. **a**, **b** and **c** is for C/S0.84A, C/S1.5A, and C/S2.0A, respectively

[38, 39]. While due to the geometric differences of silicate tetrahedra in the dreierketten chains and the presence of Ca^{2+} ions and protons that bound with Si–O groups, the silicate tetrahedra environments of Q^1, Q^2 and Q^3 are further distinguished as Q^1(OH), Q^1(Ca), Q^2p, Q^2i, Q^2b, Q^2(Ca), Q^2v, Q^3(OH) and Q^3(Ca) [14, 37, 40–43]. Specifically, Q^1(Ca) and Q^1(OH) [40] refer to the Q^1 units bound with Ca^{2+} ions and protons, respectively. The bridging tetrahedra bound with two protons is represented as Q^2p [41–43], and that bound with one proton and one Ca^{2+} ion is represented as Q^2i [37, 41, 42]. The paired tetrahedra coordinated to two main-layer Ca^{2+} ions, and the bridging tetrahedra bound with two interlayer Ca^{2+} ions are represented as Q^2(Ca) and Q^2b, respectively [41–43]. Although Q^2(Ca) and Q^2b are in different chemical states, and the presence of Ca^{2+} ions in bridging sites for charge compensation significantly influences its hydrogen bonding, the chemical shifts for the two types of silicate tetrahedra are similar to each other according to the NMR experiments in Ref. [10]. Therefore, here we considered the Q^2(Ca) and Q^2(b) as the same

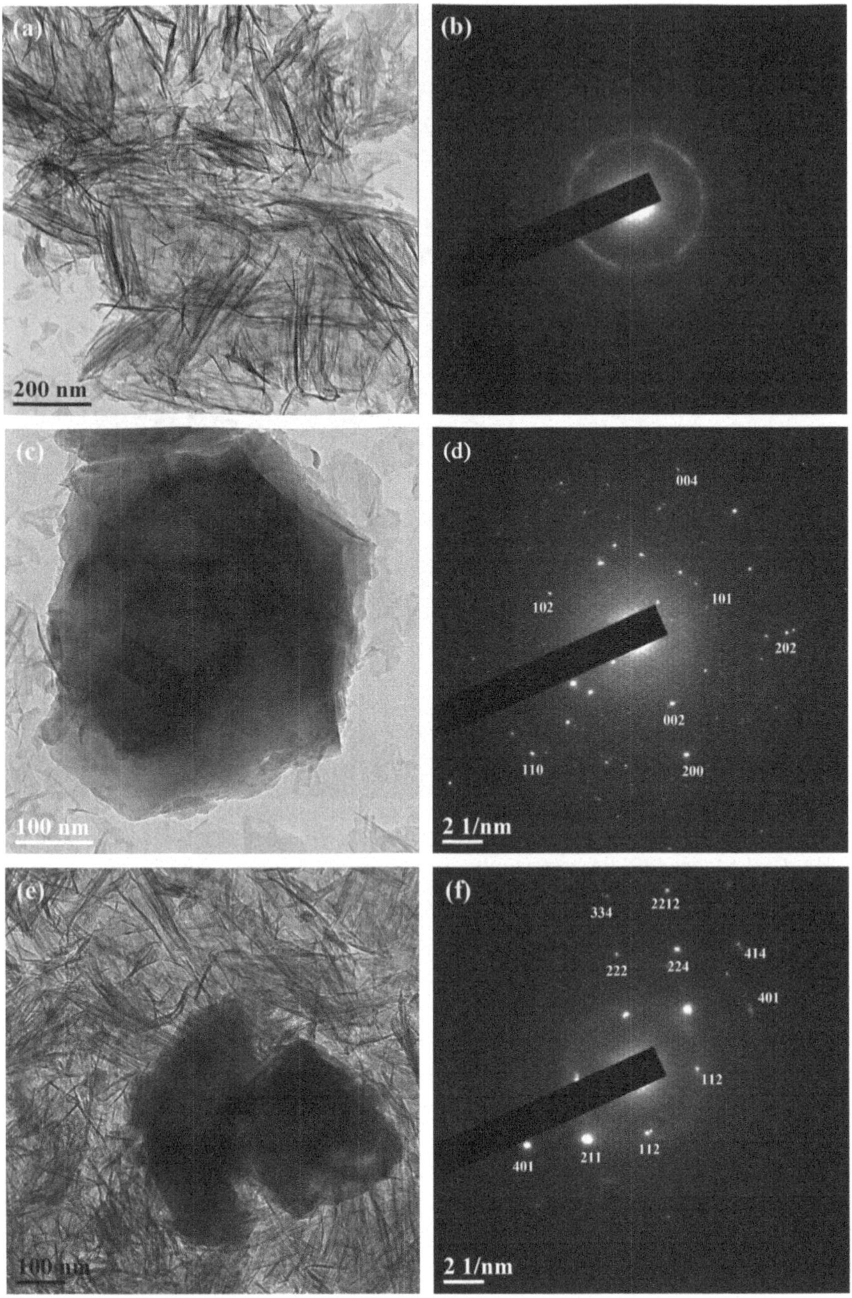

Fig. 5.11 TEM observations and SAED patterns for C-S-H in C/S1.5 after the cryogenic attack. **a** Shows the presence of C-S-H, **b** shows $Ca(OH)_2$ (portlandite), and **c** for $CaCO_3$ (vaterite)

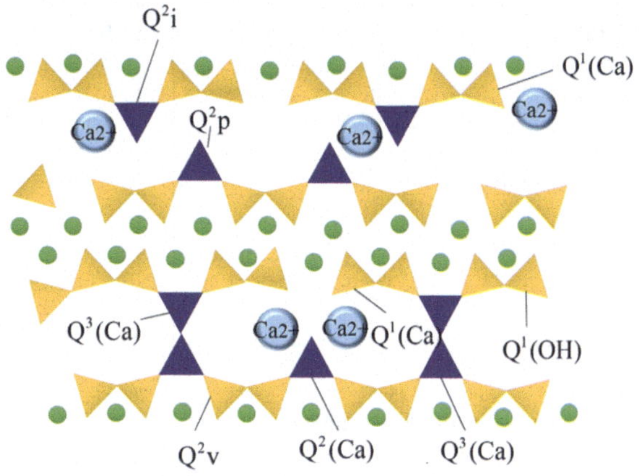

Fig. 5.12 Schematic diagram of atomistic structure of C-S-H after Ref. [42]

species, using Q^2(Ca) to represent both moieties. For the paired tetrahedra neighboring to Q^3, it is named with Q^2v [41]. Similarly, the bridging tetrahedra at the Q^3 site bound with Ca^{2+} ion and proton are called Q^3(Ca) [37, 40] and Q^3(OH) [41, 42], respectively. A schematic diagram of the C-S-H structure is shown in Fig. 5.12.

According to the above-mentioned statement, the deconvolution results of these six ^{29}Si NMR spectrums are illustrated in Fig. 5.13 and Table 5.3. The fractions of the different chemical states of silicate anions are shown in Table 5.4. Usually, the mean chain length (MCL) of alumino-silicate chains of C-(A)-S-H can be calculated from Eq. 5.7 [37]. In our case, it can be more specifically expressed as Eq. 5.8. The calculations for different C-S-H groups are presented in Table 5.3.

$$\mathrm{MCL} = \frac{2\left(Q^1 + Q^2(0\mathrm{Al}) + \frac{3}{2}Q^2(1\mathrm{Al}) + Q^3(0\mathrm{Al}) + Q^3(1\mathrm{Al})\right)}{Q^1} \tag{5.7}$$

$$\mathrm{MCL} = \frac{2\left[Q^1(\mathrm{OH}) + Q^1(\mathrm{Ca}) + Q^2\mathrm{p} + Q^2\mathrm{i} + Q^2(\mathrm{Ca}) + Q^2\mathrm{v} + Q^3(\mathrm{Ca}) + Q^3(\mathrm{OH})\right]}{Q^1(\mathrm{OH}) + Q^1(\mathrm{Ca})} \tag{5.8}$$

It is clear from Fig. 5.13 that the atomic structure of C-S-H can be significantly affected by the C/S ratio. Q^3(Ca) and Q^3(OH) peaks were found in the C/S0.84 group, indicating the branching silicate tetrahedrons were formed in the C-S-H silicate chains. These two peaks were not observed in the other two groups. The total fraction of Q^1 units (including both Q^1(Ca) and Q^1(OH)) rose with the increasing C/S ratio. Therefore, a lower MCL value was obtained at a higher C/S ratio, as presented in Table 5.4, which means the polymerization degree of silicate chains decreased. These observations are strictly consistent with other studies [14, 41].

In comparison, considerable changes were observed for the atomic structure of C-S-H after the cryogenic attack, and the structure variations resulting from the cryogenic attack are different, with the bulk C/S ratio increasing from 0.84 to 2.0. The calculated increments in the fractions of different Q^n sites listed in Table 5.3

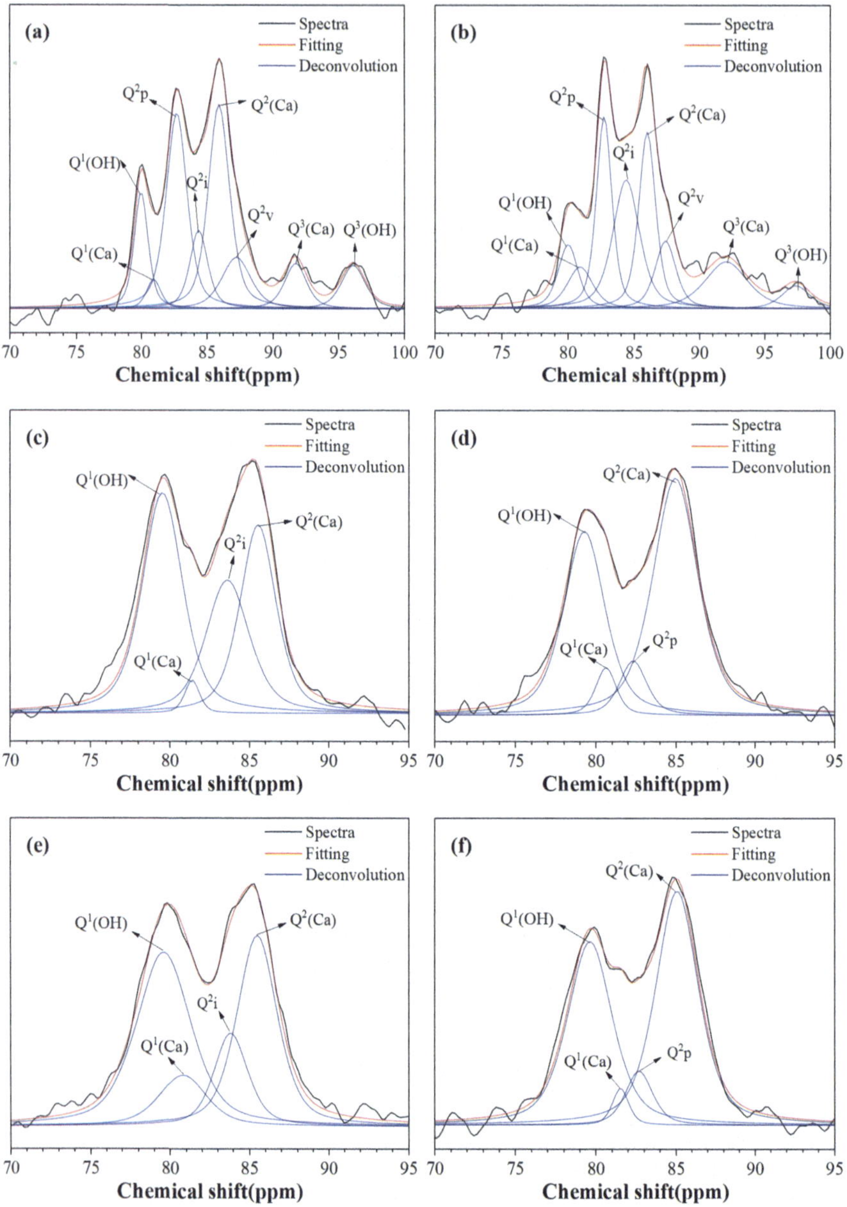

Fig. 5.13 Deconvolutions of ^{29}Si NMR spectrum. **a**, **b** illustrates for C/S0.84B and C/S0.84A1, respectively. **c**, **d** for C/S1.5B and C/S1.5A1, **e**, **f** for C/S2.0B and C/S2.0A1. Label B represents before the cryogenic attack, Label A1 represents after the cryogenic attack. Detailed deconvolution information for these spectra is shown in Table 5.4

Table 5.3 Deconvolution results of ^{29}Si NMR spectra of C-(A)-S-H

Symbol	Descriptions (Si–O–X)	Typical chemical shift (ppm)	Deconvoluted chemical shift (ppm)					
			C/S0.84		C/S1.5		C/S2.0	
			B	A1	B	A1	B	A1
Q^1(OH)	End SiO_4 tetrahedra bonded to protons	– 79 ~ – 80 [2–4]	– 79.95	– 79.90	– 79.50	– 79.25	– 79.60	– 79.60
Q^1(Ca)	End SiO_4 tetrahedra bonded to Ca^{2+} ions	– 81 [5]	– 80.90	– 80.80	– 81.35	– 80.60	– 80.80	– 81.50
Q^2p	Bridging SiO_4 tetrahedra bonded to two protons	– 82.1 [2–4]	– 82.65	– 82.62	– 83.60	–	– 83.85	–
Q^2i	Bridging SiO_4 tetrahedra bonded to one proton and one Ca^{2+} ion	– 83.9 [2–4]	– 84.35	– 84.30	–	– 82.30	–	– 82.65
Q^2(Ca)	Within SiO_4 tetrahedra bonded to two Ca^{2+} ions	– 85.3 [2–4]	– 85.90	– 85.90	– 85.50	– 84.95	– 85.50	– 85.04
Q^2v	Paired SiO_4 tetrahedra neighboring to Q^3	– 87.3 [2–4, 6]	– 87.20	– 87.30	–	–	–	–
Q^3(Ca)	Branching SiO_4 tetrahedra bonded to one Ca^{2+} ion	– 91 [5, 6]	– 91.70	– 91.95	–	–	–	–
Q^3(OH)	Branching SiO_4 tetrahedra bonded to one proton	– 102.7 [2, 3]	– 96.10	– 97.28	–	–	–	–

Table 5.4 Fractions of different chemical states of silicate anions in C-S-H together with MCL values

Q(n)	Fraction (%)		Increment (%)	Fraction (%)		Increment (%)	Fraction (%)		Increment (%)
	C/S0.84B	C/S0.84A1		C/S1.5B	C/S1.5A1		C/S2.0B	C/S2.0A1	
Q^1(OH)	10.61	7.95	− 2.66	40.38	36.34	− 4.04	41.70	40.30	− 1.41
Q^1(Ca)	2.65	6.65	4.00	2.32	4.71	2.38	10.68	2.96	− 7.73
Q^2p	26.75	18.80	− 7.95	–	7.16	7.16	–	7.28	7.28
Q^2i	9.28	21.77	12.48	26.47	–	− 26.47	14.00	–	− 14.00
Q^2(Ca)	27.84	18.86	− 8.99	30.83	51.79	20.96	33.62	49.47	15.85
Q^2v	9.57	9.01	− 0.56	–	–	–	–	–	–
Q^3(Ca)	6.60	12.89	6.30	–	–	–	–	–	–
Q^3(OH)	6.69	4.07	− 2.62	–	–	–	–	–	–
Mean chain length (MCL) (unit: 1)									
	15.08	13.69	− 1.36	4.68	4.87	0.19	3.82	4.62	0.82

Note "–" means that the corresponding value is 0

present more detailed information for the structure transition. For the C/S0.84 group, the total fraction of Q^1 units had a small increase after the cryogenic attack, leading to the MCL slightly decreasing from 15.08 to 13.69 (Table 5.4), which indicates that the long silicate chains were depolymerized into shorter chains due to the cryogenic treatment. It is notably from Table 5.4 that the variations of Q^1(OH), Q^1(Ca), Q^2p, Q^2i, Q^2(Ca) and even Q^3(Ca) and Q^3(OH) fractions were all significant, whereas the fractions of Q^1 (total fraction of Q^1(OH) and Q^1(Ca)), Q^2 (total fraction of Q^2p, Q^2i and Q^2v) and Q^3 (total fraction of Q^3(Ca) and Q^3(OH)) were not significantly varied. It can be deduced from these observations that the silicate chains of C-S-H in C/S0.84 were generally stable, while the bonding ions (Ca^{2+} and H^+) for charge balance were perhaps simultaneously reconstructed under the cryogenic attack.

As for C/S1.5 and C/S2.0, it is easy to identify some common features in the variations of atomic structures for these two C-S-H samples from Table 5.4. Specifically, the contents of Q^2(Ca) and Q^2p had the most predominant increases in both samples, while the disappearance of Q^2i resulted in its fraction reversely and significantly decreased. The difference is that there was another considerable decrease with an increment of − 7.73% in the Q^1(Ca) fraction for C/S2.0, which led to its slightly decreased MCL. In contrast, the MCL of C/S1.5 maintained steadily. This means the silicate chains of C-S-H in C/S2.0 were perhaps slightly polymerized due to the cryogenic attack. These results suggest that the skeleton of silicate chains was still stable even under extremely cryogenic conditions. However, the bonding between silicate tetrahedra and different cations, such as H^+ and Ca^{2+}, may be broken.

These results present clear evidence for the structure stability of C-S-H under the influence of ultra-low temperatures. The atomic structure variations revealed by ^{29}Si NMR spectrums indicate that the atomistic structure of C-S-H is not exactly unchangeable. Further, these variations can be another proof of calcium leaching.

In order to show the atomistic variations more clearly, the structural-chemical formulae of these C-S-H phases were calculated based on Richardson's C-(A)-S-H(I) model [8] which has been discussed and applied to characterize the synthetic C-(A)-S-H in Ref. [5]. Although this model seems somewhat simplistic in representing a realistic C-S-H structure, discussing some of the parameters is still valuable. According to this model, the structural-chemical formula of single-chain tobermorite or C-(A)-S-H(I) can be represented as Eq. 5.9 [8].

$$Ca_4\left[Si_{1-f-v}Al_f\square_v O_{3-2v}\right]H_{2i}Ca_{2-i}(Ca,Na_2,K_2)_{3f}\cdot mH_2O \tag{5.9}$$

where the aluminum substitution situation is considered. □ is the vacant silicate tetrahedra site, *f*, and *v* denote the fraction of the Al-substituted site and the fraction of the vacant site, respectively. In this study, aluminum ions were not introduced into the C-S-H phases; thus, *f* equals 0 for all groups. *i* represents the extent of charge balance by H^+ or Ca^{2+}. The alkali cations (Ca^{2+}, Na^+, K^+) in the round bracket are mainly aimed to maintain the charge balance as Al^{3+} substitutes for Si^{4+}, and these cations mainly exist in the interlayers. The value of m is calculated from TGA results, mainly referring to the bound water.

The fraction of vacant sites v and charge balance extent i in C-S-H can be obtained from Eqs. 5.10 and 5.11, respectively.

$$v = \frac{1}{1 + \text{MCL}} \tag{5.10}$$

$$i = 6[1 - (1 - v) \cdot \text{C/S}] \tag{5.11}$$

The C/S values have been stoichiometrically determined from TGA results in Sect. 5.3.2. In particular, the protonation value of i only ranges from 0 to 2 [8]. Once the calculation result is out of this range, the maximum or minimum value (2 or 0) will be given to i. These parameters, the number of interlayer Ca2 + ions per tetrahedral site and the structure-chemical formulae are shown in Table 5.5. The formulae calculated here refer to 6 tetrahedral sites [5, 8].

Although the calculation results shown in Table 5.5 seem to be not entirely consistent with the ^{29}Si NMR spectrums in Fig. 5.13, the main trend is still similar. Thus, it is generally reliable for both results. The calculated number of interlayer Ca^{2+} per tetrahedral site is 0 for C/S0.84, though the fingerprints for the participation of Ca^{2+} in charge balance were observed in ^{29}Si NMR spectrums (the presence of Q^1(Ca) and Q^2i). In combination with the ^{29}Si NMR spectrums and the above calculations, a basic consensus is that the interlayer charges are mainly balanced by protons instead of Ca^{2+} ions at a low C/S ratio (typically C/S0.84 group). This is consistent with the conclusion made in Ref. [44]. The structural-chemical formula of C-S-H in C/S0.84 was much more stable and without significant changes in its formula parameters compared with other samples, indicating that the C-S-H's atomic structure was more stable under the cryogenic attack. The parameters of C/S1.5 saw obvious variations. A significant reduction in the number of interlayer Ca^{2+} ions per tetrahedral site suggested that the interlayer Ca^{2+} ions for charge compensation were partially removed from the structure. This further validates the calcium leaching phenomenon observed by our TGA and SEM–EDS measurements. In addition, the increase in the number of m indicated that more water molecules had been hosted in a building block of the C-S-H in C/S1.5 after the cryogenic attack. As for the C-S-H in C/S2.0, the most significant change was the apparent reduction in the number of vacant sites from 0.208 to 0.178, indicating slight polymerization and the interlayer Ca^{2+} ions were the predominant cation species for charge compensation in C-S-H of this sample.

5.3.5 *Chemical Bonds*

The infrared transmittance spectrums of C-S-H phases before and after the cryogenic attack are shown in Fig. 5.14. The characteristic bands generated by the symmetric and asymmetric stretching vibrations (v_1 and v_3) and in-plane and out-of-plane bending vibrations (v_2 and v_4) of Si–O bonds in silicate chains are positioned in

Table 5.5 Calculated parameters of the structural-chemical formulae of C-S-H together with interlayer Ca^{2+} ions per tetrahedral site

Samples	C/S	*v*	*i*	m	Interlayer Ca^{2+} per tetrahedral site	Structural-chemical formulae
C/S0.84B	0.642	0.062	2.000	3.479	0.000	$Ca_4[Si_{0.938}\square_{0.062}O_{2.876}]_6H_{4.000}Ca_{0.000}\cdot 3.479H_2O$
C/S0.84A1	0.644	0.068	2.000	3.316	0.000	$Ca_4[Si_{0.932}\square_{0.068}O_{2.864}]_6H_{4.000}Ca_{0.000}\cdot 3.316H_2O$
C/S1.5B	1.052	0.176	0.797	3.582	0.201	$Ca_4[Si_{0.824}\square_{0.176}O_{2.648}]_6H_{1.594}Ca_{1.203}\cdot 3.600H_2O$
C/S1.5A1	0.903	0.170	1.504	3.933	0.083	$Ca_4[Si_{0.830}\square_{0.170}O_{2.659}]_6H_{3.007}Ca_{0.496}\cdot 3.933H_2O$
C/S2.0B	1.374	0.208	0.000	3.414	0.333	$Ca_4[Si_{0.792}\square_{0.208}O_{2.585}]_6H_{0.000}Ca_{2.000}\cdot 3.414H_2O$
C/S2.0A1	1.334	0.178	0.000	3.406	0.333	$Ca_4[Si_{0.822}\square_{0.178}O_{2.644}]_6H_{0.000}Ca_{2.000}\cdot 3.406H_2O$

the range of 1300–400 cm^{-1} [45]. The bands at around 450 cm^{-1} and 670 cm^{-1} are assigned to v_2 (O-Si–O) [46, 47] and v_4 (Si–O-Si) [46, 47] vibrations, respectively, which were observed in all spectrums. The v_3 (Si–O) vibration is also assigned at around 970 cm^{-1}, witnessed in all spectrums. The characteristic band at around 740 cm^{-1} was only seen in C/S0.84, attributing to the v_1(Si–O) vibration in Q^3 units [31]. These observations are consistent with the atomic structure information obtained from the ^{29}Si NMR spectrums shown in Fig. 5.13. The adsorption band at 3643 cm^{-1} was only observed in C/S2.0, which is attributed to the stretching vibration of the O–H bond in calcium hydroxide [31, 46].

From the comparisons shown in Fig. 5.14, the v_3(Si–O) vibration band in C/S0.84 shifted to a lower wavenumber after the cryogenic attack, while this band in C/S1.5 shifted to a higher wavenumber, the variations were not significant. The fingerprint evidence for the polymerization of derieketten chains can be reflected on the higher wavenumber of the characteristic band associated with v_3(Si–O) stretching vibration in Q^2 units, as suggested in [45–47]. From this perspective, the silicate chains in C/

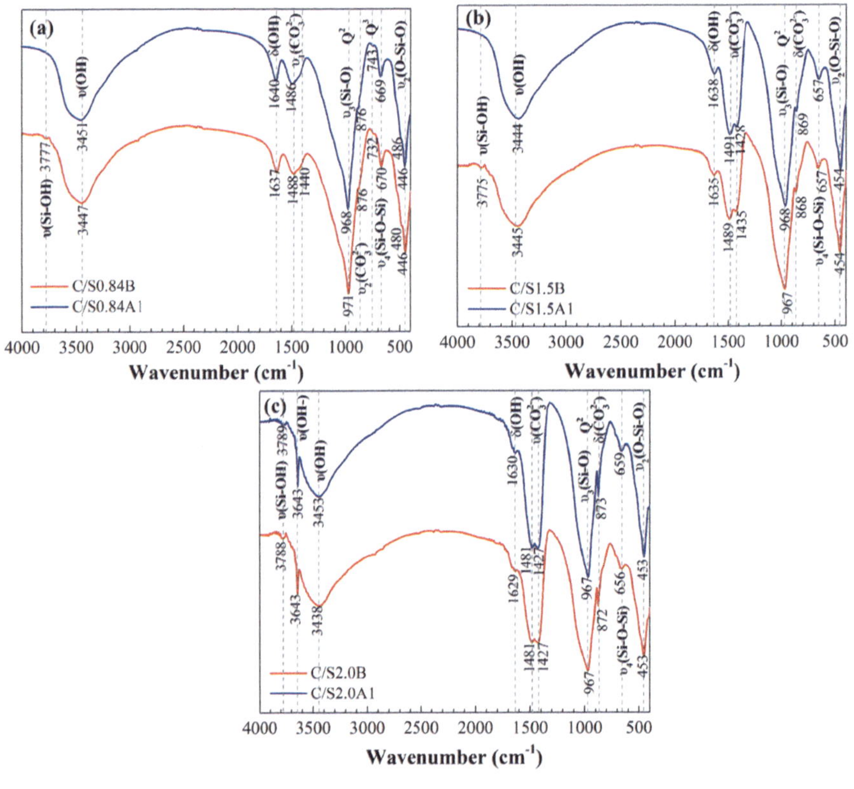

Fig. 5.14 Comparisons for FTIR spectra of C-S-H before and after the cryogenic attack. Label B represents before the cryogenic attack, Label A1 represents after the cryogenic attack

S0.84 and C/S1.5 were slightly depolymerized and polymerized, further supporting the ^{29}Si NMR results.

In Fig. 5.14, the adsorption bands at around 1640 cm^{-1} and 3450 cm^{-1} are attributed to the bending vibrations of δ(H–O–H) and stretching vibrations of *ν*(HO–H), respectively [48, 49]. The δ(H–O–H) and *ν*(HO–H) bands experienced the most significant variations under the influences of cryogenic attack. After the attack, the bands for δ(H–O–H) shifted to a higher wavenumber in all groups, irrespective of the varying C/S ratios. This change is interpreted as a stronger bonding between water molecules and the host structure [47]. In other words, the hydrophilic nature of C-S-H was perhaps slightly increased after the cryogenic attack. Apart from C/S1.5, the bands for *ν*(HO–H) vibration in C/S0.84 and C/S2.0 also shifted to a higher wavenumber. Before the cryogenic attack, a weak band for the stretching vibration of Si–OH bonds at around 3775 cm^{-1} [46] was observed in all groups. However, this characteristic band vanished or weakened after the attack. Although the Si–OH band was very weak, the disappearance or attenuation of this band was coherent with the small polymerization observed in ^{29}Si NMR results for C/S1.5 and C/S2.0, which is probably a sign of the slight polymerization.

5.3.6 A DNA Code Description

5.3.6.1 C-S-H Atomistic Variation

The results presented above clearly show that the cryogenic attack may lead to C-S-H leaching interlayer Ca^{2+} ions, while this calcium leaching process is highly controlled by its C/S ratio. According to Ref. [44], our ^{29}Si NMR experimental results and structural-chemical formulae calculations, Ca^{2+} ions are freed from the interlayer of C-S-H at C/S = 2/3, the omission of silicate tetrahedra in single silicate "dreieketten" chains is competitive within 2/3 < C/S < 1, and Ca^{2+} ions begin to incorporate in the interlayer, leading to a more disordered structure and a reduction in layer thickness [44]. This can explain why calcium leaching was not observed in the C/S0.84 samples after the cryogenic attack, as revealed by TGA. As the C/S ratio increased to a higher level, more Ca^{2+} ions are present in the interlayer and contribute to the charge balance. The amount of interlayer Ca^{2+} ions per tetrahedral site increased, making it more unstable and more likely to leach Ca^{2+} ions [4]. This makes it reasonable to observe a more significant increase in calcium hydroxide fraction in C/S1.5 after the cryogenic attack.

From the above explanation, the calcium leaching should have been more significant in the C/S2.0 group, while the observed increase of calcium hydroxide fraction in this group was not as significant as in group C/S1.5, which somewhat conflicts with the above analysis. However, it should be noted that the actual calcium-to-silica ratio of the dried synthetic C-S-H products in group C/S2.0 was only at around 1.4, as shown in the calculation results, instead of at the high bulk ratio 2.0. Since the synthetic C-S-H products are easily decomposed once they are no longer in the mother liquor or the pore solution, according to Refs. [50, 51], the drying treatment may have already led to the decomposition of C-S-H, resulting in the formation of C-S-H/$Ca(OH)_2$ nanocomposite [10]. In other words, some interlayer calcium ions for the C-S-H in C/S2.0 may have already leached before the cryogenic attack, which our XRD and TGA results can support that a considerable amount of $Ca(OH)_2$ and $CaCO_3$ were found in the products. Therefore, the increase of calcium hydroxide fraction induced by cryogenic attack in C/S2.0 was milder. Fortunately, this increase was still identifiable, as shown in Figs. 5.8 and 5.9. Meanwhile, common features of atomistic variations in the fractions of Q^2p, Q^2(Ca), and Q^2i were found in both C/S1.5 and C/S2.0. These results provide evidence of calcium leaching for the C-S-H in the C/S2.0 group.

The calcium leaching process was closely related to the atomic structure variations of C-S-H. Therefore, it can be shown by a back analysis of the changes in atomic structure. It is noted that the total increment of different silicate anions is always equal to the decrement of other anions, which means the increased silicate anions must derive from the decreased ones. This is easy to understand according to the mass conservation constraint. Likewise, the total fraction of Q^2(Ca) and Q^2p was generally equal to the decrement of Q^2i fraction in C/S1.5, and the total fraction of Q^2(Ca) and Q^2p was nearly equal to the total decrement of Q^2i and Q^1(Ca) in C/S2.0 as shown by the ^{29}Si NMR deconvolution results in Table 5.4. As stated before, the silicate tetrahedra of Q^2(Ca) and Q^1(Ca) are bound with Ca^{2+} ions for charge balance, and the silicate tetrahedra of Q^2p is bound with protons for charge compensation, while that of Q^2i is concurrently bound with both Ca^{2+} ion and proton to keep charge balance. It is also noteworthy that the protons became the predominant species for interlayer charge balance after calcium leaching, particularly in C/S1.5, as presented in Table 5.5, while the number of interlayer Ca^{2+} ions per tetrahedra site decreased from 0.201 to 0.083. Thus, it can be deduced that the bonding ions, namely Ca^{2+} and proton, in Q^2i sites were reconstructed in the calcium leaching process, forming Q^2(Ca) and Q^2p.

5.3.6.2 A DNA Code Rule for C-S-H Nanostructure

To understand this process more clearly, a Brick C-S-H model based on an atomistic building block description proposed by Mohamed et al. [9] was adopted to analyze the reconstruction of bonding ions. Kunhi Mohamed et al. [9] developed an atomistic structure for C-S-H that can describe the nanostructures. This is considered a more reliable model for C-S-H and can be used to represent the C-S-H with varying Ca/Si ratios. We adopted this description method to discuss the variations in the atomistic structure of our C-S-H samples. The definitions and notations of different atom groups are illustrated in Fig. 5.15. Here we have defined two types of interlayer water molecules. One with dark gray ball represents the original water molecule in the interlayer. The other with red ball stands for newly formed water molecule. The other definitions are consistent with those defined in the literature [9]. Here, we briefly describe these definitions: < and > represent the dimer silicate associated with

<table>
<tr><th>Code</th><th>Unit</th><th>stoichiometry</th><th>Charge</th><th>Code</th><th>Unit</th><th>stoichiometry</th><th>Charge</th></tr>
<tr><td colspan="4">Ca-Si sheet above interlayer</td><td colspan="4">Ca-Si sheet below interlayer</td></tr>
<tr><td><</td><td></td><td>CaSiO3.5</td><td>-1</td><td>></td><td></td><td>CaSiO3.5</td><td>-1</td></tr>
<tr><td>S</td><td></td><td>SiO2</td><td>0</td><td>S</td><td></td><td>SiO2</td><td>0</td></tr>
<tr><td>[S]'</td><td></td><td>[SiO]H</td><td>+1</td><td>[S]'</td><td></td><td>[SiO]H</td><td>+1</td></tr>
<tr><td>/</td><td></td><td>Ca(OH)2</td><td>0</td><td>\</td><td></td><td>Ca(OH)2</td><td>0</td></tr>
<tr><td>(</td><td></td><td>CaSiO4</td><td>-2</td><td>)</td><td></td><td>CaSiO4</td><td>-2</td></tr>
<tr><td>S|</td><td></td><td>SiO1.5</td><td>+1</td><td>|S</td><td></td><td>SiO1.5</td><td>+1</td></tr>
<tr><td>[</td><td></td><td>Ca(OH)1.5</td><td>+0.5</td><td>]</td><td></td><td>Ca(OH)1.5</td><td>+0.5</td></tr>
<tr><td colspan="8">Interlayer</td></tr>
<tr><td>C</td><td></td><td>Ca</td><td>+2</td><td>°</td><td></td><td>OH</td><td>-1</td></tr>
<tr><td>1</td><td></td><td>H2O</td><td>0</td><td>1</td><td></td><td>H2O</td><td>0</td></tr>
</table>

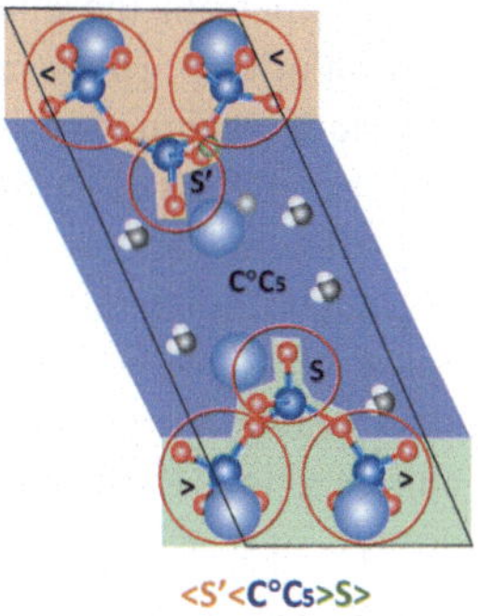

Fig. 5.15 Definition and notation of different atom groups used for describing the atomistic structure of C-S-H. After the atomistic structure represented by Kunhi Mohamed et al. [9]. H_2O unit with a red ball represents the newly formed water molecule

calcium above and below the interlayer. The bridging silicate with a stoichiometry of SiO_2 is represented by S. The protonated silanol group is denoted by an apostrophe ('), representing a proton H^+. C and ° denote the interlayer species of calcium ion and hydroxyl respectively. The interlayer water molecules are represented by a number. (and) denote the end silicate above and below the interlayer, respectively. Here, a calcium hydroxide unit is symbolized by /.

5.3.6.3 DNA-Code Rule Descriptions for the Atomistic Variations

Based on these definitions and the experimental results analyzed above, we proposed three possible reactions in Eqs. 5.12, 5.13, and 5.14 that are responsible for the calcium leaching and atomistic structure variations of C-S-H with high C/S ratio.

$$< S' < C^{o}C^{o}_{5} > S' > \rightarrow <' S' </C_5 > S > \tag{5.12}$$

$$< S' < C^{o}C^{o}_{5} > S' > \rightarrow <' S' < CC_7 > S > \tag{5.13}$$

$$< S\,('C_7 > S)'('S < C_7')S > \rightarrow < S('C_8 > S >< S < C_7')S > \tag{5.14}$$

The corresponding sketches of Eqs. 5.12 and 5.13 are shown in Fig. 5.16a–c respectively. In the first reaction process of Eq. 5.12, $Ca(OH)_2$ is formed from two nearby $(CaOH)^+$ units that participate in compensating the charge of bridging silicate in Q^2i, and the remaining interlayer Ca^{2+} ion has to be charge compensated by one of bridging silicate, consequently forming Q^2(Ca). Meanwhile, its original proton will participate in the charge balance of another bridging silicate, leading to the formation of Q^2p. However, this reaction was probably not the only process during the calcium leaching since the increments of Q^2p and Q^2(Ca) fractions should have been generally equal to each other if the reaction of Eq. 5.12 was the only process during calcium leaching. However, the Q2(Ca) fraction increment was far beyond that of Q^2p according to the deconvolution results of C/S1.5. Meanwhile, a nonnegligible increase in bound water was also observed for the C-S-H in C/S1.5. Therefore, another reaction process of Eq. 5.13 is also proposed, as illustrated in Fig. 5.16b. In this reaction, the protons in the two bridging silicates of Q^2i are bonded with hydroxyls in $(CaOH)^+$ units, producing two water molecules. The silicates are then charge-compensated by the two remaining Ca^{2+} ions, therefore forming Q^2p. These two reactions are perfectly consistent with our experimental observations, including the increases of $Ca(OH)_2$ and bound water fractions as well as the atomistic structure variations in C-S-H of C/S1.5.

However, the reaction processes in Eqs. 5.12 and 5.13 are still unable to fully cover the atomic structure variations of C-S-H in C/S2.0. Apart from the Q^2(i), a decrease was also observed for the fraction of Q^1(Ca) units, accompanied by a slight polymerization of the C-S-H in this group. Therefore, Eq. 5.14 is additionally proposed to describe this atomistic variation. The formation of the Si–O–Si bond in two adjacent Q^1(Ca) units leads to the reduction of vacant sites and the increase of the Q^2(Ca) fraction.

It is admitted that these proposed reactions are plausible interpretations of our experiment results and calculations. In the future, more stable C-S-H products with different C/S ratios should be synthesized and investigated for their cryogenic stability to validate our deductions, and further studies such as DFT calculations and refined NMR tests are still needed to confirm their validity. Also, it is urgent to detect the interlayer calcium leaching using other techniques with high resolution, such as synchrotron X-ray diffraction. These are beyond the scope of the current paper.

5.4 Cryogenic Stability of C-S-H Physical Stability

In addition to focusing on the impact of cryogenic attack on the nanostructure of C-S-H, we further investigated the influence of cryogenic attack on the physical properties of C-S-H and the degradation mechanism of its microstructure. We evaluated the physical stability of C-S-H under cryogenic attack, primarily referring to its nanomechanical properties, C-S-H particle pore structure, and the volume stability of its fundamental gel grain units.

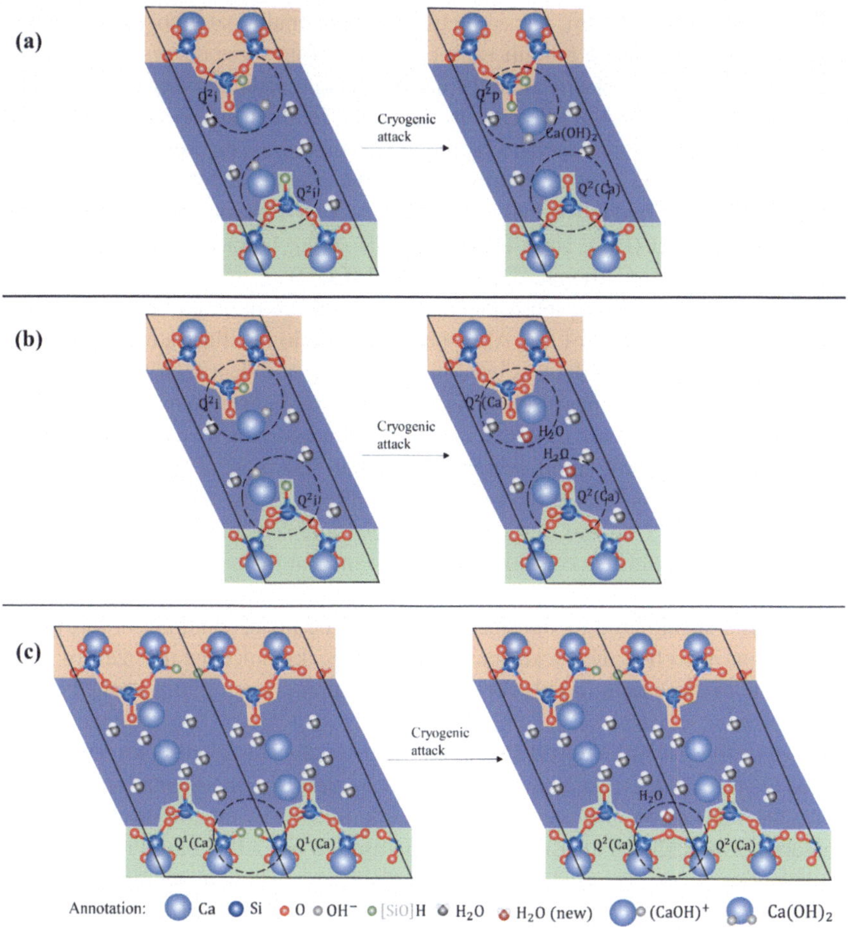

Fig. 5.16 Schematic diagrams for the proposed three reactions. **a** Illustrates the atomistic variations of Eq. 5.12, Interlayer $Ca(OH)_2$ is formed in this reaction. **b** Is for Eq. 5.13, accompanied by the formation of interlayer water. **c** Is for Eq. 5.14, showing the slight polymerization of C-S-H in C/S2.0. These models are constructed according to Ref. [9]

5.4.1 Nanomechenical Stability of C-S-H

5.4.1.1 Calibration, PDF Estimation and Deconvolution

The nanomechanical property analyzed here mainly refers to the elastic modulus obtained from AFM-based indentation. As noted, 200 AFM-based indentations were performed on ten randomly selected locations of each C-S-H sample. Therefore, an overall of 200 force curves were obtained for each sample measurement. In analyzing these raw data, the deflection sensitivity S and spring constant k were calibrated to be 76.2 0 nN/V and 32 N/m, respectively. Using the analytical theory described in Sect. 2.6.1, a set of elastic moduli were calculated from these curves. KDE model further analyzed these elastic moduli, and the distinguishable sets of the results corresponding to individual phases were separated by the Gaussian mixture model deconvolution method as presented before. Based on these data processing managements, the probability density distributions of the elastic modulus are depicted in Fig. 5.17. The identified individual phases include Very-Low-density C-S-H (VLD), Low-Density C-S-H (LD), High-Density C-S-H (HD), Ultra-High-Density C-S-H (UHD), and Calcium-Carbonate-Density C-S-H (CCD).

Before further analyzing these indentation results, some key points need to be stressed here. The scheme of this AFM-based indentation is illustrated in Fig. 5.18. The typical sizes of the indenters used in common nanoindentation [52] and other AFM-based indentation measurements [53] usually range from tens to hundreds of nanometers. In contrast, the spherical radius of the indentation tip used in this indentation operation was only 8 nm, which was smaller than most of the indenters used previously. As a result, this led to two essential effects on the AFM-based indentation measurements. One was that the tip became more sensitive to the variations of phase property and surface condition, which meant that the indentations were more likely to be influenced by gel pores and the phase differences. The other was that the measured elastic moduli were better to reveal the intrinsic nature of individual C-S-H grains as well as the packed basic building blocks, since the tip size was more comparable to those building blocks, allowing the tip to be straightly applied on the C-S-H grains.

Based on these facts analyzed above, it is understandable to identify more than one form of VLD and CCD phases in some C-S-H samples. In nanoindentation measurements, the VLD phases were viewed as zones dominated by macroporosity, namely capillary porosity [54]. Due to the smaller tip used in our AFM-based indentations, they ought to be zones dominated by gel porosity. According to Jennings [55], the gel pores were classified into small and large gel pores characterized in sizes at 1–3 nm and 3–12 nm, respectively. Therefore, the two VLD phases (VLD1 and VLD2) identified in sample C/S1.5B may simply result from the intrinsic difference of this gel pore system. As for those elastic moduli values higher than 44 GPa, they were apparently not associated with the C-S-H phase compared with other measurements. Herein, they were ascribed to crystalline phases like calcite. As revealed by XRD and TGA results, apart from C-S-H, the calcium carbonate existed in all those samples. The AFM observations even straightly proved its presence in the test sample of C/S2.0. In addition, the elastic modulus of calcite was measured at around 75 ± 3 GPa [56]. Therefore, they were assigned to be Calcium-Carbonate-Density C-S-H, abbreviated as CCD, for convenience. It is noteworthy that the tip used in this study may be difficult to indent on this hard phase. As a result, the measured elastic moduli corresponding to CCD were not as stable as that of other phases. Hence, the CCD1, CCD2, CCD3, and CCD4 identified in the C-S-H samples were all attributed to the CCD phase.

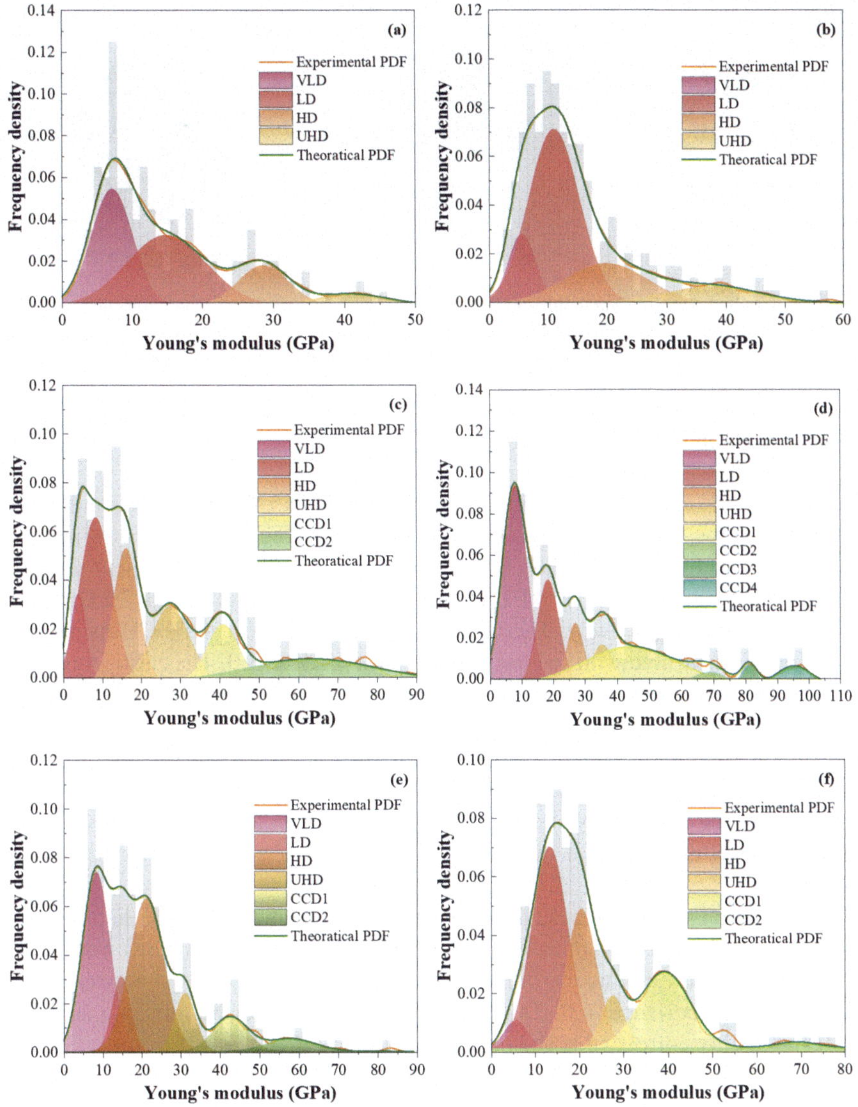

Fig. 5.17 Probability density distributions of elastic modulus of C-S-H before and after cryogenic attack. **a**, **b** Are for C/S0.84 before and after cryogenic attack, respectively; **c**, **d** are for C/S1.5 before and after cryogenic attack, respectively; **e**, **f** are fore C/S2.0 before and after cryogenic attack, respectively

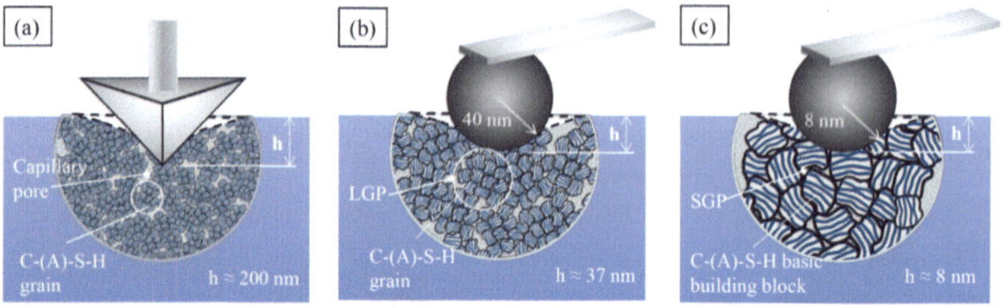

Fig. 5.18 Contact area between the indenter and the sample in different indentation methods: **a** Constantinides et al., nanoindentation [52]; **b** Jones et al., indentation based on atomic force microscopy [57]; **c** indentation based on atomic force microscopy used in this study

5.4.1.2 Elastic Moduli Comparisons

The deconvolution results of the elastic moduli of VLD, LD, HD, UHD, and CCD and their volume fractions are shown in Table 5.6 and Fig. 5.19. Previous nanoindentation measurements [58, 59] on the hydrated ordinary Portland cement showed that the elastic moduli of LD and HD have an elastic modulus of 21.7 ± 2.2 GPa and 29.4 ± 2.4 GPa, respectively. The UHD phase, identified as C-S-H/CH nanocomposite [60], is measured with an elastic modulus of 45 GPa [58]. The elastic modulus of VLD has a characteristic value lower than 11 GPa according to the nanoindentation measurements by Kim et al. and Emmy et al. [61, 62] who used a similar compaction procedure for the synthetic C-S-H solids. As shown in Table 5.6, the elastic moduli of these phases obtained from the AFM-based indentation measurement in this study were very similar to the previous nanoindentation results, which validates the effectiveness of this nanomechanical property measurement method adopted here.

Before the cryogenic attack, it was shown that a higher C/S ratio had a positive effect on the nanomechanical property of C-S-H. Its elastic modulus was found to be uplifting with the increase of the C/S ratio, which is exactly consistent with the preceding experience [63]. The results presented here further indicate that this uplift was highly correlated to the increasing amount of HD C-S-H as the C/S ratio increased. Almost every individual phase (VLD, LD, HD, and UHD) had a proportional increase with the increasing C/S ratio, though not significant. This, however, is not precisely consistent with the conclusions made from previous nanoindentation measurements [64, 65], which tended to attribute the elastic modulus to the intrinsic signature of each individual phase. Therefore, it was usually believed that this intrinsic signature is unchangeable. As analyzed above, the AFM-based indentation utilized here was theoretically closer to reflecting the intrinsic mechanical nature of C-S-H basic building blocks. Therefore, it is believed that the increase of the elastic modulus of the individual phase with the increasing C/S ratio was due to the different intrinsic natures of those building blocks. This also means that the so-called intrinsic signatures of LD and HD are not truly intrinsic and not stereotypical as they are on

Table 5.6 Deconvolution results for AFM-based indentation performed on C-S-H discs

C/S	C-S-H phase characterization	E for B[a] (GPa)		E for A1[b] (GPa)		Increment (%)	Average E (GPa)	
		μ	σ	μ	σ		B[a]	A1[b]
0.84	VLD1	7.1	0.07	5.6	0.08	– 21.1	15.7	14.9
	LD	14.9	0.64	11.1	0.14	– 25.5		
	HD	28.6	0.2	19.8	3.8	– 30.8		
	UHD	40.9	0.71	37.5	2.48	– 8.3		
1.5	VLD1	3.9	0.15	–	–	5.4[d]	17.6	15.3
	VLD2	8.3	0.83	7.8	0.14			
	LD	16.1	0.41	18.2	0.17	13.0		
	HD	27.4	0.25	26.7	0.18	– 2.6		
	UHD	40.9	0.27	35.1	0.43	– 14.1		
	CCD1	63.3	1.22	44.1	5.01	– 12.5[e]		
	CCD2	–	–	68.8	0.61			
	CCD3	–	–	81.1	0.22			
	CCD4	–	–	95.3	0.37			
2.0	VLD1	8.3	0.08	5.5	2.01	– 33.7	18.6	21.5
	VLD2	14.7	0.12	13.3	3.1	– 9.5		
	LD	21.1	0.22	20.5	2.91	– 2.8		
	HD	31.1	0.15	27.6	1.26	– 11.3		
	UHD	42.5	0.17	39.3	0.22	– 7.5		
	CCD1	57.9	0.59	69.1	1.47	19.3		

[a] Before cryogenic attack,
[b] After cryogenic attack,
[c] Increment I = $\frac{\left(\mu_{A1}^{phase} - \mu_{B}^{phase}\right)}{\mu_B} \times 100\%$,
[d] To the increment I of VLD in C/S0.84, I = $\frac{(\mu_{A1}^{VLD1} \times V_{VLD1} + \mu_{A1}^{VLD2} \times V_{VLD2} - \mu_{B}^{VLD2})}{\mu_{B}^{VLD2}} \times 100\%$,
[e] To the increment I of CCD in C/S0.84, I = $\frac{(\mu_{A1}^{CCD1} \times V_{CCD1} + \mu_{A1}^{CCD2} \times V_{CCD2} + \mu_{A1}^{CCD3} \times V_{CCD3} + \mu_{A1}^{CCD4} \times V_{CCD4} - \mu_{B}^{CCD1})}{\mu_{B}^{CCD1}}$ $\times 100\%$, V denotes the volume fraction of the corresponding phase

a smaller scale and closer to the size of C-S-H basic building blocks. Besides, the deconvolution results were not always the same, even in previous nanoindentation measurements from different studies.

The cryogenic attack degradation to the nanomechanical properties of C-S-H was also obviously shown in Table 5.6. The elastic modulus increments of the individual phases are calculated and listed. These increments are basically negative. In comparison, the cryogenic attack had generally decreased the elastic modulus to each individual phase, except for VLD and LD in C/S1.5. Overall, the average weight elastic modulus E of C-S-H decreased. Since the CCD phases were not associated with C-S-H, they were excluded from the calculation of average E. As shown in

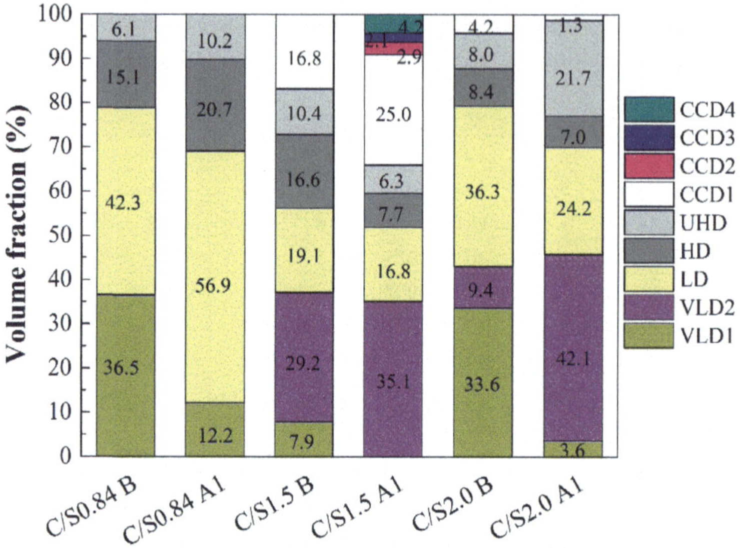

Fig. 5.19 Volume fractions of different phases in C-S-H discs before and after cryogenic attack

Table 5.6, The average E of C/S0.84 decreased from 15.7 GPa to 14.9 GPa, and that of C/S1.5 decreased from 17.6 GPa to 15.3 GPa. Although the calculated average E of C/S2.0 increased from 17.6 GPa to 21.5 GPa, this increase was mainly due to the dramatic increase in UHD fraction, as shown in Fig. 5.19. Therefore, the average E attributed to C-S-H in C/S2.0 also decreased after the cryogenic attack. It seems that the nanoscale degradation of mechanical properties in C-S-H was irrelevant to its C/S ratio. This degradation occurred in all C-S-H samples. In contrast to some macroscopic mechanical measurements under the cryogenic attack [66], the nanoscale mechanical degradation of C-S-H is consistent with the macroscopic mechanical degradation of cement-based concrete. This provides a new insight into the macroscale degradation of cement-based materials after exposure to cryogenic attack. It is suggested that apart from the freezing pressure [67, 68] and inhomogeneous thermal deformation [69], the macroscale degradation of cement-based material under cryogenic attack may also result from the intrinsically nanoscale mechanical degradation of C-S-H.

5.4.2 Pore Structure Stability

Figure 5.20 displays the cumulative pore volume and the pore size distribution of C-S-H samples before and after cryogenic attack which were obtained from nitrogen adsorption measurements. Since the nitrogen adsorption method is only effective in showing the pores with maximum diameters no more than 100 nm, the pore structure information obtained from this method can reveal the packing conditions of C-S-H grains. According to Jenning et al. [55, 70], the C-S-H grains are composed of a number of basic building blocks with a smaller size at around 5 nm. Apart from the mesopores or capillary pores formed between C-S-H grains, gel pores also exist in C-S-H grains. They are formed between those packed basic building blocks and are characterized into two types according to their sizes: small gel pores and large gel pores with a characteristic size of 1–3 nm and 3–12 nm, respectively. As shown in Fig. 5.20, the conditions of these pore systems were clearly exposed.

The comparison of the non-cryogenic-attack samples (that before the cryogenic attack) shown in Fig. 5.20 shows that the pore structure conditions of C-S-H were significantly related to C/S ratios. As indicated by nitrogen adsorption, the most probable pore sizes of C-S-H in C/S0.84, C/S1.5, and C/S2.0 were 25.77 nm, 17.86 nm, and 14.68 nm, respectively. The accumulative pore volume of C/S0.84 was higher than the other two samples. This difference was mainly due to the higher volume of mesopores between 20 and 35 nm in C/S0.84. However, for those pores smaller than 20 nm, their corresponding derivative pore volumes in C/S0.84 were apparently lower than that in C/S1.5 and C/S2.0, particularly for small and large gel pores. The

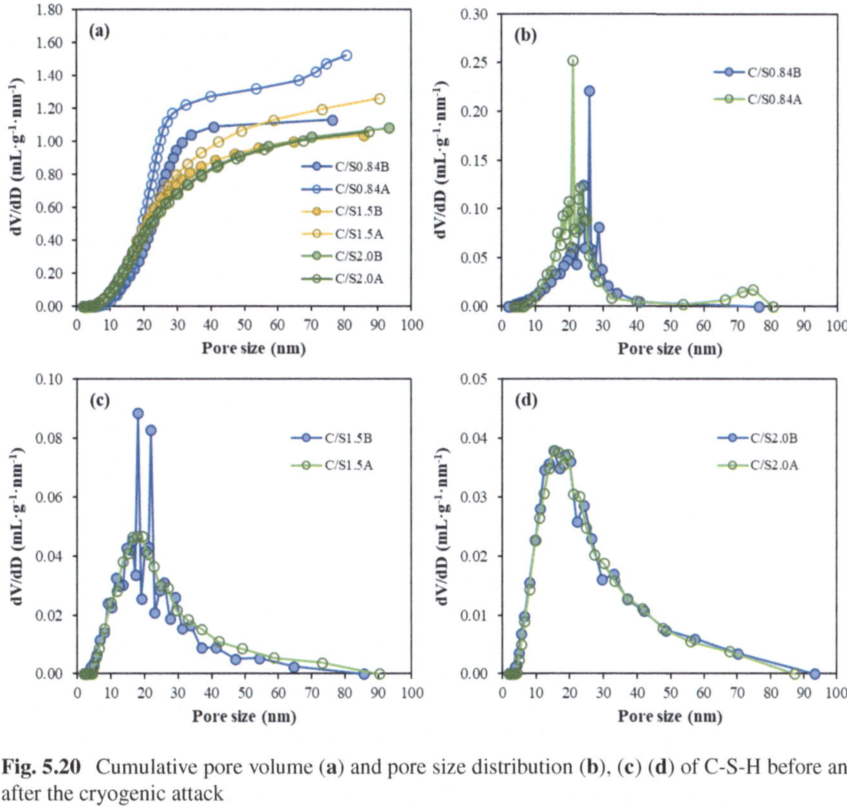

Fig. 5.20 Cumulative pore volume (**a**) and pore size distribution (**b**), (**c**) (**d**) of C-S-H before and after the cryogenic attack

pore structure conditions, including pore volume and pore size distribution of C-S-H in C/S1.5 and C/S2.0, were very similar. It is worth mentioning that the larger C-S-H grains were observed in C/S0.84 by AFM (Fig. 5.4), which has been attributed to the increasing structure ordering of C-S-H with decreasing C/S ratio. It can be deduced that larger pores were more likely to be formed between those larger grains, therefore leading to a higher pore volume in the C-S-H sample overall. Further, an interesting correlation revealed by these observations is that the C/S ratio is predominant in regulating the structure ordering and the packing conditions of C-S-H.

The cryogenic attack induced volumetric variations of these C-S-H samples are significant. As shown in Fig. 5.20, the accumulative pore volumes of C-S-H in C/S0.84 and C/S1.5 all experienced significant increases due to the cryogenic attack. Nonetheless, their increases were attributed to different reasons, as depicted in Fig. 5.20. For C/S0.84, its most probable pore size reduced to 20.89 nm, and the derivative pore volumes of pores in 12–25 nm increased a lot, while the derivative pore volumes of other pores out of this pore size range generally maintained at their original levels. This indicates that those arising pores may be derived from microcracks formed in C-S-H grains and are named with small microcracks here. For C/S1.5, the increase of accumulative pore volume was mainly ascribed to the increase of the derivative pore volume of pores with a diameter of 35–80 nm, while those smaller pores (< 20 nm) remained stable. Therefore, those increasing pores in C/S1.5 were mainly assigned to mesopores and macropores. Considering that these arising pores were even larger than some C-S-H grains, as revealed in Fig. 5.4, they were believed to be large microcracks formed between C-S-H grains instead of inside the C-S-H grains. Since a considerable amount of macropores (55–80 nm) were witnessed to emerge in the C/S0.84 group after the cryogenic attack (Fig. 5.20b), it is further deduced that these large microcracks were present in this group as well. As depicted in Fig. 5.20, the C-S-H in C/S2.0 had the most stable pore system among the three groups. Its accumulative pore volume and pore size distribution all stayed steady without much difference under the cryogenic attack. According to these facts, the schemes of the variations of C-S-H packing conditions and the formation of microcracks in the C-S-H gel systems are illustrated in Fig. 5.21. The configurations of small gel pores, large gel pores, small microcracks, and large microcracks are depicted in these schemes.

As analyzed above, the small gel pores and large gel pores of C-S-H in every group were able to stabilize themselves under cryogenic attack, which means that the cryogenic attack did not destroy the packing conditions of those C-S-H basic building blocks. However, the formation of microcracks indicated that the binding force between those blocks was likely to be broken at the interference of cryogenic attack. From this perspective, it can be deduced from these results that the binding force between C-S-H basic building blocks was probably increased with the increase of the C/S ratio. Hence, a most stable pore system was obtained for the C-S-H in C/S2.0, while microcracks were observed inside C-S-H grains of C/S0.84 and macropores (large microcracks) emerged out between C-S-H grains in both C/S0.84 and C/S1.5 under the cryogenic attack. Notably, these emerging microcracks were very destructive to cement-based materials, particularly in the presence of pore water.

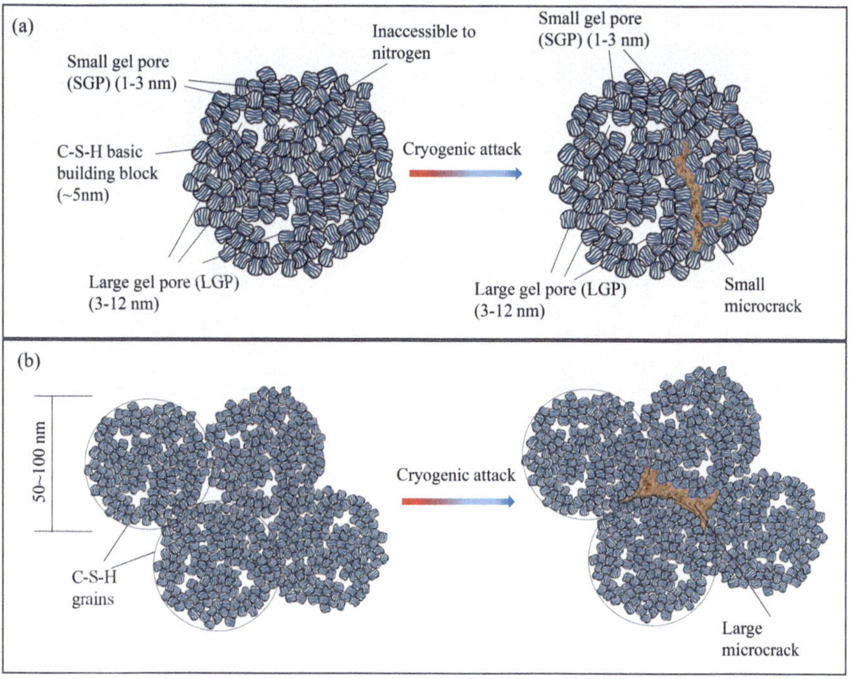

Fig. 5.21 Schemes of microcrack formation in C-S-H. **a** Scheme of the formation of small microcracks inside the C-S-H grain. This mainly happens to C-S-H with a low C/S ratio (C/S0.84 in this study). **b** Scheme of the formation of large microcracks among C-S-H grains. This is more common in C-S-H despite its C/S ratio (mainly C/S0.84 and C/S1.5 in this study)

The previous studies [67, 71] showed that the freezing of water at these microcracks produced great crystallization pressure, which was very destructive to the matrix. This finding provides another insight to understand the degradation of cement-based materials.

5.4.3 Micromorphology Stability of C-S-H

SEM micro-morphologies of C-S-H with different C/S ratios are shown in Fig. 5.22. Platelet-like morphology is observed for all C-S-H phases. It seems that the platelet-like characteristics of C-S-H are more obvious as the C/S ratio increases. This is particularly obvious for C-S-H in C/S2.0. Therefore, the cryogenic attack induced variations for the micro-morphologies of C-S-H in C/S1.5 and C/S2.0 are insignificant. While for the C-S-H phase in C/S0.84, it generally shows more porous after cryogenic attack as depicted in Fig. 5.22 a and b, which is basically compatible with the nitrogen adsorption results. Overall, a single cryogenic attack has an insignificant impact on the micro-morphology of C-S-H.

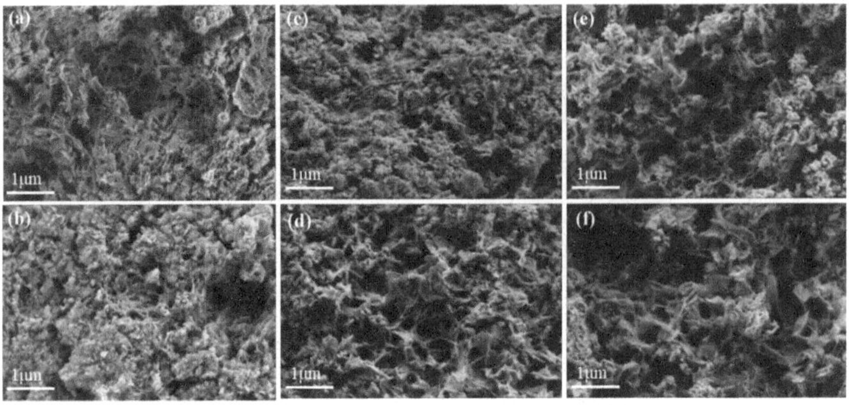

Fig. 5.22 SEM micro-morphology of C-S-H before and after cryogenic attack. **a**, **b** Are for C/S0.84 before and after the cryogenic attack, respectively; **c**, **d** are for C/S1.5 before and after the cryogenic attack, respectively; **e**, **f** are for C/S2.0 before and after the cryogenic attack, respectively

5.4.4 Interlayer Stability of C-S-H

As mentioned before, the basic globular unit of C-S-H is called C-S-H basic building blocks or bricks, according to Jennings [55]. Those small building blocks pack together and form large C-S-H grains and clusters. Obviously, the packing conditions of the blocks are significantly influential to the mechanical properties of C-S-H, as revealed by nanoindentation [58] and our AFM-based indentation. Along with the packing conditions, the properties of those blocks also play a vital role in the overall mechanical properties of the C-S-H phase. The volumetric conditions of C-S-H basic building blocks were therefore assessed here by short-term helium inflow. Since the helium molecule has a very tiny size at around 0.260 nm [72], thus it is capable of entering into small gel pores (1–3 nm). Although the interlayer spaces of the dried C-S-H are in several angstroms, they are still unreachable to the helium molecules in a short-term test [28]. Hence, the skeletal density obtained from the short-term helium inflow is assigned to the average bulk density of C-S-H basic building blocks. This further reveals the volumetric characteristics of the building blocks.

The skeletal density of C-S-H obtained from previous helium pycnometry measurements [28] was 2.450 g/cm^3, which did not consider the C/S ratio. Those measurements [28] were usually performed on hydrated cement paste to obtain the skeletal density of C-S-H, which made it difficult to control the C/S ratio of C-S-H and to evaluate its influences. The results shown in this study provided a new insight. The short-term helium inflow measurement results of C-S-H samples with different C/S ratios are displayed in Fig. 5.23. Before the cryogenic attack, it was shown that the skeletal density of C-S-H was highly proportional to the C/S ratio. A minimum value of 2.135 g/cm^3 was reached by C-S-H in C/S0.84. For C/S1.5, it had a middle skeletal density value of 2.298 g/cm^3. The C-S-H in C/S2.0 obtained a maximum density of 2.366 g/cm^3. Since the C-S-H samples were vacuum-dried, the bulk water was considered completely removed from the C-S-H gel pores and the interlayer spaces. These results reveal that the C-S-H building blocks were denser at a higher C/S ratio. Due to the increasing tightness of the basic building blocks, it can be further deduced that those building blocks gained a better capability of resisting deformation. Thus, a higher rigidity was obtained. This aligns with the elastic moduli characteristics revealed by the AFM-based indentations in Sect. 3.3. The C-S-H with a higher C/S ratio was inclined to obtain higher elastic moduli, which may result from a higher tightness of the basic building blocks.

A significant increase of the skeletal density was observed for all C-S-H samples under the influence of cryogenic attack. From Fig. 5.23, the skeletal densities of C-S-H in C/S0.84, C/S1.5, and C/S2.0 increased to 2.240 g/cm^3, 2.353 g/cm^3, and 2.502 g/cm^3, respectively. The skeletal density was calculated by $\rho_s = m_b/V_b$, where m_b is the average mass of the basic building blocks, and V_b is the average volume. Again, these cryogenic-attack-treated C-S-H samples were all dried before short-term helium inflow measurements, which means that the influence of bulk water on the density measurement results was eliminated. Hence, m_b was considered as a constant. In this case, the increases in densities were highly attributed to the decrease of V_b, which therefore indicated that those C-S-H basic building blocks had experienced volume shrinkage. According to the building block model proposed by Jennings [55], intraglobular pores were formed between the silicate chain layers. The results obtained here indicate that these intraglobular pores collapsed under the cryogenic attack, leading to the volume shrinkage of the building blocks, therefore measured with a higher bulk density value. This can be observed in all C-S-H samples. A scheme depicting such a volumetric variation of these blocks is illustrated in Fig. 5.24a.

In addition to the influence on the bulk density of those building blocks themselves, their volume shrinkage may initiate some secondary damage to the microstructures of C-S-H. As shown in Fig. 5.24b, when the intraglobular pores of two neighboring

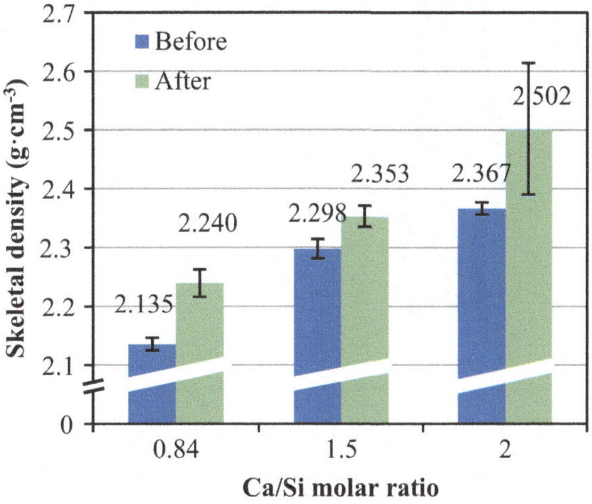

Fig. 5.23 Skeletal density of C-S-H with varying C/S ratios before and after cryogenic attack

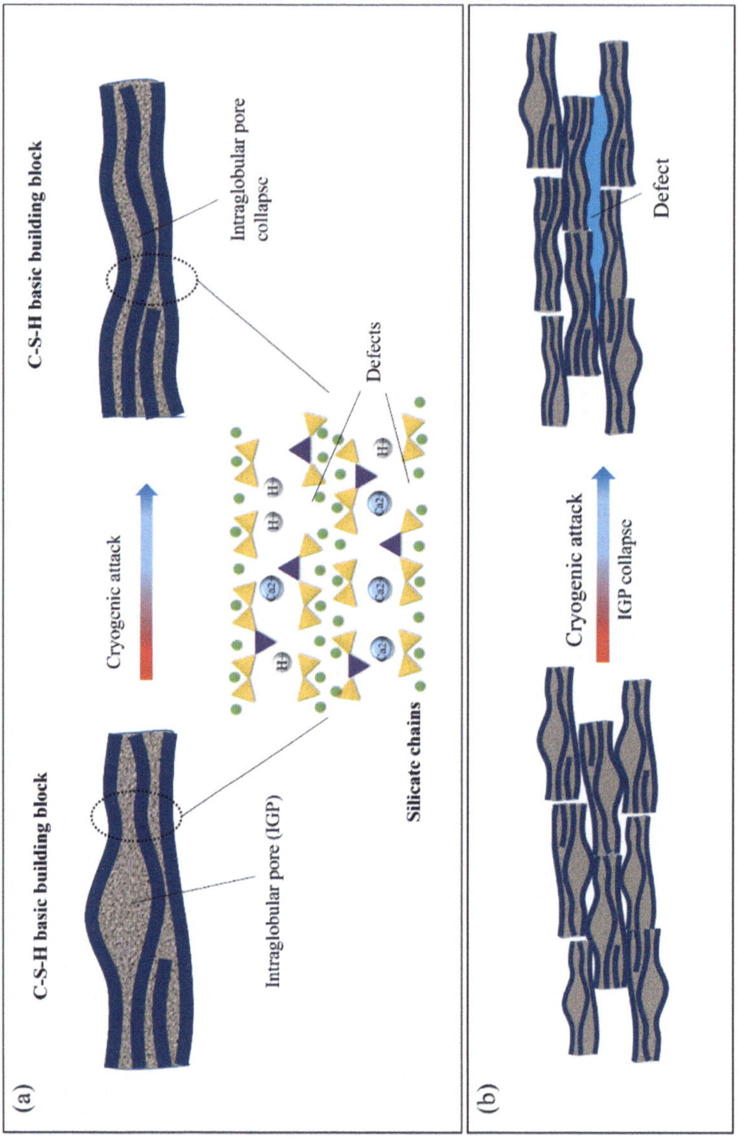

Fig. 5.24 Volume stability of C-S-H basic building block under cryogenic attack. **a** Scheme of the introglobular pore collapse of single C-S-H basic building block. **b** Scheme of the local defect formation among C-S-H basic building blocks due to the collapse of their introglobular pores. These local defects may connect and form microcracks

basic building blocks collapse, they are likely to debond from each other, therefore inducing local defects in the C-S-H clusters. The original defects formed between basic building blocks are particularly humble, though. However, once these defects expand to a large scale and connect to each other, they turn into microcracks, which are exactly negative to the C-S-H. Interestingly, these microcracks were observed in all C-S-H phases under the cryogenic attack as analyzed in Sect. 5.4.2. The scheme shown in Fig. 5.24b clearly explains the formation process of these microcracks, which shows that the formation of the small and large microcracks may initiate from the collapse of intraglobular pores.

5.4.5 Degradation Mechanism of C-S-H Under Cryogenic Attack

Based on the preceding analysis of the evolution of C-S-H nanostructure, interlayer porosity within C-S-H, and pore structures between C-S-H particles, we can further analyze the deteriorating pathway of C-S-H microstructure. The degradation of their microstructure is common for cement-based materials operating in low or ultra-low-temperature conditions. The primary debate lies in whether this degradation is driven by external forces or caused by changes in the inherent structure. These external forces are typically attributed to hydraulic, osmotic, and crystallization pressures. These forces are significant and play a dominant role in the degradation of cement-based materials under ultra-low-temperature or normal freezing–thawing conditions. In this study, we artificially eliminated the influence of these external forces by removing bulk water content from the samples before subjecting them to cryogenic attack. However, we still observed the occurrence of microstructural deterioration. Furthermore, the impact of thermal contraction was reduced by controlling the slow temperature rise and fall rate of 0.5 °C/min. Therefore, the observed structural collapses and microcracks may be attributed to changes in the internal structure.

Herein, we propose a bottom-up, multiscale deterioration pathway for C-S-H under cryogenic attack, as depicted in Fig. 5.25.

As per the analysis of C-S-H nanostructures in Sect. 5.3, it is evident that under cryogenic attack, certain primary silica-oxygen chains within C-S-H undergo de-aggregation, fracture, or interlayer ionic bond reorganization, leading to the formation of new defective sites at the nanoscale level. The formation of these defects results in the collapse and shrinkage of interlayer pores within the fundamental gel grain units of C-S-H, as demonstrated in the preceding Sect. 5.4.4.

Moreover, the volume contraction of these fundamental gel grain units might cause secondary damage to the microstructure of C-S-H. As depicted in Fig. 5.25, when interlayer pores of adjacent C-S-H gel grain units collapse, it is probable that they detach from each other, forming localized defects. Although the initial localized defects formed between fundamental gel grain units may seem inconsequential, once

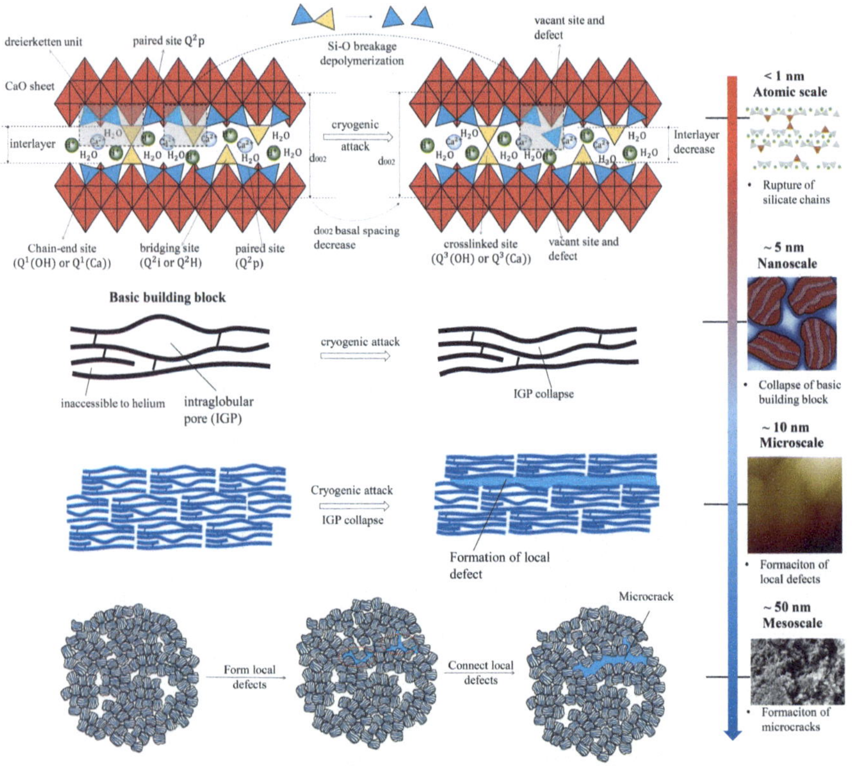

Fig. 5.25 Upscaling degradation of C-S-H under cryogenic attack

these local defects enlarge and interconnect, they further evolve into microcracks, which are detrimental to the stability of C-S-H's microstructure and properties. It is noteworthy that the nitrogen adsorption pore structure test results discussed in Sect. 5.4.2 indicated the formation of these microcracks. Figure 5.25 describes the bottom-up formation process of these microcracks, suggesting that alterations in the primary silica-oxygen chains of C-S-H may gradually initiate the formation of small and large microcracks within C-S-H.

5.5 Concluding Remarks

In this chapter, we present a comprehensive study of the stability characteristics of the nanostructure and physical properties of C-S-H under cryogenic attack. We analyze microscopic morphological differences in C-S-H phases with varying Ca/Si ratios, elucidate the transformation processes and mechanisms governing C-S-H nanostructures during cryogenic attack, and explore how Ca/Si ratios influence the stability of

C-S-H nano-mechanical properties under ultra-low-temperature conditions. Additionally, we reveal the evolution of multiscale pore structures in C-S-H under cryogenic exposure and propose a bottom-up deterioration pathway to explain cryogenically driven microstructural degradation. Based on the research in this chapter, the main conclusions can be summarized as follows:

(1) The structural ordering of C-S-H decreases as the Ca/Si ratio increases. Under low Ca/Si conditions, C-S-H tends to form larger aggregated particles, meaning that highly ordered C-S-H structures promote the aggregation and growth of C-S-H, resulting in larger particle structures. When Ca/Si = 0.84, the size of aggregated C-S-H particles can reach 60–70 nm; at Ca/Si = 1.5, typical C-S-H particle sizes are around 30–40 nm, significantly smaller than those under low Ca/Si conditions. This typical structural characteristic will significantly impact the microstructural volume stability of C-S-H under ultra-low temperatures.
(2) C-S-H nanostructures in systems with lower Ca/Si ratios are more stable under cryogenic attack. The atomic structure of C-S-H in the C/S0.84 system is the most stable, while in the C/S1.5 and C/S2.0 systems, the atomic structure of C-S-H undergoes more noticeable changes, especially at Ca/Si = 1.5. In the C/S1.5 system, the average number of Ca^{2+} ions per tetrahedral silicon-oxygen bonding layer in C-S-H significantly decreases from 0.201 to 0.083. This reduction indicates a decrease in Ca^2 + ions involved in charge balancing the primary silica-oxygen chain structure, leading to the leaching of interlayer Ca^{2+} ions. In the C/S2.0 system, the average vacancy ratio within the silicon chains decreases from 0.208 to 0.178, while the average chain length increases from 3.82 to 4.62, indicating slight aggregation. Additionally, the observed leftward shift in the v(OH) peak position in the FTIR results suggests an enhancement in the hydrophilicity of C-S-H after cryogenic attack.
(3) Young's modulus of C-S-H measured using atomic force microscopy-based nanoindentation testing is closer to the inherent nanomechanical properties of the fundamental gel grain units of C-S-H. As the Ca/Si ratio increases, the average Young's modulus of C-S-H also increases. This increase is attributed to the higher content of the HD C-S-H phase with the increasing Ca/Si ratio, alongside the rise in Young's moduli for VLD C-S-H, LD C-S-H, and HD C-S-H phases. However, cryogenic attack decreases the Young's moduli of various phases of C-S-H, consequently lowering the average Young's modulus of C-S-H. Notably, this degradation trend in the nanoscale mechanical properties of C-S-H is not correlated with the compactness of C-S-H aggregation. The observed degradation in nanoscale mechanical properties of C-S-H aligns with the degradation trend in the macroscopic mechanical performance of cement-based materials under ultra-low temperatures.
(4) The volume stability of C-S-H under cryogenic attack is closely related to the Ca/Si ratio. Comparatively, the internal structure of C-S-H with higher Ca/Si ratios exhibits more excellent stability under cryogenic attack. After the cryogenic attack, secondary microcracks form within the C-S-H, and their formation locations are associated with the particle sizes of the C-S-H aggregates. In the

case of low Ca/Si ratio systems such as C/S0.84, small microcracks ranging from 10 to 25 nm form within the aggregated C-S-H particles, while larger microcracks form between these aggregated particles. Conversely, for high Ca/Si ratio C-S-H, as their aggregated particle sizes are smaller than those of low Ca/Si ratio C-S-H particles, only larger microcracks form between these smaller aggregated particles.

(5) A bottom-up deterioration pathway of C-S-H microstructural breakdown proceeds as follows: Initially, de-polymerization or interlayer ion recombination occurs in a few silicate chains, resulting in the formation of new defects at the atomistic level. This formation of defects leads to the collapse and contraction of interlayer pores within the fundamental gel grain units of C-S-H, creating localized flaws at nanoscale. The extension and interconnection of these localized flaws within the fundamental gel grain units of C-S-H result in the formation of microcracks, and the continued propagation of these microcracks induces more extensive microstructural damage.

References

1. I.G. Richardson, G.W. Groves, Microstructure and microanalysis of hardened ordinary Portland-cement pastes. J. Mater. Sci. **28**(1), 265–277 (1993)
2. I.G. Richardson, The nature of C-S-H in hardened cements. Cem. Concr. Res. **29**(8), 1131–1147 (1999)
3. I.G. Richardson., The nature of the hydration products in hardened cement pastes. Cem. Concr. Compos. **22**, 97–113 (2000).
4. H. Taylor, Hydrated calcium silicates. Part I. Compound formation at ordinary temperatures. J. Chem. Soc. **3**, 3682–3690 (1950)
5. E.T. Rodriguez, K. Garbev, D. Merz, L. Black, I.G. Richardson, Thermal stability of C-S-H phases and applicability of Richardson and Groves' and Richardson C-(A)-S-H(I)models to synthetic C-S-H. Cem. Concr. Res. **93**, 45–56 (2017)
6. C.L. Hu, S. Yao, F.B. Zou, S. Nie, Z.C. Liu, F.Z. Wang, Insights into the influencing factors on the micro-mechanical properties of calcium-silicate-hydrate gel. J. Am. Ceram. Soc. **102**(4), 1942–1952 (2019)
7. I.G. Richardson, J. Skibsted, L. Black, R.J. Kirkpatrick, Characterisation of cement hydrate phases by TEM, NMR and Raman spectroscopy. Adv. Cem. Res. **22**(4), 233–248 (2010)
8. I.G. Richardson, Model structures for C-(A)-S-H(I), Acta Crystallogr. B: Struct. Sci. Cryst. Eng. Mater. **70**(6), 903–923 (2014)
9. A. Kunhi Mohamed, S.C. Parker, P. Bowen, S. Galmarini, An atomistic building block description of C-S-H—towards a realistic C-S-H model. Cem. Concr. Res. **107**, 221–235 (2018)
10. A. Kumar, B.J. Walder, A. Kunhi Mohamed, A. Hofstetter, B. Srinivasan, A.J. Rossini, K. Scrivener, L. Emsley, P. Bowen, The atomic-level structure of cementitious calcium silicate hydrate. J. Phys. Chem. C **121**(32), 17188–17196 (2017)
11. E. Pustovgar, R.P. Sangodkar, A.S. Andreev, M. Palacios, B.F. Chmelka, R.J. Flatt, J.B. d'Espinose de Lacaillerie, Understanding silicate hydration from quantitative analyses of hydrating tricalcium silicates. Nat. Commun. **7**, 10952 (2016)
12. H.F.w. Taylor., Proposed struture for calcium silicate hydrate gel. J. Am. Ceram. Soc. **69**(6), 464–67 (1986)

13. I.G. Richardson., G.W. Groves., Models for the composition and structure of calcium silicate hydrate(C-S-H) gel in hardened tricalcium silicate pastes. Cem. Concr. Res. **22**, 1001–1010 (1992)
14. Y. He, L. Lu, L.J. Struble, J.L. Rapp, P. Mondal, S. Hu, Effect of calcium–silicon ratio on microstructure and nanostructure of calcium silicate hydrate synthesized by reaction of fumed silica and calcium oxide at room temperature. Mater. Struct. **47**(1–2), 311–322 (2014)
15. F. Brunet, P. Bertani, T. Charpentier, A. Nonat, J. Virlet, Application of ^{29}Si homonuclear and 1H–29Si heteronuclear NMR correlation to structural studies of calcium silicate hydrates. J. Phys. Chem. B **108**, 15494–15502 (2004)
16. X. Cong., R.J. Kirkpatrick., ^{29}Si MAS NMR study of the structure of calcium silicate hydrate. Adv. Cem. Based Mater. **3**, 144–156 (1996)
17. A. Nonat, X. Lecoq, The structure, stoichiometry and properties of C-S-H prepared by C_3S hydration under controlled condition, in *Nuclear Magnetic Resonance Spectroscopy of Cement-Based Materials*. ed. by P. Colombet, H. Zanni, A.-R. Grimmer, P. Sozzani (Springer, Berlin Heidelberg, Berlin, Heidelberg, 1998), pp.197–207
18. P. Faucon., J.M. Delaye., J. Virlet., Molecular dynamics simulation of the structure of calcium silicate hydrates I. $Ca_{4+x}Si_6O_{14+2x}(OH)_{4-2x}(H_2O)_2$ ($0 \leq x \leq 1$). J. Solid State Chem. **127**, 92–97 (1996)
19. J.S. Dolado, M. Griebel, J. Hamaekers, A molecular dynamic study of cementitious calcium silicate hydrate (C-S-H) gels. J. Am. Ceram. Soc. **90**(12), 3938–3942 (2007)
20. J.A. Gard, H.F.W. Taylor, Calcium silicate hydrate (II) ("C-S-H(II)"). Cem. Concr. Res. **6**(5), 667–677 (1976)
21. H. Viallis., P. Faucon., J.-C. Petit., A. Nonat., Interaction between salts (NaCl, CsCl) and calcium silicate hydrates (C−S−H). J. Phys. Chem. B **103**, 5212–5219 (1999)
22. W.T. Li, B.Q. Dong, Z.X. Yang, J. Xu, Q. Chen, H.X. Li, F. Xing, Z.W. Jiang, Recent advances in intrinsic self-healing cementitious materials. Adv. Mater. **30**(17), 1705679 (2018)
23. Z.W. Jiang, W.T. Li, Z.C. Yuan, Influence of mineral additives and environmental conditions on the self-healing capabilities of cementitious materials. Cem. Concr. Compos. **57**, 116–127 (2015)
24. S. Grangeon, F. Claret, C. Lerouge, F. Warmont, T. Sato, S. Anraku, C. Numako, Y. Linard, B. Lanson, On the nature of structural disorder in calcium silicate hydrates with a calcium/silicon ratio similar to tobermorite. Cem. Concr. Res. **52**, 31–37 (2013)
25. E. Gallucci, X. Zhang, K.L. Scrivener, Effect of temperature on the microstructure of calcium silicate hydrate (C-S-H). Cem. Concr. Res. **53**, 185–195 (2013)
26. M.S. Kim, Y. Jun, C. Lee, J.E. Oh, Use of CaO as an activator for producing a price-competitive non-cement structural binder using ground granulated blast furnace slag. Cem. Concr. Res. **54**, 208–214 (2013)
27. P. Trtik, J. Kaufmann, U. Volz, On the use of peak-force tapping atomic force microscopy for quantification of the local elastic modulus in hardened cement paste. Cem. Concr. Res. **42**(1), 215–221 (2012)
28. H.M. Jennings., A model for the microstructure of calcium silicate hydrate in cement paste, Cem. Concr. Res. **30**, 101–116 (2000)
29. E. L'Hopital, B. Lothenbach, K. Scrivener, D.A. Kulik, Alkali uptake in calcium alumina silicate hydrate (C-A-S-H). Cem. Concr. Res. **85**, 122–136 (2016)
30. B. Lothenbach, D. Nied, E. L'Hopital, G. Achiedo, A. Dauzeres, Magnesium and calcium silicate hydrates. Cem. Concr. Res. **77**, 60–68 (2015)
31. E. Kapeluszna, L. Kotwica, A. Rozycka, L. Golek, Incorporation of Al in C-A-S-H gels with various Ca/Si and Al/Si ratio: microstructural and structural characteristics with DTA/TG, XRD, FTIR and TEM analysis. Constr. Build. Mater. **155**, 643–653 (2017)
32. J.J. Kim, E.M. Foley, M.M. Reda Taha, Nano-mechanical characterization of synthetic calcium–silicate–hydrate (C–S–H) with varying CaO/SiO_2 mixture ratios. Cem. Concr. Compos. **36**, 65–70 (2013)
33. P.J. McDonald, V. Rodin, A. Valori, Characterisation of intra- and inter-C-S-H gel pore water in white cement based on an analysis of NMR signal amplitudes as a function of water content. Cem. Concr. Res. **40**(12), 1656–1663 (2010)

34. L. Alarcon-Ruiz, G. Platret, E. Massieu, A. Ehrlacher, The use of thermal analysis in assessing the effect of temperature on a cement paste. Cem. Concr. Res. **35**(3), 609–613 (2005)
35. J.J. Thomas, S.A. FitzGerald, D.A. Neumann, R.A. Livingston, State of water in hydrating tricalcium silicate and portland cement pastes as measured by quasi-elastic neutron scattering. J. Am. Ceram. Soc. **84**(8), 1811–1816 (2001)
36. M.B. Pinson, E. Masoero, P.A. Bonnaud, H. Manzano, Q. Ji, S. Yip, J.J. Thomas, M.Z. Bazant, K.J. Van Vliet, H.M. Jennings, Hysteresis from multiscale porosity: Modeling water sorption and shrinkage in cement paste. Phys. Rev. Appl. **3**(6), 064009 (2015)
37. X. Guo, M. Song, Micro-nanostructures of tobermorite hydrothermal-synthesized from fly ash and municipal solid waste incineration fly ash. Constr. Build. Mater. **191**, 431–439 (2018)
38. E.T. Rodriguez, I.G. Richardson, L. Black, E. Boehm-Courjault, A. Nonat, J. Skibsted, Composition, silicate anion structure and morphology of calcium silicate hydrates (C-S-H) synthesised by silica-lime reaction and by controlled hydration of tricalcium silicate (C_3S). Adv. Appl. Ceram. **114**(7), 362–371 (2015)
39. S. Thomas., K. Meise-Gresch., W. Müller-Warmuth., I. Odler., MAS NMR studies of partially carbonated Portland cement and tricalcium silicate pastes. J. Am. Ceram. Soc. **76**(8), 1998–2004 (1999)
40. H. Noma., Y. Adachil., M. Takushi., Y. Hideo., Y. Tadashi., N. Yoshihisa., ^{29}Si MAS NMR spectroscopy of poorly-crystalline calcium silicate hydrates (C-S-H), in *Nuclear Magnetic Resonance Spectroscopy of Cement-Based Materials* (Springer, Berlin, Heidelberg, 1998), pp. 159–168
41. I. Klur., B. Pollet., J. Virlet., A. Nonat., C-S-H structure evolution with calcium content by multinuclear NMR, in *Nuclear Magnetic Resonance Spectroscopy of Cement-Based Materials* (Springer, Verlag Berlin Heidelberg, 1998), pp. 119–141
42. F. Matsushita, Y. Aono, S. Shibata, Calcium silicate structure and carbonation shrinkage of a tobermorite-based material. Cem. Concr. Res. **34**(7), 1251–1257 (2004)
43. K. Kurumisawa, T. Nawa, H. Owada, M. Shibata, Deteriorated hardened cement paste structure analyzed by XPS and 29Si NMR techniques. Cem. Concr. Res. **52**, 190–195 (2013)
44. K. Garbev, G. Beuchle, M. Bornefeld, L. Black, P. Stemmermann, Cell dimensions and composition of nanocrystalline calcium silicate hydrate solid solutions. Part 1: synchrotron-based X-ray diffraction. J. Am. Ceram. Soc. **91**(9), 3005–3014 (2008)
45. I.F. Sáez del Bosque, S. Martínez-Ramírez, M.T. Blanco-Varela, FTIR study of the effect of temperature and nanosilica on the nano structure of C-S-H gel formed by hydrating tricalcium silicate. Constr. Build. Mater. **52**, 314–323 (2014)
46. P. Yu, R.J. Kirkpatrick, B. Poe, P.F. McMillan, X.D. Cong, Structure of calcium silicate hydrate (C-S-H): Near-, mid-, and far-infrared spectroscopy. J. Am. Ceram. Soc. **82**(3), 742–748 (1999)
47. M.Y.A. Mollah, W.H. Yu, R. Schennach, D.L. Cocke, A Fourier transform infrared spectroscopic investigation of the early hydration of Portland cement and the influence of sodium lignosulfonate. Cem. Concr. Res. **30**(2), 267–273 (2000)
48. R. Ylmén, L. Wadsö, I. Panas, Insights into early hydration of Portland limestone cement from infrared spectroscopy and isothermal calorimetry. Cem. Concr. Res. **40**(10), 1541–1546 (2010)
49. T. Richard, L. Mercury, F. Poulet, L. d'Hendecourt, Diffuse reflectance infrared Fourier transform spectroscopy as a tool to characterise water in adsorption/confinement situations. Journal of Colloid and Interface **304**(1), 125–136 (2006)
50. J.E. Rossen, B. Lothenbach, K.L. Scrivener, Composition of C-S-H in pastes with increasing levels of silica fume addition. Cem. Concr. Res. **75**, 14–22 (2015)
51. T.T.H. Bach, C.C.D. Coumes, I. Pochard, C. Mercier, B. Revel, A. Nonat, Influence of temperature on the hydration products of low pH cements. Cem. Concr. Res. **42**(6), 805–817 (2012)
52. G. Constantinides, F.J. Ulm, The nanogranular nature of C-S-H. J. Mech. Phys. Solids **55**(1), 64–90 (2007)
53. C. Plassard, E. Lesniewska, I. Pochard, A. Nonat, Investigation of the surface structure and elastic properties of calcium silicate hydrates at the nanoscale. Ultmi **100**(3–4), 331–338 (2004)

54. V. Zanjani Zadeh, C.P. Bobko, Nanoscale mechanical properties of concrete containing blast furnace slag and fly ash before and after thermal damage. Cem. Concr. Compos. **37**, 215–221 (2013)
55. H.M. Jennings, Refinements to colloid model of C-S-H in cement: CM-II. Cem. Concr. Res. **38**(3), 275–289 (2008)
56. C. Plassard, E. Lesniewska, I. Pochard, A. Nonat, Intrinsic elastic properties of calcium silicate hydrates by nanoindentation, in *12th International Congress on the Chemistry of Cement* (Cement Association of Canada, Montreal, Canada, 2007), p. 44
57. C.A. Jones, Z.C. Grasley, J.A. Ohlhausen, Measurement of elastic properties of calcium silicate hydrate with atomic force microscopy. Cem. Concr. Compos. **34**(4), 468–477 (2012)
58. G. Constantinides, F.J. Ulm, K. Van Vliet, On the use of nanoindentation for cementitious materials. Mater. Struct. **36**(3), 191–196 (2003)
59. G. Constantinides, F.-J. Ulm, The effect of two types of C-S-H on the elasticity of cement-based materials: results from nanoindentation and micromechanical modeling. Cem. Concr. Res. **34**(1), 67–80 (2004)
60. J.J. Chen, L. Sorelli, M. Vandamme, F.-J. Ulm, G. Chanvillard, A coupled nanoindentation/SEM-EDS study on low water/cement ratio Portland cement paste: Evidence for C-S-H/$Ca(OH)_2$ nanocomposites. J. Am. Ceram. Soc. **93**(5), 1484–1493 (2010)
61. W. Shen, G. Gan, R. Dong, Y. Tan, H. Chen, L. Xiao, Fabrication and characterization of nano colloid surfaced concrete. Mater. Struct. **44**(9), 1559–1564 (2011)
62. E.M. Foley, J.J. Kim, M.M. Reda Taha, Synthesis and nano-mechanical characterization of calcium-silicate-hydrate (C-S-H) made with 1.5 CaO/SiO_2 mixture. Cem. Concr. Res. **42**(9), 1225–1232 (2012)
63. J. Li, W. Zhang, P.J.M. Monteiro, Structure and intrinsic mechanical properties of nanocrystalline calcium silicate hydrate. ACS Sustainable Chem. Eng. **8**(33), 12453–12461 (2020)
64. M. Vandamme, F.J. Ulm, P. Fonollosa, Nanogranular packing of C-S-H at substochiometric conditions. Cem. Concr. Res. **40**(1), 14–26 (2010)
65. I.G. Richardson., The nature of C-S-H in hardened cements. Cem. Concr. Res. **29**, 1131–1147 (1999)
66. J. Zhengwu., D. Zilong., L. Wenting., Z. Lei., Freeze-thaw effect at ultra-low temperature on properties of mortars. J. Chin. Ceram. Soc. 42(5), 596–600 (2014)
67. Z. Sun, G.W. Scherer, Effect of air voids on salt scaling and internal freezing. Cem. Concr. Res. **40**(2), 260–270 (2010)
68. L. Liu, G. Ye, E. Schlangen, H. Chen, Z. Qian, W. Sun, K. van Breugel, Modeling of the internal damage of saturated cement paste due to ice crystallization pressure during freezing. Cem. Concr. Compos. **33**(5), 562–571 (2011)
69. K. Wang, P.J. Monteiro, B. Rubinsky, A. Arav, Microscopic study of ice propagation in concrete. ACI Mater. J. **93**(4) (1996)
70. A.J. Allen, J.J. Thomas, H.M. Jennings, Composition and density of nanoscale calcium–silicate–hydrate in cement. Nat. Mater. **6**(4), 311–316 (2007)
71. K.H. Khayat., M. Polivka., Cryogenic frost resistance of lightweight concrete containing silica fume. ACI Spec. Publ. **114**, 915–928 (1989)
72. X. Yang, Z.P. Sun, L.L. Shui, Y.L. Ji, Characterization of the absolute volume change of cement pastes in early-age hydration process based on helium pycnometry. Constr. Build. Mater. **142**, 490–498 (2017)

Chapter 6
Cryogenic Stability of Calcium-Aluminosilicate-Hydrate

6.1 Introduction

In Chap. 5, we elaborate on the influence of cryogenic attack on the nanomechanical stability and micro/nanostructure of C-S-H, systematically studying the differences in cryogenic stability under various Ca/Si ratios. The results indicate that the physical properties and nanostructure of C-S-H are challenging to maintain under cryogenic attack, regardless of changes in the Ca/Si ratio. The nanomechanical performance and microscopic structure of C-S-H degrade, aligning with the macroscopic deterioration trend of cryogenic concrete. It is noteworthy that when the Ca/Si ratio is lower, the cryogenic stability of C-S-H is slightly better. This suggests that the chemical structural characteristics of C-S-H contribute to stabilizing the physical properties and micro/nanostructure of concrete. However, we have limited knowledge about enhancing and reinforcing the stability of C-S-H structure and physical properties under cryogenic attack.

Cryogenic concrete was initially developed to build all-concrete liquefied natural gas (LNG) tanks due to an increasing need for green natural gas energy consumption [1–4]. Due to global warming, the Paris Agreement calls for urgent and worldwide collaborative actions to lower carbon emissions using cleaner energies, further accelerating the demand for LNG and cryogenic storage tanks. In addition, the ambitious space exploration projects proposed by the U.S, China, and Russia plan to build permanent habitable bases on the Moon and Mars, which also motivates the development of cryogenically stable construction materials since the environmental temperature can be as low as − 200 °C on the Moon [5, 6] and − 140 °C on Mars [7]. The chemical analysis shows that both Lunar and Martian soils contain significant CaO, Al_2O_3, Fe_2O_3/FeO, and SiO_2 [8–10], making it possible to develop cement-related materials. As the main binding phase in a hydrated cement system, calcium-(alumino-)silicate-hydrate (C-(A-)S-H) plays a predominant role in regulating the strength and durability of cement-based materials [11–13]. Therefore, understanding the C-(A-)S-H nanostructure under a cryogenic attack environment is crucial to

Z. Jiang and X. Zhu, *Cryogenic Stability of Cement-Based Materials at Micro/Nano Scale*, https://doi.org/10.1007/978-981-96-1836-1_6

advance our knowledge and develop more cryogenically stable cement-based materials. Unfortunately, the C-(A-)-S-H nanostructure is still not well understood.

After comprehensive studies [12, 14–21], the nanostructures of Al-free calcium-silicate-hydrate (C-S-H) have been resolved more clearly with less controversy than C-A-S-H. The silicate chains are made of three types of silicate tetrahedra, including end-chain tetrahedra (or chain-terminating tetrahedra), pairing tetrahedra, and bridging tetrahedra. Silicate chains are arranged in layers, and the intralayer space hosts a CaO sheet that coordinates the silicon atoms. In addition, there is a water-rich interlayer space between two layers, which is considered to host charge-balancing species like Ca^{2+} ions or $Ca(OH)^{+}$ units [15, 22, 23]. The pairing silicate tetrahedra are coordinated with the intralayer CaO, while the bridging silicate tetrahedra are coordinated to two neighboring pairing sites by sharing oxygen atoms. Unlike the quasi-infinite silicate chains in tobermorite, the silicate chains in C-S-H are defective due to missing bridging tetrahedra within dreierketten units, present in multimer forms, e.g., dimer, pentamer, and octamer [22, 24–27].

In practice, one of the most widely accepted ways to improve performance and reduce carbon footprint is replacing a portion of cement clinker by using supplementary cementitious materials (SCMs) such as metakaolin, fly ash, and slag [28–30]. These Al-rich materials introduce additional aluminum to the hydration environment, resulting in the uptake of Al in C-S-H. The Al/Si ratio in hydrated Portland cement is about 0.1 due to the incorporation of Al. This value can reach 0.3 with an Al-rich supplementary cementitious material, e.g., with metakaolin or with alkali-activated systems [31]. The mean aluminosilicate chain length increases as the Al/Si ratio increases [31, 32], and the uptake of Al incorporation in C-S-H is highly dependent on the Al concentration in the solution, as indicated in Ref. [33, 34].

Figure 6.1 shows the multiscale structure of C-A-S-H. Regarding the atomic structure of Al-substituted C-A-S-H, two primary issues remain unclear. The first issue is the substitution site of aluminum for silicon in the dreierketten chains of tobermorite-like C-S-H. One can reasonably consider that the possible substitutions take place at the end-chains (Q^1), pairing sites (Q^2p), bridging sites (Q^2b), and branching sites (Q^3) of the dreierketten chains. However, it is widely accepted that the substitution of Al^{3+} for Si^{4+} mainly occurs at the bridging sites (including Q^2 and branching Q^3 sites) of the dreierketten chain [35]. The possibility of substitution at pairing sites remains highly debated. The pairing substitution was first proposed by Faucon et al. [36], who assigned the Al(IV) signal at 58–67 ppm to the pairing Al-tetrahedra. However, such an assignment conflicted with the presence of Al-branching substitution [37]. The ^{27}Al and ^{29}Si NMR study done by Pardal et al. [38] further tries to provide experimental evidence for this possibility. According to Pardal et al. [38], a narrow Al resonance signal of around 72 ppm was assigned to the Al resonance of pairing tetrahedral Al(IV). Although this conclusion was cited by Bach et al. [39] in their NMR analysis of C-A-S-H structure, the theoretical studies [40–42] have provided critical insight in showing that aluminum is energetically unfavorable to incorporate in the pairing sites since the pairing Al has higher energy than the bridging or branching Al. In addition, this Al(IV) resonance at 72 ppm observed by Pardal et al. [38] probably corresponds to bridging tetrahedral aluminates neighboring Ca^{2+}

cations according to the atomic structure calculations shown in [43]. Note that the Loewenstein rule [44] suggests avoiding the Al–O–Al linkage in aluminosilicate zeolite materials since this moiety can increase the kinetic energy, making the structure unstable [45]. Following this rule, the presence of Al–O–Al linkage in C-A-S-H also leads to a high-energy and less stable structure [46]. Therefore, Al cannot simultaneously substitute the neighboring bridging and pairing silicon atoms. The very recent C-A-S-H brick model [43] does not consider the pairing Al(IV) substitution and considers that the Al substitution only occurs at the bridging sites. Accordingly, more evidence is still needed to determine whether the Al substitution is likely to happen at the pairing sites. At present, it is more widely accepted to assume that Al substitution occurs at the bridging sites only [34, 47–50].

Another issue is the aluminum coordination in C-A-S-H, which is related to the present forms of aluminate in aluminosilicate chains, interlayer, or at the surface. The coordination of aluminum in the C-A-S-H structure has long been misunderstood and debated. Conventionally, there are three types of possible aluminum coordination present in an aluminate phase: four-fold Al(IV) (tetrahedral), five-fold Al(V) (pentahedral), and six-fold Al(VI) (octahedral). A well-recognized fact is that Al is primarily incorporated in the silicate chains as Al(IV) [51, 52]. Initially, the observed Al(VI) by ^{27}Al NMR spectra was ascribed to an AFm-type structure layer within C-S-H [53, 54]. Later on, the Al(VI) resonance at 5 ppm was identified as a signal for Al(VI) in C-A-S-H silicate chains by Faucon et al. [36]. However, this point of view was challenged by Andersen et al. [55, 56] afterward. They found that the disappearance of the Al(VI) signal under an elevated temperature did not change the framework of the aluminosilicate chains. The intense peak at 5 ppm was then explicitly considered as the signal for the presence of a so-called "third aluminum hydroxide" (TAH) phase [56]. The TAH phase is considered a layer double hydroxide (LDH) phase by Taylor et al. [57]. In their description, the Ca-Al LDH layers are probably present in the outer-product C-S-H, and the Mg–Al LDH or $Al(OH)_3$-based structure layers are plausibly included in the inner-product. For a long time, the possibilities for the presence of Al(V) and Al(VI) in C-A-S-H were dismissed [58]. In a recent study, the TAH phase was proven nonexistent, and the Al resonance at 5 ppm was identified in the bridging Al(VI) species by applying DFT calculations and DNP NMR measurements [43]. This study suggests that the bridging Al(V) and Al(VI) complexes are present in C-A-S-H. Actually, the bridging Al(VI) can also be confirmed by previous experimental data for precipitated C-A-S-H phases provided by Stade et al. [53, 54]. They pointed out a linear relation between the Ca/Si ratio and the relative incorporation ratio Al(IV)/(Al(IV) + Al(VI)). Specifically, the measured minimum and maximum Al(IV)/(Al(IV) + Al(VI)) ratios are 0 and 1 at a Ca/Si ratio of 0.9 and 1.5, respectively. However, at that time, they ascribed this phenomenon to the preferential incorporation of octahedral trivalent ions into AFm-type layers at high Ca/Si ratios. Nonetheless, the presence of interlayer Al(V) and Al(VI) is still controversial. It is considered that the formation of this pentahedral Al is probably due to the replacement of sevenfold coordinated interlayer calcium ions [32, 55]. The Al(VI) aluminates are also likely to exist in the interlayer or at the particle surface

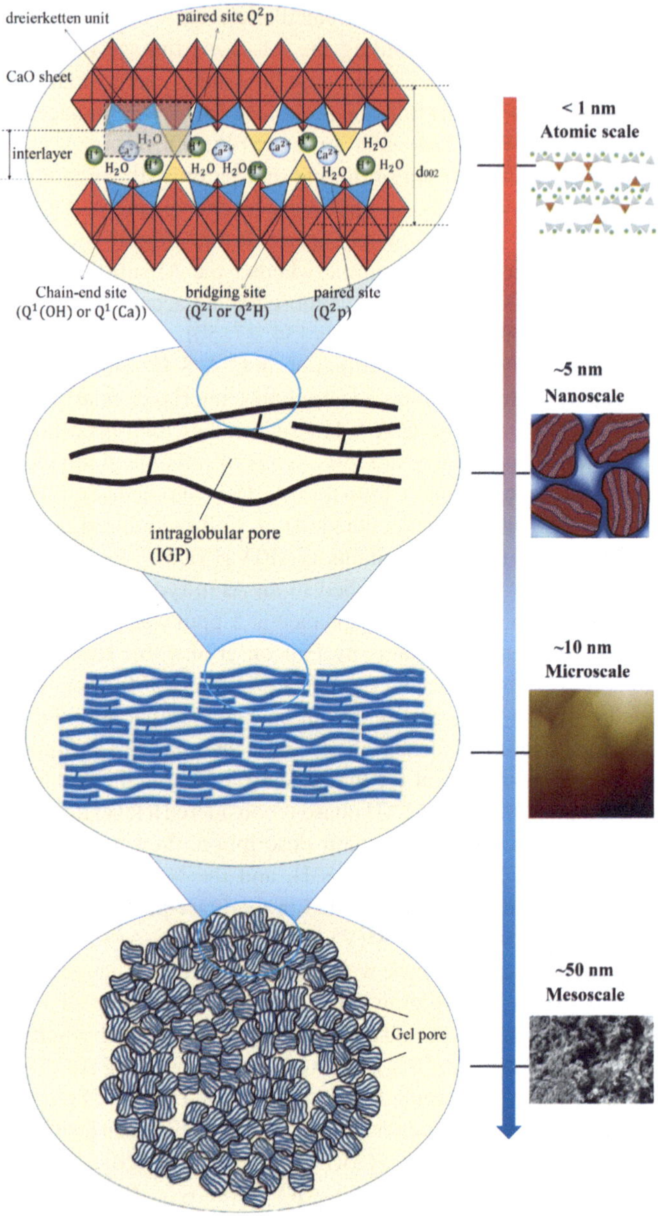

Fig. 6.1 Multiscale structure of C-A-S-H

[37, 59]. It was experimentally observed that the fraction of Al(V) in C-A-S-H has a slight increase with the Al/Si ratio [32].

In this chapter, we show the potential of Al incorporation in enhancing the cryogenic stability of the C-A-S-H. In particular, we will discuss how the chemical composition alters the cryogenic stability of C-A-S-H. Some findings shed new light on the presence of bridging and interlayer aluminosilicates in C-A-S-H. Regarding aluminum substitution sites, we assume here that substitution occurs solely at bridging and branching sites, consistent with previous research [34, 47, 50, 60]. Furthermore, we explored the stability of the nanomechanical properties of the nanostructured stable C-A-S-H system and their degradation pathways, evaluating the role of aluminum incorporation in enhancing the cryogenic stability of C-S-H.

6.2 C-A-S-H Synthesis and Characterization

6.2.1 Hydrothermal Synthesis and Cryogenic Attack Regime

The precipitations of C-A-S-H phases were implemented using a hydrothermal reaction method. Calcined CaO powders (Aladdin, 95% purity), nano SiO_2 (Aladdin, 98% purity), nano Al_2O_3 (Aladdin, 98% purity), and ultra-pure water were used to synthesize C-A-S-H phases with a bulk Al/(Al + Si) ratio of 0.1 and 0.2, and bulk Ca/(Al + Si) ratio ranging from 0.84 to 1.75, a constant water/solid ratio of 12. The mix proportions for all the precipitations are shown in Table 6.1. There were two main successive procedures for each precipitation: (i) the mixtures were constantly stirred in a nitrogen-protected reactor at a constant temperature of 60 °C for 10 h, (ii) the mixed slurry was transferred to an air-tight Teflon bottle constrained by a thick stainless steel layer, and then hydrothermally reacted at 120 °C for another 10 h. Similar precipitation procedures can be found in our previous works [4, 12, 19], where one can find more details about the procedures for preparing samples. The precipitated solids were obtained by filtration in a 45 μm filter, freeze-dried for 7 days, and then stored in an air-tight and nitrogen-protected desiccator containing saturated $CaCl_2$ solution and NaOH pellets.

The prepared C-A-S-H samples were cryogenically treated within a temperature-controllable air-tight box. By simultaneously regulating the releasing rate of liquefied nitrogen and the heating rate of the heater, the temperature was controlled to cool down slowly from ambient temperature to − 170 °C at a constant rate of − 0.5 °C/min. This cooling process took 380 min. The samples were then held at a constant temperature of − 170 °C for 180 min, enabling the C-A-S-H samples to experience a full cryogenic attack. Then, the temperature was increased to 20 °C at a ramping rate of 0.5 °C/min for another 380 min and then kept at ambient temperature for an additional 60 min. After the treatment, the samples were all moved to the air-tight and nitrogen-protected storage desiccators.

Table 6.1 The mix proportion of synthetic C-A-S-H solids

Name	Ca/(Al + Si)	Al/(Al + Si)	W/S[1]	Ca/Si	Al/Si	CaO (g)	Nano SiO_2 (g)	Nano Al_2O_3 (g)	Ultra-pure water (g)
CA1B	0.84	0.10	12	0.93	0.11	12.31	13.88	1.310	330
CA2B	0.84	0.20	12	1.05	0.25	12.42	12.44	2.642	330
CA3B	1.25	0.10	12	1.67	0.11	16.26	10.27	0.969	330
CA4B	1.25	0.20	12	1.88	0.25	16.37	9.18	1.950	330
CA5B	1.50	0.10	12	1.39	0.11	15.04	11.39	1.075	330
CA6B	1.50	0.20	12	1.56	0.25	15.14	10.19	2.165	330
CA7B	1.75	0.10	12	1.94	0.11	17.27	9.35	0.882	330
CA8B	1.75	0.20	12	2.19	0.25	17.37	8.35	1.774	330

Note [1] W/S is the water to solid ratio

6.2.2 *Phase Assemblage of Synthetic Products*

XRD patterns of the precipitated solids before and after the cryogenic attack are shown in Fig. 6.2. The chemical compositions are listed in Table 6.2. C-A-S-H phases are the leading products in the precipitates, indicating the effectiveness of our synthesis strategy. The intensity of the (020) and ($\bar{2}20$) peak at ~ 3.1 Å [47, 49, 61] decreases with the increase of Al content, suggesting that the increased Al uptake leads to more structural disorder in the C-A-S-H nanostructure. Regarding the basal spacing (002) peak at ~ 12.3 Å observed in the precipitates with a Ca/(Al + Si) ratio of 0.84 and 1.5, the intensity significantly decreases, and the basal spacing distance increases with a higher Al/Si ratio. Katoite phases are more likely to precipitate under a higher Al availability condition [48], particularly when the Al/(Al + Si) ratio is above 0.1 [34, 50]. However, they are less likely to precipitate at a low Ca/(Al + Si) ratio system below 1.25, even at a high Al/(Al + Si) ratio of 0.2, which indicates that Al uptake by C-S-H is more preferential at a relatively lower Ca/Si ratio. Portlandite is observed in the precipitates with a high bulk Ca/(Al + Si) ratio $\geq$ 1.5. Portlandite formation probably occurs during the drying procedure when the precipitates leave away the synthetic mother liquor [58, 62]. As a result, the formation of high Ca/(Al + Si) C-(A-)S-H is hindered by the precipitation of portlandite [63].

The phase composition was not altered under the cryogenic attack. No phase formed or disappeared, as indicated by the XRD patterns in Fig. 6.2b. Regarding the crystallinity of C-A-S-H, except the tobermorite-like C-A-S-H (C/S0.84-A/S0.1), the (020) & ($\bar{2}20$) peak of precipitates showed a slight increase in intensity after the cryogenic attack, indicating a more structurally ordered nanostructure. For C-A-S-H with Ca/(Al + Si) = 0.84, Al/(Al + Si) = 0.1, the basal spacing distance shortened slightly from 12.5 Å to 12.3 Å. While the basal spacing distance of C-A-S-H with Ca/(Al + Si) = 1.5, Al/(Al + Si) = 0.1 increased slightly from 13.6 Å to 13.8 Å. Albeit the (002) peaks were not observed for other precipitates except for the abovementioned two, the results shown here, together with our previous XRD

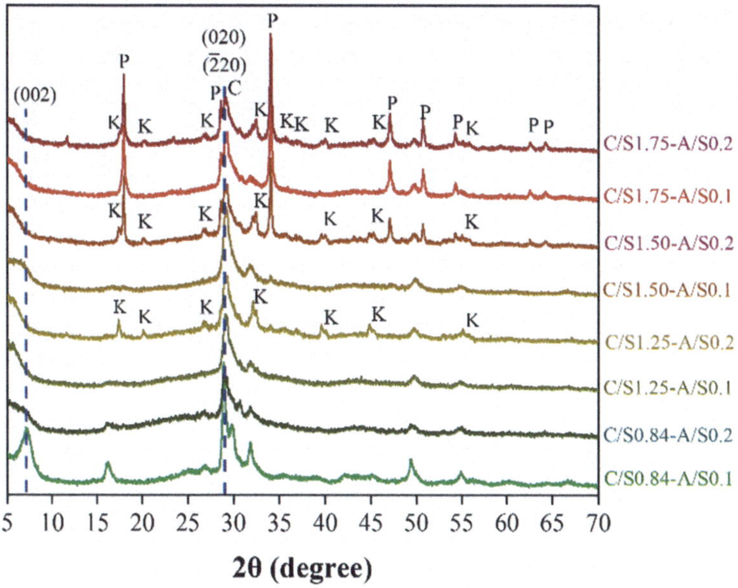

Fig. 6.2 XRD patterns of precipitated solids. P: portlandite, K: katoite, C: calcite. C/S represents Ca/(Al + Si) ratio, A/S stands for Al/(Al + Si) ratio, e.g., C/S0.84-A/S0.1 means the precipitate with Ca/(Al + Si) = 0.84, Al/(Al + Si) = 0.1

measurements for Al-free C-S-H [19], indicate that the interlayer basal spacing of low-Ca C-(A-)S-H is likely to decrease under the cryogenic attack.

6.2.3 SEM Micromorphology

The SEM microstructure of the synthesized C-A-S-H product is shown in Fig. 6.3. In the case of aluminum-doped C-S-H, it exhibits a highly disordered microscopic morphology. Comparing the surface morphology shown in Fig. 6.3, both calcium content and aluminum content significantly influence the appearance of C-A-S-H. Surfaces of low Ca/(Al + Si) ratio C-A-S-H exhibit a more porous characteristic compared to systems with higher Ca/(Al + Si) ratios, displaying a plate-like morphology. As the Ca/(Al + Si) ratio increases, the surface of C-A-S-H becomes denser. When Ca/(Al + Si) > 1.5, the plate-like structures nearly disappear, replaced by agglomerates of C-A-S-H. Another noteworthy feature with increasing Ca/(Al + Si) ratio is the gradual observation of the presence of $Ca(OH)_2$ or $CaCO_3$ in the synthesized product.

As the amount of aluminum incorporation increases, the size of the plate-like structures formed on the surface of the C-A-S-H product becomes larger, a feature particularly evident in the system with Ca/(Al + Si) = 0.84. Based solely on surface

Table 6.2 Chemical compositions of precipitated solids

Bulk Al/(Si + Al)	Bulk Ca/(Si + Al)	Al/Si	Ca/Si	Al/(Al + Si)	Ca/(Si + Al)	Solids
0.1	C/S0.84B[a]	0.102	1.142	0.092	1.038	CASH
	C/S0.84A[b]	0.256	0.899	0.204	0.716	CASH
	C/S1.25B	0.243	1.336	0.195	1.075	CASH
	C/S1.25A	0.267	1.307	0.210	1.032	CASH
	C/S1.50B	0.243	1.435	0.194	1.156	CASH
	C/S1.50A	0.187	1.357	0.155	1.150	CASH
	C/S1.75B	0.266	2.135	0.209	1.692	CASH + Portlandite
	C/S1.75A	0.196	1.703	0.163	1.424	CASH + Portlandite
0.2	C/S0.84B	0.553	1.044	0.354	0.675	CASH
	C/S0.84A	0.301	1.034	0.231	0.795	CASH
	C/S1.25B	0.310	1.437	0.236	1.099	CASH + Katoite
	C/S1.25A	0.213	1.409	0.175	1.162	CASH + Katoite
	C/S1.50B	0.171	1.514	0.145	1.294	CASH + Portlandite + Katoite
	C/S1.50A	0.298	1.531	0.229	1.181	CASH + Portlandite + Katoite
	C/S1.75B	0.282	1.873	0.220	1.461	CASH + Potlandite + Katoite + Calcite
	C/S1.75A	0.152	1.789	0.131	1.556	CASH + Potlandite + Katoite + Calcite

Note [a] Before cryogenic attack, [b] After cryogenic attack

morphology observations, the increased aluminum incorporation seems to create a more compact C-A-S-H surface structure, as shown in Fig. 6.3e and f. Additionally, with the increase in Ca/(Al + Si) ratio, spherical substances can be observed on the surface of high-Al-doped C-A-S-H products, as depicted in Fig. 6.3f and g. Combined with the XRD results illustrated in Fig. 6.2, these spherical substances are likely to be impurities of Katoite formed in high-Al C-A-S-H systems.

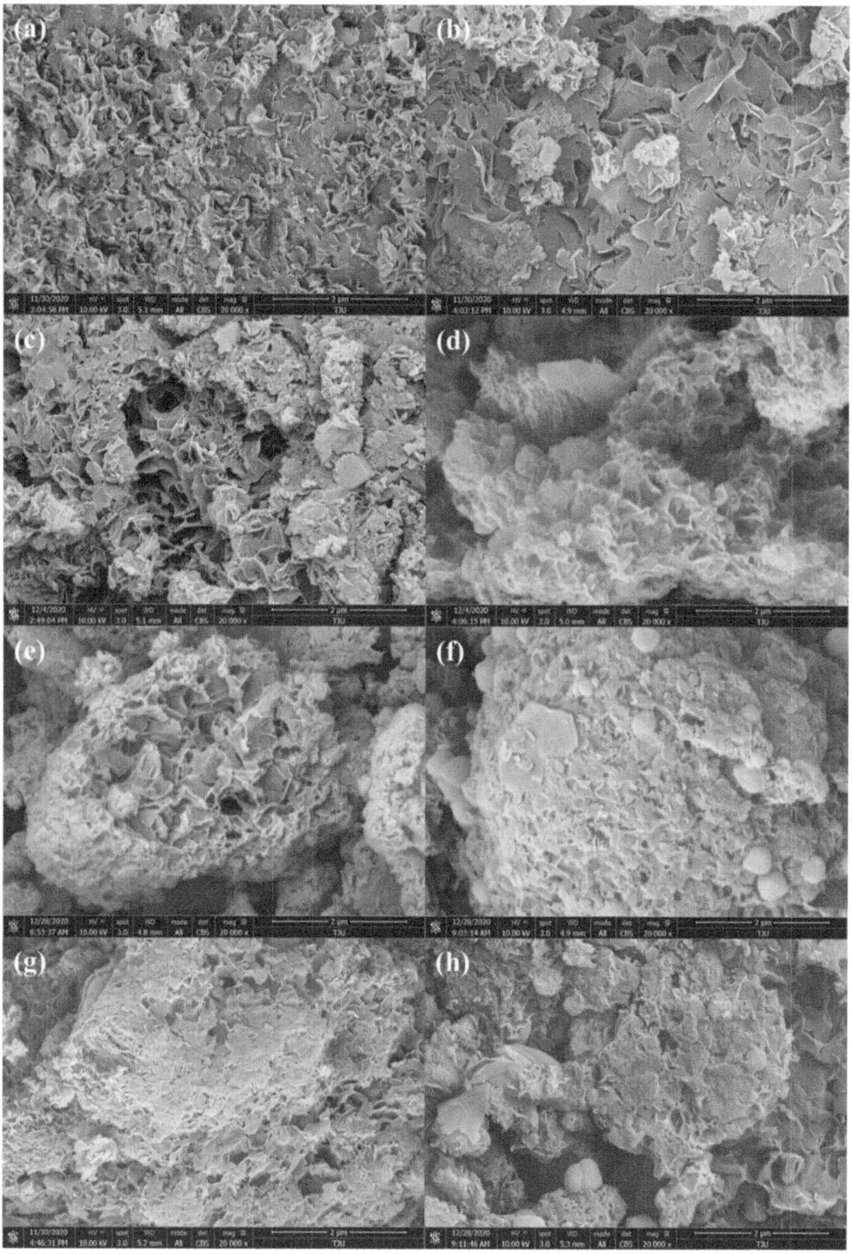

Fig. 6.3 The SEM microstructural characteristics of C-A-S-H: **a** C/S0.84-A/S0.1; **b** C/S0.84-A/S0.2; **c** C/S1.25-A/S0.1; **d** C/S1.25-A/S0.2; **e** C/S1.50-A/S0.1; **f** C/S1.50-A/S0.2; **g** C/S1.75-A/S0.1; **h** C/S1.75-A/S0.2

6.3 Cryogenic Stability of C-A-S-H Nanostructure

6.3.1 *Phase Assemblage*

The phase composition was not altered under the cryogenic attack. No phase formed or disappeared, as indicated by the XRD patterns in Fig. 6.4. Regarding the crystallinity of C-A-S-H, except the tobermorite-like C-A-S-H (C/S0.84-A/S0.1), the (020) and ($\bar{2}$20) peak of precipitates showed a slight increase in intensity after the cryogenic attack, indicating a more structurally ordered nanostructure. For C-A-S-H with Ca/(Al + Si) = 0.84, Al/(Al + Si) = 0.1, the basal spacing distance shortened slightly from 12.5 Å to 12.3 Å. While the basal spacing distance of C-A-S-H with Ca/(Al + Si) = 1.5, Al/(Al + Si) = 0.1 increased slightly from 13.6 Å to 13.8 Å. Albeit the (002) peaks were not observed for other precipitates except for the abovementioned two, the results shown here, together with our previous XRD measurements for Al-free C-S-H [19] indicate that the interlayer basal spacing of low-Ca C-(A-)S-H is likely to decrease under the cryogenic attack.

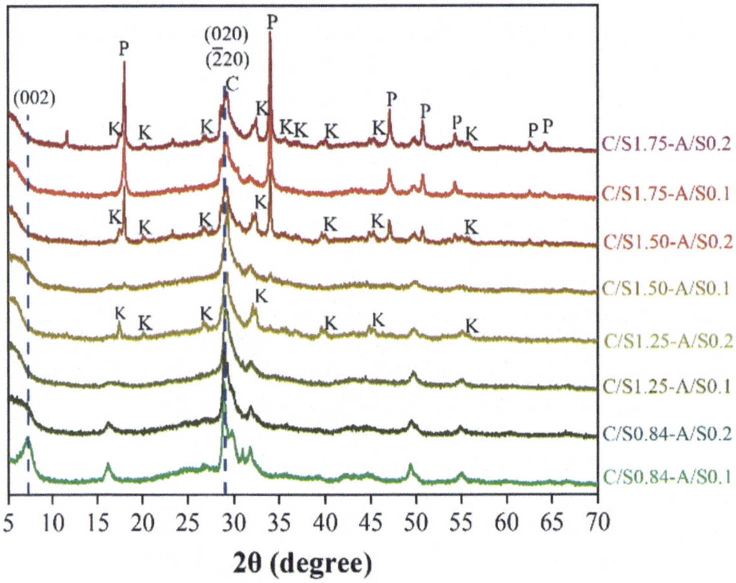

Fig. 6.4 The XRD patterns of C-A-S-H synthesized products after cryogenic attack. P represents $Ca(OH)_2$, K represents Katoite, and C represents calcite

6.3.2 *Composition Variation*

6.3.2.1 Dehydration/de-Hydroxylation/de-Carbonation Stages

Figure 6.5 displays the mass loss and corresponding derivative values of cryogenically treated and non-treated precipitates recorded in TGA. Four prominent peaks were observed in all DTG curves. The primary mass loss peak at ~ 120 °C is related to the dehydration of interlayer C-A-S-H water. The broad peaks at ~ 410 °C observed in Fig. 6.5a–e are ascribed to the de-hydroxylation of C-A-S-H, which is mainly involved in the de-hydroxylation of interlayer Ca-OH according to previous measurements on Al-free C-S-H [19, 64]. At the same time, it is probably followed with partial de-hydroxylation of Si–OH and Al–OH (if any) regarding Al-incorporated C-A-S-H in the current study. For those precipitates intermixed with portlandite, the de-hydroxylation of crystallized portlandite $Ca(OH)_2$ is sharp within 400–450 °C, as shown in Fig. 6.5f–h. The peaks centered at ~ 680 °C are attributed to the de-carbonation of calcium carbonates $CaCO_3$. Although protection measures have been taken to prevent carbonation, mild carbonation is inevitable during the synthesis, transfer, cryogenic attack, and even drying process. According to Ref. [60], the peaks at ~ 820 °C are associated with C-A-S-H decomposition as it is transformed to wollastonite. Sharp peaks at ~ 310 °C are additionally observed in high Al/(Al + Si) systems except that with Ca/(Al + Si) = 0.84, which are relevant to the dehydration of katoite [60, 65].

Combining the tangent method and previous TGA studies [19, 64, 66], there is good confidence that one can assign the de-hydroxylation and de-carbonation stages at the range of 350–500 °C and 600–750 °C, respectively. For clarity we use hydroxyl water to describe the water lost in the de-hydroxylation stage. Regarding the mass loss of interlayer bound water, the corresponding dehydration stage is considered within 100–350 °C, consistent with the assignment adopted in Ref. [19, 64]. Note that the dehydration of katoite present in some precipitates partially overlaps with the dehydration of interlayer bound water within 275–350 °C. Nonetheless, the mass loss of interlayer bound water predominantly occurs between 100 and 275 °C.

6.3.2.2 Carbonate and Corrected Hydroxyl Water

The carbonates observed in the precipitates are mainly $CaCO_3$-type phases. Carbonation reaction occurs as: $Ca(OH)_2 + CO_2 \rightarrow CaCO_3 + H_2O$. It consumes 1 mol of $Ca(OH)_2$ to produce 1 mol of $CaCO_3$ along with 1 mol of H_2O. Therefore, the variations in the carbonate content are closely related to the amount of $Ca(OH)_2$ and interlayer water. In particular, the carbonation that occurred during the cryogenic attack can lead to underestimating hydroxyl water content. For this reason, the corrected hydroxyl water content corresponding to the consumed $Ca(OH)_2$ during carbonation was estimated according to the carbonation reaction. Figure 6.6 compares calcium carbonate content and the corrected hydroxyl water present in precipitates before

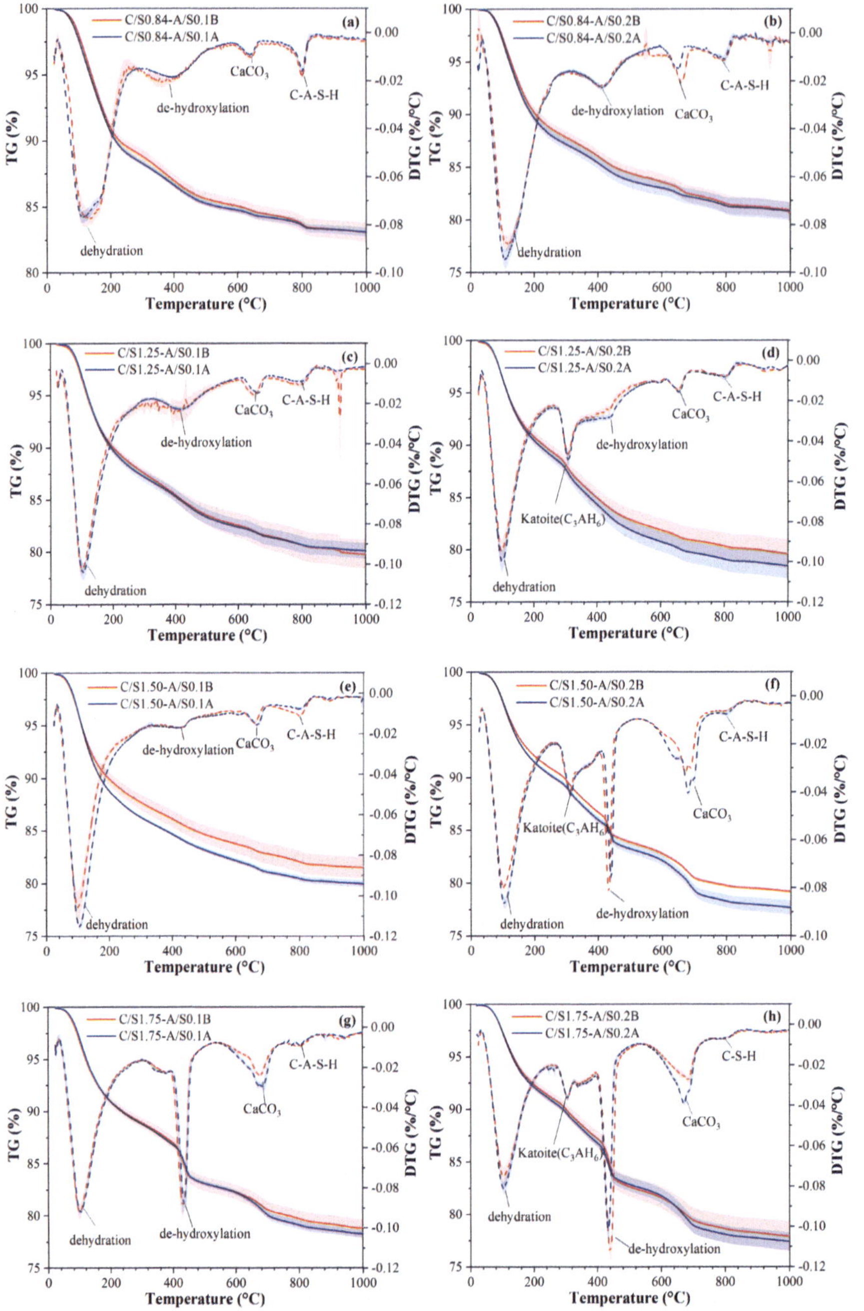

Fig. 6.5 Thermogravimetric analysis curves of precipitation solids before and after cryogenic attack. B represents before the cryogenic attack, and A represents after the cryogenic attack

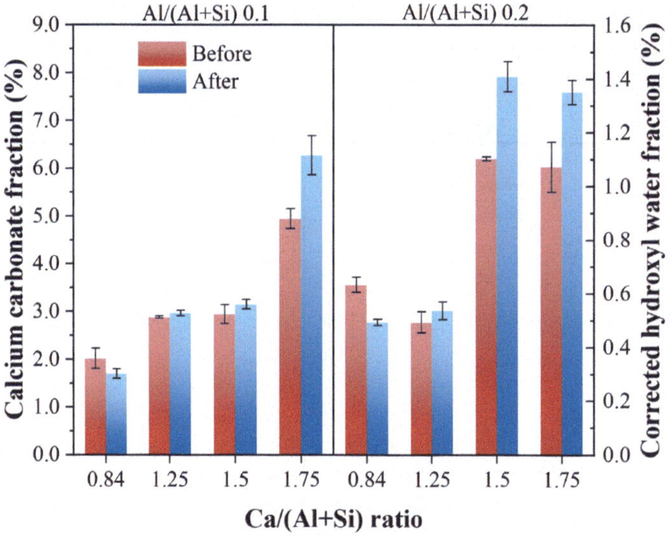

Fig. 6.6 The mass fraction of calcium carbonate and calculations for the content of corrected hydroxyl water produced during carbonation

and after the cryogenic attack. Carbonation is more severe in the precipitates with a higher Ca/(Al + Si) ratio and a higher Al/(Al + Si) ratio than those with low ratios. The content of carbonates generally shows a slight increase after the cryogenic attack, which means that $Ca(OH)_2$ phases were partially carbonated during the attack. Correspondingly, the corrected hydroxyl water content slightly increased after the cryogenic attack. The carbonate variations in precipitates containing portlandite are more significant, showing that the crystallized $Ca(OH)_2$ phases are more likely to be carbonated than interlayer $Ca(OH)_2$.

6.3.2.3 Interlayer Bound Water

Figure 6.7 illustrates the mass loss in the range of 100–350 °C corresponding to the dehydration of interlayer water. For those precipitates containing katoite, the measured interlayer water contents are not exact due to the partial overlap with katoite dehydration within 275–350 °C. However, the overlap does not change the variation trend for interlayer water content according to the mass loss within 100–275 °C and 275–350 °C. As shown in Fig. 6.8, a predominant amount of interlayer water is driven out of C-A-S-H within 100–275 °C, while the mass loss for katoite is less than 1.5%. The variation of interlayer water depicted in Figs. 6.7 and 6.8a exhibits the same trend. Therefore, one can reasonably estimate the variation of interlayer water content from Fig. 6.7. The cryogenic attack led to a slight increase of less than 1.5% in interlayer water content for all precipitates, irrespective of the Ca/(Al + Si) and Al/(Al + Si) ratios, reflecting that a portion of water was adsorbed or condensed into the C-A-S-H solids during the cryogenic attack, which is consistent with our previous study of Al-free C-S-H [19].

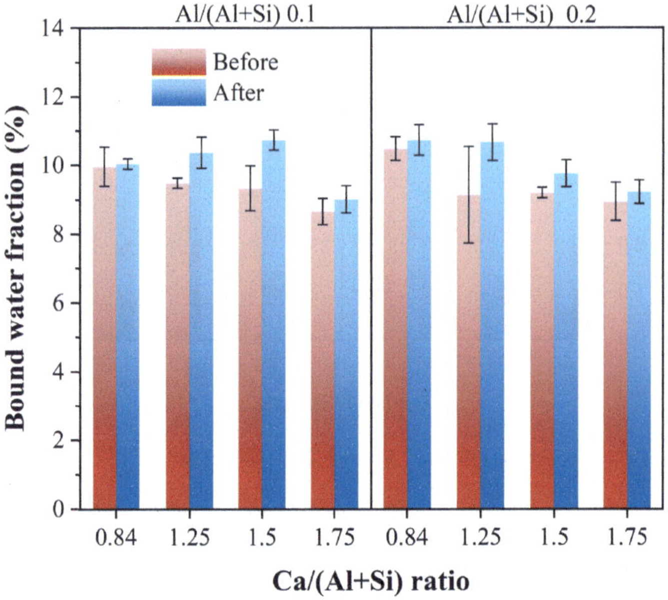

Fig. 6.7 Comparisons of interlayer bound water weight fractions of precipitates before and after cryogenic attack

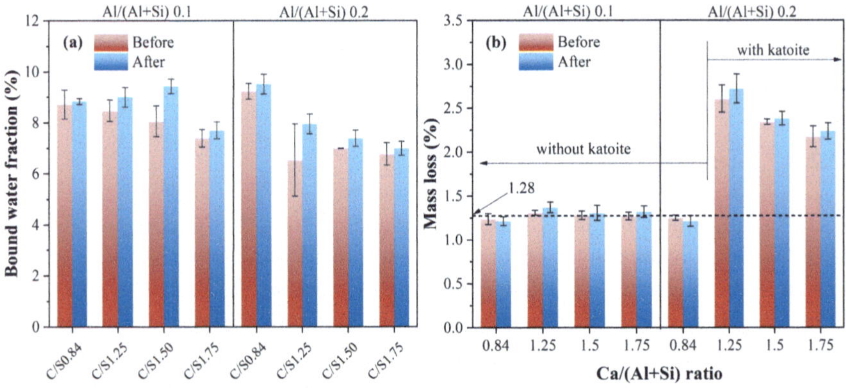

Fig. 6.8 Measurements of interlayer water weight fraction from mass loss within 100–275 °C (**a**), and mass loss within 275–350 °C (**b**)

6.3.2.4 Hydroxyl Water

Figure 6.9a displays the hydroxyl water contents directly measured from mass loss within 350–500 °C. Herein, hydroxyl water is related to the de-hydroxylation of R-OH where R can be Ca, Al, and Si, which means that the higher the hydroxyl water content obtained from TGA, the higher the amount of hydroxyl groups host in C-A-S-H solids. The de-hydroxylation of Ca-OH contributes predominantly to the hydroxyl water lost during thermogravimetric analysis, as indicated by similar measurements on Al-free C-S-H [19]. As noted before, carbonation consumes the $Ca(OH)_2$. In order to eliminate the influence of carbonation, the equivalent hydroxyl water content was added to the corresponding measured hydroxyl water content, yielding the initial hydroxyl water content, i.e., the content of hydroxyl water in the absence of carbonation, as shown in Fig. 6.9b.

From Fig. 6.9, the amount of hydroxyl water present in C-A-S-H generally increases with the Ca/(Al + Si) ratio. Such a relation is mainly associated with the larger amount of calcium hydroxide in a higher Ca/Si ratio system. Apart from the increased Ca content, the increase in Al uptake content leads to an increase in hydroxyl water content as well. The apparent growth in the amount of hydroxyl water is observed for all precipitates as the Al/(Al + Si) ratio increased from 0.1 to 0.2, suggesting that Al-rich systems have a higher propensity to form hydroxyl groups. Comparing the precipitates before and after the cryogenic attack, the amounts of hydroxyl water variations are insignificant. A slight increase is seen for precipitates with a Ca/(Al + Si) ratio of 1.25 and 1.5. In contrast, it seems that C-A-S-H, with a low Ca/(Al + Si) ratio of 0.84, has the best cryogenic stability in hydroxyl water content (about 2.7%), regardless of the Al/(Al + Si) ratio. Similar observations are also found in Al-free C-S-H at a low Ca/Si ratio [19]. It is also noteworthy that the fraction of hydroxyl water in C-A-S-H is more stable under cryogenic attack than for Al-free C-S-H submitted to the same cryogenic treatment [19].

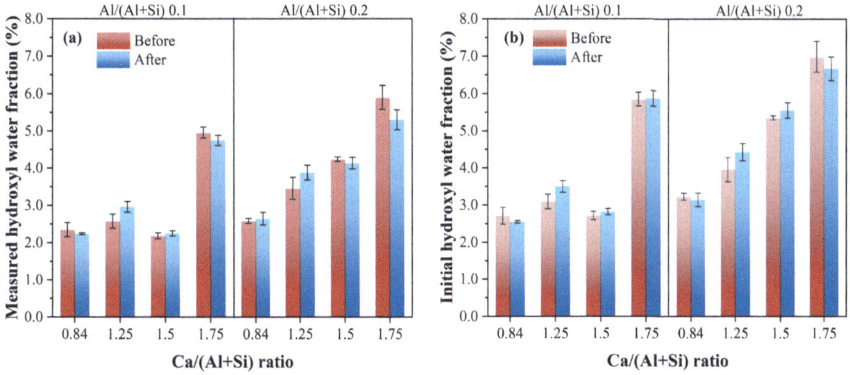

Fig. 6.9 Measurements of hydroxyl water weight fraction (**a**), and calculations for the initial hydroxyl water weight fraction including corrected hydroxyl water formed during carbonation (**b**)

6.3.3 Description of C-A-S-H Atomistic Structure

The (alumino)silicate chemical environments are characterized by a notation of the type Q^n(mAl, R) where Q denotes a silicate tetrahedra unit, n represents the number of corresponding linked tetrahedra, m stands for the number of linked aluminate structures, and R is the charge-balanced interlayer species, e.g., H^+ and Ca^{2+}. Since m is either equal to 0 or 1, the notation is further abbreviated as Q^n(R) and Q^n(Al, R), respectively. This notation is convenient for describing the silica at different positions within the dreierketten chains, such as the bridging, the pairing, and the end-chain sites. However, the chemical environments of silicate tetrahedra are more complex with the presence of interlayer species. It is worth mentioning that the silica environments in the (alumino)silicate chains of C-(A-)S-H are highly influenced by the interlayer charge-compensated ions such as Ca^{2+} and H^+, which further leads to a shift for the corresponding Si resonance according to previous measurements for bridging Q^2 tetrahedra [22, 67–71]. Similar shielding behavior has been observed for the end-chain Q^1 sites due to different charge-compensated species like Ca^{2+} and H^+ [71, 72]. According to the theoretical study [71], the hydroxylated Q^1 sites (Q^1(OH)) are less shielded than Q^1 charge-compensated by Ca cations only (Q^1(Ca)). Myers et al. [72] resolved the above two Q^1 sites with a chemical shift of – 78 ppm and – 80 ppm in their ^{29}Si MAS NMR analysis. However, the 2D ^{1}H{^{29}Si} NMR measurement results obtained by Klur et al. [70] were against such assignments. Therefore, the current knowledge regarding this issue is still not well understood. Herein, we assigned a single Q^1 peak at ~ 79 ppm, as we normally do.

Following the above rule and the postulation that the aluminum is substituted at the bridging and branching sites only, the positions of the different peaks were referenced from Ref. [19, 50, 60, 70–73]. The schematic diagrams for crosslinked and non-crosslinked C-A-S-H atomic structures are illustrated in Fig. 6.10. Q^3(OH) and Q^3(Ca) denote branching silicate tetrahedra neighboring interlayer H^+ and Ca^{2+}, respectively; Q^1(OH) and Q^1(Ca) represent end-chain silicate tetrahedra neighboring interlayer H^+ and Ca^{2+}, respectively; Q^2H stands for bridging silicate tetrahedra charge-compensated by two interlayer H^+; Q^2(Ca) means the bridging silicate tetrahedra charge-compensated by Ca^{2+} only; Q^2i represents the bridging silicate tetrahedra charge-compensated by H^+ and $Ca(OH)^+$; Q^2p stands for the pairing tetrahedra; and Q^2p(Al) and Q^3(Al) denote the pairing and branching tetrahedra neighboring one substituted aluminate, respectively. Note that Q^2p(Al(VI)) and Q^2p(Al(IV)) have different chemical shifts according to the experiments and simulations in Ref. [43]. It turns out that the chemical shift of Q^2p(Al(VI)) is at ~ 77 ppm. Since the number of pairing sites is twice as many as bridging/branching sites, when performing the decomposition, the relative intensity of bridging (together with branching) and pairing tetrahedra should satisfy relation Eq. 6.1.

$$\frac{\begin{array}{c}Q^3(\mathrm{Al}) + Q^3(\mathrm{Ca}) + Q^3(\mathrm{OH}) + \left[Q^2\mathrm{p}(\mathrm{Al}) + Q^2\mathrm{p}(\mathrm{Al(VI)})\right]/2 \\ + Q^2\mathrm{H} + Q^2(\mathrm{Ca}) + Q^2\mathrm{i}\end{array}}{Q^2\mathrm{p} + Q^2\mathrm{p}(\mathrm{Al}) + Q^2\mathrm{p}(\mathrm{Al(VI)})} = \frac{1}{2} \tag{6.1}$$

According to the atomistic descriptions for C-A-S-H illustrated in Fig. 6.10, branching silicate tetrahedra, e.g., Q^3, and $Q^3(Al)$, are distinctive fingerprints for the presence of crosslinked aluminosilicate chains. The fraction of crosslinked chain d can be additionally determined according to Eq. 6.), which shows the relative intensity of branching sites to bridging/branching sites. The total fraction of the branching site is $Q^3(OH) + Q^3(Ca) + 2Q^3(Al)$.

$$d = \frac{Q^3(OH) + Q^3(Ca) + 2Q^3(Al)}{Q^2H + Q^2(Ca) + Q^2i + Q^3(OH) + Q^3(Ca) + Q^3(Al) + \left[Q^2p(Al) + Q^2p(Al(VI))\right]/2} \tag{6.2}$$

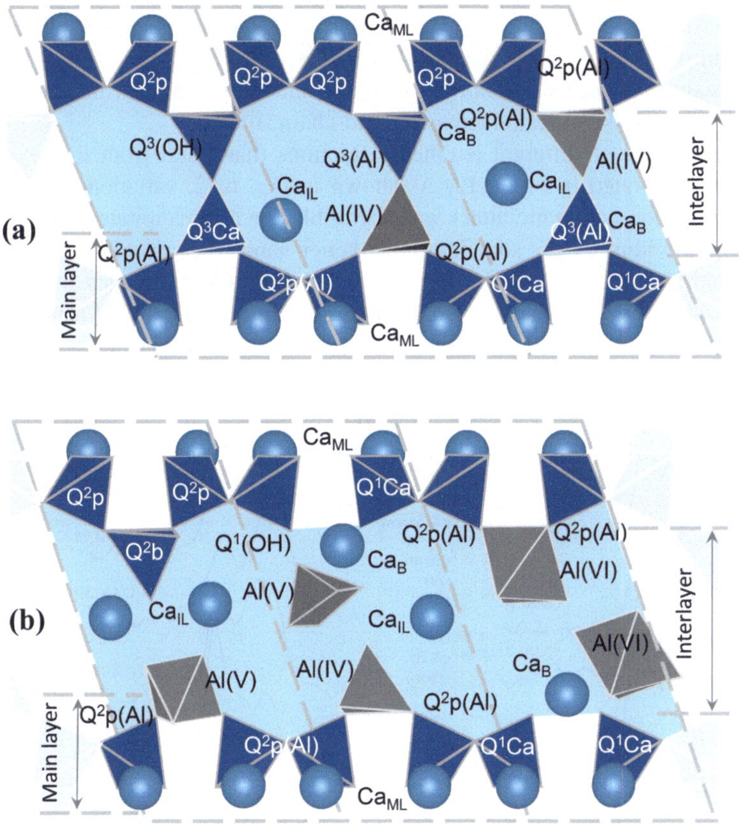

Fig. 6.10 Schematic diagrams for crosslinked (**a**) and non-crosslinked C-A-S-H structure (**b**). The dark gray polyhedrons stand for aluminates in the Al-substituted sites of the dreierketten chains as well as in the interlayer. The dark gray tetrahedra, pentahedrons, and octahedra are Al(IV), Al(V), and Al(VI) aluminates, respectively. The presence of bridging Al(V) and Al(VI) is indicated by Kunhi Mohamed et al. [43]. The dark blue tetrahedra represents silicate tetrahedra. The cyan balls denote calcium cations. The main-layer calcium cations are labeled as Ca_{ML}, the interlayer calcium cations are labeled Ca_{IL}, and the bridging calcium is labeled Ca_B. (**a**) and (**b**) are adapted from [43]

6.3.4 Silicate Tetrahedra in Low-Al System

Figure 6.11 shows the ^{29}Si NMR spectra of the cryogenically treated and non-treated precipitates with an Al/(Al + Si) ratio of 0.1. The fractions of different silicate tetrahedra sites decomposed from the spectra are presented in Table 6.3.

From Fig. 6.11, the ^{29}Si NMR spectra tend to shift towards less negative values as the Ca/(Al + Si) ratio increases from 0.84 to 1.75. Following this shift, the fraction of end-chain silicate tetrahedra goes up from about 18% to over 50%, as shown in Table 6.3. Crosslinked aluminosilicate chains are merely formed at a low Ca/(Al + Si) ratio of 0.84 since the resonance of branching silicate was only detected in this precipitate. The crosslinked chains occupy a percentage of 36.5% out of the total amount of aluminosilicate chains. In crosslinked C-A-S-H, the observed Q^3(Al) and Q^2p(Al) resonances reveal Al substitution occurring at branching and bridging sites, confirming previous findings [34, 47]. Once the Ca/Si ratio is over 1.0, no more crosslink is observed, leading to a significant decrease in pairing silicates and correspondingly a significant increase in end-chain silicates.

The variations of different tetrahedra fractions that result from the cryogenic attack are illustrated in Fig. 6.12. As shown in Fig. 6.12, variations in fractions of Q^n sites upon cryogenic attack are negligible for all precipitates. Most of the changes are within ± 3%, showing much better nanostructure stability under this harsh condition than Al-free C-S-H [19]. Albeit the fraction changes are insignificant, it is still possible to describe the general transition of C-A-S-H nanostructure from Table 6.3 and Fig. 6.12. Regarding the crosslinked C-A-S-H with Ca/(Al + Si) =

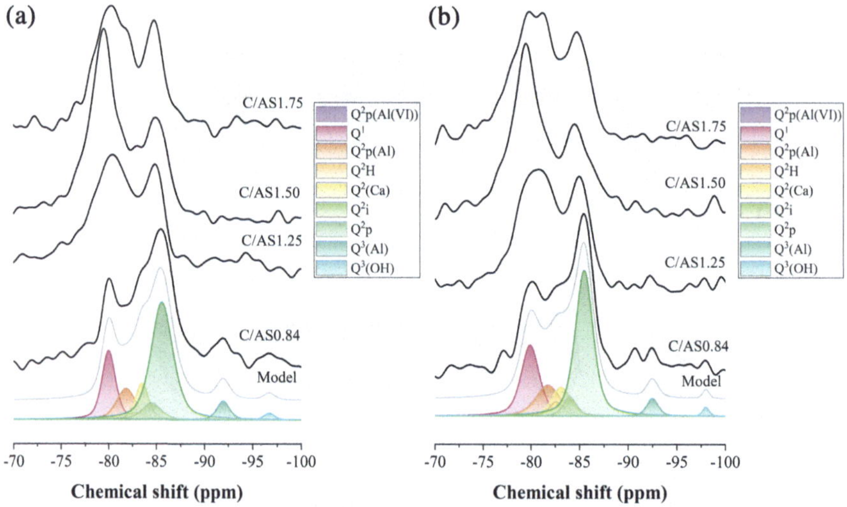

Fig. 6.11 ^{29}Si MAS NMR deconvolution results of C-A-S-H with Al/(Al + Si) ratio of 0.1. **a** Before cryogenic attack, **b** after cryogenic attack

Table 6.3 The fractions (%) of different silicate tetrahedral sites and the crosslinked chain fraction in low-Al C-A-S-H samples

Al/(Al + Si)	Ca/(Al + Si)	$I_{Q2p(Al(VI))}$ (~− 77.0 ppm)	I_{Q1} (~− 79.5 ppm)	$I_{Q2p(Al)}$ (~− 81.5 ppm)	I_{Q2H} (~− 82.5 ppm)	$I_{Q2(Ca)}$ (~− 83.5 ppm)	I_{Q2i} (~− 84.3 ppm)	I_{Q2p} (~− 85.2 ppm)	$I_{Q3(Al)}$ (~− 92.0 ppm)	$I_{Q3(OH)}$ (~− 97.0 ppm)	d[1]
0.1	C/S0.84B[2]	0.00	18.22	11.44	2.10	9.24	6.08	46.95	4.65	1.32	0.365
	C/S0.84A[2]	0.00	22.08	12.85	3.50	9.18	4.44	43.39	3.55	1.00	0.288
	C/S1.25B	8.09	41.50	16.53	1.72	6.51	3.10	22.55	0.00	0.00	0.000
	C/S1.25A	4.79	39.33	18.58	2.40	6.61	3.36	24.93	0.00	0.00	0.000
	C/S1.50B	5.11	55.04	5.66	2.46	5.50	3.53	22.69	0.00	0.00	0.000
	C/S1.50A	7.54	54.34	5.13	2.83	3.77	4.54	21.86	0.00	0.00	0.000
	C/S1.75B	4.87	50.12	9.08	3.91	2.14	5.83	24.06	0.00	0.00	0.000
	C/S1.75A	3.46	49.95	11.33	2.68	4.48	4.73	23.37	0.00	0.00	0.000

Note [1]. d is the fraction of crosslinked chain in C-A-S-H. [2]. B represents before the cryogenic attack, A represents after the cryogenic attack

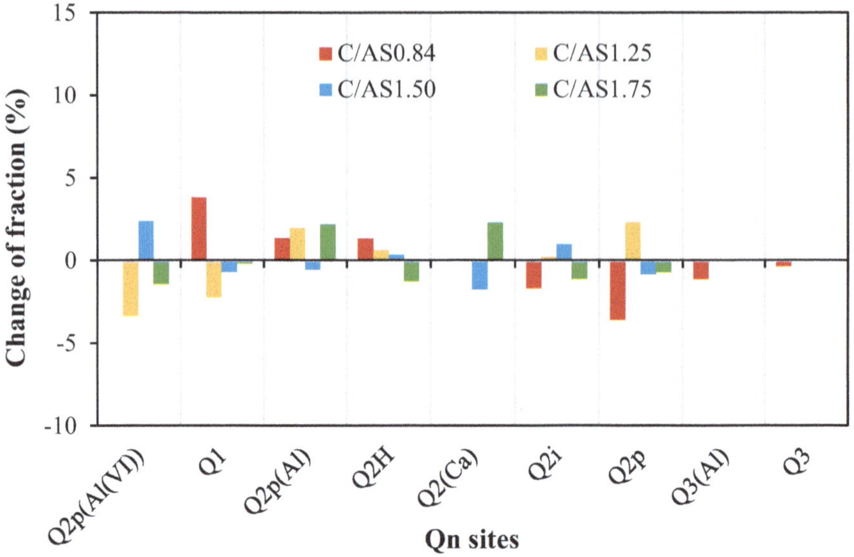

Fig. 6.12 The variations of Si environments after the cryogenic attack for C-A-S-H with Al/(Al + Si) = 0.1. The values are calculated by $I_{after} - I_{before}$ (I_{after} and I_{before} are the fractions of silicate species before and after the attack, respectively) for each site

0.84, Al/(Al + Si) = 0.1, it is noteworthy that the fractions of branching silicate tetrahedra, i.e., Q^3(Al) and Q^3(OH), decrease after the cryogenic attack. Consequently, the crosslinked chain fraction decreased to 28.8%, suggesting that the crosslinks in crosslinked aluminosilicate chains were partially broken. A similar observation was also found in crosslinked Al-free C-S-H [19]. Moreover, the subsequent decrease of Q^2p indicates that the broken branching sites were further ruptured at the bridging positions, yielding more end-chain silicate tetrahedra. For non-crosslinked C-A-S-H nanostructures, modest changes can be found at Q^2p(Al), Q^2p(Al(VI)), and end-chain sites. Note that all the above-mentioned nanostructure transitions are relatively modest in crosslinked and non-crosslinked C-A-S-H in comparison with Al-free C-S-H [19].

6.3.5 *Silicate Tetrahedra in High-Al System*

Figure 6.13 displays the ^{29}Si NMR spectra of C-A-S-H phases with an Al/(Al + Si) ratio of 0.2 before and after the cryogenic attack. Table 6.4 shows the detailed fraction values of varying silicate types in C-A-S-H phases. Similar to what we observed in C-A-S-H with a lower Al/(Al + Si) ratio of 0.1, the spectra are shifted to less negative values when the Ca/(Al + Si) ratio increases from 0.84 to 1.75, consistently with the previous findings [34, 74]. This change is probably associated with the shielding effect of interlayer Ca^{2+} ions [71]. However, Q^3 resonances disappeared in all these Al-rich precipitates, even for that with a Ca/(Al + Si) ratio of 0.84, suggesting that none of these phases is crosslinked. Regarding C-A-S-H phases with a Ca/(Al + Si) ratio of 0.84, along with the absence of Q^3, we observe that the pairing Q^2p fraction drops down and that the end-chain Q^1 fraction rises as the Al/(Al + Si) ratio increases from 0.1 to 0.2. This observation proves that the Al incorporation in crosslinked tobermorite cuts down the aluminosilicate chains [47], which is linked to the lower capability of Al substitution at the crosslinking sites. Therefore, when the substitution occurs in a low Ca/Si system, increasing Al content can decrease the formation of crosslinked species [72]. When it comes to a C-A-S-H system at Ca/Si ratio above 1, no cross-linking is observed, but incorporating Al promotes longer chains by occupying more bridging sites, as evidenced by a higher amount of pairing Q^2(Al) (Q^2p(Al) + Q^2p(Al(VI)) and a lower number of end-chain Q^1 (see Table 6.4). This further confirms the previous study [75].

Figure 6.14 illustrates the variations of Q^n species fraction after the cryogenic attack. Unlike the precipitates with Al/(Al + Si) = 0.1, the nanostructure of these Al-rich precipitates with Al/(Al + Si) = 0.2 exhibits much more severe variations due to the cryogenic attack. More specifically, apparent variations mainly occurred at end-chain Q^1 sites and pairing Q^2p(Al) sites. The Q^1 fraction increased by approximately 5–10%, while the fraction of Q^2p(Al) species generally decreased by almost the same value. Based on the atomistic structure shown in Fig. 6.10, a rational explanation is that these changes in Q^n species are related to the amount of incorporated Al at the bridging sites. The increase in end-chain species fraction suggests that the aluminosilicate chains were ruptured within the dreierketten chains, thus exposing

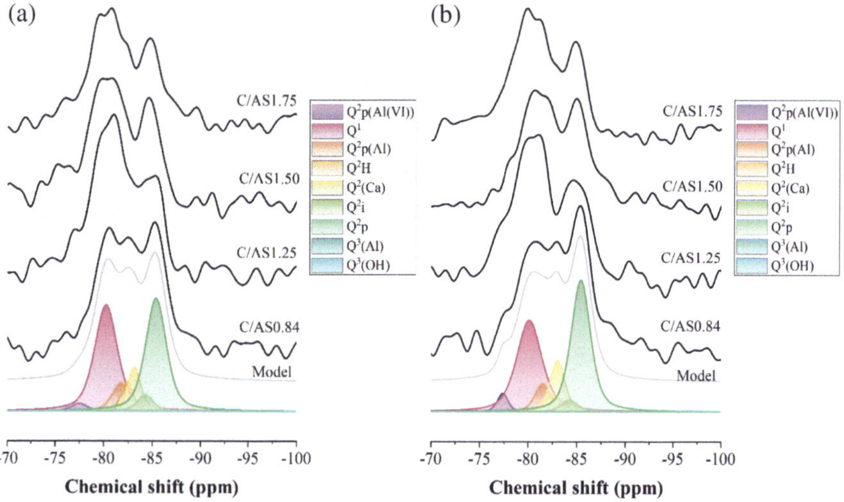

Fig. 6.13 ^{29}Si MAS NMR deconvolution results of C-A-S-H with Al/(Al + Si) ratio of 0.2. **a** Before cryogenic attack, **b** after cryogenic attack

Table 6.4 The fractions (%) of different silicate tetrahedral sites and the crosslinked chain fraction in high-Al C-A-S-H samples

Al/(Al + Si)	Ca/(Al + Si)	$I_{Q2p(Al(VI))}$ (~– 77.0 ppm)	I_{Q1} (~– 79.5 ppm)	$I_{Q2p(Al)}$ (~– 81.5 ppm)	I_{Q2H} (~– 82.5 ppm)	$I_{Q2(Ca)}$ (~– 83.5 ppm)	I_{Q2i} (~– 84.3 ppm)	I_{Q2p} (~– 85.2 ppm)	$I_{Q3(Al)}$ (~– 92.0 ppm)	$I_{Q3(OH)}$ (~– 97.0 ppm)	d
0.2	C/AS0.84B[2]	1.88	36.34	7.54	5.33	8.96	3.96	35.99	0.00	0.00	0.00
	C/AS0.84A[2]	2.98	33.95	6.59	4.38	11.36	2.92	37.82	0.00	0.00	0.00
	C/AS1.25B	4.60	29.44	30.09	1.83	5.93	4.43	23.67	0.00	0.00	0.00
	C/AS1.25A	7.00	37.89	21.50	1.10	3.76	6.35	22.41	0.00	0.00	0.00
	C/AS1.50B	7.09	30.31	31.76	2.58	0.61	7.29	20.36	0.00	0.00	0.00
	C/AS1.50A	3.99	34.46	25.60	0.97	1.61	9.63	23.74	0.00	0.00	0.00
	C/AS1.75B	3.95	35.87	20.48	7.34	3.67	2.32	26.36	0.00	0.00	0.00
	C/AS1.75A	6.16	46.10	12.22	4.08	5.18	2.59	23.67	0.00	0.00	0.00

Note [1]. d is the fraction of crosslinked chain in C-A-S-H. [2]. B represents before the cryogenic attack, A represents after the cryogenic attack

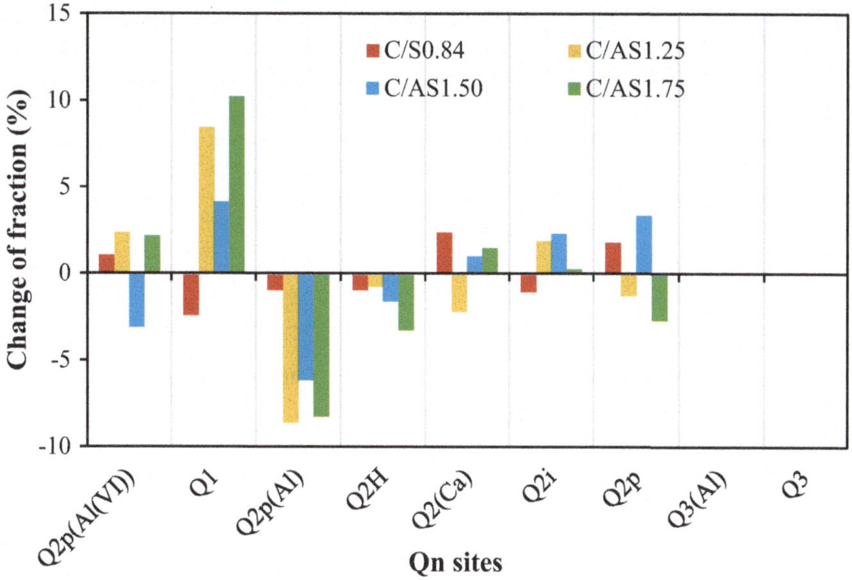

Fig. 6.14 The variations of Si environments after the cryogenic attack with Al/(Al + Si) = 0.2. The values are calculated by I_{after}—I_{before} (I_{after} and I_{before} are the fractions of silicate species before and after the attack, respectively) for each site

more end-chain units. The subsequent decrease of pairing Q^2p(Al) fraction further indicates that the number of Al-substituted bridges decreased after the cryogenic attack. In other words, the initial bridging aluminates present in the dreierketten chains were partially released from the chains, and the associated Q^2p(Al) units were converted to end-chain Q^1 units. The released isolated aluminates were probably incorporated as a part of interlayer species. Our investigation of the ^{27}Al NMR spectra will provide more evidence for this transition (see Sect. 6.3.8). Notably, the nanostructure of the C-A-S-H phase with a low Ca/(Al + Si) ratio of 0.84 did not experience these extreme variations, with fractions of Q^n species almost constant under the cryogenic attack. This C-A-S-H phase shows the best cryogenic stability among the Al-rich C-A-S-H precipitates studied here. We may deduce that the short- or long-range ordering may also contribute to the enhanced stability of C-(A)-S-H with Ca/Si between 0.84 and 1.0.

6.3.6 Al Resonance Assignment in C-A-S-H

Two typical ^{27}Al NMR spectra for crosslinked and non-crosslinked C-A-S-H are displayed in Fig. 6.15, which shows all Al resonances decomposed in this study. There are three domains shown in the spectra centered at approximately 70 ppm, 30 ppm,

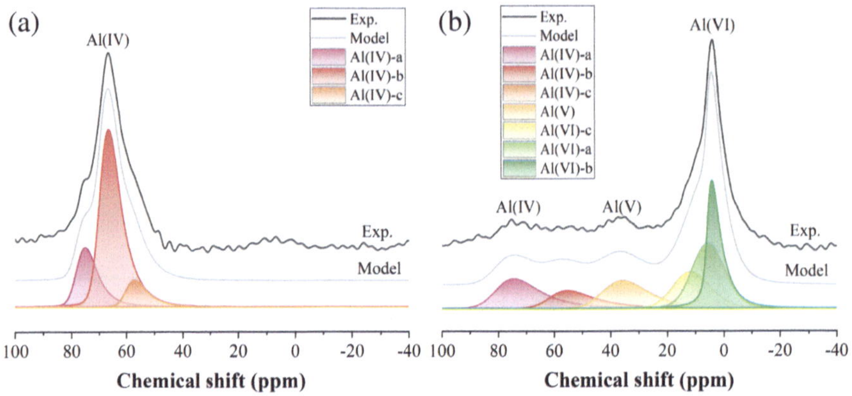

Fig. 6.15 Typical deconvolution for ^{27}Al NMR spectra of crosslinked C-A-S-H (Al/(Al + Si) = 0.1, Ca/(Al + Si) = 0.84) (**a**) and high-Al non-crosslinked C-A-S-H (Al/(Al + Si) = 0.2, Ca/(Al + Si) = 1.75) with katoite (**b**)

and 5 ppm, attributed to tetrahedral (AlO_4, Al(IV)), pentahedral (AlO_5, Al(V)), and octahedral (AlO_6, Al(VI)) aluminates, respectively. The tetrahedral aluminate has a broad resonance from 40 to 80 ppm, and this is ascribed to the presence of varying disordered Al(IV) forms. Three refined resonances at ~ 55 ppm, ~ 65 ppm, and ~ 75 ppm are decomposed from the tetrahedral aluminate domain, labeled as Al(IV)-c, Al(IV)-b, Al(IV)-a, respectively. The broad resonance at around 55 ppm is also observed in other ^{27}Al NMR studies [34, 76–79], and the latter two peaks were reported in Refs. [34, 37, 38, 72, 80]. Herein, these tetrahedral Al resonances are all assigned to the bridging Al-tetrahedra in aluminosilicate chains of C-A-S-H. While we are not going to identify them explicitly here, further discussion will be presented later.

A general assignment to Al(V) is attributed as an interlayer species replacing an interlayer cation Ca^{2+} [34, 37, 81]. Although the latest theoretical study indicates that bridging Al(V) can be present in the aluminosilicate chains of C-A-S-H [43], there is no experimental evidence so far to support this point of view. Herein, it is assigned to a single Al(V) without further deconvolution, which, therefore, includes both interlayer and bridging pentahedral aluminates (if any). Regarding the Al(VI) resonance region, three refined resonances Al(VI)-a, Al(VI)-b, and Al(VI)-c are decomposed from the spectra. Al(VI)-c at ~ 14 ppm is assigned to octahedral Al in katoite as reported previously, which is only observed in three Ca-rich and Al-rich precipitates, consistent with previous XRD and TGA observations. The correlated signals observed for ^{27}Al at 9 ppm in 2D ^{27}Al{^{1}H} HETCOR measurement by Aditya et al. [82] indicated the presence of Al^{VI}-OH species in the C-A-S-H phase. A similar signal at 7 ppm was also observed for the hydrated oilwell cement, which was also associated with Al(VI) species in C-A-S-H or aluminoferrite [83]. This signal is resolved in our measurement and assigned to Al(VI)-a at ~ 9 ppm. As summarized before, Al(VI)-b at ~ 5 ppm could be the bridging octahedral aluminate [43] as well as the interlayer species [37, 59], analogously to Al(V) assigned before.

6.3.7 Aluminate in Low-Al System

The ^{27}Al NMR spectra for each precipitate with a low Al/(Al + Si) ratio of 0.1 are displayed in Fig. 6.16. Deconvolutions of these spectra are summarized in Table 6.5. The shapes of these spectra vary significantly with the Ca/(Al + Si) ratio. For the crosslinked C-A-S-H with Ca/(Al + Si) = 0.84, the spectrum displays a broad high-intensity Al(IV) resonance only. This reveals a key feature: the incorporated Al is merely allowed to exist in tetrahedral coordination in the aluminosilicate chains of crosslinked C-A-S-H. The broad Al(IV) resonance region is due to branching tetrahedral Al, and Al(IV)-c was observed in this low-Ca precipitate only. It is noteworthy that Al(IV)-b represents an overwhelming fraction of over 70% of all Al-tetrahedra. At higher Ca/(Al + Si) ratio ($\geq$ 1), Al(IV), Al(V), and Al(VI) resonances are all present. The predominantly high intensity of Al(VI) resonance suggests that Al was mainly octahedral-coordinated in those high-Ca precipitates. Al(VI)-c was not observed for all the low-Al precipitates, indicating the inexistence of katoite, which is consistent with XRD and TGA observations. Al(V) has a low intensity for all precipitates with varying Ca/(Al + Si) ratios.

Figure 6.17 depicts the fractional change upon cryogenic attack of each type of aluminate. After the cryogenic attack, the changes in most ^{27}Al NMR spectra are

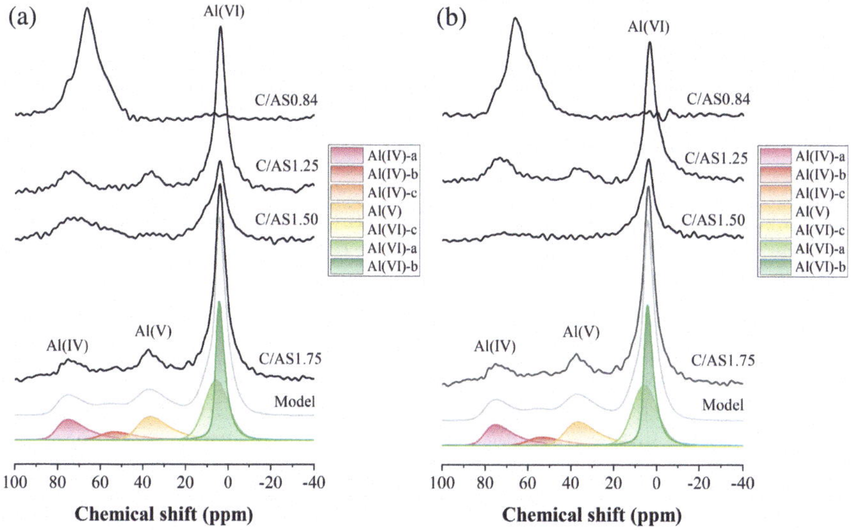

Fig. 6.16 ^{27}Al MAS NMR deconvolution results of C-A-S-H with Al/(Al + Si) ratio of 0.1. **a** Before cryogenic attack, **b** after cryogenic attack

Table 6.5 The fractions (%) of Al in synthetic solids in low-Al C-A-S-H samples

Al/(Al + Si)	Ca/(Al + Si)	$I_{Al(IV)\text{-}a}$	$I_{Al(IV)\text{-}b}$	$I_{Al(IV)\text{-}c}$	$I_{Al(V)}$	$I_{Al(VI)\text{-}a}$	$I_{Al(VI)\text{-}b}$	$I_{Al(VI)\text{-}c}$
0.1	C/S0.84B	19.17	70.47	10.36	0.00	0.00	0.00	0.00
	C/S0.84A	17.68	71.60	10.73	0.00	0.00	0.00	0.00
	C/S1.25B	6.49	7.92	0.00	9.10	21.41	55.09	0.00
	C/S1.25A	6.58	11.15	0.00	6.97	20.22	55.08	0.00
	C/S1.50B	26.42	11.00	0.00	4.29	32.73	25.55	0.00
	C/S1.50A	7.84	5.46	0.00	0.00	32.27	54.43	0.00
	C/S1.75B	14.10	6.07	0.00	16.80	31.99	31.04	0.00
	C/S1.75A	14.89	3.60	0.00	12.63	32.25	36.63	0.00

modest, and little redistribution can be found in the portions of aluminate species. Significant variations are only seen for the precipitate with Ca/(Al + Si) = 1.5. In this system, the intensity of tetrahedral Al(IV)-a, Al(IV)-b, and pentahedral Al(V) decreases by ~ 19%, ~ 5.5%, and ~ 4.5%, respectively, while the intensity of octahedral Al(VI)-c increases significantly by ~ 29%. Apparently, the tetrahedral and pentahedral aluminates are partly converted to octahedral aluminates under the cryogenic attack. For other precipitates, the changes in aluminates composition are much more minor, particularly for the crosslinked C-A-S-H system (with Ca/(Al + Si) = 0.84) in which the proportion of each aluminate is negligibly altered, which confirms the superior cryogenic stability for this phase. Note that the stable proportions of different aluminates in the low-Ca system can only reflect the stability of aluminates and not that of the dreierketten chains. A transformation of a smaller magnitude of Al(IV)-b and Al(V) to Al(VI)-b is also observed for the high-Ca precipitate (with Ca/(Al + Si) = 1.75).

6.3.8 Aluminate in High-Al System

The ^{27}Al NMR spectra for the high-Al precipitates with Al/(Al + Si) = 0.2 before and after the cryogenic attack are displayed in Fig. 6.18, and the deconvolution results for the intensities of Al resonances are given in Table 6.6. Similar to low-Al precipitates, the incorporated Al is mainly tetrahedral at a Ca/(Al + Si) ratio below 1 and octahedral at a Ca/(Al + Si) ratio above 1. However, all the spectra exhibit Al(IV), Al(V), and Al(VI) resonances, even for the low-Ca system. The shapes of these spectra of high-Al precipitates differ slightly from those of low-Al precipitates shown in Fig. 6.16. For the precipitate with Ca/(Al + Si) = 0.84 Al/(Al + Si) = 0.2 (non-crosslinked), the Al(IV) resonance region has two prominent peaks, which are related to the Al(IV)-a and Al(IV)-b resonances. The corresponding deconvolution results in Table 6.6 reveal that Al(IV)-a and Al(IV)-b account for comparable proportions of ~ 44% and ~ 30%, respectively. These observations are

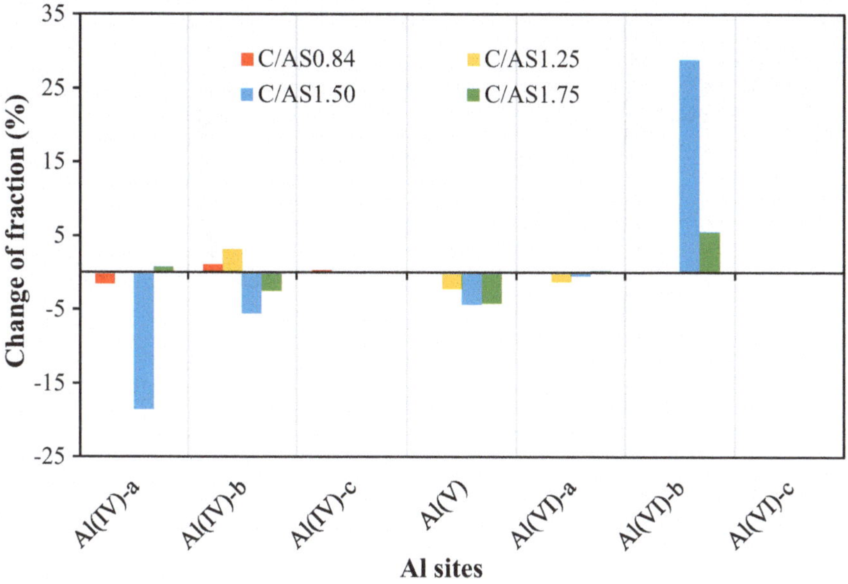

Fig. 6.17 The variations of Al environments after the cryogenic attack with Al/(Al + Si) = 0.1. The values are calculated by $I_{after} - I_{before}$ (I_{after} and I_{before} are the fractions of aluminate species before and after the attack, respectively) for each site

significantly different from the crosslinked C-A-S-H at low Al/(Al + Si), which has a predominant proportion of Al(IV)-b and a small proportion of Al(IV)-a and Al(IV)-c. In addition, the presence of katoite broadens the Al(VI) resonance from ~ 20 ppm to ~ − 20 ppm. Apart from the low-Ca system, Al(VI)-c resonance related to katoite is observed for all these precipitates.

The changes upon cryogenic attack of the different aluminates are displayed in Fig. 6.19. As before, the low-Ca precipitates (with Ca/(Al + Si) = 0.84) exhibit negligible changes, showing excellent stability under cryogenic attack. For other high-Ca precipitates, a common trend can be identified: the fraction of Al(V) decreases significantly, Al(IV) (Al(IV)-a and Al(IV)-b) decreases slightly, while Al(VI)-b increases significantly. Note that the decrease (Al(V) + Al(VI)) generally compensates for the increase (Al(VI)-b), which provides evidence for the conversion from pentahedral and tetrahedral aluminates to octahedral aluminates under the cryogenic attack. The conversion mainly occurs between Al(V) and Al(VI)-b. Combined with the ^{29}Si NMR results, these conversions can be understood better, which will be discussed later. The decomposed intensities of Al(VI)-c for katoite have the same values after the cryogenic attack, indicating that the amount of katoite was unchanged. Katoite stability confirms that our calculation for the interlayer water content proposed in Sect. 6.3.3 is valid.

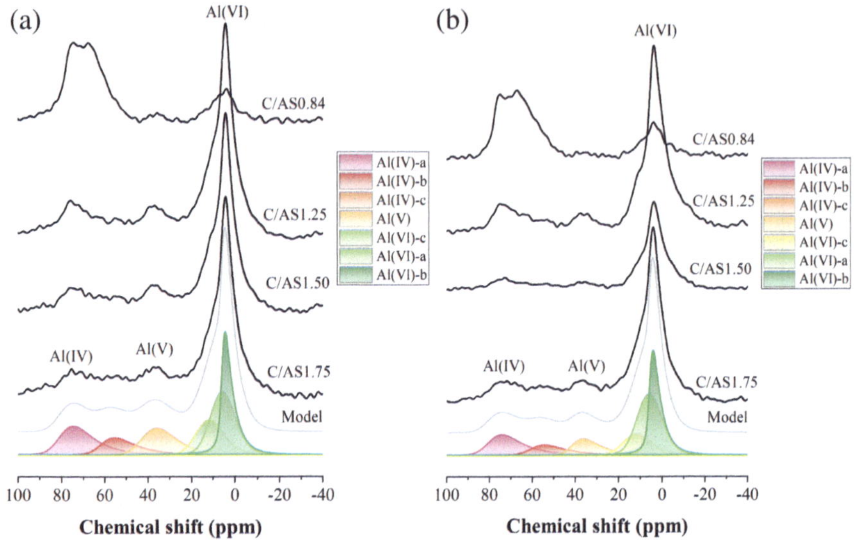

Fig. 6.18 ^{27}Al MAS NMR deconvolution results of C-A-S-H with Al/(Al + Si) ratio of 0.2. **a** Before cryogenic attack, **b** after cryogenic attack

Table 6.6 The fractions (%) of Al in synthetic solids in high-Al C-A-S-H samples

Al/(Al + Si)	Ca/(Al + Si)	$I_{Al(IV)\text{-}a}$	$I_{Al(IV)\text{-}b}$	$I_{Al(IV)\text{-}c}$	$I_{Al(V)}$	$I_{Al(VI)\text{-}a}$	$I_{Al(VI)\text{-}b}$	$I_{Al(VI)\text{-}c}$
0.2	C/S0.84B	43.63	29.68	3.37	3.38	9.21	10.73	0.00
	C/S0.84A	43.29	30.89	4.84	1.61	7.36	12.01	0.00
	C/S1.25B	15.38	4.07	0.00	12.78	15.25	35.86	16.66
	C/S1.25A	15.17	3.38	0.00	6.20	14.40	44.33	16.52
	C/S1.50B	12.29	4.72	0.00	12.66	30.45	32.06	7.82
	C/S1.50A	10.36	2.66	0.00	4.68	30.43	44.50	7.37
	C/S1.75B	12.82	7.68	0.00	13.69	26.91	26.26	12.63
	C/S1.75A	13.66	7.31	0.00	10.01	27.69	28.94	12.38

6.3.9 Evolution of Aluminosilicate Chain Under Cryogenic Attack

From the deconvolution results, quantitative analysis of the aluminosilicate chains is possible, such as the mean chain length (MCL), the fraction of vacant position, the fraction of aluminum substitution, and the Al/Si ratio. The mean chain length represents the average total number of Si-tetrahedra and aluminate in a single aluminosilicate chain along ***b*** axis. It can be calculated using Eq. 6.3 [84].

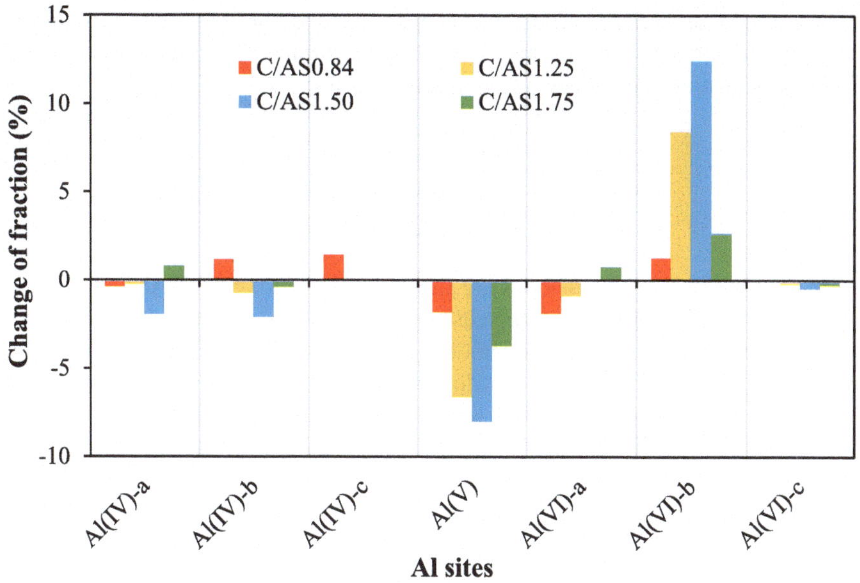

Fig. 6.19 The variations of Al environments after the cryogenic attack with Al/(Al + Si) = 0.2. The values are calculated by $I_{after} - I_{before}$ (I_{after} and I_{before} are the fractions of aluminate species before and after the attack, respectively) for each site

$$\mathrm{MCL} = \frac{2\left[\mathrm{Q}^1 + \mathrm{Q}^2(0\mathrm{Al}) + \frac{3}{2}\mathrm{Q}^2(1\mathrm{Al}) + \mathrm{Q}^3(0\mathrm{Al}) + \mathrm{Q}^3(1\mathrm{Al})\right]}{\mathrm{Q}^1} \tag{6.3}$$

Here, it can be further developed as Eq. 6.4.

$$\mathrm{MCL} = \frac{2\left[\begin{array}{c}\mathrm{Q}^1 + \mathrm{Q}^2\mathrm{H} + \mathrm{Q}^2\mathrm{Ca} + \mathrm{Q}^2\mathrm{i} + \frac{3}{2}\left[\mathrm{Q}^2\mathrm{p(Al)} + \mathrm{Q}^2\mathrm{p(Al(VI))}\right] \\ + \mathrm{Q}^2\mathrm{p} + \mathrm{Q}^3\mathrm{(OH)} + \mathrm{Q}^3\mathrm{(Al)}\end{array}\right]}{\mathrm{Q}^1} \tag{6.4}$$

The fraction of vacant sites v can be determined by Eq. 6.5.

$$v = \frac{1/2\mathrm{Q}^1}{\begin{array}{c}3/2\mathrm{Q}^1 + \mathrm{Q}^2\mathrm{H} + \mathrm{Q}^2\mathrm{Ca} + \mathrm{Q}^2\mathrm{i} + \frac{3}{2}\left[\mathrm{Q}^2\mathrm{p(Al)} + \mathrm{Q}^2\mathrm{p(Al(VI))}\right] \\ + \mathrm{Q}^2\mathrm{p} + \mathrm{Q}^3\mathrm{(OH)} + \mathrm{Q}^3\mathrm{(Al)}\end{array}} \tag{6.5}$$

The fraction of aluminum substitution in aluminosilicate chains f can be determined using Eq. 6.6.

$$f = \frac{1/2\left[\mathrm{Q}^2\mathrm{p(Al)} + \mathrm{Q}^2\mathrm{p(Al(VI))}\right]}{\begin{array}{c}3/2\mathrm{Q}^1 + \mathrm{Q}^2\mathrm{H} + \mathrm{Q}^2\mathrm{Ca} + \mathrm{Q}^2\mathrm{i} + \frac{3}{2}\left[\mathrm{Q}^2\mathrm{p(Al)} + \mathrm{Q}^2\mathrm{p(Al(VI))}\right] \\ + \mathrm{Q}^2\mathrm{p} + \mathrm{Q}^3\mathrm{(OH)} + \mathrm{Q}^3\mathrm{(Al)}\end{array}} \tag{6.6}$$

It is noted that the Al/Si ratio calculated according to the ^{29}Si NMR decompositions refers explicitly to the relative intensity of substituted Al and silicate tetrahedra in aluminosilicate chains. Herein, it is specified as Eq. 6.7.

$$Al_{bridging}/Si = \frac{1/2\left[Q^2p(Al) + Q^2p(Al(VI))\right]}{Q^1 + Q^2H + Q^2Ca + Q^2i + Q^2p(Al) + Q^2p(Al(VI)) + Q^2p + Q^3(OH) + Q^3(Al)} \quad (6.7)$$

The results of these calculations are shown in Fig. 6.20. As Al/(Al + Si) = 0.1, the mean chain length tends to decrease as the Ca/(Al + Si) ratio increases from 0.84 to 1.75. Meanwhile, the fraction of vacant sites increases. Compared with the non-crosslinked precipitates, the crosslinked C-A-S-H (Ca/(Al + Si) = 0.84, Al/(Al + Si) = 0.1) exhibits a significantly longer aluminosilicate chain and fewer vacant sites. As Al/(Al + Si) increases to 0.2, the non-crosslinked C-A-S-H phase shows a longer mean aluminosilicate chain length, while the original crosslinked C-A-S-H is converted to a non-crosslinked phase with the MCL value decreasing from 11.60 to 5.76. From Fig. 6.20c, d, for the precipitates with a low Ca/(Al + Si) of 0.84, a higher Al content does not lead to a higher Al substitution fraction or a higher $Al_{bridging}/Si$ ratio, which confirms that Al uptake is less favorable for a low Ca/(Al + Si) system. However, for the precipitates with a higher Ca/(Al + Si) above 1, Al uptake in dreierketten chains is more favorable as the Al/(Al + Si) ratio increased from 0.1 to 0.2. The corresponding Al substitution fraction and $Al_{bridging}/Si$ ratio show a significant increase. It is noted that the Al substitution fraction and $Al_{bridging}/Si$ ratio are lower than the target bulk values, suggesting that Al was not completely incorporated in the aluminosilicate chains.

The quantitative results in Fig. 6.20 reveal clearer evolutions of the aluminosilicate chains upon cryogenic attack (see plain lines versus dashed lines in Fig. 6.20). The aluminosilicate chains of C-A-S-H with Al/(Al + Si) = 0.1 exhibit superior stability than those with Al/(Al + Si) = 0.2. As inferred before, the mean chain length of crosslinked C-A-S-H (Ca/(Al + Si) = 0.84, Al/(Al + Si) = 0.1) slightly declined from 11.60 to 9.64 due to the breakage of silicate chains. For these low Al/(Al + Si) C-A-S-H phases (Ca/(Al + Si) > 1.25), it is interesting to observe that they show a slightly higher Al substitution fraction and $Al_{bridging}/Si$ ratio after the cryogenic attack, which is due to the increased amount of pairing $Q^2(Al)$ ($Q^2p(Al)$ + $Q^2p(Al(VI))$) units. When the Al/(Al + Si) increases to 0.2, the C-A-S-H precipitates exhibit shorter aluminosilicate chains and more vacant sites after cryogenic attack, except for that with Ca/(Al + Si) = 0.84. Meanwhile, Al substitution fraction and $Al_{bridging}/Si$ ratio are decreased, which can be attributed to the release of bridging aluminates. Overall, we can safely deduce from these observations that C-A-S-H nanostructure has better cryogenic stability with a low Al/(Al + Si) ratio and/or a low Ca/(Al + Si) ratio.

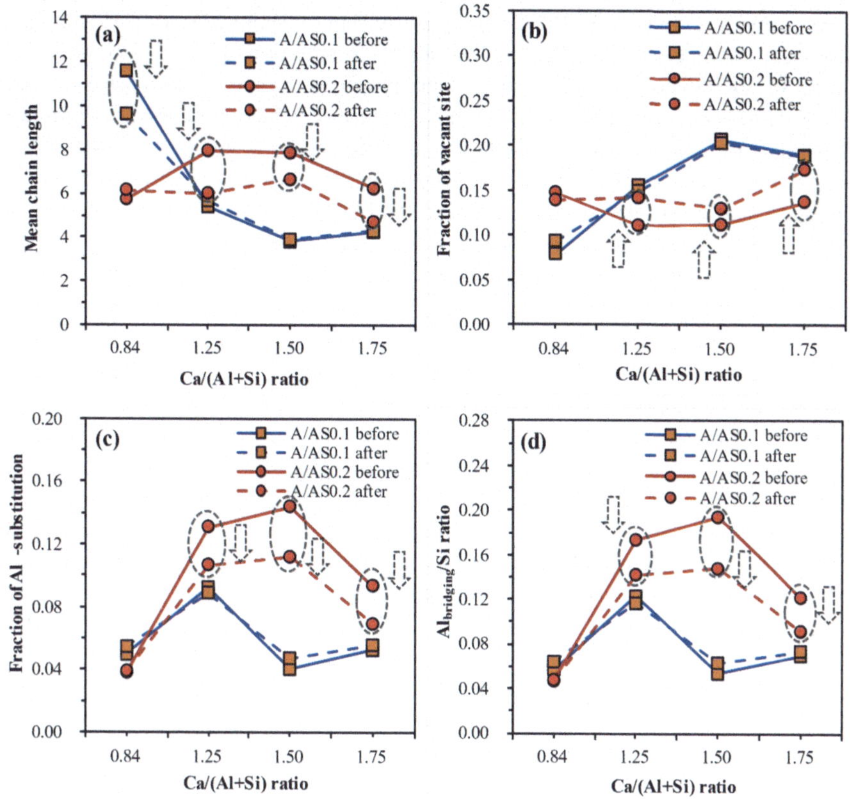

Fig. 6.20 The mean chain length (**a**), the fraction of vacant sites (**b**), the fraction of Al substitutions (**c**), and bridging Al-structures to Si-tetrahedra ratio (**d**) along the aluminosilicate chains. These results are calculated from the ^{29}Si MAS decomposed results

6.4 Extended DNA Code Rule for C-A-S-H

6.4.1 Tetrahedral Aluminate: Bridging Forms

It seems clear that Al(IV) is predominant in the low-Ca C-A-S-H system, which is confirmed by our ^{27}Al NMR measurements and a previous study [81]. Our ^{27}Al NMR results further suggest that the incorporated Al in crosslinked C-A-S-H networks can only be coordinated in a tetrahedral form, predominantly in an Al(IV)-b complex. Another general view of Al(IV) is that it is incorporated as a part of aluminosilicate chains of C-A-S-H and not as an interlayer species [38]. What remains unclear is the form that Al(IV) takes in the chains. The three resonance peaks, i.e., Al(IV)-a at ~ 75 ppm, Al(IV)-b at ~ 65 ppm, and Al(IV)-c at ~ 55 ppm, must exist in distinctive forms with different complexes. In the C-A-S-H brick model [43], Al(IV) is formed by adding AlO(OH) unit after the removal of bridging SiO_2.

As shown previously, Al(IV)-c is only present in crosslinked C-A-S-H samples. This Al resonance was also detected in other investigations [34, 76–79]. We note that these observations [34, 76–79] are mainly targeted towards C-(N-)A-S–H and N-A-S–H gels contained in alkali-activated materials, which are known to be more crosslinked networks than C-(A-)S-H gels in ordinary Portland cement pastes [79]. For a synthetic C-A-S-H solid, the Al(IV)-c signal at 55 ppm is clearly observed for a more crosslinked system at a low Ca/Si of 0.6 [34]. Combined with our ^{27}Al NMR results in Sect. 6.3.6, it is rational to assign this signal to the tetrahedral aluminates at branching sites linked to Q^3 Si-tetrahedra, which is present in crosslinked C-A-S-H only. According to Kunhi Mohamed et al. [43], this branching Al(IV)-c aluminate forms the complex $[AlO_3(OH)]^{4-}$.

The Al resonances Al(IV)-b at ~ 65 ppm and Al(IV)-a at ~ 75 ppm are also observed or decomposed for C-A-S-H in previous ^{27}Al NMR measurements [34, 37, 38, 72, 80]. There are essentially three different points of view to assign those two resonances: (1) Al resonance at ~ 65 ppm is assigned to bridging Al(IV), and Al resonance at ~ 75 ppm is associated with pairing Al(IV) [38]; (2) the former ~ 65 ppm peak is related to bridging Al(IV), and the latter ~ 75 ppm peak is attributed to bridging Al(IV) charge-compensated by surface/interlayer pentahedral and octahedral aluminates [34, 72]; (3) both of them are attributed to bridging Al(IV) but charge-compensated by H^+ and Ca^{2+} [72], respectively, analogously to Si-tetrahedra as stated in the preceding analysis. Since pairing aluminate is energetically unfavored and has not made a consensus, as stated before, the first assignment is excluded here. The mismatched proportions of Al(V) + Al(VI) and Al(IV)-b indicate that the second assignment is not correct, either. Regarding the third assignment, the latest DFT study of C-A-S-H atomic-level structure [43] can provide additional evidence for the shielding effect of H^+ and Ca^{2+} on the bridging Al(IV) aluminates. Generally, the small H^+ cations charge-compensating the aluminates have relatively small effects on the spherical symmetry of the electric field around the nucleus, leading to a lower frequency of Al resonance. In comparison, the large cationic species lead to a higher frequency [85–87]. Herein, the Al(IV)-a at the high chemical shift of ~ 75 ppm are assigned to tetrahedral aluminates $[AlO_3(OH)]^{4-}$ at bridging sites of dreierketten chains which are neighboring the large cations, e.g., Ca^{2+}, and the Al(IV)-b at a relatively lower chemical shift of ~ 65 ppm is attributed to the bridging tetrahedral aluminates $[AlO_2(OH)_2]^{3-}$ charge-balanced by small cations only, e.g., H^+. According to their DFT calculations, the bridging tetrahedral $[AlO_4]^{5-}$ unit constructed by Kunhi Mohamed et al. [43] at a high Ca/Si ratio of 1.75 has a high chemical shift of 79 ppm. This aluminate was not observed in our ^{27}Al NMR spectra nor in the previous measurements [34, 37, 38, 72, 80]. It is inferred that tetrahedral aluminates $[AlO_4]^{5-}$ may not exist in C-A-S-H since they are energetically less favorable than other complexes [43].

6.4.2 Pentahedral Aluminate: Bridging and Interlayer Forms

Al resonance at around 38 ppm is usually assigned to Al(V). According to the calculation done by Kunhi Mohamed et al. [43], the energy of Al(V) is always at the intermediate of Al(IV) and Al(VI). For this reason, it is generally considered that the Al(V) species are present in the C-A-S-H at any given Ca/Si ratio. However, our results indicate that Al(V) is not present in the crosslinked low-Ca C-A-S-H. Al(V) is generally considered to exist in the interlayer space of C-A-S-H rather than at the bridging sites [37, 88], and the formation of this Al species originates from the substitution of Al for Ca present in the interlayer [36, 55, 56]. However, a recent theoretical study [43] reveals that bridging pentahedral aluminates are present in C-A-S-H as well, which has not been experimentally proved so far. Fortunately, the variations of C-A-S-H nanostructure detected by our ^{29}Si and ^{27}Al NMR spectra may provide some evidence.

As indicated by ^{29}Si NMR results, the low-Al high-Ca C-A-S-H phases have slightly higher mean chain length, Al substitution ratio, and $Al_{bridging}$/Si ratio after the cryogenic attack (see Fig. 6.20), accompanied by a slightly higher fraction of Q^2p(Al(VI)) (see Fig. 6.12). These observations suggest that the corresponding aluminosilicate chains are incorporated with more aluminates. Interestingly, ^{27}Al NMR shows, for the same samples, that the fraction of Al(V) decreases slightly, while the fraction of Al(VI)-b increases (see Fig. 6.17). Therefore, this suggests that some interlayer Al(V) are converted to bridging octahedral aluminates (which will be discussed later) occupying sites in the dreierketten chains that were previously vacant. For the high-Al C-A-S-H precipitates (with Al/(Al + Si) = 0.2, Ca/(Al + Si) > 1), the significant decrease in mean chain length and Q^2p(Al) fraction observed by ^{29}Si NMR reveals that the aluminosilicate chains are ruptured, releasing the bridging aluminates from dreierketten chains under the cryogenic attack. Moreover, the significant decrease of Al(V) is followed by the increase of Al(VI)-b, as shown in ^{27}Al NMR results, which indicates that the released bridging aluminates are mainly pentahedral ones and are further converted to octahedral aluminates. One should note that both the internal energy and entropy of separated interlayer aluminates are higher than those of bridging aluminates [43]. When considering the contribution of both internal energy and entropy, it is still possible that some interlayer aluminates may be more stable than bridging ones. This is probably the reason for the observed conversion from released Al(V) to Al(VI). These observations provide evidence for the presence of bridging pentahedral aluminates in aluminosilicate chains of C-A-S-H.

As for the complex related to the interlayer and bridging Al(V), Kunhi Mohamed et al. [43] proposed a $[AlO(OH)_4]^{3-}$ complex for interlayer Al(V), and two complexes, i.e., $[AlO_3(OH)_2]^{5-}$ and $[AlO_2(OH)_3]^{4-}$, for the bridging Al(V). The calculated chemical shift is 37 ppm for the interlayer Al(V) $[AlO(OH)_4]^{3-}$, 37 ppm for the bridging Al(V) $[AlO_3(OH)_2]^{5-}$, and 40–46 ppm for the bridging Al(V) $[AlO_2(OH)_3]^{4-}$. The latter chemical shift is somehow higher than our ^{27}Al NMR measurements. Accordingly, we exclude the $[AlO_2(OH)_3]^{4-}$ complex and consider that bridging pentahedral aluminate corresponds to the $[AlO_3(OH)_2]^{5-}$ complex, while the $[AlO(OH)_4]^{3-}$ complex is considered for the interlayer pentahedral aluminate.

6.4.3 Octahedral Aluminate: Bridging and Interlayer Forms

The Al(VI)-b signal at 5 ppm observed by ^{27}Al NMR was previously assigned to a new silica-free phase, i.e., the so-called "third aluminate hydrate" (TAH). The latest study [43] contradicts this point of view by performing DNP NMR experiments and DFT calculations, revealing that the TAH is a bridging Al(VI) species of C-A-S-H dreierketten chains. When the Al substitution occurs at the bridging sites of C-A-S-H with a high Ca/Si ratio, it is suggested that both Al(V) and Al(VI) can be found, even with lower energies than Al(IV) [43]. The nanostructure variations of low-Al C-A-S-H (Al/(Al + Si) = 0.1) and high-Al C-A-S-H (Al/(Al + Si) = 0.2) observed in this study indicate the presence of bridging Al(VI) and interlayer Al(VI), respectively.

The ^{29}Si NMR spectra show that the aluminosilicate chains of low-Al C-A-S-H are generally stable under the cryogenic attack. The fractional changes (see Fig. 6.14) for almost all the Q^n species are insignificant, including Q^2p(Al) and Q^2p(Al(VI)) tetrahedra, which stabilizes the mean chain length and $Al_{bridging}$/Si ratio (see Fig. 6.20). These observations suggest that bridging aluminate is present in the aluminosilicate chains and the skeleton remains stable under cryogenic attack. However, a significant conversion (over 15%) from Al(IV)-a to Al(VI)-b was observed for the C-A-S-H with Ca/(Al + Si) = 1.5, see Fig. 6.17. As we have discussed before, the tetrahedral aluminates Al(IV) belong to bridging units, while such a conversion did not lead to shorter aluminosilicate chains and lower $Al_{bridging}$/Si ratio. Therefore, the octahedral aluminates Al(VI)-b formed must belong to bridging units as well. The increase of Q^2p(Al(VI)) in the C-A-S-H sample with Ca/(Al + Si) = 1.5, Al/(Al + Si) = 0.1 further confirms this deduction (see Fig. 6.12).

While unlikely for low-Al C-A-S-H, the aluminosilicate chains of high-Al C-A-S-H are ruptured at the bridging positions occupied by aluminates, leading to a shorter chain length and a lower $Al_{bridging}$/Si ratio, as shown in Figs. 6.14 and 6.20. The decrease of Q^2p(Al) and increase of Q^1 tetrahedra both confirm the release of bridging aluminates from aluminosilicate chains. The changes in aluminate species shown in Fig. 6.19 further reveal that the released bridging aluminates are mainly pentahedral units Al(V) converted to octahedral units Al(VI)-b. Therefore, the Al(VI)-b formed must belong to an interlayer species out of the skeleton of aluminosilicate chains. From the above discussion, our results suggest the presence of both bridging and interlayer aluminates, with Al resonances assigned at ~ 5 ppm.

Regarding the atomistic structure of bridging Al(VI) in C-A-S-H, three possible complexes, i.e. $[AlO_2(OH)_4]^{5-}$, $[AlO_2(OH)_{(4-x)}(H_2O)_x]^{5-x}$, and $[AlO(ØH)(OH)_4]^{6-}$ (ØH represents a Brønsted acid hydroxysilicate bond, Al – OH – Si), were theoretically studied in [43]. Comparing the energies and ^{27}Al chemical shifts, it is suggested that the primary Al(VI) structure in C-A-S-H is $[AlO_2(OH)_4]^{5-}$. The conversion from Al(IV)-a to Al(VI)-b observed in the present study further confirms this conclusion since such a conversion usually follows an equation of the form $[AlO_3OH]^{4-}$ + $CaOH^+ + H_2O \rightleftharpoons [AlO_2(OH)_4]^{5-} + Ca^{2+}$ according to [43]. As stated previously, the bridging Al(IV)-a has a complex of $[AlO_3(OH)]^{4-}$, so it seems reasonable to have a $[AlO_2(OH)_4]^{5-}$ complex for bridging octahedral aluminates. Similarly, since the bridging Al(V) has a complex of $[AlO_3(OH)_2]^{5-}$, the observed conversion from bridging Al(V) to interlayer Al(VI)-b leads to suggest a complex of $[Al(OH)_6]^{5-}$ for interlayer octahedral aluminates. There is another Al(VI) signal at ~ 9 ppm, i.e., Al(VI)-a, resolved for the ^{27}Al NMR spectra. It appears to have superior cryogenic stability since the corresponding fractional changes upon cryogenic attack were insignificant for all C-A-S-H samples. This means that Al(VI)-a is probably more energetically favorable than bridging Al(VI)-b. Based on the simulation [43], Al(VI)-a is probably another form of bridging octahedral aluminate with complex of $[AlO_2(OH)_3(H_2O)]^{4-}$ (relative energy of – 0.27 eV and ^{27}Al chemical shift of 10 ppm).

6.4.4 Extended DNA-Code Rule for C-A-S-H Brick

Kunhi Mohamed et al. [15] developed a DNA-code rule for describing the atomistic structure of Al-free C-S-H brick, which is elegant to generate a C-S-H molecular model. The deconvolutions of ^{29}Si and ^{27}Al NMR spectra presented above offer the possibility to develop a similar DNA code rule to describe C-A-S-H brick for generating a realistic C-A-S-H model. Compared with the DNA-code rule for Al-free C-S-H, the current C-A-S-H DNA-code rule introduces and specifies the bridging and interlayer aluminates in C-A-S-H bricks. The DNA codes, related units, corresponding stoichiometries, and charges are displayed in Fig. 6.21. Herein, Al^{IV}, Al^{V}, and Al^{VI} represent the AlO_2, AlO_3, and AlO_4 for bridging tetrahedral, pentahedral, and octahedral aluminates, respectively, and ${Al_{IL}}^{V}$ and ${Al_{IL}}^{VI}$ denote AlO_5 and AlO_6 for interlayer pentahedral and octahedral aluminates, respectively. Since the geometries of aluminates and silicates have been identified, typical atomistic C-A-S-H bricks that include bridging and interlayer aluminates can be constructed, as shown in Fig. 6.22. Regarding the typical Al-free C-S-H bricks, one can find detailed descriptions in Ref. [15]. Herein, the charge-compensated species are mainly H^+, Ca^{2+}, and $Ca(OH)^+$. $Ca(OH)_2$ units are also considered for the mass balance once the Ca/Si ratio is too high. Figure 6.22a illustrates a typical Al-free C-S-H brick with six silicate tetrahedra, four main layer calcium atoms, several interlayer charge-balanced calcium ions, hydroxyl groups, and water molecules. Figure 6.22b–d display the presence of bridging tetrahedral aluminates. Albeit the bridging $[AlO_4]^{5-}$ complex may not exist in C-A-S-H brick as analyzed before, it is still shown here as a possible complex. Figure 6.22e, f depict the presence of bridging and interlayer pentahedral aluminates, respectively. Figure 6.22g, h show two forms of bridging octahedral aluminates, i.e. $[AlO_2(OH)_4]^{5-}$ and $[AlO_2(OH)_3(H_2O)]^{4-}$, respectively. A brick with interlayer octahedral aluminate is presented in Fig. 6.22i. The last two figures

Code	Unit	Stoichiometry	Charge	Code	Unit	Stoichiometry	Charge
Ca-Si sheet above interlayer				Ca-Si sheet below interlayer			
<		$CaSiO_{3.5}$	-1	>		$CaSiO_{3.5}$	-1
S		SiO_2	0	S		SiO_2	0
A^{IV}		AlO_2	-1	A^{IV}		AlO_2	-1
A^{V}		AlO_3	-3	A^{V}		AlO_3	-3
A^{VI}		AlO_4	-5	A^{VI}		AlO_4	-5
[S/A]'		[SiO/AlO]H	+1	[S/A]'		[SiO/AlO]H	+1
/		$Ca(OH)_2$	0	\		$Ca(OH)_2$	0
(		$CaSiO_4$	-2	)		$CaSiO_4$	-2
S|		$SiO_{1.5}$	+1	|S		$SiO_{1.5}$	+1
A|		$AlO_{1.5}$	0	|A		$AlO_{1.5}$	0
[		$Ca(OH)_{1.5}$	+0.5	]		$Ca(OH)_{1.5}$	+0.5
Interlayer							
C		Ca	+2	°		OH	-1
A_{IL}^{V}		AlO_5	-7	A_{IL}^{VI}		AlO_6	-9
1		H_2O	0	[AO]'		[AlO]H	+1

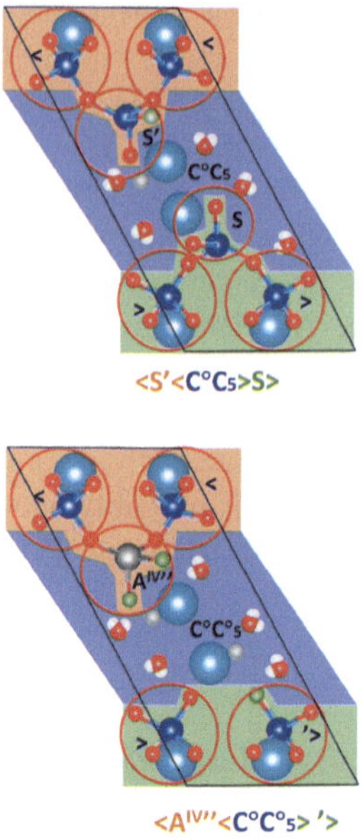

Fig. 6.21 An extended DNA-code description for the C-A-S-H brick model. The table illustrates detailed atomic structures and the corresponding chemical stoichiometry, charges, and codes. This atomistic description for C-A-S-H building blocks is inherited from the C-S-H brick model proposed by Kunhi Mohamed et al. [15]. In addition, we add codes to describe the bridging Al-structures (Al(IV), Al(V), and Al(VI)) and interlayer aluminates (Al(V) and Al(VI)). Two examples on the right side are given for better understanding

describe the crosslinked bricks with branching tetrahedral aluminates and silicates, respectively.

6.4.5 *Extended Characterization of C-A-S-H Model Based on DNA-Code Rule*

Now that we have constructed DNA codes for aluminates in the last section, characterizations such as the stoichiometry and MCL of C-A-S-H can be done using these notations, as Kunhi Mohamed et al. [15] did for C-S-H. Note that the stoichiometry and MCL equations given by Kunhi Mohamed et al. [15] are adapted to non-crosslinked C-S-H only, although they have proposed DNA codes for crosslinked units. Herein, besides introducing the notations of aluminates, we further extended those equations to characterize non-crosslinked and crosslinked C-(A)-S-H. The stoichiometry equation is given as Eq. 6.8, where n_X represents the number of unit X (see Fig. 6.21). In order to be compatible with their original equations, we highlight that n_X is the total number of unit X with different orientations, i.e., that above and

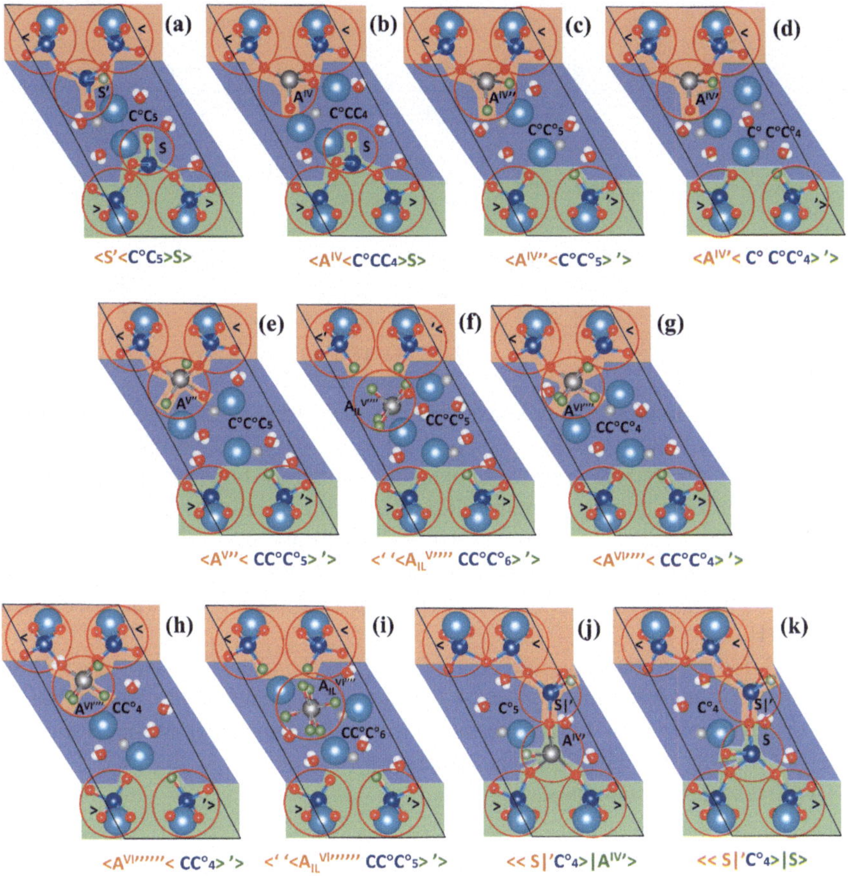

Fig. 6.22 Atomistic descriptions for the representative Al-containing C-A-S-H building blocks. These diagrams are based on the silicate and aluminate species decomposed from ^{29}Si NMR and ^{27}Al NMR spectra. The purpose is to describe the forms of different silicate and aluminate species. Here we list all the possible forms of aluminate species. Then, one can construct the rest of similar building blocks simply by combining these silicate and aluminate species

below the interlayer, e.g., < and >, (and).

$$Ca_{\kappa_1}Al_{\kappa_2}SiO_{\kappa_3} \cdot \kappa_4H_2O \tag{6.8}$$

where,

$$\kappa_1 = \frac{n_< + n_C + n_/ + n_(}{n_< + n_S + n_(+ n_{S|}} \tag{6.9}$$

$$\kappa_2 = \frac{n_{A^{IV}} + n_{A^V} + n_{A^{VI}} + n_{A^V_{IL}} + n_{A^{VI}_{IL}}}{n_< + n_S + n_(+ n_{S|}} \tag{6.10}$$

$$\kappa_3 = \frac{\begin{array}{c}3.5n_C + 2n_S + 4n_{(} + 1.5n_{S|} + 2n_{A^{IV}} + 3n_{A^{V}} \\ +4n_{A^{VI}} + 5n_{A^{V}_{IL}} + 6n_{A^{VI}_{IL}} - 0.5n' + 0.5n_{\circ} + n_{/}\end{array}}{n_{<} + n_S + n_{(} + n_{S|}} \tag{6.11}$$

$$\kappa_4 = \frac{0.5 \cdot \left(n' + n_{\circ}\right) + n_{/} + n_{H_2O}}{n_{<} + n_S + n_{(} + n_{S|}} \tag{6.12}$$

The equation for mean chain length MCL is extended as Eq. 6.13:

$$MCL = \frac{n_{<} + n_S + n_{(} + n_{A^{IV}} + n_{A^{V}} + n_{A^{VI}} + n_{S|} + n_{A|}}{0.5n_{<} - n_S + n_{(}} \tag{6.13}$$

Note that the monomers are considered in the original stoichiometry and MCL equations for C-S-H using DNA-code rule. To reconcile with the original equations, the monomers are also included in our extended equations shown above. As summarized in the Introduction, the aluminosilicate chains are usually present in a multimer form. Silicate monomers are considered in absence from C-(A)-S-H according to the $3n - 1$ rule, which can be confirmed by many NMR experimental studies [34, 89]. When no silicate monomer exists, $n_{(} = 0$, then Eqs. 6.8–6.13 characterize usual C-(A)-S-H.

In addition, we further give the equations for calculating the fraction of vacant site v and the fraction of Al-substitution f, which are not presented in the original model of Kunhi Mohamed et al. [15]. Note that silicate monomers shall not be considered when calculating the two properties. Equation 6.14 and 6.15 are proposed for usual C-(A)-S-H without monomers.

$$v = \frac{0.5 \cdot (n_{<} - 2n_S)}{1.5n_{<} + n_{A^{IV}} + n_{A^{V}} + n_{A^{VI}} + n_{S|} + n_{A|}} \tag{6.14}$$

$$f = \frac{n_{A^{IV}} + n_{A^{V}} + n_{A^{VI}} + n_{A|}}{1.5n_{<} + n_{A^{IV}} + n_{A^{V}} + n_{A^{VI}} + n_{S|} + n_{A|}} \tag{6.15}$$

6.4.6 DNA-Code Description for Main Reactions

With the DNA-code rule described above, the conversions that occurred in C-A-S-H during the cryogenic attack can be clearly illustrated. As analyzed before, there are two primary conversions that occur for low-Al and high-Al C-A-S-H systems. For low-Al C-A-S-H, the apparent conversion from bridging Al(IV)-a to bridging Al(VI)-b was mainly observed for the precipitates with Ca/(Al + Si) = 1.5, which can be illustrated as Fig. 6.23a. Our TGA results indicate that the content of interlayer water increased after the cryogenic attack, which means that a portion of water was absorbed into the interlayer during the cryogenic attack. A plausible reaction is that part of the interlayer water molecule was dissociated into hydroxyl groups and protons, further re-coordinated with bridging aluminum. The bridging Al(IV)-a has a complex of $[AlO_3(OH)]^{4-}$ charge-compensated by two $Ca(OH)^+$ units, while the bridging Al(VI)-b has a complex of $[AlO_2(OH)_4]^{5-}$ charge-compensated by one Ca^{2+} and one $Ca(OH)^+$ unit. Therefore, the conversion from Al(IV) to Al(VI) occurs as Eq. 6.16, which does not change the skeleton of aluminosilicate chains. This water driven inter conversion between the two moieties was first discovered by Kunhi Mohamed [43]. Here we confirm that it can also occur during cryogenic attack.

$$[AlO_3(OH)]^{4-} + CaOH^+ + H_2O \rightleftharpoons \left[AlO_2(OH)_4\right]^{5-} + Ca^{2+} \tag{6.16}$$

For high-Al C-A-S-H, we observed a conversion from bridging Al(V) to interlayer Al(VI)-b, which is illustrated in Fig. 6.23b and Eq. 6.17. The bridging Al(V) has a complex of $[AlO_3(OH)_2]^{5-}$ charge-compensated by Ca^{2+} and $Ca(OH)^+$, while the interlayer Al(VI)-b is constructed as $[Al(OH)_6]^{3-}$ which is charge-balanced by Ca^{2+} and $Ca(OH)^+$ as well. In addition, this conversion leads to the formation of Q^1(OH) silicate tetrahedra. We assume that Eq. 6.17 is irreversible. Note that this reaction shortens the aluminosilicate chains.

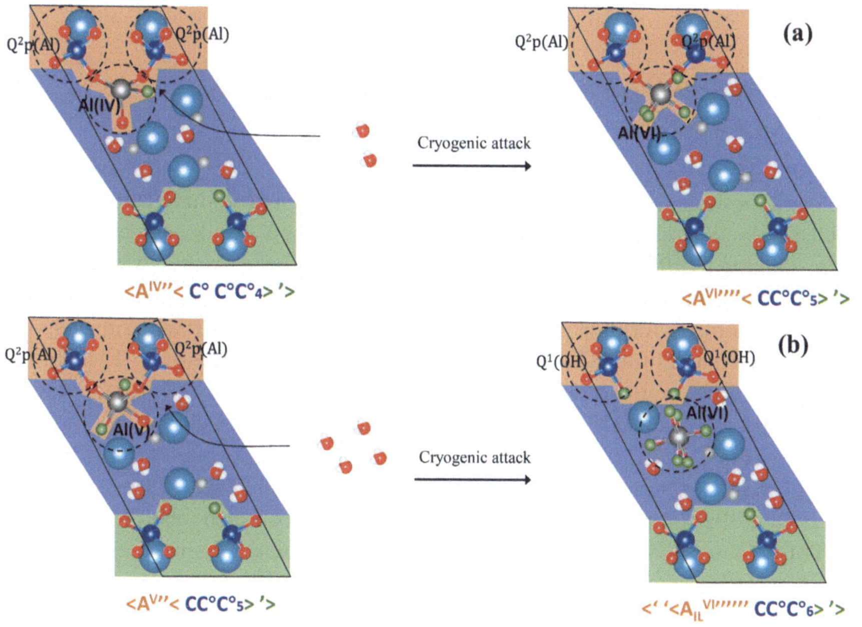

Fig. 6.23 The main changes during the cryogenic attack. **a** The hydration of bridging Al(IV) in low Al/(Al + Si) ratio C-A-S-H. **b** The rupture of Al(V)-structure in aluminosilicate chain and the hydration into interlayer Al(VI)-structure

$$\left[AlO_3(OH)_2\right]^{5-} + 2[CaSiO_3] + 3H_2O \rightarrow \left[Al(OH)_6\right]^{3-} + 2[CaSiO_3(OH)]^- \quad (6.17)$$

It is noted that both reactions produce more hydroxyl groups, consistent with our TGA results for hydroxyl water content shown in Fig. 6.6. However, we admit that the driving force for these reactions is still unclear. Perhaps, the reactions can simultaneously occur in both low-Al and high-Al C-A-S-H systems, while their intensities are associated with the Ca/Si and Al/Si ratio.

6.5 General Structural-Chemical Formula of C-A-S-H

In the previous section, we clarified the specific forms of aluminum salts within C-A-S-H and their corresponding structural configurations. Additionally, we proposed the DNA encoding description rules for the C-A-S-H nanostructure. In this section, we will further present the generalized structural chemical formula of C-A-S-H. We categorize C-A-S-H into three distinct structural forms: cross-linked, non-cross-linked, and hybrid types.

6.5.1 General Structural-Chemical Formula Model of C-A-S-H

Although the stoichiometry of C-A-S-H has already been given in Sect. 6.4.5 based on the DNA-code rule, the number of n_X of each unit type X is unknown in the experiment, making it difficult to apply the model. Herein, with the clear view of chemical species shown in our proposed extended DNA-code rule, we proposed a general structural-chemical formula model which is consistent with NMR experiments.

A General model for C-S-H in hydrated C_3S paste was proposed by Richardson and Groves [18] in 1992, which assumes that silicate chains of C-S-H follow a $3n - 1$ rule and contain $Ca(OH)_2$ units. Considering the presence of Al^{3+} or Fe^{3+} substitution in C-S-H, Richardson and Groves [90] further developed a Substituted General model excluding the so-called AFm-type structure layers from C-A-S-H. Myers et al. [91] proposed the so-called cross-linked substituted tobermorite model (CSTM) to describe the structural model of calcium-sodium aluminosilicate hydrate gel. They made a sound description of the derivation of the generalized model and the associations of Ca/Si, Al/Si, and MCL correlations by combining the non-crosslinked and crosslinked tobermorite structure. This model leads to a possibility for describing the crosslinked C-A-S-H structure. Based on the inspection of a number of C-S-H and related phases, Richardson [84] proposed structural-chemical formulae for single-chain and double-chain C-(A)-S-H(I) phase, which can be used to explain some experimental observations such as mean chain length, layer spacing, H_2O/Si ratio. Al

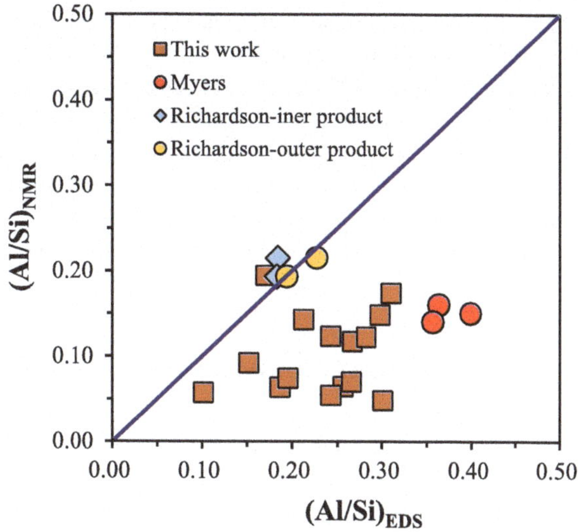

Fig. 6.24 Comparisons for the Al/Si ratio determined by EDS measurement and ^{29}Si MAS NMR analysis. Both Myers's EDS results [91] and ours are determined by ESEM-EDS. Richardson's EDS results [92] are obtained from TEM-EDS measurement

substitution for Si is considered to occur at the bridging site. The charge-compensated interlayer species for substituted Al tetrahedra are Ca^{2+}, Na^{+}, or K^{+}. However, as Richardson stated in his paper, this model cannot explain the high amount of Al incorporated into C-A-S-H in real hydrated cement and the Ca-rich C-(A)-S-H.

None of these structural-chemical models can consider the interlayer Al species as parts of the C-A-S-H phase. For example, in the CSTM model, although they noted that the interlayer Al impacts the formulae, the significance of the interlayer Al(VI) for the interlayer Ca^{2+} was still underestimated, and they failed to consider this interlayer aluminate in this model. As a result, there is a substantial mismatch between the calculated Al/Si ratio and the measured Al/Si ratio by SEM–EDS (see Fig. 6.24), albeit it may not be the only reason.

Herein, we propose a general structural-chemical formulae model for crosslinked and non-crosslinked C-A-S-H based on the atomistic bricks described in Sect. 4.2.1. As for the previous structural-chemical models [84, 90, 91], the current model should be supported by ^{29}Si NMR data, and ^{27}Al NMR data are also needed. The general model is given in Eq. 6.18, representing the average structural-chemical formula values for a brick containing six Si-tetrahedra, four main-layer calcium, and several interlayer species. Note that the charge-compensated species used for chemical units is consistent with that in the DNA-code rule. Hence the general structural-chemical formula model proposed here is compatible with Eq. 6.8 determined from the DNA-code rule when it is applied to C-S-H. Moreover, one can easily reformulate Eqs. 6.8 and 6.18 in the same format. Although the proposed model is a bit complicated, it provides a clearer image of the C-A-S-H atomic structure with more detailed information than Eq. 6.8, such as the fraction of Al-substitution, the extent of charge-compensation of proton and calcium, and the content of interlayer aluminates, etc.

$$Ca_4\left[Si_{1-f-v}Al_fV_vO_\alpha\right]_6H_\beta Ca_\gamma(OH)_\delta A^{V}_{\varepsilon}A^{VI}_{\theta}\cdot mH_2O\cdot zCa(OH)_2 \tag{6.18}$$

where the main layer dreierketten chain is included in the square bracket, V represents vacant sites of aluminosilicate chains, A^{V} denotes the interlayer pentahedral aluminate $[AlO(OH)_4]^{3-}$, A^{VI} stands for the interlayer octahedral aluminates $[Al(OH)_6]^{3-}$, f is the Al substitution ratio, v is the vacancy, α is the number of oxygen present in dreierketten chain, β, γ, and δ are the amount of H^+, Ca^{2+} and OH^-, respectively, ε and θ represent the amount of interlayer pentahedral and octahedral aluminates, respectively, m defines the amount of interlayer water, and z $Ca(OH)_2$ units are added to the interlayer for reaching a high Ca/Si ratio in a Ca-rich phase.

6.5.2 *Normalized Parameters of C-A-S-H Brick Model*

Table 6.7 presents the structural-chemical formula values and normalized values for silicate and aluminate species in a C-A-S-H brick. C represents the total Al/Si ratio including both interlayer and bridging aluminates, φ_n values are the fractions of aluminate species, which can be determined from ^{27}Al NMR spectra deconvolution following the method described in this work. Similarly, Q_n values are fractions of silicate species calculated from ^{29}Si NMR spectra deconvolution. It is noted that the fraction values φ_n for aluminates are normalized from fractions for bridging and interlayer aluminate of C-A-S-H, i.e. Al(IV)-a, Al(IV)-b, Al(IV)-c, Al(V), Al(VI)-a, and Al(VI)-b, the fractions for other Al-phases e.g. katoite and AFm should be excluded. Specifically, φ_n and Q_n values follow the following rule.

$$\varphi_1 + \varphi_2 + \varphi_3 + \varphi_4 + \varphi_5 + \varphi_6 = 1 \tag{6.19}$$

$$Q_1 + Q_2 + Q_3 + Q_4 + Q_5 + Q_6 + Q_7 + Q_8 + Q_9 + Q_{10} + Q_{11} = 1 \tag{6.20}$$

Table 6.8 presents the details for the complexes, values, and charge-balanced species of different silicates and aluminates.

6.5.3 *Structural-Chemical Formula Model of Non-crosslinked C-A-S-H*

The general structural-chemical formula model for C-A-S-H brick is shown in Eq. 6.21.

$$\begin{aligned} &Ca_4\left[Si_{1-f-v}Al_f V_v O_\alpha\right]_6 H_\beta Ca_\gamma (OH)_\delta \left[AlO(OH)_4\right]_\varepsilon \left[Al(OH)_6\right]_\theta \\ &\cdot mH_2O \cdot zCa(OH)_2 \end{aligned} \tag{6.21}$$

Table 6.7 The fractions of the aluminate and silicate sites of C-A-S-H that are occupied by Si, Al or vacancy

Sites/item	Notes	Values for crosslinked C-A-S-H	Value for non-crosslinked C-A-S-H
Total Al/Si ratio	Bridging and interlayer Al	C	C
Al substituted ratio	–	f	f
Ca/Si ratio	C/S	x	x
Al(IV)-a	Bridging tetrahedral aluminate	φ_1	φ_1
Al(IV)-b	Bridging tetrahedral aluminate	φ_2	φ_2
Al(IV)-c	Branching tetrahedral aluminate	φ_3	$\varphi_3 = 0$
Al(V)	φ_{41}: bridging pentahedral aluminate φ_{42}: interlayer pentahedral aluminate	$\varphi_4 = \varphi_{41} + \varphi_{42} = 0$	$\varphi_4 = \varphi_{41} + \varphi_{42}$
Al(VI)-a	Bridging octahedral aluminate	$\varphi_5 = 0$	φ_5
Al(VI)-b	φ_{61}: bridging octahedral aluminate φ_{62}: interlayer octahedral aluminate	$\varphi_6 = \varphi_{61} + \varphi_{62} = 0$	$\varphi_6 = \varphi_{61} + \varphi_{62}$
Vacancy	Denoted by V	v	v
Q^1(OH)	–	Q_1	Q_1
Q^1(Ca)	–	Q_2	Q_2
Q^2H	–	Q_3	Q_3
Q^2(Ca)	–	Q_4	Q_4
Q^2i	–	Q_5	Q_5
Q^3(OH)	Present in crosslinked system only	Q_6	$Q_6 = 0$
Q^3(Ca)	Present in crosslinked system only	Q_7	$Q_7 = 0$
Q^3(Al)	Present in crosslinked system only	Q_8	$Q_8 = 0$
Q^2p	–	Q_9	Q_9
Q^2p(Al)	–	Q_{10}	Q_{10}
Q^2p(Al(VI))	–	Q_{11}	Q_{11}
Q^1(0Al)Si	Q^1(OH)/ Q^1(Ca)	$2v$	$2v$
Q^2(0Al)Si	Q^2H /Q^2(Ca)/ Q^2i/ Q^2p	$2/3 - 2f - v(2 - p)$	$1 - 3(f + v)$
Q^2(1Al)Si	Q^2p(Al)	$2f$	$2f$

(continued)

Table 6.7 (continued)

Sites/item	Notes	Values for crosslinked C-A-S-H	Value for non-crosslinked C-A-S-H
$Q^3(0Al)Si$	Q^3	$1/3 - v(1 + p - q) - 2f$	–
$Q^3(1Al)Si$	$Q^3(Al)$	$f - qv$	–

For non-crosslinked C-A-S-H, the multiplicities of structural units in main layer and interlayer can be obtained according to Table 6.9. Note that branching species, i.e. $Q^3(OH)$, $Q^3(Ca)$, $Q^3(Al)$, and Al(IV)-c, are not present in non-crosslinked C-A-S-H, therefore, their corresponding multiplicities are assigned to be 0 by default, i.e. Q_6, Q_7, Q_8, and $\varphi_3 = 0$. The values of the global multiplicities in the structural-chemical formula (6.21) are deduced from those of the structural units as follows:

O_α:

$$\alpha = 3 - 2v + y \cdot C \cdot (\varphi_{41} + 2\varphi_5 + 2\varphi_{61}) \tag{6.22}$$

H_β:

$$\begin{aligned}\beta &= 6y \cdot (Q_1 + 2Q_3 + Q_5) + 6y \cdot C \cdot (\varphi_1 + 2\varphi_2 + 2\varphi_{41} + 5\varphi_5 + 4\varphi_{61}) \\ &= 6y \cdot \left[(Q_1 + 2Q_3 + Q_5) + C \cdot (\varphi_1 + 2\varphi_2 + 2\varphi_{41} + 5\varphi_5 + 4\varphi_{61})\right]\end{aligned} \tag{6.23}$$

Ca_γ:

$$\begin{aligned}\gamma &= 6y \cdot Q_4 + 6y \cdot (Q_2 + Q_5) + 6y \cdot C \cdot (\varphi_{41} + \varphi_5 + \varphi_{61}) \\ &\quad + 6y \cdot C \cdot (2\varphi_1 + \varphi_2 + \varphi_{41} + \varphi_{61}) + 2 \times 6y \cdot C \cdot \varphi_{42} + 2 \times 6y \cdot C \cdot \varphi_{62} \\ &= 6y \cdot \left[(Q_2 + Q_4 + Q_5) + C \cdot (\varphi_{41} + \varphi_5 + \varphi_{61} + 2\varphi_1 + \varphi_2 \right. \\ &\quad \left. + \varphi_{41} + \varphi_{61} + 2\ \varphi_{42} + 2\varphi_{62})\right] \\ &= 6y \cdot \left[(Q_2 + Q_4 + Q_5) + C \cdot (2\varphi_1 + \varphi_2 + 2\varphi_4 + \varphi_5 + 2\varphi_6)\right] \\ &= 6y \cdot \left[(Q_2 + Q_4 + Q_5) + C \cdot (1 + \varphi_1 + \varphi_4 + \varphi_6)\right]\end{aligned} \tag{6.24}$$

$(OH)_\delta$:

$$\begin{aligned}\delta &= 6y \cdot (Q_2 + Q_5) + 6y \cdot C \cdot (2\varphi_1 + \varphi_2 + \varphi_{41} + \varphi_{61}) \\ &\quad + 6y \cdot C \cdot \varphi_{42} + 6y \cdot C \cdot \varphi_{62} \\ &= 6y \cdot \left[(Q_2 + Q_5) + C \cdot (2\varphi_1 + \varphi_2 + \varphi_{41} + \varphi_{61} + \varphi_{42} + \varphi_{62})\right] \\ &= 6y \cdot \left[(Q_2 + Q_5) + C \cdot (2\varphi_1 + \varphi_2 + \varphi_4 + \varphi_6)\right]\end{aligned} \tag{6.25}$$

${A^V}_\varepsilon$:

$$\varepsilon = 6y \cdot C \cdot \varphi_{42} \tag{6.26}$$

Table 6.8 The values of the structural-chemical formula for a C-A-S-H brick unit with six silicate/aluminate units

Items	Complex	Values	Symbols	Charge-balanced species
Total Al/Si ratio	–	C	C	–
Ratio of substituted Al to Si	–	$f/(1-f-v)$	f/y	–
Vacancy	–	v	v	–
Ca/Si ratio	–	x	x	–
Si atoms in unit	–	$6{\cdot}(1-f-v)$	$6y$	–
Bridging Al atoms	–	$6f$	$6f$	–
Total Al (bridging and interlayer)	–	$6{\cdot}C{\cdot}(1-f-v)$	$6y{\cdot}C$	–
Interlayer Al	–	$6{\cdot}[C{\cdot}(1-f-v)-f]$	$6(y{\cdot}C{\text -}f)$	–
Al(IV)-a[1]	$[AlO_3(OH)]^{4-}$	$6{\cdot}C{\cdot}(1-f-v){\cdot}\varphi_1$	$6y{\cdot}C{\cdot}\varphi_1$	$2Ca(OH)^+$
Al(IV)-b[1]	$[AlO_2(OH)_2]^{3-}$	$6{\cdot}C{\cdot}(1-f-v){\cdot}\varphi_2$	$6y{\cdot}C{\cdot}\varphi_2$	$Ca(OH)^+$
Al(IV)-c[1]	$[AlO_3(OH)]^{4-}$	$6{\cdot}C{\cdot}(1-f-v){\cdot}\varphi_3$	$6y{\cdot}C{\cdot}\varphi_3$	$Ca(OH)^+$
Al(V)-briging[1,2]	$[AlO_3(OH)_2]^{5-}$	$6{\cdot}C{\cdot}(1-f-v){\cdot}\varphi_{41}$	$6y{\cdot}C{\cdot}\varphi_{41}$	Ca^{2+},$Ca(OH)^+$
Al(V)-interlayer[1,2]	$[AlO(OH)_4]^{3-}$	$6{\cdot}C{\cdot}(1-f-v){\cdot}\varphi_{42}$	$6y{\cdot}C{\cdot}\varphi_{42}$	Ca^{2+},$Ca(OH)^+$
Al(VI)-a[1]	$[AlO_2(OH)_3(H_2O)]^{4-}$	$6{\cdot}C{\cdot}(1-f-v){\cdot}\varphi_5$	$6y{\cdot}C{\cdot}\varphi_5$	Ca^{2+}
Al(VI)-b-bridging[1,3]	$[AlO_2(OH)_4]^{5-}$	$6{\cdot}C{\cdot}(1-f-v){\cdot}\varphi_{61}$	$6y{\cdot}C{\cdot}\varphi_{61}$	Ca^{2+},$Ca(OH)^+$
Al(VI)-b-interlayer[1,3]	$[Al(OH)_6]^{3-}$	$6{\cdot}C{\cdot}(1-f-v){\cdot}\varphi_{62}$	$6y{\cdot}C{\cdot}\varphi_{62}$	Ca^{2+},$Ca(OH)^+$
Q^1(OH)	$[SiO_3(OH)]^{3-}$	$6{\cdot}(1-f-v){\cdot}Q^1(OH)$	$6y{\cdot}Q_1$	Neutral
Q^1(Ca)	$[SiO_4]^{4-}$	$6{\cdot}(1-f-v){\cdot}Q^1(Ca)$	$6y{\cdot}Q_2$	$Ca(OH)^+$
Q^2H	$[SiO_2(OH)_2]^{2-}$	$6{\cdot}(1-f-v){\cdot}Q^2H$	$6y{\cdot}Q_3$	Neutral
Q^2(Ca)	$[SiO_4]^{4-}$	$6{\cdot}(1\text{-}f\text{-}v){\cdot}Q^2(Ca)$	$6y{\cdot}Q_4$	Ca^{2+}
Q^2i	$[SiO_3(OH)]^{3-}$	$6{\cdot}(1-f-v){\cdot}Q^2i$	$6y{\cdot}Q_5$	$Ca(OH)^+$
Q^3(OH)	$[SiO_3(OH)]^{3-}$	$6{\cdot}(1-f-v){\cdot}Q^3(OH)$	$6y{\cdot}Q_6$	Neutral

(continued)

Table 6.8 (continued)

Items	Complex	Values	Symbols	Charge-balanced species
Q^3(Ca)	$[SiO_4]^{4-}$	$6\cdot(1-f-v)\cdot Q^3(Ca)$	$6y\cdot Q_7$	$Ca(OH)^+$
$Q^3(Al)^4$	$[SiO_3(OH)]^{3-}$	$6\cdot(1-f-v)\cdot Q^3(OH)$	$6y\cdot Q_8$	Neutral

Note

1. The fractions of aluminates are normalized, the intensities for other Al resonances related to other Al-type phases are excluded
2. Al(V) includes bridging and interlayer Al(V). Therefore, the fraction of Al(V) and that of bridging φ_{41} and interlayer Al(V) φ_{42} have a relation of $\varphi_4 = \varphi_{41} + \varphi_{42}$. Specifically, $0 \le \varphi_{41} \le \varphi_4$, $0 \le \varphi_{42} \le \varphi_4$
3. Similarly, Al(VI)-b includes bridging and interlayer Al(V). Therefore, the fraction of Al(VI) φ_6 and that of bridging φ_{61} and interlayer Al(V)-b φ_{62} have a relation of $\varphi_6 = \varphi_{61} + \varphi_{62}$. Specifically, $0 \le \varphi_{61} \le \varphi_6$, $0 \le \varphi_{62} \le \varphi_6$
4. Since Q^2p(Al(VI)) can be resolved from ^{29}Si NMR spectra, we have

$$\varphi_{61} = \frac{Q^2p(Al(VI))}{Q^2p(Al(VI))+Q^2p(Al)} = \frac{Q_{11}}{Q_{10}+Q_{11}}$$

5. Since the Q^3(Al) species has a very small amount in crosslinked C-A-S-H, herein, we assign a complex of $[SiO_3(OH)]^{3-}$ to this Si-tetrahedron, which will not significantly change the structural values for crosslinked C-A-S-H

Table 6.9 Multiplicities for the structural-chemical formula of non-crosslinked C-(A)-S-H brick

Site	Main chain layer					
Species	Ca	Si_6	Al_6	V_6	$[Si/A^{IV}]O_6$	$[Al^V/Al^V]O_6$
Values	4	$1-f-v$	f	v	$3-2v$	$y\cdot C\cdot(\varphi_{41}+2\varphi_5+2\varphi_{61})$
Site	**For main-layer SiO_x charge balance**			**For main-layer AlO_x charge balance**		
Species	H^+	Ca^{2+}	$Ca(OH)^+$	H^+	Ca^{2+}	$Ca(OH)^+$
Values	$6y(Q_1+2Q_3+Q_5)$	$6y\cdot Q_4$	$6y\cdot(Q_2+Q_5)$	$6y\cdot C(\varphi_1+2\varphi_2+2\varphi_{41}+5\varphi_5+4\varphi_{61})$	$6y\cdot C(\varphi_{41}+\varphi_5+\varphi_{61})$	$6y\cdot C(2\varphi_1+\varphi_2+\varphi_{41}+\varphi_{61})$
Site	**Interlayer Al(V) charge balance**			**Interlayer Al(VI) charge balance**		
Species	$[AlO(OH)_4]^{3-}$	Ca^{2+}	$Ca(OH)^+$	$[Al(OH)_6]^{5-}$	Ca^{2+}	$Ca(OH)^+$
Values	$6y\cdot C\cdot\varphi_{42}$	$6y\cdot C\cdot\varphi_{42}$	$6y\cdot C\cdot\varphi_{42}$	$6y\cdot C\cdot\varphi_{62}$	$6y\cdot C\cdot\varphi_{62}$	$6y\cdot C\cdot\varphi_{62}$
Site	**Interlayer free species**					
Species	H_2O			$Ca(OH)_2$		
Values	m			z		

$A^{VI}{}_{\theta}$:

$$\theta = 6y \cdot C \cdot \varphi_{62} \tag{6.27}$$

6.5.4 *Structural-Chemical Formula Model of Crosslinked C-A-S-H*

Similarly, the multiplicities of the structural units in main layer and interlayer are obtained according to Table 6.10. One should note that the crosslinked C-A-S-H only contains tetrahedral aluminates in the bridging and branching sites, which means the fractions of Al(V) and Al(VI), i.e. φ_{41}, φ_{42}, φ_5, φ_{61}, φ_{62}, are at 0. The global multiplicities are then obtained as follows:

O_α:

$$\alpha = 17/6 - 2v \tag{6.28}$$

H_β:

$$\begin{aligned}\beta &= 6y \cdot (Q_1 + 2Q_3 + Q_5 + Q_6 + Q_8) + 6y \cdot C \cdot (\varphi_1 + 2\varphi_2 + \varphi_3)\\ &= 6y \cdot \left[(Q_1 + 2Q_3 + Q_5 + Q_6 + Q_8) + C \cdot (1 + \varphi_2)\right]\end{aligned} \tag{6.29}$$

Ca_γ:

$$\begin{aligned}\gamma &= 6y \cdot Q_4 + 6y \cdot (Q_2 + Q_5 + Q_7) + 6y \cdot C \cdot (2\varphi_1 + \varphi_2 + \varphi_3)\\ &= 6y \cdot \left[(Q_2 + Q_4 + Q_5 + Q_7) + C \cdot (1 + \varphi_1)\right]\end{aligned} \tag{6.30}$$

$(OH)_\delta$:

$$\begin{aligned}\delta &= 6y \cdot (Q_2 + Q_5 + Q_7) + 6y \cdot C \cdot (2\varphi_1 + \varphi_2 + \varphi_3)\\ &= 6y \cdot \left[(Q_2 + Q_5 + Q_7) + C \cdot (1 + \varphi_1)\right]\end{aligned} \tag{6.31}$$

Table 6.10 Multiplicities for the structural-chemical formula of crosslinked C-(A)-S-H brick

Site	**Main chain layer**					
Species	Ca	Si_6	Al_6	V_6	$[Si/Al^{IV}]O_6$	$[Al^V/Al^{VI}]O_6$
Values	4	$1-f-v$	f	v	$17/6-2v$	0
Site	**Main-layer SiO_x charge balance**			**Main-layer AlO_x charge balance**		
Species	H^+	Ca^{2+}	$Ca(OH)^+$	H^+	Ca^{2+}	$Ca(OH)^+$
Values	$6y(Q_1+2Q_3+Q_5+Q_6+Q_8)$	$6y{\cdot}Q_4$	$6y{\cdot}(Q_2+Q_5+Q_7)$	$6y{\cdot}C(\varphi_1+2\varphi_2+\varphi_3)$	0	$6y{\cdot}C(2\varphi_1+\varphi_2+\varphi_3)$
Site	**Interlayer Al(V) charge balance**			**Interlayer Al(VI) charge balance**		
Species	$[AlO(OH)_4]^{3-}$	Ca^{2+}	$Ca(OH)^+$	$[Al(OH)_6]^{5-}$	Ca^{2+}	$Ca(OH)^+$
Values	0	0	0	0	0	0
Site	**Interlayer free species**					
Species	H_2O			$Ca(OH)_2$		
Values	m			z		

$A^{V}{}_{\varepsilon}$:

$$\varepsilon = 0 \tag{6.32}$$

$A^{VI}{}_{\theta}$:

$$\theta = 0 \tag{6.33}$$

6.5.5 General Structural-Chemical Formula Model of Mixed C-A-S-H

In a mixed C-A-S-H system contains both non-crosslinked and crosslinked precipitates. The multiplicities for the structural units in the main layer and interlayer are listed in Table 6.11. The global multiplicities for general mixed C-A-S-H are obtained as follows:

O_{α}:

$$\alpha = (18 - \mathrm{d})/6 - 2v + y \cdot C \cdot (\varphi_{41} + 2\varphi_5 + 2\varphi_{61}) \tag{6.34}$$

H_{β}:

$$\beta = 6y \cdot (Q_1 + 2Q_3 + Q_5 + Q_6 + Q_8) + 6y \cdot C \cdot (\varphi_1 + 2\varphi_2 + \varphi_3 + 2\varphi_{41} + 5\varphi_5 + 4\varphi_{61})$$

Table 6.11 Multiplicities for the structural-chemical formula of mixed C-(A)-S-H brick

Site	Main chain layer					
Species	Ca	Si_6	Al_6	V_6	$[Si/A^{IV}]O_6$	$[Al^{V}/ Al^{V}]O_6$
Values	4	$1-f-v$	f	v	$(18-d)/6 - 2v$	$y{\cdot}C{\cdot}(\varphi_{41} + 2\varphi_5 + 2\varphi_{61})$
Site	**Main-layer SiO_x charge balance**			**Main-layer AlO_x charge balance**		
Species	H^+	Ca^{2+}	$Ca(OH)^+$	H^+	Ca^{2+}	$Ca(OH)^+$
Values	$6y(Q_1 + 2Q_3 + Q_5 + Q_6 + Q_8)$	$6y{\cdot}Q_4$	$6y{\cdot}(Q_2 + Q_5 + Q_7)$	$6y{\cdot}C(\varphi_1 + 2\varphi_2 + \varphi_3 + 2\varphi_{41} + 5\varphi_5 + 4\varphi_{61})$	$6y{\cdot}C(\varphi_{41} + \varphi_5 + \varphi_{61})$	$6y{\cdot}C(2\varphi_1 + \varphi_2 + \varphi_3 + \varphi_{41} + \varphi_{61})$
Site	**Interlayer Al(V) charge balance**			**Interlayer Al(VI) charge balance**		
Species	$[AlO(OH)_4]^{3-}$	Ca^{2+}	$Ca(OH)^+$	$[Al(OH)_6]^{5-}$	Ca^{2+}	$Ca(OH)^+$
Values	$6y{\cdot}C{\cdot}\varphi_{42}$	$6y{\cdot}C{\cdot}\varphi_{42}$	$6y{\cdot}C{\cdot}\varphi_{42}$	$6y{\cdot}C{\cdot}\varphi_{62}$	$6y{\cdot}C{\cdot}\varphi_{62}$	$6y{\cdot}C{\cdot}\varphi_{62}$
Site	**Interlayer free species**					
Species	H_2O			$Ca(OH)_2$		
Values	m			z		

$$
\begin{aligned}
&= 6y \cdot \left[(Q_1 + 2Q_3 + Q_5 + Q_6 + Q_8) + C \right. \\
&\quad \cdot (\varphi_1 + 2\varphi_2 + \varphi_3 + 2\varphi_{41} + 5\varphi_5 + 4\varphi_{61}))
\end{aligned}
\tag{6.35}
$$

Ca_γ:

$$
\begin{aligned}
\gamma &= 6y \cdot Q_4 + 6y \cdot (Q_2 + Q_5 + Q_7) \\
&\quad + 6y \cdot C \cdot (\varphi_{41} + \varphi_5 + \varphi_{61}) + 6y \cdot C \cdot (2\varphi_1 + \varphi_2 + \varphi_3 + \varphi_{41} + \varphi_{61}) \\
&\quad + 2 \times 6y \cdot C \cdot \varphi_{42} + 2 \times 6y \cdot C \cdot \varphi_{62} \\
&= 6y \cdot \left[(Q_2 + Q_4 + Q_5 + Q_7) + C \cdot (\varphi_{41} + \varphi_5 + \varphi_{61} + 2\varphi_1 \right. \\
&\quad + \varphi_2 + \varphi_3 + \varphi_{41} + \varphi_{61} + 2\varphi_{42} + 2\varphi_{62})] \\
&= 6y \cdot \left[(Q_2 + Q_4 + Q_5 + Q_7) \right. \\
&\quad + C \cdot (2\varphi_1 + \varphi_2 + \varphi_3 + 2\varphi_4 + \varphi_5 + 2\varphi_6)] \\
&= 6y \cdot \left[(Q_2 + Q_4 + Q_5 + Q_7) + C \cdot (1 + \varphi_1 + \varphi_4 + \varphi_6) \right]
\end{aligned}
\tag{6.36}
$$

$(OH)_\delta$:

$$
\begin{aligned}
\delta &= 6y \cdot (Q_2 + Q_5 + Q_7) + 6y \cdot C \cdot (2\varphi_1 + \varphi_2 + \varphi_3 + \varphi_{41} + \varphi_{61}) \\
&\quad + 6y \cdot C \cdot \varphi_{42} + 6y \cdot C \cdot \varphi_{62} \\
&= 6y \cdot \left[(Q_2 + Q_5 + Q_7) + C \cdot (2\varphi_1 + \varphi_2 + \varphi_3 + \varphi_{41} \right. \\
&\quad + \varphi_{61} + \varphi_{42} + \varphi_{62})] \\
&= 6y \cdot \left[(Q_2 + Q_5 + Q_7) + C \cdot (2\varphi_1 + \varphi_2 + \varphi_3 + \varphi_4 + \varphi_6) \right] \\
&= 6y \cdot \left[(Q_2 + Q_5 + Q_7) + C \cdot (1 + \varphi_1 - \varphi_5) \right]
\end{aligned}
\tag{6.37}
$$

$A^{V}{}_{\varepsilon}$

$$
\varepsilon = 6y \cdot C \cdot \varphi_{42} \tag{6.38}
$$

$A^{VI}{}_{\theta}$:

$$
\theta = 6y \cdot C \cdot \varphi_{62} \tag{6.39}
$$

6.5.6 Summary of C-A-S-H General Structural-Chemical Formula Model

For the purpose of reference and comparison among the generalized structural chemical formula models of the various types of C-A-S-H discussed above, this section provides a summary. The generalized structural chemical formula models for non-cross-linked, cross-linked, and hybrid C-A-S-H brick models can be uniformly represented by Eq. 4.39. The specific meanings can be found in Sect. 6.5.1, and the computed values for each parameter are summarized in Table 6.12.

$$Ca_4\left[Si_{1-f-v}Al_fV_vO_\alpha\right]_6H_\beta Ca_\gamma(OH)_\delta A_\varepsilon^{V}A_\theta^{VI}\cdot mH_2O\cdot zCa(OH)_2 \tag{6.40}$$

It is noted that the total Al/Si ratio of C-A-S-H is a critical parameter in our model. The presence of interlayer aluminates in C-A-S-H cannot be determined from ^{29}Si NMR spectra. Unfortunately, it is still controversial to measure an accurate Al/Si ratio for C-A-S-H through general techniques like SEM–EDS and TEM-EDS. As indicated in Fig. 6.24, the Al/Si ratio detected from SEM–EDS are always higher than that obtained from TEM-EDS. It is unknown whether SEM–EDS overestimates

Table 6.12 The values of the structural-chemical formula for a C-A-S-H brick with six silicate/aluminate

Types	α	β	γ	δ	ε	θ
Non-crosslinked	$3-2v+y\cdot C\cdot(\varphi_{41}+2\varphi_5+2\varphi_{61})$	$6y[(Q_1+2Q_3+Q_5)+C\cdot(\varphi_1+2\varphi_2+2\varphi_{41}+5\varphi_5+4\varphi_{61})]$	$6y\cdot[(Q_2+Q_4+Q_5)+C\cdot(1+\varphi_1+\varphi_4+\varphi_6)]$	$6y\cdot[(Q_2+Q_5)+C\cdot(1-\varphi_5+\varphi_1)]$	$6y\cdot C\cdot\varphi_{42}$	$6y\cdot C\cdot\varphi_{62}$
Crosslinked	$17/6-2v$	$6y\cdot[(Q_1+2Q_3+Q_5+Q_6+Q_8)+C\cdot(1+\varphi_2)]$	$6y\cdot[(Q_2+Q_4+Q_5+Q_7)+C\cdot(1+\varphi_1)]$	$6y\cdot[(Q_2+Q_5+Q_7)+C\cdot(1+\varphi_1)]$	0	0
General mixed	$(18-d)/6-2v+y\cdot C\cdot(\varphi_{41}+2\varphi_5+2\varphi_{61})$	$6y\cdot[(Q_1+2Q_3+Q_5+Q_6+Q_8)+C\cdot(\varphi_1+2\varphi_2+\varphi_3+2\varphi_{41}+5\varphi_5+4\varphi_{61})]$	$6y\cdot[(Q_2+Q_4+Q_5+Q_7)+C\cdot(1+\varphi_1+\varphi_4+\varphi_6)]$	$6y\cdot[(Q_2+Q_5+Q_7)+C\cdot(1+\varphi_1-\varphi_5)]$	$6y\cdot C\cdot\varphi_{42}$	$6y\cdot C\cdot\varphi_{62}$

Note $y=1-f-v$, C represents the total Al/Si ratio including both interlayer and bridging aluminates, Q_n values are the fractions of silicate Q^n species which can be determined from ^{29}Si NMR spectra deconvolution followed with the method in the current study. More precisely, Q_1, Q_2, Q_3, Q_4, Q_5, Q_6, and Q_7 stand for the fractions of Q^1(OH), Q^1(Ca), Q^2H, Q^2(Ca), Q^2i, Q^3(OH), and Q^3(Ca), respectively. Similarly, φ_n values are fractions of aluminate species calculated from ^{27}Al NMR spectra deconvolution. That is, φ_1, φ_2, φ_3, and φ_5, are the fractions of Al(IV)-a, Al(IV)-b, Al(IV)-c, and Al(VI)-a, respectively; φ_{41} and φ_{42} denote the fraction of bridging and interlayer Al(V), respectively; φ_{61} and φ_{62} refer to the fraction of bridging and interlayer Al(VI)-b, respectively. According to our results, crosslinked C-A-S-H has Al-tetrahedra only. Therefore, the pentahedral and octahedral aluminates are excluded from the crosslinked C-A-S-H model. For a general C-A-S-H model mixed with crosslinked and non-crosslinked networks, the fraction of crosslinked chain *d* is introduced here, which can be determined from Eq. 6.2

the Al/Si ratio or TEM-EDS underestimates it, which was already noticed by Myers et al. [72] as well. More experimental work is needed in this direction.

6.6 Physical Stability of C-A-S-H Under Cryogenic Attack

In the previous analysis, we examined the stability characteristics of C-A-S-H nanostructures under different chemical composition conditions at ultra-low temperatures. It has been established that the nanostructure of low-Ca C-A-S-H exhibits superior stability at ultra-low temperatures compared to high-Ca systems. Based on this, this section will investigate the stability characteristics of the physical properties of low-Ca C-A-S-H systems under cryogenic attack, elucidating the enhancing effect of aluminum incorporation on the cryogenic stability of C-A-S-H. To achieve this, we designed three categories of low-Ca C-(A)-S-H systems, controlling Ca/Si at 1, with Al/Si ratios of 0, 0.11, and 0.22.

6.6.1 Characteristics of Low-Al C-A-S-H

Figure 6.25a depicts the XRD patterns of synthetic samples before and after the cryogenic attack. C-(A)-S-H is the primary solid precipitated in the hydrothermal reactions. The difference is that the aluminum incorporation additionally introduces vaterite and calcite in synthetic solids, namely AS0.11 and AS0.22; the relevant diffraction peaks are relatively weak, though. No crystalline portlandite peaks are found, which is usually observed in the re-precipitated method [93]. The re-precipitated method is time-consuming in obtaining the target sample, usually taking months, which is far beyond the time spent in our hydrothermal method (only about 20 h). As shown in Fig. 6.25b, the main peak ((020) and (− 220)) at around 3.08 Å is sharper with the presence of aluminum. The corresponding full width at half maximum (FWHM) value of this peak decreases when aluminum is incorporated, showing that it can promote the crystallization of C-A-S-H. This result confirms the conclusion made by Geng et al. [49]. While in the observation of Kapluszna et al. [63], they found that aluminum increased the FWHM value, implying a decrease in crystallinity. Such a difference probably results from temperature [94]. A higher synthesis temperature can also promote C-(A)-S-H crystallization, according to Ref. [49]. Apart from C-A-S-H, other aluminate phases like C_2AH_8 and C_3AH_6 are not observed in the products. The primary reason for this observation is probably associated with the low Ca/Si ratio of the Al-incorporated systems. The aluminosilicate chains of the low-Ca C-A-S-H system are less defective than the high-Ca system, which provides more substitution sites for Al. The chemical compositions of these samples are presented in Table 6.13. As expected, the samples have a low Ca/Si ratio of around 1.

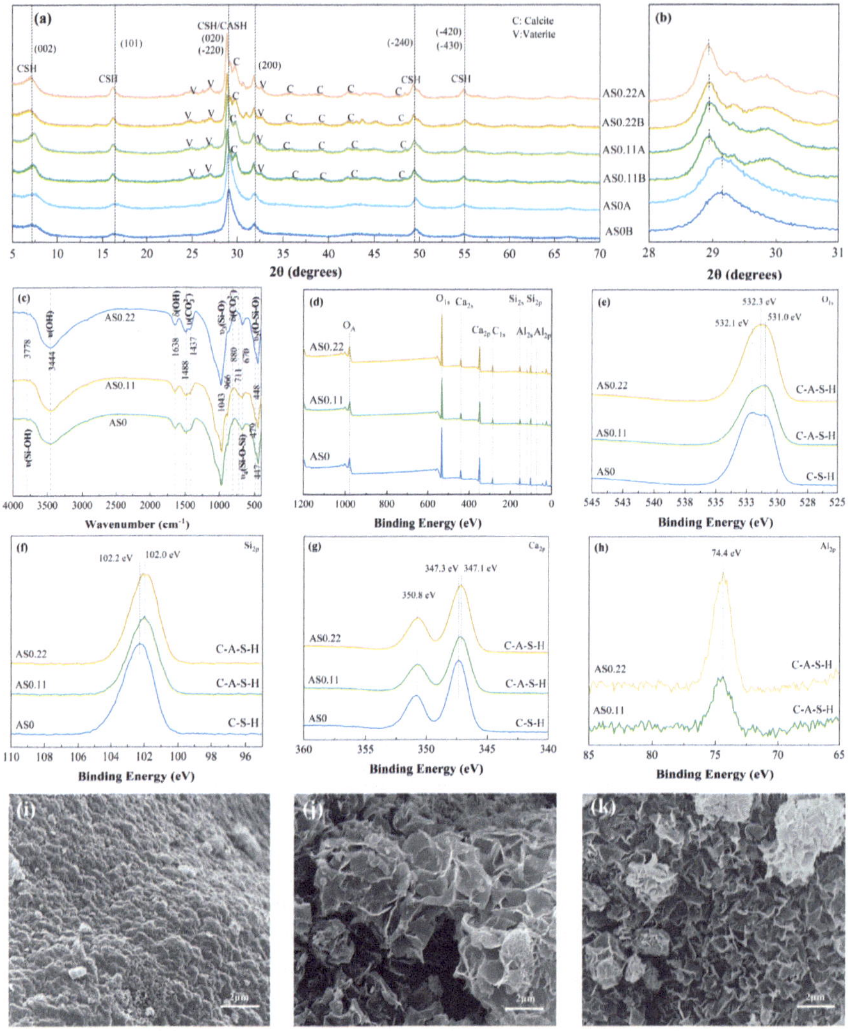

Fig. 6.25 Characterizations of synthetic samples. XRD patterns (**a**, **b**); FTIR spectra (**c**); XPS spectra (**d**–**h**), survey spectra (**d**), O_{1s} (**e**), Si_{2p} (**f**), Ca_{2p} (**g**), Al_{2p} (**h**); SEM micromorphology of AS0 (**i**), AS0.11 (**j**), and AS0.22 (**k**). AS represents Al/Si ratio, the same below

Figure 6.25c exhibits the FTIR spectra of synthetic samples. Narrow bands at 1645 cm^{-1} and 3500 cm^{-1} are assigned to bending stretching vibrations of hydroxyl [63], which implies the existence of poorly crystalline products or confined interlayer water. Typical adsorption bands for Si–O-Si and Si–O–Al bonds of aluminosilicate structures are present at 1300–850 cm^{-1} [63]. As shown in Fig. 6.25c, the wavenumbers of the main 1300–850 cm^{-1} band show lower values with aluminum

Table 6.13 Chemical compositions of synthetic C-A-S-H products from EDS analysis

Samples	Ca (At.%)	Si (At.%)	Al (At.%)	Ca/Si	Al/Si	Average Ca/Si	Average Al/Si
AS0	9.1	10.1	0.00	0.90	0.00	0.92	0.00
	10.8	11.6	0.00	0.93	0.00		
	11.1	11.9	0.00	0.93	0.00		
AS0.11	7.3	7.3	1.3	1.00	0.18	1.03	0.17
	9.3	9.1	1.1	1.02	0.12		
	9.1	8.5	1.7	1.07	0.20		
AS0.22	11.5	11.5	2.3	1.00	0.20	1.04	0.21
	8.0	7.5	1.7	1.07	0.23		
	9.7	9.3	1.8	1.04	0.19		

incorporation. This is due to the partial substitution of Si–O–Al bonds for Si–O-Si bonds. In addition, the aluminum incorporation can be reflected by the weak signal intensity observed at around 1044 cm^{-1}, which is considered a fingerprint for double Si=O bonds or Al=O bonds [63]. In comparison, this weak band has a slightly more vigorous intensity in AS0.11 and AS0.22. The band at 711 cm^{-1} observed in Al-incorporated AS0.11 and AS0.22 samples is associated with vibrations of three-membered rings [95], and the band for vibrations of four-membered rings is at 670 cm^{-1} [95]. In a high Ca/Si ratio system, the $Si–O^{-}$ terminal bonds that link with interlayer calcium ions may form. Its characteristic band shows at around 943 cm^{-1}, according to Ref .[63].

Figure 6.25d–h displays the XPS spectra of synthetic samples. Elements of oxygen and calcium are collected in the survey spectra (Fig. 6.25d), with O_A at ~ 980 eV, O_{1s} at ~ 529 eV, Ca_{2s} at ~ 441 eV, and Ca_{2p} at ~ 349 eV. As expected, elements Si (Si_{2s} and Si_{2p}) are detected in all samples. Al_{2s} at ~ 119 eV and Al_{2p} at ~ 76 eV are detected on AS0.11 and AS0.22 incorporated with aluminum. Figure 6.25e–h plots the spectra details for each element. Although calcite and vaterite make it challenging to analyze O_{1s} spectra in Fig. 6.25e quantitatively, the trend is still apparent. AS0.11 and AS0.22 are detected with a lower O_{1s} binding energy than AS0. This shift is more likely related to the Al substitution rather than depolymerization [96]. The Al-O has a longer bond length than Si–O, resulting in lower binding energy [97]. From Fig. 6.25f, the Si_{2p} binding energy slightly decreases with aluminum incorporated into the solid. Considering that these samples are synthesized with the same bulk Ca/Si ratio, the decrease is reasonably attributed to the aluminum substitution in silicate chains. The aluminum substitution for bridging silicon leads to an increase in net charge, consequently promoting the shielding of Si nuclei, which may initiate such a shift in Si_{2p} binding energy [98]. In contrast to O_{1s} and Si_{2p}, Ca_{2s} binding energy does not sufficiently change with the aluminum content since Ca atoms are mainly coordinated seven-fold. As expected, a higher aluminum content incorporated in the chains increases the intensity of Al_{2p} (see Fig. 6.25h).

The Micro-morphology of synthetic phases is illustrated in Fig. 6.25i–k. The morphology of Al-free AS0 (Fig. 6.25i) is significantly different from that of AS0.11 and AS0.22 (Fig. 6.25j, k). AS0 shows amorphous features and has a much denser surface. In comparison, foil-like morphology features can be observed for C-A-S-H with aluminum incorporated, making the structure more porous than AS0.

6.6.2 *Nanostructural Stability of C-A-S-H*

This section focuses on elucidating the stability characteristics of the nanomechanical properties of low-Ca C-A-S-H systems under cryogenic attack, employing the atomic force microscopy-based nanoindentation testing and analysis method. Below, we will analyze the overall statistical results of the tests and proceed with deconvolution analysis step by step.

6.6.2.1 Calibration and Experimental PDF

According to previous reports [12, 99], the Poisson's ratio of C-(A)-S-H is assumed to be 0.3. Regarding the calibrated parameters for the cantilever, it has a deflection sensitivity of 76.2 nN/V and a spring constant of 32 N/m. The probability density distributions of elastic moduli obtained from indentation are shown in Fig. 6.26. The experimental PDFs for samples before and after the cryogenic attack are overlapped to identify the difference. Those PDFs are all multimodal with more than one peak. Another common feature is that the predominant peak is located below 10 GPa for all samples. Nonetheless, AS0 still responds differently from the other two C-A-S-H samples regarding the cryogenic attack. As shown in Fig. 6.26a, the predominant peak of the Al-free phase shits to a lower elastic modulus after the attack. In comparison, the predominant peaks for AS0.11 and AS0.22 shift to a reverse direction with higher elastic moduli. Furthermore, the frequency density also evolves differently among these samples. Although the frequency density of high elastic moduli (> 20 GPa) increases in both Al-free and Al-containing phases, that of low elastic moduli presents different evolution trends. For AS0, the decrease of the frequency density mainly occurred in the low elastic moduli region within 7 ~ 20 GPa, whereas it mainly occurred in the very low elastic moduli region (< 7 GPa) for AS0.11 and AS0.22.

6.6.2.2 Deconvolution of AFM-Based Indentation

Although the compacted C-(A)-S-H gels differ from the hydrated cement paste with the inner and outer product, the heterogeneous nature of C-(A)-S-H can still lead to the difference in the nanomechanical properties. Herein, we named the various phases LD, HD, and UHD C-S-H by mechanical analogy with the phases identified in hydrated cement systems. Previous nanoindentation studies [64, 100] on compacted

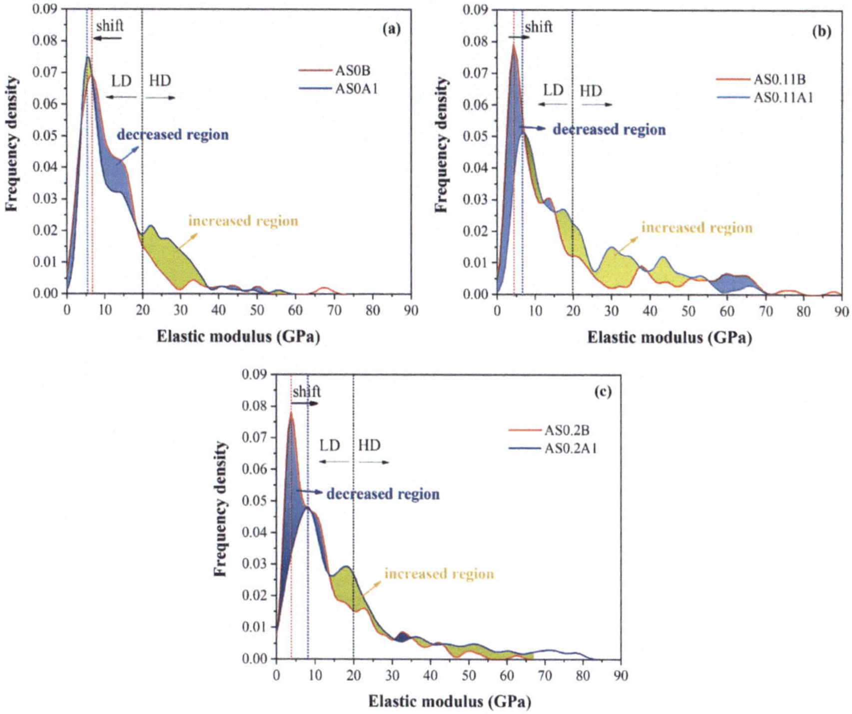

Fig. 6.26 Overlap of experimental probability density distributions before (B) and after (Al) cryogenic attack. AS0 (**a**), AS0.11 (**b**), AS0.22 (**c**)

synthetic C-S-H samples used the same philosophy, i.e., decomposing the mechanical response of their compacts into LD, HD, and UHD C-S-H. Figure 6.27 displays the deconvolution results for the experimental PDFs of elastic moduli. As mentioned before, the different packing nature of gel globules leads to the variability of LD and HD C-S-H [101], which can be identified from the different elastic moduli by using nanoindentation [102, 103]. However, the embedment of crystalline phases such as portlandite and calcium carbonates could also influence the measured mechanical results, as indicated in Ref. [99, 104, 105]. The insertion of portlandite crystals results in a higher value of measured elastic modulus, identified as ultra-high-density (UHD) C-S-H [103]. The loosely packed globules are also identified by nanoindentation [64, 100], named very-low-density (VLD) C-S-H, which is highly associated with gel pores. Especially the AFM tip used in this study has a small tip, which makes it sensitive to the loosely packed region.

According to previous nanoindentation results [64, 103], the characteristic elastic moduli for VLD, LD, HD, and UHD are < 15 GPa, 20 ± 5.0 GPa, 30 ± 5.0 GPa, and 42 ± 7.0 GPa, respectively. These characteristics can be found in our measurement systems as well. From Fig. 6.27, the elastic moduli of these samples range widely from 2 to 90 GPa. Those values over 49 GPa were obtained for all samples, which

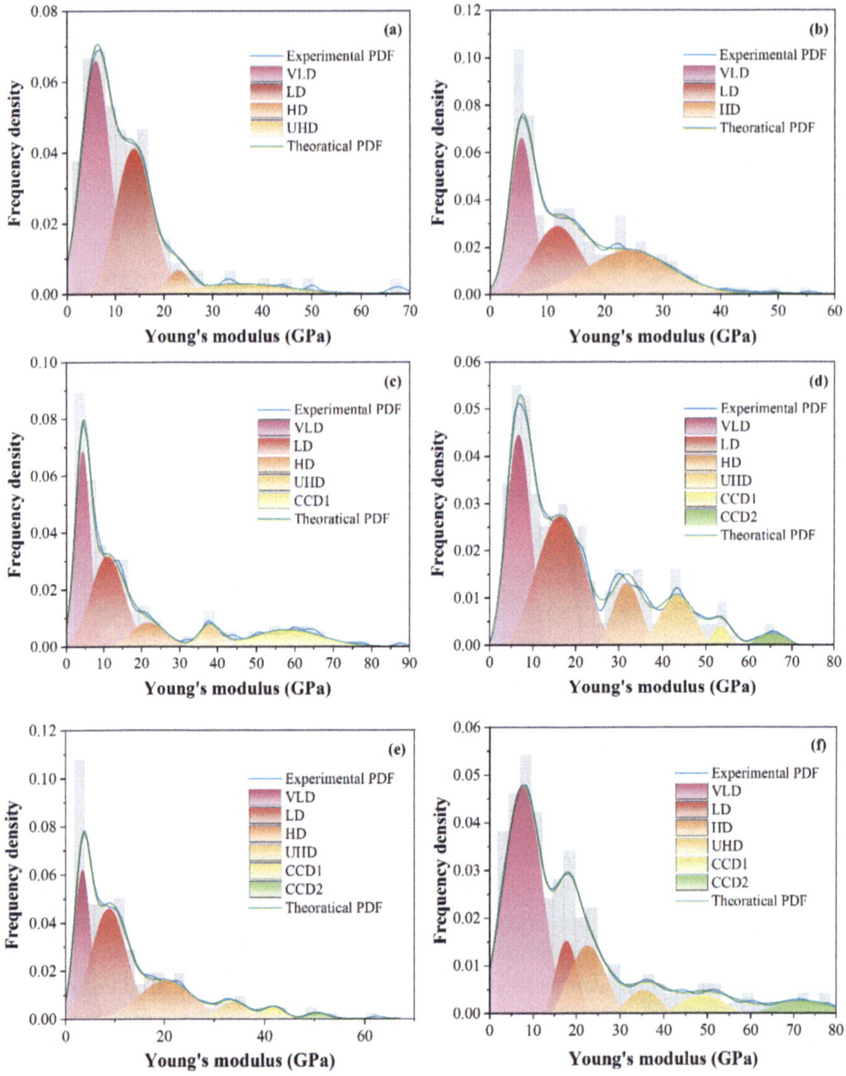

Fig. 6.27 Statistical results for elastic moduli obtained from AFM-based indentations and deconvolution results for AS0B (**a**), AS0A1 (**b**), AS0.11B (**c**), AS0.11A1 (**d**), AS0.22B (**e**), AS0.22A1 (**f**)

herein are speculated to originate from crystalline phases. The XRD (Fig. 6.25) results revealed that calcite is the predominant crystalline phase in the synthetic solids. Therefore, the phases with a high elastic modulus over 60 GPa are assigned to calcite crystals or C-(A)-S-H embedded with calcite, denoted as CCD1 and CCD2 C-S-H, analogously to the descriptions in Ref. [12]. Based on these principles, the elastic moduli for measured discs were statistically analyzed. It is noteworthy that multimodal distributions of elastic moduli are observed for discs sustained cryogenic attack, opposite to the less multimodal distributions for those before the attack.

6.6.2.3 Elastic Moduli of Decomposed Phases

Figure 6.28a compares the average elastic modulus of three samples before and after the cryogenic attack, and Fig. 6.28b–d show the statistically analyzed elastic moduli for decomposed C-(A)-S-H phases from VLD to CCD. Before the attack, the average elastic moduli of Al-containing phases for AS0.11 and AS0.22 are 17.9 GPa and 12.6 GPa, respectively, higher than 11.6 GPa for Al-free phase. The corresponding standard deviations for the average elastic moduli of AS0, AS0.11, and AS0.22 are 8.0 GPa, 18.2 GPa, and 10.3 GPa, respectively, simply showing that those C-(A)-S-H phases are very heterogeneous.

From Fig. 6.28a, C-A-S-H samples were observed with significant increases in their average elastic moduli, while the average elastic modulus of Al-free C-S-H had a relatively small increase. Specifically, the average elastic moduli of AS0, AS0.11,

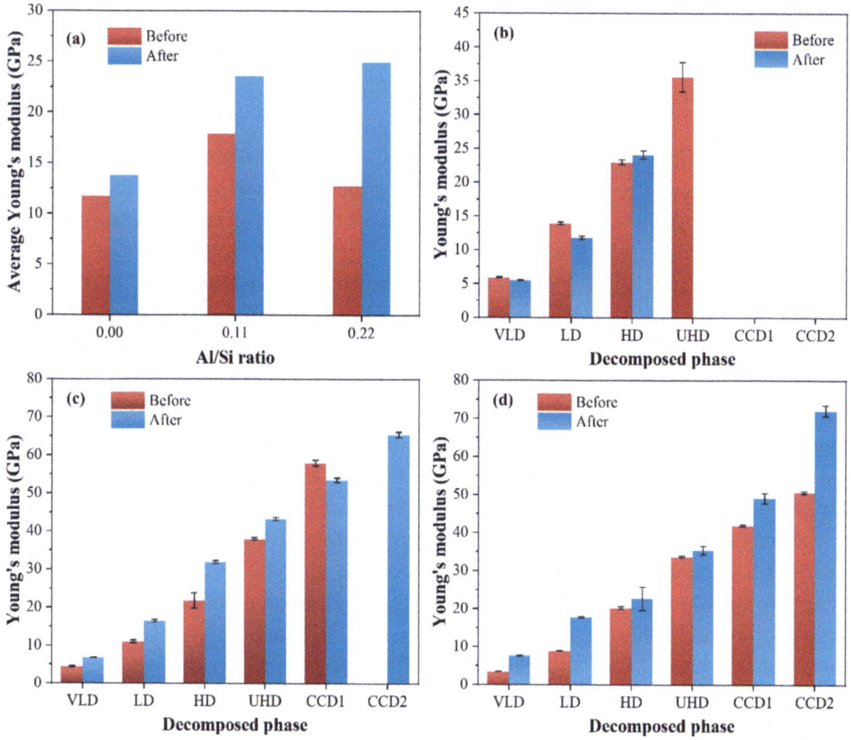

Fig. 6.28 Elastic moduli of decomposed phases (**a–c**), and average elastic moduli of C-(A)-S-H samples (**d**). (**a–c**) for AS0, AS0.11, AS0.22, respectively

and AS0.22 increased to 13.4 GPa, 24.3 GPa, and 25.0 GPa, respectively, due to the cryogenic attack. The corresponding standard deviations are 7.7 GPa, 17.6 GPa, and 21.9 GPa, respectively. Notably, although both Al-free and Al-containing samples obtained higher average elastic moduli after the attack, the evolution of elastic moduli for their corresponding decomposed phases was surprisingly reversed. The variations on the elastic moduli of different decomposed phases are highly affected by incorporating aluminum. Regarding the C-S-H in AS0, the elastic modulus of its decomposed VLD and LD phases had a nonnegligible decrease of 6.8% and 15.1%, respectively, and the resolved peak for the UHD phase even disappeared. Generally, the decomposed phases of C-S-H tended to obtain a smaller elastic modulus after the attack, parallel to the observations in our previous study [12]. As aluminum was incorporated into C-S-H (i.e., AS0.11 and AS0.22), unlike the Al-free C-S-H, the elastic moduli of decomposed Al-containing phases reversely increased after the cryogenic attack, even for the UHD, CCD1, and CCD2. Particularly, the elastic moduli of VLD, LD, and HD significantly increased by 52.3%, 50.0%, and 46.3% for AS0.11, 117.1%, 100.0%, and 11.8% for AS0.22. These differences reflect a different resistance to cryogenic attack between C-A-S-H and C-S-H. Incorporating aluminum in C-S-H positively prevents its nanomechanical property from severe degradation under the cryogenic attack.

6.6.2.4 Volume Fraction of Decomposed Phases

The volume fractions of different decomposed C-(A)-S-H phases were obtained from the KED analysis on the indentation measurements, as shown in Fig. 6.29. The statistical analysis suggests that CCD1 and CCD2 were present in Al-containing phases only, albeit in a relatively low volume fraction. This is perfectly coherent with the facts revealed by XRD (Fig. 6.25) that calcite and vaterite exist in AS0.11 and AS0.22 only. Although the fraction of carbonate obtained from the TG measurement is a mass fraction, we can still roughly compare the carbonate volume fractions between the deconvolution and the TG measurement since the density of calcium carbonate is close to C-(A)-S-H. In contrast, the CCD fraction of AS0.11 before the cryogenic attack approaches the fraction of carbonates obtained from TG, while the CCD fraction of AS0.22 is higher than its carbonate fraction determined by the TG. Such a bias probably results from the inhomogeneous distribution of the solids since the crystalline calcite and vaterite may have orientation preferences during the compression [106]. Fortunately, this will not influence the C-(A)-S-H analysis. Before the cryogenic attack, the VLD and LD occupied the predominant volume fraction among those decomposed phases. Note that the total volume fractions of VLD and LD for AS0, AS0.11, and AS0.22 were 80.6%, 70.6%, and 68.4%, respectively, showing a decreasing trend with increasing aluminum content. Surprisingly, their HD phases occupied 3.9%, 9.7%, and 22.3%, respectively, reversely showing a rising trend with the increasing aluminum content. To our best knowledge, such a relation between decomposed phases and aluminum content has never been reported before.

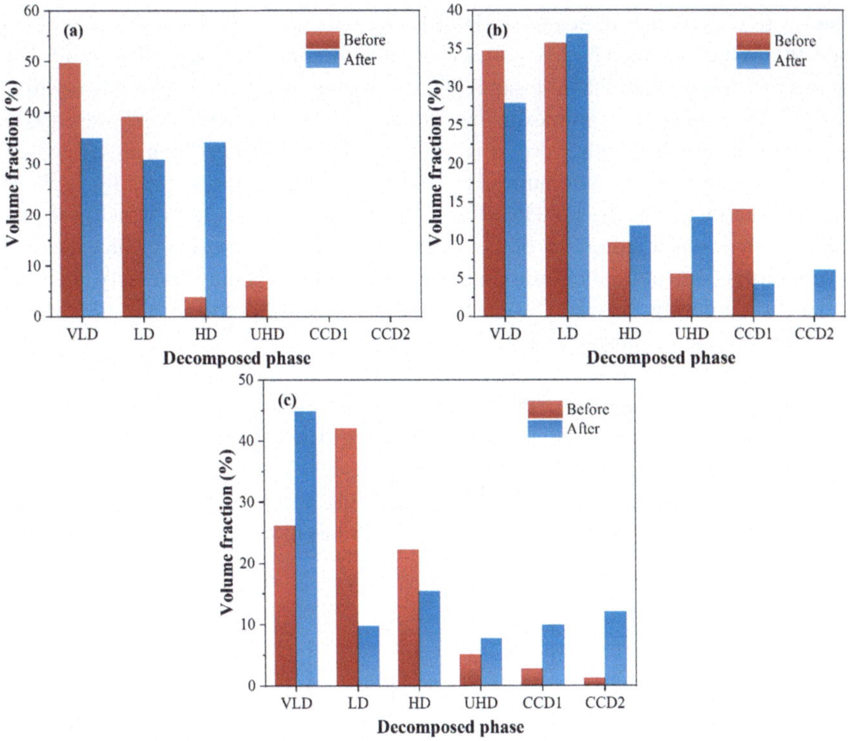

Fig. 6.29 Fractions of decomposed phases in C-(A)-S-H. **a**–**c** for AS0, AS0.11, AS0.22, respectively

The volume fractions of decomposed phases in each sample were significantly reorganized upon the cryogenic attack. The total fraction of VLD and LD phases for AS0, AS0.11, and AS0.22 decreased by 23.2%, 5.7%, and 13.7%, respectively, due to the attack. In contrast, the volume fractions of those high-density phases, i.e., HD, UHD, and CCD, tended to increase. Specifically, the volume fraction of HD of AS0 had a significant increase after the cryogenic attack. The volume fraction of VLD for AS0.22 significantly increased, opposite to the variations observed in the other two samples. The C-A-S-H sample of AS0.11 generally had the best cryogenic stability in maintaining the volume fractions of VLD, LD, HD phases under the cryogenic attack among these three samples.

6.6.3 Interlayer Space of C-A-S-H

A high-resolution XRD measurement was performed to evaluate the interlayer basal spacing of C-(A)-S-H upon cryogenic attack. The peaks in the range 5° ~ 10° 2θ are relevant to d_{002}, which indicates a 3D-ordering structure. The mean basal spacing and FWHM are averaged by five patterns on each sample, as shown in Fig. 6.30. The intensity of d_{002} peak of AS0.11 and AS0.22 is higher than that of AS0. Meanwhile, the FWHM of these two samples is lower than AS0, suggesting that the presence of aluminum results in a higher ordering structure for cross-linked phases. However, note that the uptake of aluminum also enlarges the basal spacing, and a high Al content even increases the FWHM (Fig. 6.30h) and decreases the intensity of d_{002} peak, as can be observed in Fig. 6.30g. These observations indicate that an excessive incorporation content of aluminum in cross-linked C-A-S-H may reversely decrease its structure ordering. Nevertheless, a high Al content still produces a more ordered phase than that in the absence of Al.

In agreement with the atomic C-(A)-S-H structure, the basal spacing is also altered upon the cryogenic attack (see Fig. 6.30g). The basal spacing of AS0 has an apparent decrease, which indicates that the distance between two silicate chains, e.g., d_{002} interlayer space, is shortened. In contrast, an increase is observed for the basal spacing of the Al-containing sample in AS0.22, suggesting an expansion of its interlayer space. The variation of the basal spacing of AS0.11 lies in between these two samples

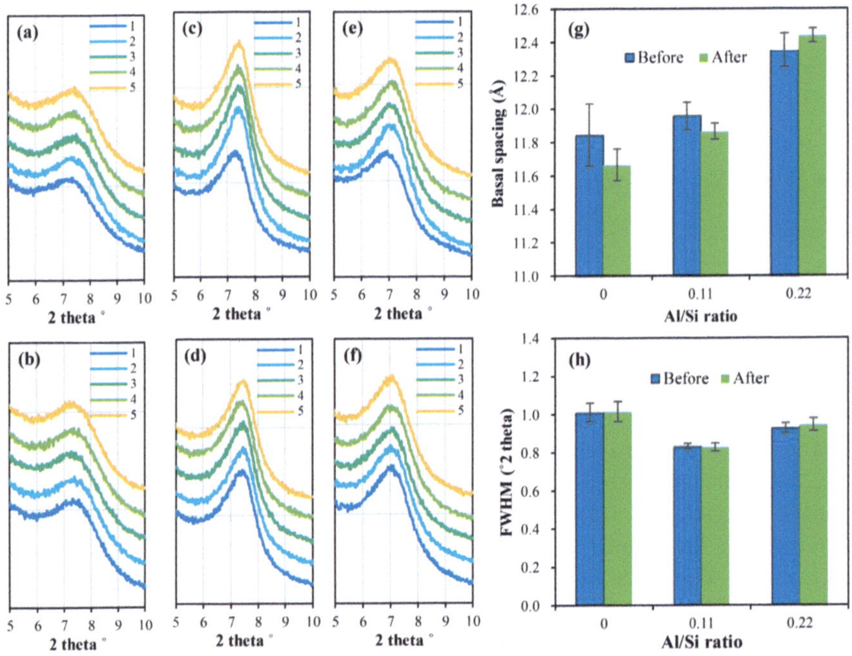

Fig. 6.30 Comparisons for the basal spacing and FWHM of C-(A)-S-H products before and after cryogenic attack. The XRD measurements for d_{002} peak were performed five times for each sample. **a**, **c** and **e** are the results of the non-cryogenically treated samples of AS0, AS0.11, and AS0.22, respectively; **b**, **d** and **f** are the results of cryogenically treated samples of the corresponding groups; **g** and **h** show the results of basal spacing and FWHM, respectively

mentioned above, only showing a mild decrease. Regarding the FWHM, we did not observe significant variations. Since these C-(A)-S-H phases are not entirely crystallized or ordered, the measured d_{002} peaks are broadened, making it difficult to resolve a tiny variation of the FWHM. However, the variation for the FWHM is consistent with the alteration observed for the interlayer basal spacing. The AS0.11 has a stable FWHM value upon the attack, just like its basal spacing, while the expanded interlayer basal spacing of AS0.22 follows with a slight increase in its FWHM.

6.6.4 Volumetric Variations of Intraglobular Pores

In the previous section, we assessed the stability of the interlayer space in C-(A)-S-H, regarded as characterization of the nanoscale space of approximately 1 nm. Herein, helium pyconometry is adopted to measure the bulk volume of basic building blocks that are usually considered at a scale of five nm. Compared with water vapor and helium, nitrogen is more likely to be blocked outside the smallest pores by the "ink bottle" [107]. The average minimum size of inaccessible pore for nitrogen is 1.1 nm which is in the range of SGP [107]. Thus, nitrogen is hindered from penetrating small gel pores (SGP) and interlayer space. Fortunately, the tiny SGP can be theoretically reached by helium flow since the helium can penetrate a confined pore with a diameter of 0.260 nm. Note that the interlayer space of a basic building block is assumed inaccessible for short-term helium inflow [101]. Therefore, the measured skeletal density from helium pycnometry is considered the average bulk density of the C-(A)-S-H basic building blocks [12].

Figure 6.31 shows the C-(A)-S-H skeletal density variations upon the attack. The impact of evaporable water on the internal space of C-(A)-S-H has been eliminated since the evaporable water had been removed from samples before sustaining the attack and helium pycnometry measurement. Before the attack, the Al-free C-S-H had the lowest skeletal density of 2.199 g/cm^3, while AS0.11 and AS0.22 were 2.387 g/cm^3 and 2.307 g/cm^3, respectively, which indicates that the incorporation of aluminum leads to a higher gel density. We should note that the internal space cannot hold constant under the cryogenic attack, and the variations depend on the aluminum content. A non-negligible increase is observed for the density of AS0, which means that the internal microstructure, including interlayer space and intra-globular pore (IGP) of C-S-H collapsed upon the attack, forming a denser structure. Reversely, skeletal densities of AS0.11 and AS0.22 decrease after the attack, indicating a volume expansion of the building blocks. These observations reflect different volumetric stability of Al-free C-S-H and C-A-S-H under ultra-cold conditions. Al-free building blocks are inclined to sustain a volume shrinkage, whereas Al-containing building blocks are more likely to expand. Note that the density of AS0.22 decreased much more significantly than that in AS0.11. In comparison, the density of AS0.11 merely experienced a slight decrease from 2.387 g/cm^3 to 2.368 g/cm^3, while an evident reduction from 2.307 g/cm^3 to 2.203 g/cm^3 was witnessed for

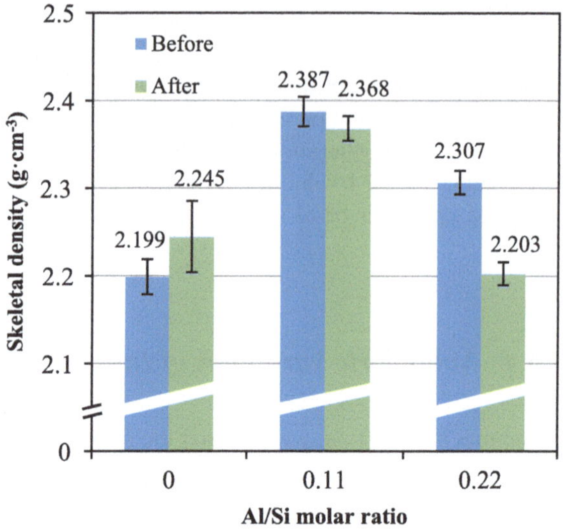

Fig. 6.31 The skeletal density of C-(A)-S-H before and after the cryogenic attack

AS0.22. Under cryogenic attack, the C-A-S-H building blocks showed better volumetric stability in a relatively lower Al/Si ratio. Among these samples, the building blocks in the Al-containing products of AS0.11 have the most stable structure since they are able to maintain their IGP structure upon the cryogenic attack. In contrast, basic building blocks of AS0 are incapable of remaining stable and preventing the IGP from collapsing.

6.6.5 Variations of Pore Networks

This section will assess the cryogenic stability of C-(A)-S-H mesoporous structure. The BJH analysis for the nitrogen adsorption measurement is depicted in Fig. 6.32. In contrast, AS0.11 and AS0.22 are occupied with a prominently higher pore volume than Al-free C-S-H before the attack. Such a significant difference between C-A-S-H and Al-free C-S-H mainly accounts for those large mesopores (> 20 nm) and macropores (> 50 nm), as shown in Fig. 6.32a. The measured cumulative pore volume of AS0, AS0.11, and AS0.22 presents an insignificant difference between 0 and 20 nm. Once the pore diameter is over 20 nm, the cumulative pore volume of AS0.11 and AS0.22 exceeds that of AS0. One should note that although incorporating Al may lead to a more porous microstructure, the higher incorporation content of aluminum does not necessarily yield a more porous structure, as indicated by our results. Herein, AS0.11 has the most significant cumulative pore volume instead of AS0.22.

From Fig. 6.32, even a single cryogenic attack can make a conspicuous reorganization to the C-(A)-S-H pore networks. The cumulative pore volume of AS0 has a significant increase, whereas AS0.11 sees a reduction, and the pore volume of

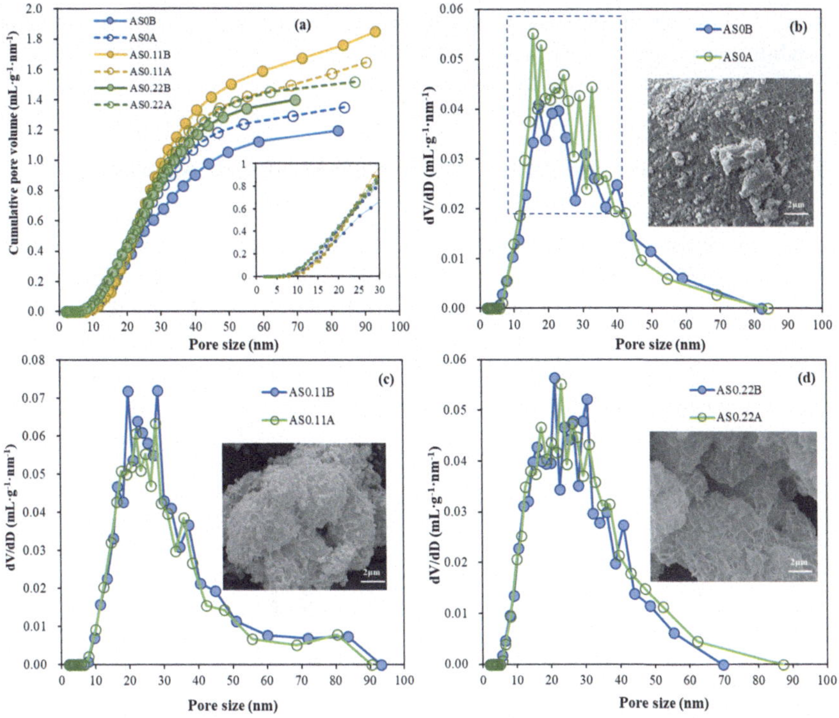

Fig. 6.32 Comparisons of pore structure of synthetic samples upon the cryogenic attack. **a** Shows the results of the cumulative pore volume. **b**, **c**, and **d** Show the pore size distribution of AS0, AS0.11, and AS0.22, respectively

AS0.22 generally remains steadily, merely a slight increase. Among these phases, the pore network of C-S-H in AS0 has been reconstructed violently. Upon the cryogenic attack, the cumulative pore volume of AS0 rises when the pore size is over 15 nm, while the other two Al-containing samples are more stable, particularly when the pore size is below 30 nm. From Fig. 6.32b, the significant variation observed for Al-free C-S-H is mainly derived from mesopores within 10–40 nm. It is believed that these increased pores are micro-cracks forming in C-S-H floccules. Consequently, the attack resulted in a more porous surface. Micro-cracks are widely distributed. The increased pore volume measured for AS0 after the attack is ascribed to the formation of micro-cracks. In contrast, the mesopores in Al-containing C-A-S-H solids represented a more stable trait. The result of AS0.11 presented in Fig. 6.32c suggests that the pores larger than 16 nm may sustain a slight collapse after the attack, consequently decreasing the pore volume. These pores in AS0.22 were reshaped and reconstructed mildly without significant variations. Nonetheless, the pore size distributions of measurable mesopores in AS0.11 and AS0.22 generally remain steadily, reflecting that these mesopores are more stable under the cryogenic attack compared with AS0. These results indicate that the incorporation of aluminum also strengthens the stability of the pore networks of C-(A)-S-H grains.

6.6.6 Geometric Characteristics of Pore Structure Networks

This section performs quantitative analysis on the pore networks of those Al-free and Al-containing phases. The CPSM theory is adopted here to quantify the geometric characteristics of pore structure networks, from which we can assess the evolution of nominal pore length, overall pore shape, and tortuosity upon cryogenic attack. The CPSM model fits the experimental adsorption and desorption isotherms, as plotted in Fig. 6.33, yielding the target key parameters, i.e., N_S, $D_{max.eff}$, $D_{min.eff}$, D_{mean}, and τ_{CPSM} (see details in Table 6.14).

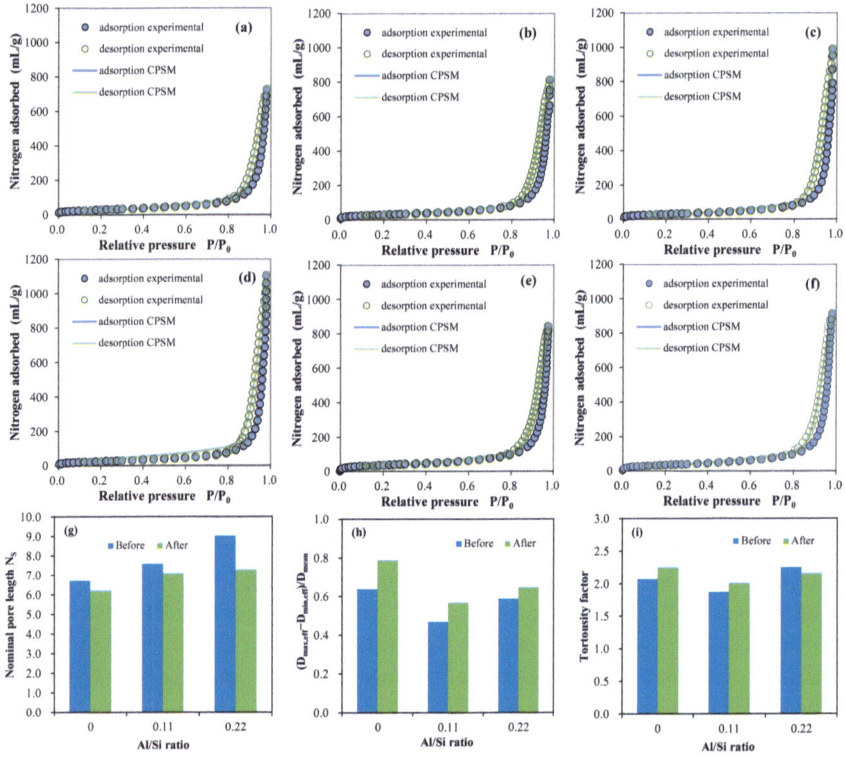

Fig. 6.33 CPSM simulation for nitrogen adsorption and desorption measurements. **a**, **b**, and **c** are the sorption isotherms of the non-cryogenically treated samples of AS0B, AS0.11B, and AS0.22B, respectively; **d**, **e**, and **c** are the results of cryogenically treated samples AS0A1, AS0.11A1, and AS0.22A1, respectively; **g** the nominal pore length Ns; **h** the pore shape factor $(D_{\max,eff} - D_{\min,eff})/D_{mean}$; **i** the tortuosity factor

Table 6.14 The nominal pore length and CPSM-tortuosity factors of C-(A)-S-H samples

Samples	N_S	$D_{\min,\text{eff}}$ (nm)	$D_{\max,\text{eff}}$ (nm)	D_{mean} (nm)	$D_{\max,eff} - D_{\min,eff}$ (nm)	$\frac{D_{\max,eff} - D_{\min,eff}}{D_{mean}}$	τ_{CPSM}
AS0B	6.73	17.07	37.68	32.44	20.62	0.64	2.08
AS0A1	6.24	16.37	41.07	31.23	24.70	0.79	2.26
AS0.11B	7.61	22.31	42.38	42.99	20.08	0.47	1.88
AS0.11A1	7.12	19.93	41.84	38.35	21.91	0.57	2.02
AS0.22B	9.05	17.14	35.76	31.70	18.62	0.59	2.26
AS0.22A1	7.32	18.00	39.82	33.83	21.83	0.65	2.17

From Fig. 6.33, the nominal pore length N_S of AS0, AS0.11, and AS0.22 before cryogenic attack is 6.73, 7.61, and 9.05, respectively. The nominal pore length results imply that the incorporation of aluminum in C-A-S-H promotes its pore length, and a higher Al/Si ratio may breed a longer pore tunnel. Nonetheless, since the pore size distributions of AS0.11 and AS0.22 are more accumulational, as indicated by the pore shape factor, i.e., $(D_{\max,eff} - D_{\min,eff})/D_{mean}$ (Fig. 6.33i), the increasing nominal pore length does not necessarily give rise to a more tortuous pore network with a higher tortuosity. In comparison, the pore sizes of the AS0.11 and AS0.22 are more homogenous than AS0, and the AS0.11 has the most uniform pore size distribution.

Apparently, the cryogenic attack reshaped the pore networks. The N_S values (Fig. 6.33g) of AS0, AS0.11, and AS0.22 declined to 6.24, 7.12, and 7.32, respectively. At the same time, their corresponding pore shape factors shifted upward to 0.79, 0.57, and 0.69, respectively. Consequently, the tortuosity factors of AS0 and AS0.11 are calculated with higher values of 2.26 and 2.02, respectively, while the tortuosity of AS0.22 reversely moves slightly toward a lower value of 2.17. According to the above comparisons, AS0.11 has better stability in preserving its pore geometry. These results quantitively describe the evolution of pore geometries under the cryogenic attack. The harsh attack decreased the nominal pore length and the homogeneity of the pore size distribution for all samples, regardless of Al-free C-S-H or C-A-S-H. The decreased nominal pore length suggests that the pore networks are split and broken into shorter pore tunnels under the cryogenic attack, which is highly likely due to the propagation of microcracks, particularly the large microcrack. The formation of microcracks cuts down the pore tunnels, shortening the nominal pore length and widening the pore size distribution (see Fig. 6.33).

6.6.7 The Role of Aluminum

Based on the stability assessment and analysis of C-(A)-S-H structures at different scales as mentioned above, we found that the nano-mechanical properties and cryogenic stability of the low-Ca C-A-S-H system are superior to the Al-free C-S-H system. Particularly, the low-Al and low-Ca cross-linked C-A-S-H system (Ca/Si

= 1, Al/Si = 0.11) maintains a highly stable micro-nanostructure under cryogenic attack.

For the low-Ca C-(A)-S-H system, aluminum incorporation promotes cross-linking between the silicate chains. According to Geng et al. [49], these connections formed in the cross-linking areas act as pillars, strengthening the interlayer structure of C-(A)-S-H and increasing its rigidity. Hence, the volume stability of C-A-S-H basic gel units is superior to C-S-H gel units. When the interlayer space extends to interlayer pores, there might be a slight expansion. However, in the case of cross-linked C-S-H, the average chain length is shorter than that of cross-linked C-A-S-H [32, 73], implying that C-S-H silicate chains inherently contain more defect vacancies than cross-linked C-A-S-H. Moreover, notably, cross-linked silicate chains undergo partial breakage at bridging sites, resulting in more defect vacancies within the chains, which is detrimental to the stability of the interlayer structure, making it more sensitive to external or internal forces. Figure 6.34 illustrates the schematic evolution of the C-(A)-S-H basic gel units under cryogenic attack. The changes in nanostructure might correspond to irreversible deformation. For Al-free C-S-H, the collapsed basic gel units create a denser micro-nanostructure, further increasing its true density. The noticeable expansion of the internal space within the C-A-S-H basic gel units in AS0.22 leads to structural looseness, forming a micro-nanostructure with a lower true density. The relatively well-preserved silicate chains in the AS0.11 sample might explain its outstanding cryogenic stability in maintaining the interlayer spacing and internal pore structure.

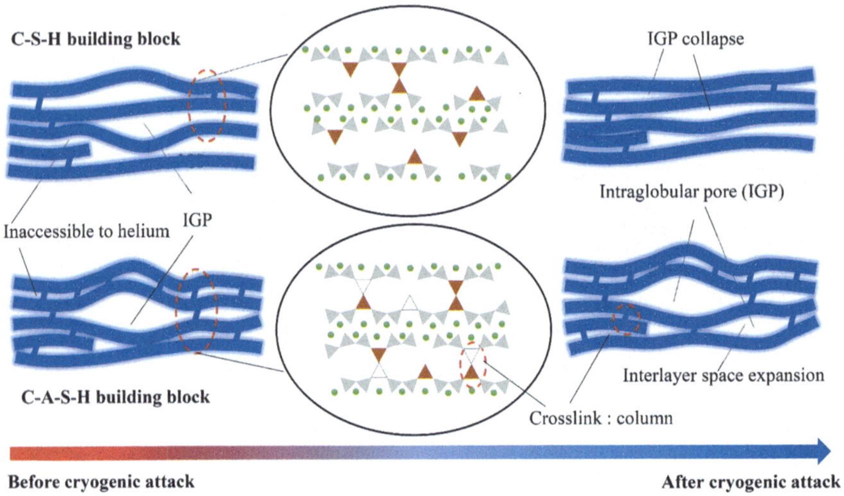

Fig. 6.34 Schemes for the volumetric variations of Al-free C-S-H and Al-containing C-A-S-H building blocks

6.7 Concluding Remarks

In this chapter, we show the potential of Al incorporation of C-A-S-H in enhancing its cryogenic stability. We systematically analyze the composition and atomic structure of C-A-S-H, revealing the stability of low-Al and high-Al C-A-S-H nanostructures under cryogenic attack. For the first time, we experimentally demonstrate the presence of bridging and interlayer aluminates in C-A-S-H, further expanding the understanding of the existing C-A-S-H nanostructures. We introduce a DNA-code rule to describe and construct molecular models of C-A-S-H, and propose a generalized structural-chemical formula for C-A-S-H models that can describe both cross-linked and non-cross-linked structures, considering bridging and interlayer aluminates. Furthermore, we explore the cryogenic stability characteristics of C-A-S-H nano-mechanical properties and multiscale micro-nanostructures. The main conclusions are as follows:

(1) Under cryogenic attack, the nanostructure of low-Al C-A-S-H systems ($Al/(Al + Si) = 0.1$) exhibits better cryogenic stability compared to Al-free C-S-H systems and also surpasses the nanostructure stability of high-Al C-A-S-H ($Al/(Al + Si) = 0.2$). Low aluminum incorporation enhances the cryogenic stability of C-(A)-S-H nanostructures.

(2) Under cryogenic attack, the primary silica chain structure of low-Al C-A-S-H systems remains generally more stable. However, in low-Ca cross-linked systems, the network might experience minor fractures at branched aluminate sites, resulting in a reduction in the average chain length. In contrast, other high-Ca systems tend to incorporate the interlayer pentahedral aluminates as bridging octahedral units, causing slightly longer chain lengths and aluminum substitution ratios. When $Ca/(Al + Si) > 1.0$, under cryogenic attack, high-Al C-A-S-H releases bridging pentahedral aluminates from aluminosilicate chains, transforming into interlayer octahedral aluminates, significantly reducing the average chain length and Al/Si ratio.

(3) The presence forms of tetrahedral, pentahedral, and octahedral aluminates in cross-linked and non-cross-linked C-A-S-H have been explicitly identified. Tetrahedral aluminates can only exist at the bridging or branching sites of the aluminosilicate chains in C-A-S-H, and the aluminate in cross-linked C-A-S-H can only exist in a tetrahedral form. Both bridging and interlayer pentahedral aluminates coexist in C-A-S-H, characterized by a corresponding ^{27}Al chemical shift of ~ 39 ppm. Octahedral aluminates can exist in C-A-S-H at both bridging and interlayer positions, and the ~ 5 ppm ^{27}Al chemical shift is associated with both bridging and interlayer octahedral aluminates.

(4) The C-A-S-H brick model and the corresponding DNA code rules were proposed for constructing realistic C-A-S-H molecular structures, incorporating both bridging and interlayer aluminates. This DNA code rule introduces bridging tetrahedral, pentahedral, and octahedral aluminate units, as well as interlayer pentahedral and octahedral aluminate units. Building upon the C-A-S-H brick model, a generalized structural chemical formula for C-A-S-H was further

proposed, capable of describing cross-linked, non-cross-linked, and hybrid C-A-S-H systems simultaneously.

(5) The nanomechanical stability of low-Al C-A-S-H also surpasses that of Al-free C-S-H and high-Al C-A-S-H. Under cryogenic attack, the various decomposition phases of low-Al C-A-S-H remain stable, preserving the intrinsic nanomechanical properties of C-A-S-H without significant degradation. Cryogenic attack causes an increase in the average Young's modulus of C-A-S-H, which is not related to the compactness of C-A-S-H particle aggregation but should be attributed to the inherent changes in its nanomechanical properties. In contrast, the nanomechanical performance of Al-free C-S-H deteriorates, resulting in a decrease in the intrinsic Young's modulus of its decomposition phases, while the characteristic Young's modulus of the decomposition phases of Al-incorporated C-A-S-H increases after cryogenic attack. Although both Al-free C-S-H and Al-incorporated C-A-S-H achieve higher average Young's modulus due to cryogenic attack, the reasons behind this increase differ. The increase in the average Young's modulus of C-A-S-H is benefited by the increase in the characteristic Young's modulus of its decomposition phases (VLD, LD, HD, CCD), whereas the increase in the average Young's modulus of Al-free C-S-H mainly originates from the increase in HD content.

(6) Within the low-Al C-A-S-H system, the incorporation of aluminum atoms strengthens the cross-linking structure of C-A-S-H. More importantly, under cryogenic attack, the nanostructure of low-Al C-A-S-H tends to be more stable, less prone to fracture, thereby enhancing the interlayer space stability of C-A-S-H, and reducing the likelihood of collapse of the basic C-A-S-H basic building blocks. Consequently, the pore structure of C-A-S-H remains relatively stable under ultra-low temperatures, exhibiting no significant microcrack expansion or microstructural degradation as observed in C-S-H.

References

1. A.L. Marshall., Cryogenic concrete. Cryogenics **22**(11), 555–565 (1982)
2. B. He, M. Xie, Z. Jiang, C. Zhang, X. Zhu, Temperature field distribution and microstructure of cement-based materials under cryogenic freeze-thaw cycles. Constr. Build. Mater. **243**, 118256 (2020)
3. B. He, X. Zhu, Q. Ren, Z. Jiang, Effects of fibers on flexural strength of ultra-high-performance concrete subjected to cryogenic attack. Constr. Build. Mater. **265**, 120323 (2020)
4. X. Zhu, B. He, M. Xie, Y. Zhang, Z. Jiang, A comprehensive nitrogen adsorption measurement on the pore structure of calcium-silicate-hydrate subjected to cryogenic attack. Measurement **184**, 109941 (2021)
5. G.P. Hu, Y.C. Zheng, A.A. Xu, Z.S. Tang, Lunar surface temperature of global moon: preparation of database with topographic and albedo effects. IEEE Geosci. Remote Sens. Lett. **13**(1), 110–114 (2016)
6. H. Benaroya, L. Bemold, Engineering of lunar bases. Acta Astronaut. **62**(4–5), 277–299 (2008)

7. E. Sebastián, C. Armiens, J. Gómez-Elvira, M.P. Zorzano, J. Martinez-Frias, B. Esteban, M. Ramos, The rover environmental monitoring station ground temperature sensor: a pyrometer for measuring ground temperature on Mars. Sensors **10**(10) (2010).
8. M. Jiang, L. Li, Y. Sun, Properties of TJ-1 Lunar Soil Simulant. J. Aerosp. Eng. **25**(3), 463–469 (2012)
9. R. Rieder, T. Economou, H. Wänke, A. Turkevich, J. Crisp, J. Brückner, G. Dreibus, H.Y. McSween, The chemical composition of martian soil and rocks returned by the mobile alpha proton X-ray spectrometer: preliminary results from the X-ray mode. Science **278**(5344), 1771 (1997)
10. J.F. Bell Iii, H.Y. McSween Jr, J.A. Crisp, R.V. Morris, S.L. Murchie, N.T. Bridges, J.R. Johnson, D.T. Britt, M.P. Golombek, H.J. Moore, A. Ghosh, J.L. Bishop, R.C. Anderson, J. Brückner, T. Economou, J.P. Greenwood, H.P. Gunnlaugsson, R.M. Hargraves, S. Hviid, J.M. Knudsen, M.B. Madsen, R. Reid, R. Rieder, L. Soderblom, Mineralogic and compositional properties of Martian soil and dust: results from Mars pathfinder. J. Geophys. Res. Planets **105**(E1), 1721–1755 (2000)
11. C. Li, J. Li, A. Telesca, D. Marchon, K. Xu, M. Marroccoli, Z. Jiang, P.J.M. Monteiro, Effect of polycarboxylate ether on the expansion of ye'elimite hydration in the presence of anhydrite. Cem. Concr. Res. **140**, 106321 (2021)
12. X. Zhu, Z. Jiang, B. He, C. Qian, Investigation on the physical stability of calcium-silicate-hydrate with varying CaO/SiO_2 ratios under cryogenic attack. Constr. Build. Mater. **252**, 119103 (2020)
13. X. Zhu, Q. Ren, B. He, H. Zhang, Z. Jiang, Upscaling degradation of cementitious calcium (aluminate) silicate hydrate upon ultra-low temperature attack: a multiscale insight and a bottom-up enhancement route. Compos. B Eng. **243**, 110122 (2022)
14. E. Pustovgar, R.P. Sangodkar, A.S. Andreev, M. Palacios, B.F. Chmelka, R.J. Flatt, J.B. d'Espinose de Lacaillerie, Understanding silicate hydration from quantitative analyses of hydrating tricalcium silicates. Nat. Commun. **7**, 10952 (2016)
15. A. Kunhi Mohamed, S.C. Parker, P. Bowen, S. Galmarini, An atomistic building block description of C-S-H—towards a realistic C-S-H model, Cem. Concr. Res. **107**, 221–235 (2018)
16. E. Bonaccorsi, S. Merlino, A.R. Kampf, The crystal structure of tobermorite 14 A (plombierite), a C-S-H phase. J. Am. Ceram. Soc. **88**(3), 505–512 (2005)
17. I.G. Richardson, Tobermorite/jennite- and tobermorite/calcium hydroxide-based models for the structure of C-S-H: applicability to hardened pastes of tricalcium silicate, β-dicalcium silicate, Portland cement, and blends of Portland cement with blast-furnace slag, metakaolin, or silica fume. Cem. Concr. Res. **34**(9), 1733–1777 (2004)
18. I.G. Richardson., G.W. Groves., Models for the composition and structure of calcium silicate hydrate(C-S-H) gel in hardened tricalcium silicate pastes. Cem. Concr. Res. **22**, 1001–1010 (1992)
19. X. Zhu, C. Qian, B. He, Q. Chen, Z. Jiang, Experimental study on the stability of C-S-H nanostructures with varying bulk CaO/SiO_2 ratios under cryogenic attack. Cem. Concr. Res. **135**, 106114 (2020)
20. E. Gallucci, X. Zhang, K.L. Scrivener, Effect of temperature on the microstructure of calcium silicate hydrate (C-S-H). Cem. Concr. Res. **53**, 185–195 (2013)
21. Y. Zhang, Q. Zhou, J.W. Ju, M. Bauchy, New insights into the mechanism governing the elasticity of calcium silicate hydrate gels exposed to high temperature: a molecular dynamics study. Cem. Concr. Res. **141**, 106333 (2021)
22. A. Kumar, B.J. Walder, A. Kunhi Mohamed, A. Hofstetter, B. Srinivasan, A.J. Rossini, K. Scrivener, L. Emsley, P. Bowen, The atomic-level structure of cementitious calcium silicate hydrate. J. Phys. Chem. C **121**(32),17188–17196 (2017)
23. A. Nonat, The structure and stoichiometry of C-S-H. Cem. Concr. Res. **34**(9), 1521–1528 (2004)
24. H.F.w. Taylor., Proposed structure for calcium silicate hydrate gel. J. Am. Ceram. Soc. **69**(6), 464–67 (1986)

25. L.S. Dent Glasser, E.E. Lachowski, M.Y. Qureshi, H.P. Calhoun, D.J. Embree, W.D. Jamieson, C.R. Masson, Identification of some of the polysilicate components of trimethylsilylated cement paste. Cem. Concr. Res. **11**(5), 775–780 (1981)
26. K. Mohan, H.F.W. Taylor, A trimethylsilylation study of tricalcium silicate pastes. Cem. Concr. Res. **12**(1), 25–31 (1982)
27. J. Hirljac, Z.Q. Wu, J.F. Young, Silicate polymerization during the hydration of alite. Cem. Concr. Res. **13**(6), 877–886 (1983)
28. B. Lothenbach, K. Scrivener, R.D. Hooton, Supplementary cementitious materials. Cem. Concr. Res. **41**(12), 1244–1256 (2011)
29. K.L. Scrivener, A. Nonat, Hydration of cementitious materials, present and future. Cem. Concr. Res. **41**(7), 651–665 (2011)
30. Q. Ren, Z.W. Jiang, H.X. Li, X.P. Zhu, Q. Chen, Fresh and hardened properties of self-compacting concrete using silicon carbide waste as a viscosity-modifying agent. Constr. Build. Mater. **200**, 324–332 (2019)
31. C.A. Love, I.G. Richardson, A.R. Brough, Composition and structure of C-S-H in white Portland cement-20% metakaolin pastes hydrated at 25 degrees C. Cem. Concr. Res. **37**(2), 109–117 (2007)
32. Z. Dai, T.T. Tran, J. Skibsted, H. Jennings, Aluminum incorporation in the C-S-H phase of white Portland cement-metakaolin blends studied by ^{27}Al and ^{29}Si MAS NMR spectroscopy. J. Am. Ceram. Soc. **97**(8), 2662–2671 (2014)
33. X. Pardal, I. Pochard, A. Nonat, Experimental study of Si–Al substitution in calcium-silicate-hydrate (C-S-H) prepared under equilibrium conditions. Cem. Concr. Res. **39**(8), 637–643 (2009)
34. E. L'Hopital, B. Lothenbach, D.A. Kulik, K. Scrivener, Influence of calcium to silica ratio on aluminium uptake in calcium silicate hydrate. Cem. Concr. Res. **85**, 111–121 (2016)
35. P. Faucon, T. Charpentier, A. Nonat, J.C. Petit, Triple-quantum two-dimensional ^{27}Al magic angle nuclear magnetic resonance study of the aluminum incorporation in calcium silicate hydrates. J. Am. Chem. Soc. **120**(46), 12075–12082 (1998)
36. P. Faucon, A. Delagrave, C. Richet, J.M. Marchand, H. Zanni, Aluminum incorporation in calcium silicate hydrates (C−S−H) depending on their Ca/Si ratio. J. Phys. Chem. B **103**(37), 7796–7802 (1999)
37. G.K. Sun, J.F. Young, R.J. Kirkpatrick, The role of Al in C-S-H: NMR, XRD, and compositional results for precipitated samples. Cem. Concr. Res. **36**(1), 18–29 (2006)
38. X. Pardal, F. Brunet, T. Charpentier, I. Pochard, A. Nonat, ^{27}Al and ^{29}Si solid-state NMR characterization of calcium-aluminosilicate-hydrate. Inorg. Chem. **51**(3), 1827–1836 (2012)
39. T.T.H. Bach, C.C.D. Coumes, I. Pochard, C. Mercier, B. Revel, A. Nonat, Influence of temperature on the hydration products of low pH cements. Cem. Concr. Res. **42**(6), 805–817 (2012)
40. L. Pegado, C. Labbez, S.V. Churakov, Mechanism of aluminium incorporation into C-S-H from ab initio calculations. J. Mater. Chem. A **2**(10), 3477–3483 (2014)
41. P. Faucon., J.M. Delaye., J. Virlet., J.F. Jacquinot., F. Adenotg., Study of the structural properties of the C-S-H(I) by molecular dynamics simulation, Cem. Concr. Res. **27**(10), 1581–1590 (1997)
42. M.J.A. Qomi, F.J. Ulm, R.J.M. Pellenq, Evidence on the dual nature of aluminum in the calcium-silicate-hydrates based on atomistic simulations. J. Am. Ceram. Soc. **95**(3), 1128–1137 (2012)
43. A. Kunhi Mohamed, P. Moutzouri, P. Berruyer, B.J. Walder, J. Siramanont, M. Harris, M. Negroni, S.C. Galmarini, S.C. Parker, K.L. Scrivener, L. Emsley, P. Bowen, The atomic-level structure of cementitious calcium aluminate silicate hydrate. J. Am. Chem. Soc. **142**(25), 11060–11071 (2020)
44. W. Loewenstein, The distribution of aluminum in the tetrahedra of silicates and aluminates. Am. Mineral. **39**(1–2), 92–96 (1954)
45. A.V. Larin, The Loewenstein rule: the increase in electron kinetic energy as the reason for instability of Al–O–Al linkage in aluminosilicate zeolites. Phys. Chem. Miner. **40**(10), 771–780 (2013)

46. H. Manzano, J.S. Dolado, A. Ayuela, Aluminum incorporation to dreierketten silicate chains. J. Phys. Chem. B **113**(9), 2832–2839 (2009)
47. J. Li, W. Zhang, K. Garbev, G. Beuchle, P.J.M. Monteiro, Influences of cross-linking and Al incorporation on the intrinsic mechanical properties of tobermorite. Cem. Concr. Res. **136**, 106170 (2020)
48. J.Q. Li, G.Q. Geng, R. Myers, Y.S. Yu, D. Shapiro, C. Carraro, R. Maboudian, P.J.M. Monteiro, The chemistry and structure of calcium (alumino) silicate hydrate: a study by XANES, ptychographic imaging, and wide- and small-angle scattering. Cem. Concr. Res. **115**, 367–378 (2019)
49. G. Geng, R.J. Myers, J. Li, R. Maboudian, C. Carraro, D.A. Shapiro, P.J. Monteiro, Aluminum-induced dreierketten chain cross-links increase the mechanical properties of nanocrystalline calcium aluminosilicate hydrate. Sci. Rep. **7**, 44032 (2017)
50. E. L'Hôpital, B. Lothenbach, K. Scrivener, D.A. Kulik, Alkali uptake in calcium alumina silicate hydrate (C-A-S-H). Cem. Concr. Res. **85**, 122–136 (2016)
51. S. Komarneni, M. Tsuji, Selective cation exchange in substituted tobermorites. J. Am. Ceram. Soc. **72**(9), 1668–1674 (1989)
52. S. Komarneni, D.M. Roy, C.A. Fyfe, G.J. Kennedy, Naturally occurring 1.4 nm tobermorite and synthetic jennite: Characterization by ^{27}Al and ^{29}Si MASNMR spectroscopy and cation exchange properties. Cem. Concr. Res. **17**(6), 891–895 (1987)
53. H. Stade, D. Müller, G. Scheler, Zum Aufbau schlecht geordneter Calciumhydrogensilicate. V ^{27}Al-NMR-spektroskopische Untersuchungen zur Koordination des Al in C–S–H (Di, Poly). Z. Anorg. Allg. Chem. **510**(3), 16–24 (1984)
54. H. Stade, D. Müller, On the coordination of Al in ill-crystallized C-S-H phases formed by hydration of tricalcium silicate and by precipitation reactions at ambient temperature. Cem. Concr. Res. **17**(4), 553–561 (1987)
55. M.D. Andersen, H.J. Jakobsen, J. Skibsted, A new aluminium-hydrate species in hydrated Portland cements characterized by ^{27}Al and ^{29}Si MAS NMR spectroscopy. Cem. Concr. Res. **36**(1), 3–17 (2006)
56. M.D. Andersen., H.J. Jakobsen., J. Skibsted., Incorporation of aluminum in the calcium silicate hydrate (C−S−H) of hydrated Portland cements: A high-field ^{27}Al and ^{29}Si MAS NMR investigation. Inorg. Chem. **42**, 2280−2287 (2003)
57. R. Taylor, I.G. Richardson, R.M.D. Brydson, Composition and microstructure of 20-year-old ordinary Portland cement–ground granulated blast-furnace slag blends containing 0 to 100% slag. Cem. Concr. Res. **40**(7), 971–983 (2010)
58. C. Roosz, P. Vieillard, P. Blanc, S. Gaboreau, H. Gailhanou, D. Braithwaite, V. Montouillout, R. Denoyel, P. Henocq, B. Made, Thermodynamic properties of C-S-H, C-A-S-H and M-S-H phases: Results from direct measurements and predictive modelling. Appl. Geochem. **92**, 140–156 (2018)
59. I. Lognot, I. Klur, A. Nonat, NMR and infrared spectroscopies of C-S-H and Al-substituted C-S-H synthesised in alkaline solutions, in: H.Z. P. Colombet, A.-R. Grimmer, P. Sozzani (Eds.) (Ed.) *Nuclear Magnetic Resonance Spectroscopy of Cement-Based Materials* (Springer, Berlin Heidelberg, 1998), pp. 189–196
60. R.J. Myers, E. L'Hôpital, J.L. Provis, B. Lothenbach, Effect of temperature and aluminium on calcium (alumino)silicate hydrate chemistry under equilibrium conditions. Cem. Concr. Res. **68**, 83–93 (2015)
61. S. Grangeon, F. Claret, Y. Linard, C. Chiaberge, X-ray diffraction: a powerful tool to probe and understand the structure of nanocrystalline calcium silicate hydrates. Acta Crystallogr., Sect. B: Struct. Sci. **69**(5), 465–473 (2013)
62. C. Roosz, S. Gaboreau, S. Grangeon, D. Prêt, V. Montouillout, N. Maubec, S. Ory, P. Blanc, P. Vieillard, P. Henocq, Distribution of water in synthetic calcium silicate hydrates. Langmuir **32**(27), 6794–6805 (2016)
63. E. Kapeluszna, Ł Kotwica, A. Różycka, Ł Gołek, Incorporation of Al in C-A-S-H gels with various Ca/Si and Al/Si ratio: Microstructural and structural characteristics with DTA/TG, XRD, FTIR and TEM analysis. Constr. Build. Mater. **155**, 643–653 (2017)

64. J.J. Kim, E.M. Foley, M.M. Reda Taha, Nano-mechanical characterization of synthetic calcium–silicate–hydrate (C–S–H) with varying CaO/SiO_2 mixture ratios, Cem. Concr. Compos. **36**, 65–70 (2013)
65. B. Lothenbach, L. Pelletier-Chaignat, F. Winnefeld, Stability in the system CaO–Al_2O_3–H_2O. Cem. Concr. Res. **42**(12), 1621–1634 (2012)
66. X. Zhu, Q. Ren, B. He, H. Zhang, Z. Jiang, The nanomechanical nature of cementitious calcium-(alumino)silicate-hydrate under ultra-low temperature attack: an intrinsic transition and the role of aluminum for bottom-up enhancement. Cem. Concr. Compos. **134**, 104811 (2022)
67. G. Le Saout, E. Lécolier, A. Rivereau, H. Zanni, Chemical structure of cement aged at normal and elevated temperatures and pressures: Part I. Class G oilwell cement. Cem. Concr. Res. **36**(1), 71–78 (2006)
68. G. Le Saout, E. Lecolier, A. Rivereau, H. Zanni, Chemical structure of cement aged at normal and elevated temperatures and pressures, Part II: low permeability class G oilwell cement. Cem. Concr. Res. **36**(3), 428–433 (2006)
69. H. Noma., Y. Adachil., M. Takushi., Y. Hideo., Y. Tadashi., N. Yoshihisa., ^{29}Si MAS NMR spectroscopy of poorly-crystalline calcium silicate hydrates (C-S-H), in *Nuclear Magnetic Resonance Spectroscopy of Cement-Based Materials* (Springer, Berlin, Heidelberg, 1998), pp. 159–168
70. I. Klur., B. Pollet., J. Virlet., A. Nonat., C-S-H structure evolution with calcium content by multinuclear NMR, in *Nuclear Magnetic Resonance Spectroscopy of Cement-Based Materials* (Springer, Verlag Berlin Heidelberg, 1998), pp. 119–141
71. P. Rejmak, J.S. Dolado, M.J. Stott, A. Ayuela, ^{29}Si NMR in cement: a theoretical study on calcium silicate hydrates. J. Phys. Chem. C **116**(17), 9755–9761 (2012)
72. R.J. Myers, S.A. Bernal, J.D. Gehman, J.S.J. van Deventer, J.L. Provis, L. Struble, The role of Al in cross-linking of alkali-activated slag cements. J. Am. Ceram. Soc. **98**(3), 996–1004 (2015)
73. E. L'Hôpital, B. Lothenbach, G. Le Saout, D. Kulik, K. Scrivener, Incorporation of aluminium in calcium-silicate-hydrates. Cem. Concr. Res. **75**, 91–103 (2015)
74. J.J. Chen, J.J. Thomas, H.F.W. Taylor, H.M. Jennings, Solubility and structure of calcium silicate hydrate. Cem. Concr. Res. **34**(9), 1499–1519 (2004)
75. S.-Y. Hong, F.P. Glasser, Alkali sorption by C-S-H and C-A-S-H gels: Part II. Role of alumina. Cem. Concr. Res. **32**(7), 1101–1111 (2002)
76. G. Le Saoût, M. Ben Haha, F. Winnefeld, B. Lothenbach, Hydration degree of alkali-activated slags: A ^{29}Si NMR study, J. Am. Ceram. Soc. **94**(12), 4541–4547 (2011)
77. S.A. Bernal, J.L. Provis, B. Walkley, R.S. Nicolas, J.D. Gehman, D.G. Brice, A.R. Kilcullen, P. Duxson, J.S.J. van Deventer, Gel nanostructure in alkali-activated binders based on slag and fly ash, and effects of accelerated carbonation. Cem. Concr. Res. **53**, 127–144 (2013)
78. F. Bonk, J. Schneider, M.A. Cincotto, H. Panepucci, Characterization by multinuclear high-resolution NMR of hydration products in activated blast-furnace slag pastes. J. Am. Ceram. Soc. **86**(10), 1712–1719 (2003)
79. B. Walkley, R. San Nicolas, M.-A. Sani, G.J. Rees, J.V. Hanna, J.S.J. van Deventer, J.L. Provis, Phase evolution of C-(N)-A-S-H/N-A-S-H gel blends investigated via alkali-activation of synthetic calcium aluminosilicate precursors. Cem. Concr. Res. **89**, 120–135 (2016)
80. F. Puertas, M. Palacios, H. Manzano, J.S. Dolado, A. Rico, J. Rodriguez, A model for the C-A-S-H gel formed in alkali-activated slag cements. J. Eur. Ceram. Soc. **31**(12), 2043–2056 (2011)
81. G. Renaudin, J. Russias, F. Leroux, C. Cau-dit-Coumes, F. Frizon, Structural characterization of C-S-H and C–A–S–H samples—part II: local environment investigated by spectroscopic analyses. J. Solid State Chem. **182**(12), 3320–3329 (2009)
82. M.T. Janicke, C.C. Landry, S.C. Christiansen, D. Kumar, G.D. Stucky, B.F. Chmelka, Aluminum incorporation and interfacial structures in MCM-41 mesoporous molecular sieves. J. Am. Chem. Soc. **120**(28), 6940–6951 (1998)

83. A. Rawal, B.J. Smith, G.L. Athens, C.L. Edwards, L. Roberts, V. Gupta, B.F. Chmelka, Molecular silicate and aluminate species in anhydrous and hydrated cements. J. Am. Chem. Soc. **132**(21), 7321–7337 (2010)
84. I.G. Richardson, Model structures for C-(A)-S-H(I), Acta Crystallogr. B: Struct. Sci. Cryst. Eng. Mater. **70**(6), 903–923 (2014)
85. J. Brus, L. Kobera, M. Urbanová, D. Koloušek, J. Kotek, Insights into the structural transformations of aluminosilicate inorganic polymers: a comprehensive solid-state NMR study. J. Phys. Chem. C **116**(27), 14627–14637 (2012)
86. N. Katada, S. Nakata, S. Kato, K. Kanehashi, K. Saito, M. Niwa, Detection of active sites for paraffin cracking on USY zeolite by ^{27}Al MQMAS NMR operated at high magnetic field 16 T. J. Mol. Catal. A: Chem. **236**(1), 239–245 (2005)
87. Z. Yu, A. Zheng, Q. Wang, L. Chen, J. Xu, J.-P. Amoureux, F. Deng, Insights into the dealumination of zeolite HY revealed by sensitivity-enhanced ^{27}Al DQ-MAS NMR spectroscopy at high field. Angew. Chem. Int. Ed. **49**(46), 8657–8661 (2010)
88. Z. Dai, T.T. Tran, J. Skibsted, Aluminum incorporation in the C-S-H phase of white Portland cement–metakaolin blends studied by ^{27}Al and ^{29}Si MAS NMR spectroscopy. J. Am. Ceram. Soc. **97**(8), 2662–2671 (2014)
89. R.J. Myers, E. L'Hopital, J.L. Provis, B. Lothenbach, Effect of temperature and aluminium on calcium (alumino)silicate hydrate chemistry under equilibrium conditions. Cem. Concr. Res. **68**, 83–93 (2015)
90. I.G. Richardson., G.W. Groves., The incorporation of minor and trace elements into calcium silicate hydrate (C-S-H) gel in hardened cement pastes. Cem. Concr. Res. **23**, 131–138 (1993)
91. R.J. Myers, S.A. Bernal, R. San Nicolas, J.L. Provis, Generalized structural description of calcium-sodium aluminosilicate hydrate gels: the cross-linked substituted tobermorite model. Langmuir **29**(17), 5294–5306 (2013)
92. I.G. Richardson, A.R. Brough, G.W. Groves, C.M. Dobson, The characterization of hardened alkali-activated blast-furnace slag pastes and the nature of the calcium silicate hydrate (C-S-H) phase. Cem. Concr. Res. **24**(5), 813–829 (1994)
93. J. Haas, A. Nonat, From C-S-H to C–A–S–H: experimental study and thermodynamic modelling. Cem. Concr. Res. **68**, 124–138 (2015)
94. C.E. White, Effects of temperature on the atomic structure of synthetic calcium–silicate–deuterate gels: a neutron pair distribution function investigation. Cem. Concr. Res. **79**, 93–100 (2016)
95. M. Sitarz., M. Handke., W. Mozgawa., Calculations of silicooxygen ring vibration frequencies. Spectrochim. Acta, Part A **55**, 2831–2837 (1999)
96. L. Black, K. Garbev, P. Stemmermann, K.R. Hallam, G.C. Allen, Characterisation of crystalline C-S-H phases by X-ray photoelectron spectroscopy. Cem. Concr. Res. **33**(6), 899–911 (2003)
97. K. Muraoka, W. Chaikittisilp, T. Okubo, Energy analysis of aluminosilicate zeolites with comprehensive ranges of framework topologies, chemical compositions, and aluminum distributions. J. Am. Chem. Soc. **138**(19), 6184–6193 (2016)
98. L. Black, A. Stumm, K. Garbev, P. Stemmermann, K.R. Hallam, G.C. Allen, X-ray photoelectron spectroscopy of aluminium-substituted tobermorite. Cem. Concr. Res. **35**(1), 51–55 (2005)
99. P. Trtik, J. Kaufmann, U. Volz, On the use of peak-force tapping atomic force microscopy for quantification of the local elastic modulus in hardened cement paste. Cem. Concr. Res. **42**(1), 215–221 (2012)
100. E.M. Foley, J.J. Kim, M.M. Reda Taha, Synthesis and nano-mechanical characterization of calcium-silicate-hydrate (C-S-H) made with 1.5 CaO/SiO_2 mixture. Cem. Concr. Res. **42**(9), 1225–1232 (2012)
101. H.M. Jennings., A model for the microstructure of calcium silicate hydrate in cement paste. Cem. Concr. Res. **30**, 101–116 (2000)
102. G. Constantinides, F.-J. Ulm, The effect of two types of C-S-H on the elasticity of cement-based materials: results from nanoindentation and micromechanical modeling. Cem. Concr. Res. **34**(1), 67–80 (2004)

103. M. Vandamme, F.-J. Ulm, P. Fonollosa, Nanogranular packing of C-S-H at substochiometric conditions. Cem. Concr. Res. **40**(1), 14–26 (2010)
104. F.-J. Ulm, M. Vandamme, H.M. Jennings, J. Vanzo, M. Bentivegna, K.J. Krakowiak, G. Constantinides, C.P. Bobko, K.J. Van Vliet, Does microstructure matter for statistical nanoindentation techniques? Cem. Concr. Compos. **32**(1), 92–99 (2010)
105. J.J. Chen, L. Sorelli, M. Vandamme, F.J. Ulm, G. Chanvillard, A coupled nanoindentation/SEM-EDS study on low water/cement ratio Portland cement paste: Evidence for C-S-H/$Ca(OH)_2$ nanocomposites. J. Am. Ceram. Soc. **93**(5), 1484–1493 (2010)
106. H.R. Wenk, C.S. Venkitasubramanyan, D.W. Baker, F.J. Turner, Preferred orientation in experimentally deformed limestone. Contrib. Mineral. Petrol. **38**(2), 81–114 (1973)
107. H.M. Jennings, Refinements to colloid model of C-S-H in cement: CM-II. Cem. Concr. Res. **38**(3), 275–289 (2008)

Chapter 7
C-S-H Superstructure

7.1 Introduction

In the previous chapter, we investigated the strategy of aluminum incorporation to regulate the nanostructure of C-S-H, aiming to enhance its stability under ultra-low temperatures. As previously mentioned, the incorporation of aluminum atoms as an enhancement strategy is commonly observed in active aluminum-rich supplementary cementitious materials. However, the structure and properties of C-S-H are not essentially changed with aluminum incorporated. Particularly, the Al-incorporated C-A-S-H is not significantly improved in terms of its stiffness or porosity, remaining a porous material. Consequently, under ultra-low temperature freezing–thawing conditions, the C-A-S-H matrix can be still affected by the freezing and thawing of pore water. Hererin we aim to further regulate the pore structure and stiffness of C-S-H with innovative molecular editing strategy.

Globally, cement is the most used manufactured substance [1]. Approximately 10 billion cubic meters of concrete have been consumed per year over the last decade [2]. It has been over 2000 years since Roman cement was discovered [3]. The history of Portland cement is also about 200 years since it was invented [4]. To a certain extent, humankind's industry and urbanization progress over the last two centuries were nearly proportional to the amount of cement we consumed [5]. With the continuous urbanization movement in developing countries, cement will still be the most used industrial substance for a long period. Such a massive cement production is accompanied by non-negligible environmental impacts. In particular, it consumes a vast fraction of fresh water and global energy consumed by humans (~1.7% and ~ 3%, respectively) [6, 7] and releases gigatons of CO_2 (8.6% of the anthropic emissions) [8]. When comparing the chemical composition of modern cement and ancient Roman cement, it is striking to find that they are very analogous. Nonetheless, the performance of modern concrete has made a giant leap thanks to the incorporation of chemical additives [9–11] and various fibers [12, 13], as well as the fine control of the granulometry [14–16]. The compressive strength can exceed

Z. Jiang and X. Zhu, *Cryogenic Stability of Cement-Based Materials at Micro/Nano Scale*, https://doi.org/10.1007/978-981-96-1836-1_7

100 MPa, making it possible to construct skyscrapers like Khalifah Tower (828 m) and Shanghai Tower (632 m). For pursuing the long life span of engineering structures, a vast number of studies were implemented to increase the performance of concrete using nano-engineering approaches [17–19].

There are various innovative nanotechnologies reported in the literature to manipulate the C-S-H nanostructure, aiming to enhance the macroscopic properties of concrete. The reactive nanomaterials [20–24] are effective in improving strength and durability. These nano-additives exhibit high specific surface area and thus provide additional nucleation sites, which accelerates the dissolution of cement clinker and promotes the hydration process, consequently stimulating the rapid nucleation in the early hydration stage [25, 26]. At the same time, pozzolanic reactive nanoparticles can participate in the pozzolanic reaction with the dissolved Ca^{2+} cations, further producing additional C-S-H gel [26]. In addition, these additives are beneficial to forming a more robust and mechanically stable bonding framework at the nanoscale and microscale. For instance, the addition of nanosilica induces longer silicate chains [20], and the small colloidal nanosilica leads to a more compact microstructure [21], densifying the interfacial transition zone [27]. Apart from the mineral nanoparticles, nanocarbon additives, such as graphene [28–30] and carbon nanotubes [31], are also capable of improving the macroscopic mechanical properties of concrete. However, the nature of C-S-H has never been essentially altered. In particular, these additives never significantly improve stiffness and porosity compared with conventional C-S-H.

Inspired by some biominerals found in natural organisms, e.g., magnetotactic bacteria, red coral, and sea urchin spines [32, 33], organic polymers can interact with cement materials to enhance their macroscopic properties. Some innovative approaches are proposed, such as the sol-gel process and layer-by-layer self-assembly [17, 34–36]. It aims to interlock small inorganic C-S-H grains by intercalation of polymers [37], finally forming a coarsening ordered structure, similar to the above-mentioned biominerals. The intercalation is highly associated with the surface charge of polymers and the Ca/Si ratio of C-S-H gel [38]. Cationic polymers are more favorable to intercalate in the gel at a low Ca/Si ratio, while the anionic polymers favor intercalation at a high Ca/Si ratio. The polymer's molecular size may also impact the intercalation, which is not very explicit, though [39]. Pelisser et al. [40] reported that the poly(vinyl alcohol) (PVA) polymer with a small molecular size is more likely to intercalate in C-S-H than that with a large size. The PVA polymers are then used to toughen the cement-based materials [41]. Except for PVA, nitrobenzoic acid [42], poly(ethylene glycol) [18], and some biomolecules (e.g., glutamic acid and leucine) [39] are also reported to have the potential to intercalate in C-S-H. Although many organic polymers can intercalate in the gel, two primary features are required to form strong interactions, namely hydrophilic residues and anionic groups, according to ref. [9, 43]. Governed by this rule, Picker et al. [43] designed a sophisticated method to synthesize the so-called mesocrystalline C-S-H, which shows an excellent bending property. Therefore, it is explicit to conclude that organic additives show significant potential to enhance the ductility of cement concrete.

Generally, the fabrication of inorganic minerals and organic polymers offers impressive improvement for the physico-mechanical properties of the composite. The fabricated inorganic-organic materials obtain the hard and stiff nature of inorganic minerals as well as the ductile and lightweight features of organic polymers [33]. Particular interest was focused on forming a mesocrystalline structure with remarkable fracture toughness and strength. This mesocrystalline structure is commonly seen in the combination of amorphous carbonate minerals and organics [32, 44], while uncommon for the complex of nanocrystalline C-S-H and organics. Although Picker et al. [43] had successfully synthesized the so-called mesocrystalline C-S-H with excellent bending properties, the stiffness of those obtained materials was not improved, unlike the $CaCO_3$ mesocrystals. Even though Picker et al. [9] had concluded that the hydrophilic and anionic groups could have strong interactions with C-S-H, there is a lack of explicit experimental evidence to justify that they are requisites for C-S-H superstructure with mesocrystalline features. Herein, we will sequentially add hydrophilic and anionic groups into the CSH@polymer system to validate this hypothesis. We will show that we succeeded in manufacturing a hierarchical C-S-H superstructure with really high stiffness and super-low porosity. Given those features, we will investigate their potential for use in cryogenic attack conditions. The super-low porosity avoids a high water content in cement concrete, and the high stiffness enhances the resistance to the elastic deformation of the cement matrix under cryogenic attack. In this chapter, we propose a simple molecular-editing strategy to formulate hierarchical C-S-H superstructure composites, and we investigate their potential for application in cryogenic engineering.

7.2 Synthesis and Characteristics C-S-H Superstructure

7.2.1 Synthesis of C-S-H Superstructure

7.2.1.1 Materials

The raw materials used for synthesis were as follows: calcium chloride ($CaCl_2$, 95% AR, Sinopharm), sodium metasilicate ($Na_2SiO_3 \cdot 5H_2O$, 98% AR, Sinopharm), sodium hydroxide (95% AR, Sinopharm), poly(acrylic acid) (PAA, 99%, Sinopharm), poly(acrylamide) (PAAm, ~ 3 000 000 g/mol, Sinopharm), poly(acrylamide-co-acrylic salt) partial sodium salt (P(AAm-co-AA), 520 000 g/mol, acrylamide ~ 80 wt%, acrylic and acrylic salt ~ 20 wt%, Sigma-Aldrich), and ultra-pure water.

7.2.1.2 Preparation

Ultra-pure water was used to prepare the formulation, which had been boiled to remove the dissolved CO_2. After cooling down, nitrogen gas flow was used to exclude the remaining CO_2 further. The Ca/Si molar ratio was set at 1.0 for all samples. 500 mL Na_2SiO_3 solution was used in each synthesis, which was prepared at a concentration of 0.15 mol/L. The Na_2SiO_3 solution has a pH of 13.2. Then, 2.96 g PAAm, PAA, or P(AAm-co-AA) was added to the Na_2SiO_3 solution and stirred for 2 h. After that, 250 mL $CaCl_2$ solution (0.3 mol/L) was slowly added to the mixtures under stirring in a 40 °C oil bath container. After all the $CaCl_2$ solution was added, the mixtures were continuously stirred for another 2 h, then standing for three days. It was protected under an N_2 atmosphere during the whole synthesis procedure, as shown in the scheme of Fig. 7.1. Finally, synthetic solids were filtered, washed with 95% ethanol, and dried at 40 °C. Those samples were named in the form of CSH@polymer (e.g., CSH@PAA) when the polymer was added (see Fig. 7.1). The sample to which no polymer was added was our reference synthetic C-S-H and we named it CSH. The PAA offers carboxylic groups to interact with silicate chains. The PAAm donates amide groups, while the P(AAm-co-AA) provides carboxylic, amide, and deprotonated carboxylic groups. The non-bond interactions between these organic groups and silicate chains are illustrated in Fig. 7.1c–e.

7.2.2 Phase Assemblages

XRD patterns are presented in Fig. 7.2a, showing the main diffraction peaks of C-S-H for all samples. From the thermogravimetric analysis results in Fig. 7.2b, the sharp decrease of weight at ~ 500 °C is associated with the decomposition of polymers. Both the TGA and XRD results show that the obtained synthetic solids are well prepared without many impurities except for a minimal amount of calcium carbonate. A diffraction peak (002) at 2theta ~ 7° is detected for CSH, CSH@PAAm, and CSH@PAA composite. This (002) peak is related to the basal spacing between two adjacent dreierketten chains. In the reference CSH sample, a much sharper and higher intensity (002) peak at 2theta = 7.119° is observed, which corresponds to a basal spacing of 12.27 Å for the interlayer, comparable with the C-S-H obtained using pozzolanic reactions [46] and precipitation method [47]. In comparison, the peaks for CSH@PAAm and CSH@PAA have shifted to the left at 6.5°–7.1°. Unlike the reference CSH with a single intense peak, for both CSH@PAAm and CSH@PAA, two (002) peaks were automatically detected (see Fig. 7.2a). For CSH@PAAm, the two peaks are at 6.779° and 7.059°, yielding basal spacings of 13.03 and 12.51 Å, respectively. Regarding CSH@PAA, the two (002) peaks stand at 6.556° and 7.112°, corresponding to the basal spacings of 13.47 and 12.41 Å, respectively. These observations suggest that the interlayer basal spacings are expanding with the incorporation of polymers. Moreover, the split in two (002) peaks indicates that two primary nanostructures share the same ***ab*** plane but with two different basal spacings,

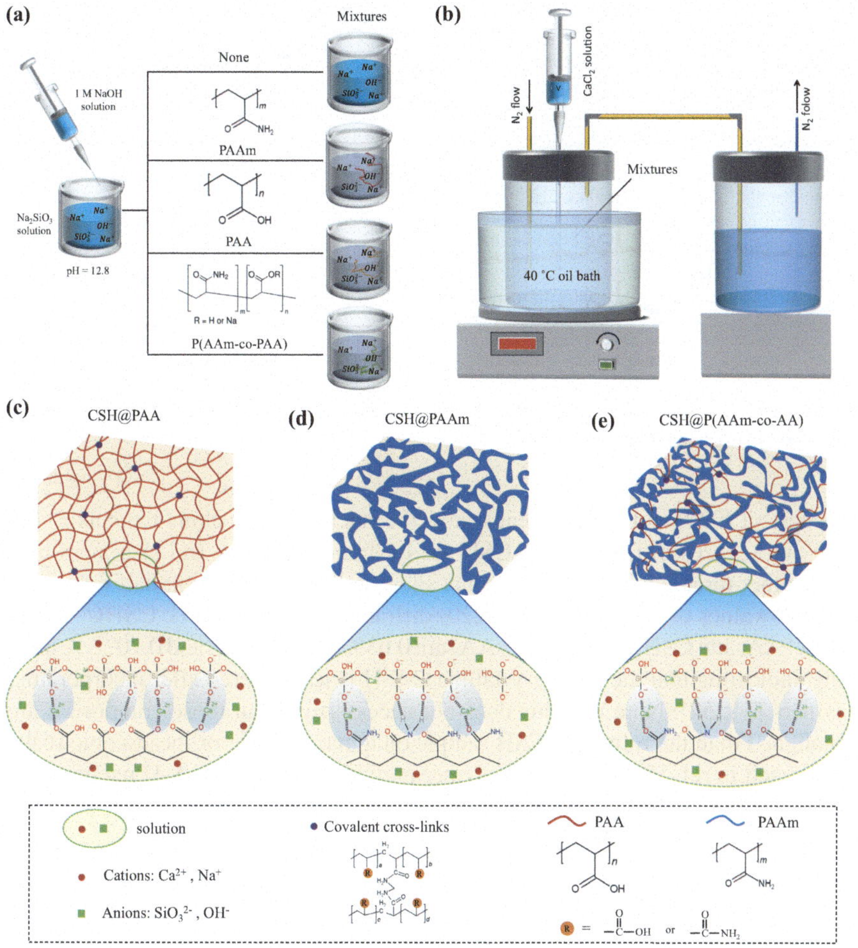

Fig. 7.1 Schemes of CSH@polymers composites preparation. **a** preparation for synthetic solution; **b** reactive condition; **c** formation of CSH@PAA; **d** formation of CSH@PAAm; **e** formation of CSH@P(AAm-co-AA). The non-bond interactions between silicate chains and organic groups are mainly present among Ca^{2+}, deprotonated silanol, amide groups, carboxylic groups, and deprotonated carboxylic groups. The schemes for bulk PAA, PAAm, and P(AAm-co-AA) are adapted from the ionogels [45]

which may also imply that the polymers have partial intercalations between silicate chains. Similar observations have also been reported for C-A-S-H due to the uptake of aluminum [46]. Note that the (002) peak disappeared in CSH@P(AAm-co-AA), suggesting that this type of anionic polymer may strongly interact with silicate chains and the intercalation of polymers between silicate chains significantly increase the heterogeneity of the stacked layers normal to ***c*** [48, 49]. X-ray diffractogram peaks with $2\theta > 29.8^\circ$ correspond to the *hk*0 reflections [18, 46, 50]. The intensity and

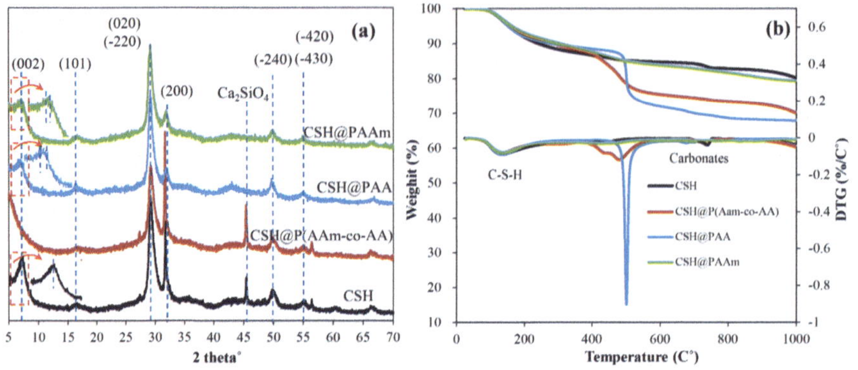

Fig. 7.2 XRD patterns **a** (red boxes and arrows indicate zoomed areas) and thermogravimetric analysis **b** of C-S-H@polymer composites

sharpness of (020) and (−220) peaks show the stacking order along the ***c***-axis. The (020) and (−220) diffractions of CSH are of higher intensity and sharpness than for the other three CSH@polymer composites, and the full width at half maximum (FWHM) values of the (020) and (−220) diffraction peaks of CSH, CSH@PAAm, CSH@PAA, and CSH@P(AAm-co-AA) are 0.639, 0.675, 0.654, and 0.707 °, respectively, indicating that the nano-crystalline C-S-H is less ordered after incorporating polymers [51, 52]. Note that this does not necessarily mean those composites are less aligned than the reference CSH sample on a macroscopic or mesoscopic scale, as indicated in ref. [53], since the polymer may interlock with C-S-H nanocrystals and form a more aligned mesostructure.

7.2.3 Chemical Bonding

The FTIR spectra for synthetic composites are shown in Fig. 7.3, and frequencies for characteristic bond vibrations are labeled in the spectra. The frequencies at 1658, 1659, and 1676 cm^{-1} are associated with the stretching vibrations of C = O moiety of CSH@PAA, CSH@PAAm, and CSH@P(AAm-co-AA), respectively, suggesting the presence of amide groups [54, 55]. The NH_2 bending vibration at 1468 cm^{-1} for CSH@PAAm, and 1461 cm^{-1} for CSH@P(AAm-co-AA) confirm the amide groups as well, showing the $-CONH_2$ group. The C-O stretch bond vibration at 1319 cm^{-1} is observed for CSH@PAA and CSH@P(AAm-co-AA) samples [55]. Let us focus on the Q^2 signal located at ~ 967 cm^{-1}, which corresponds to the stretching vibration of Si-O bonds of Q^2 silicate tetrahedra in dreierketten chains [39]. For the reference CSH sample, the Q^2 signal is located at 967 cm^{-1}, while the band shifts to lower frequencies as polymers carrying amide groups intercalate. The shift is particularly significant in CSH@P(AAm-co-AA), as shown in Fig. 7.7.3, to 963 cm^{-1}. This shift suggests that the silicate chains of CSH@P(AAm-co-AA) and CSH@PAAm

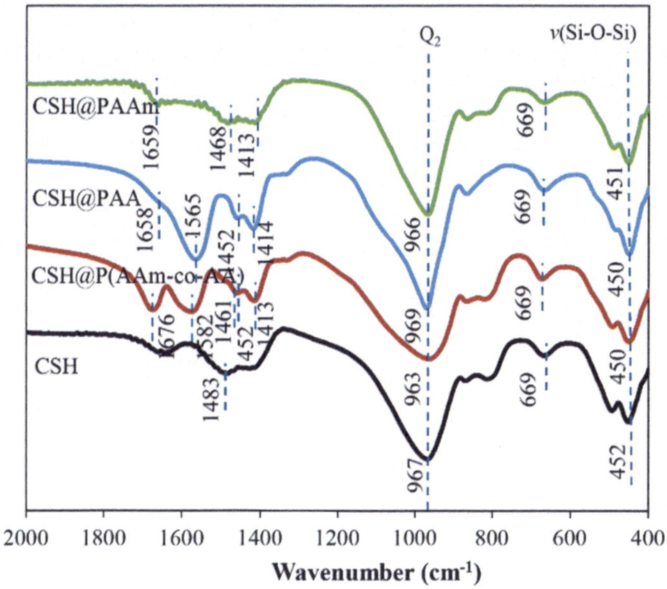

Fig. 7.3 FTIR spectrum of C-S-H@polymer composites

are less polymerized than for the reference CSH, implying that incorporating these polymers is not favorable to forming long silicate chains for nano-crystalline C-S-H. This is consistent with the observations in the XRD diffractograms analyzed before. In contrast, it is interesting to observe that the Q^2 signal of CSH@PAA has a higher frequency at 969 cm^{-1}, which means the silicate chains are more polymerized with regard to CSH. These results suggest that the amide groups may depolymerize the silicate chains, while the carboxyl groups may promote polymerization. In particular, the intercalation of anionic P(AAm-co-AA) leads to the least polymerized nanostructure for C-S-H.

7.2.4 Micromorphology

The SEM and TEM morphologies of CSH@polymers are presented in Figs. 7.4, 7.5, respectively. Significant differences are observed in the SEM micromorphologies of these samples when different polymers are added. The plate-like morphology is captured for the reference CSH sample (Fig. 7.4a), similar to the observations in previous studies [56, 57]. As the PAAm is incorporated (Fig. 7.4b), the micromorphology is shaped like a net with much more porous features than CSH. With a high magnification (Fig. 7.5), it appears more clearly that the CSH@PAAm net looks like a wicker-like fabric of modified C-S-H. Such a difference is due to the bonding between PAAm and C-S-H, as suggested by its TEM morphology in Fig. 7.6b. The polymers and foil-like C-S-H are somewhat separated but still closely intertwined. For CSH@PAA (Fig. 7.4c), the SEM morphology is more agglomerated than both CSH and CSH@PAAm. We did not observe an apparent separation between C-S-H and PAA polymers in SEM and TEM observations, which differs from CSH@PAAm. This observation suggests that PAA has more intense interaction with silicate chains than PAAm, thus forming a compact structure.

A surprisingly more aligned and compact micromorphology is observed for CSH@P(AAm-co-AA), as shown in Fig. 7.4d. Some bulk polymers are attached to the surface. The small grains are assembled in alignment, yielding a macroscopically layering structure with no microporosity, which differs significantly from the other three samples. The incorporation of P(AAm-co-AA) bridges the small nanoscrystalline C-S-H grains in the sense of long-range ordering. This hierarchical superstructure is commonly seen in natural biominerals, e.g., sea urchin spines and red coral [33], exhibiting mesocrysal features. The TEM morphology presented in Fig. 7.6d discloses a trend toward stacking order along the normal direction of the figure rather than randomly heterogeneous as CSH@PAAm and CSH@PAA composite. The P(AAm-co-AA) polymer contains two fundamental chemical features: (1) the

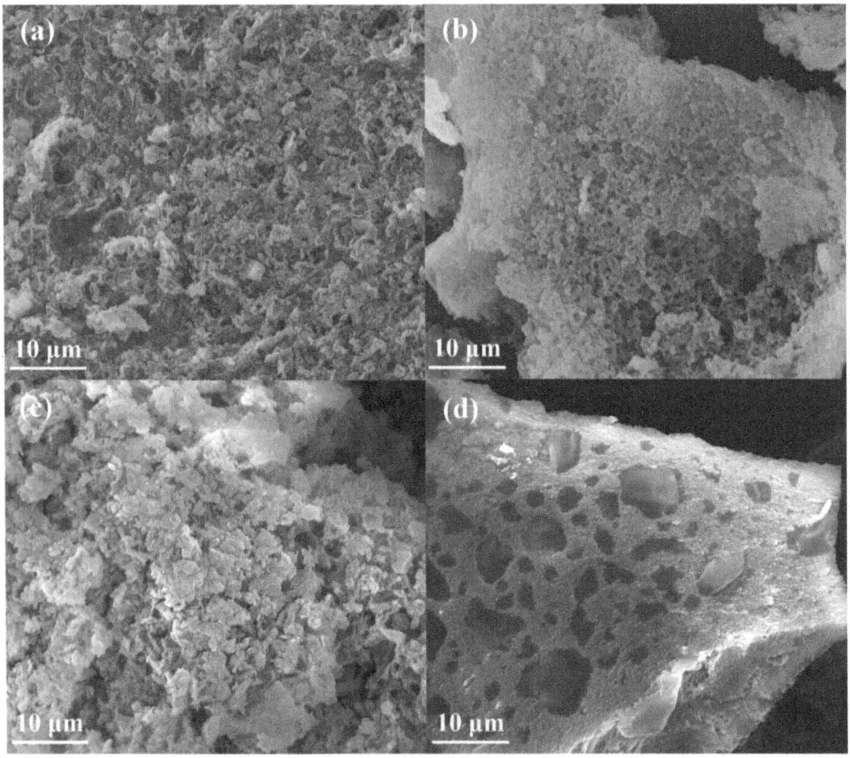

Fig. 7.4 SEM micromorphology of C-S-H@polymer composites. **a** C-S-H; **b** C-S-H@PAAm; **c** C-S-H@PAA; **d** C-S-H@P(AAm-co-AA)

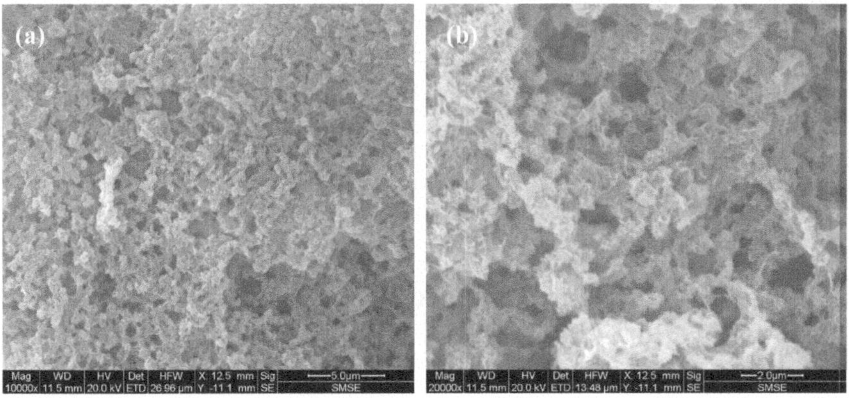

Fig. 7.5 CSH@PAAm SEM morphology in high magnification. **a** 10000x; **b** 20000x

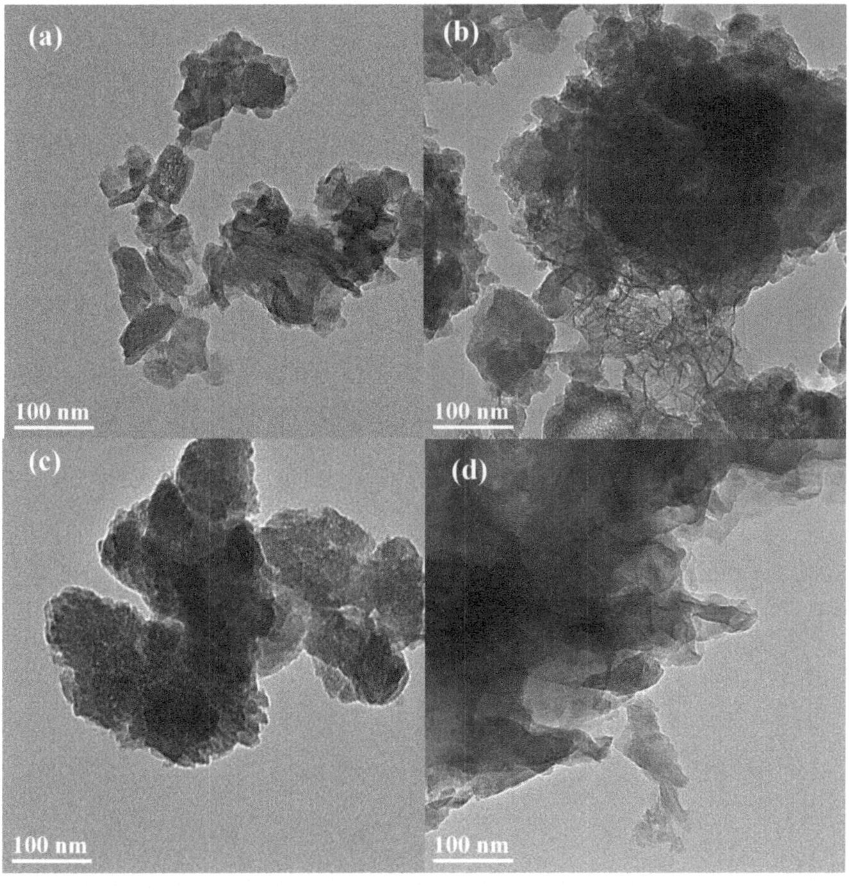

Fig. 7.6 TEM micromorphology of C-S-H@polymer composites. **a** C-S-H; **b** C-S-H@PAAm; **c** C-S-H@PAA; **d** C-S-H@P(AAm-co-AA)

hydrophilic groups, i.e., amide and carboxyl groups; (2) the anionic groups, i.e., ionized carboxylic groups. These characteristic chemical groups can form strong interactions with silicate chains, as shown in Fig. 7.2. The above comparison indicates that both hydrophilic residues and anionic groups are required to form strong interactions between inorganic C-S-H and organic polymers. The aligned hierarchical superstructure is expected to obtain improved mechanical properties.

7.3 Pore Structure and Nanomechanical Properties of C-S-H Superstructure

7.3.1 Pore Structure

The pore structure characteristics of the synthesized C-S-H organic composite material were characterized using nitrogen adsorption. The results of the adsorption isotherms and cumulative pore volume curves are presented in Fig. 5.5 for comparison. In Fig. 7.7a, each sorption isotherm corresponds to a unique porous network, yielding fruitful information regarding the pore structure and surface morphology. The first main feature is the adsorption capacity. CSH@PAAm has a far greater capacity to intake nitrogen than other samples, and CSH@P(AAm-co-AA) has the smallest intake capacity, while CSH and CSH@PAA are at an intermediate level. Consequently, we observe that the cumulative pore volumes of these CSH samples are ranked as CSH@PAAm > CSH ≈ CSH@PAA > > CSH@P(AAm-co-AA), see Fig. 7.7b. The pore volume of CSH@P(AAm-co-AA) is at 0.0077 mL/g, far lower than CSH (0.1184 ml/g), CSH@PAA (0.1181 mL/g), and CSH@PAAm (0.4045 mL/g).

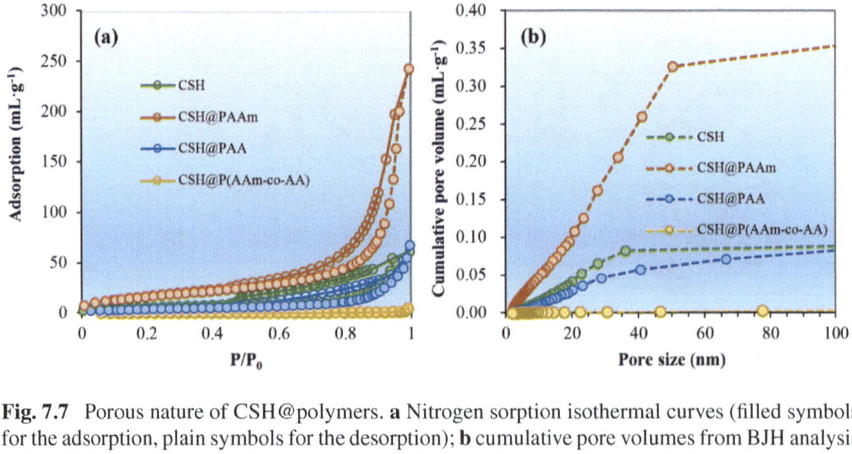

Fig. 7.7 Porous nature of CSH@polymers. **a** Nitrogen sorption isothermal curves (filled symbols for the adsorption, plain symbols for the desorption); **b** cumulative pore volumes from BJH analysis of the adsorption isotherm branch

The addition of PAAm resulted in forming a porous and loose structure in the C-S-H@PAAm organic composite material, which was not the expected outcome of this study. The introduction of PAA did not significantly reduce the porosity of C-S-H, resulting in the formation of a porous structure termed CSH@PAA. In contrast, the CSH@P(AAm-co-AA) superstructure exhibited an extremely low porosity. According to the nitrogen adsorption test results, this C-S-H superstructure contained almost no micropores, mesopores, or macropores.

7.3.2 Specific Surface Area

The specific surface areas were calculated using the BET method from the sorption isotherms; results are plotted in Fig. 7.8. For an explicit comparison, the specific surface areas of other common-seen construction materials, including hydrated cement paste, hydrated alite, ceramsite, and various rocks, are also collected and compared with our synthetic CSH@polymer composites. As expected, CSH@P(AAm-co-AA) obtains a significantly low specific surface area of 0.4509 m^2/g, two orders of magnitude less than other synthetic samples and hydrated cement pastes. It is even lower than ceramsite sand (usually used for preparing lightweight concrete) and most rocks, showing a remarkably compact microstructure.

The C-S-H@P(AAm-co-AA) superstructure depicted in Figs. 7.4 and 7.7 exhibits an extremely dense microstructure, meeting the low porosity requirements set by the research. In this scenario, its internal moisture content will be significantly reduced.

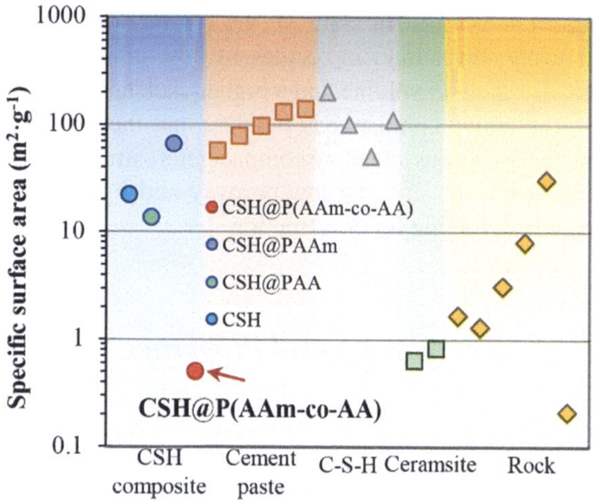

Fig. 7.8 Comparison of the specific surface area among CSH@polymers and various similar materials; data are collected from ref [58–64]

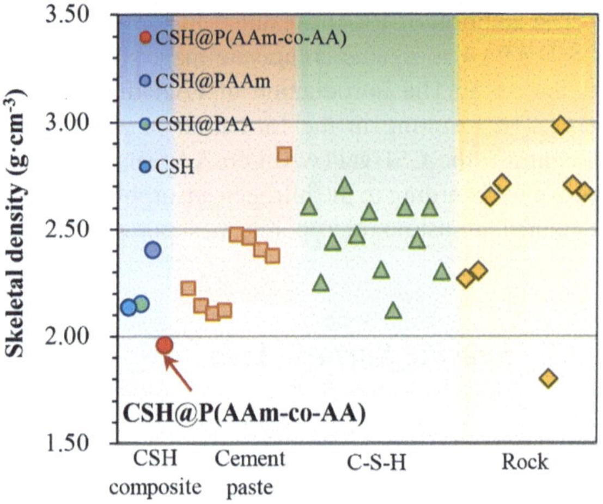

Fig. 7.9 Comparison of the skeletal density among CSH@polymers and various similar materials, data are collected from ref. [65–70], the C-S-H samples collected have different water content conditions from dry to saturated inside

7.3.3 Density

In addition, the skeletal densities of our synthetic samples are also compared with hydrated cement pastes and rocks in Fig. 7.9. The skeletal density of CSH@P(AAm-co-AA) is measured at 1.96 g/cm^3, which is also much lower than hydrated cement pastes, C-S-H, and most rocks (minerals), indicating a lightweight feature. The super-low porosity of CSH@P(AAm-co-AA) may contribute to this observation. Since the CSH@P(AAm-co-AA) is near with no microporosity, the helium molecules are hard to flow inside the superstructure, yielding a higher skeletal volume compared with other composites. In contrast, the CSH@PAAm has a significantly high porosity, allowing the helium molecules to flow efficiently inside the solids, leading to a smaller skeletal volume and a higher skeletal density than other samples.

The above comparisons indicate that the CSH@P(AAm-co-AA) sample has a much less porous and more compact microstructure than most conventional construction materials, showing low-porosity and lightweight features. These features are usually favorable for construction.

7.3.4 Nanomechanical Properties

The nanoscopic mechanical properties of these synthetic CSH@polymer composites are probed using statistical AFM-based indentions, from which Young's modulus can be obtained in each indentation spot when the Poisson's ratio is assumed. Figure 7.10a–d display the contours of Young's moduli within a region of $2 \times 2\ \mu m^2$ for CSH, CSH@PAAm, CSH@PAA, and CSH@P(AAm-co-AA), respectively, and their statistical probability density distributions (PDDs) are plotted in Fig. 7.10e–h. From the contours and PDDs, CSH@polymer composites exhibit much higher

Young's moduli than CSH, thanks to the intercalation of polymers. The PDD of CSH is less dispersed than that of the composites, as revealed by the standard deviation in Fig. 7.10j. The Young's modulus distribution is mostly within 1 ~ 20 GPa and reaches its maximum at ~ 9 GPa. The small Young's moduli (< 5 GPa) are usually ascribed to the porosity [71–73]. The moduli of CSH@PAAm are predominantly distributed within 10 ~ 45 GPa, with the highest probability at ~ 25 GPa. CSH@PAA has a broader distribution for the measured Young's moduli but mainly in 1 ~ 50 GPa and maximum at ~ 20 GPa. In contrast, CSH@P(AAm-co-AA) exhibits much higher Young's moduli on the scanning surface area than all the other samples, primarily distributed at 20 ~ 70 GPa. The different PDDs are confronted in Fig. 7.10i, offering a clear overview of the effect of polymer incorporation. The average Young's moduli are displayed in Fig. 7.10j. The average Young's modulus of CSH is at 15.7 GPa, about 34% lower than the low-density (LD) C-S-H (23.7 ± 5.9 GPa), 56% lower than the high-density (HD) C-S-H (36.1 ± 3.4 GPa) [74–76], but still in agreement with that measured for similar synthetic C-S-H [77, 78]. Such a difference may be attributed to the low Ca/Si ratio of our sample. According to [78, 79], a higher Ca/Si ratio can yield a higher modulus since the interlayer density of C-S-H increases with the Ca/Si ratio (the interlayer calcium content), and the incompressibility of the ***c*** axis is controlled by the interlayer density [79]. The higher the Ca/Si ratio, the higher the interlayer calcium content and the higher the interlayer density, thus the higher the modulus. Herein, the CSH sample has a bulk Ca/Si ratio of 1; it is, therefore, reasonable to obtain a lower average Young's modulus than the hardened cement paste with Ca/Si ≈ 1.7. In comparison with CSH, the incorporation of PAAm and PAA yield higher average Young's moduli at 28.7 GPa and 33.4 GPa, respectively. The CSH@P(AAm-co-AA) composite demonstrates remarkable stiffness, boasting the highest average Young's modulus at 47.5 GPa. This value is approximately twice that of LD C-S-H, surpasses HD C-S-H by 32%, and is three times higher than the average modulus of the reference CSH sample. Representative AFM force-separation curves for these samples are shown in Fig. 7.10k. When submitted to the same force, the deformation of the CSH@polymer composites is smaller than that of CSH, indicating an enhanced stiffness.

The positions of our CSH@polymer composites in a density-Young's modulus diagram are compared with other common materials (i.e., metal, ceramic, polymer, and carbon fiber) in Fig. 7.10l. Interestingly, the CSH@(AAm-co-AA) is located at the intersection of the domains of metals, ceramics, and polymers, and close to the carbon fiber region. These results render this C-S-H composite notable for its remarkable characteristics: super-low-porosity, lightweight, and very stiff. The other two CSH@polymer composites, i.e., CSH@PAAm and CSH@PAA, have more moderate improved stiffnesses, and their pore structures are not as dense. The porosity of CSH@PAAm is even larger than the one of CSH.

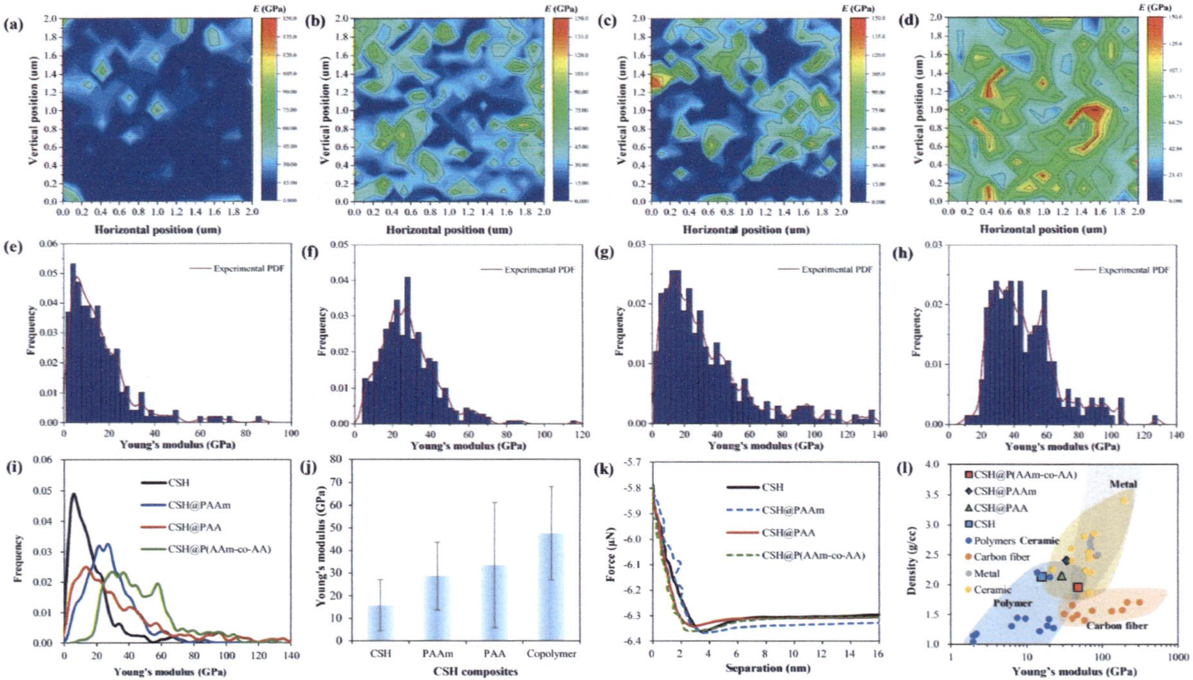

Fig. 7.10 AFM-based indentation mapping measurements and statistical analysis for synthetic CSH@polymer composites, **a**–**d** are Young's modulus contours within 2 × 2 μm^2 area for CSH, CSH@PAAm, CSH@PAA, and CSH@P(AAm-co-AA), respectively; **e**–**h** are the statistical analysis towards the modulus contours for CSH, CSH@PAAm, CSH@PAA, and CSH@P(AAm-co-AA), respectively; **i** probability density distribution of Young's moduli; **j** average Young's modulus; PAAm, PAA, and copolymer represent CSH@PAAm, CSH@PAA, and CSH@P(AAm-co-AA), respectively; **k** representative force-separation curves; **l** density-Young's modulus scatters, data are collected from Matweb [80]

7.4 Cryogenic Stability of C-S-H Superstructure

We further investigated these CSH@polymer composites regarding their cryogenic stability (−170–20 °C) to evaluate their potential for cryogenic applications. Their stabilities in terms of pore structure and nanoscopic mechanical property were evaluated.

7.4.1 *Pore Structure Stability*

The PSD and pore volume of CSH and CSH@polymer composites before and after the cryogenic attack were analyzed from the adsorption branches of gas sorption isotherms using BJH theory, as compared in Fig. 7.11. The desorption branches were also used to implement the BJH analysis; results are presented in Fig. 7.12. However, a sharp increase emerges for the volume of mesopores at 3 ~ 5 nm, as depicted in Fig. 7.12, suggesting that the network-percolation effects were strong for these samples during the gas desorption. This phenomenon is usual for pore systems with a broad distribution of interconnected pores [81, 82]. Under this circumstance, it is rational to adopt the adsorption branch to analyze the PSD [81, 83]. Note that the minimum pore sizes analyzed from both desorption and adsorption branches are greater than 15.1 Å, ensuring the validity of the Kelvin equation [84].

The comparisons show that the pore volume of CSH (Fig. 7.11a) slightly increases mainly for micropores, mesopores, and macropores, indicating the emergence of microcracks upon the attack. Unlike the CSH, the other three CSH@polymer composites have more stable micropores and mesopores. The porous CSH@PAAm (Fig. 7.11b) sample exhibits a significant increase in pore volume. The increase in pore volume is mainly attributed to the macropores > 50 nm, while the micropores and mesopores remain very stable, which is an indicator of microcrack because of the thermal contractions during the cryogenic treatment. The pores of CSH@PAA (Fig. 7.11c) and CSH@P(AAm-co-AA) (Fig. 7.11d) almost have no evident change upon the attack. The CSH@P(AAm-co-AA) maintains its low-porosity microstructure and exhibits only very little increase in its total pore volume, almost negligible compared with the variations observed for the CSH sample. Note that the size distribution of the gel pores < 10 nm of all CSH@polymer composites does not evolve much under the attack. These gel pores control the packing density and diverse the mechanical properties [74, 85]. A lower gel porosity is expected to yield enhanced mechanical properties, e.g., a higher hardness and indentation modulus [86, 87]. From this perspective, the nanoscopic mechanical properties of these CSH@polymer composites were expected to be stable since the gel porosity was not significantly altered. However, the results of AFM indentation mapping measurements are against this expectation. We will discuss it in a later section.

This comparison shows that the combination of organic polymers enhances the pore structure stability of C-S-H upon cryogenic attack. In particular, the synthetic

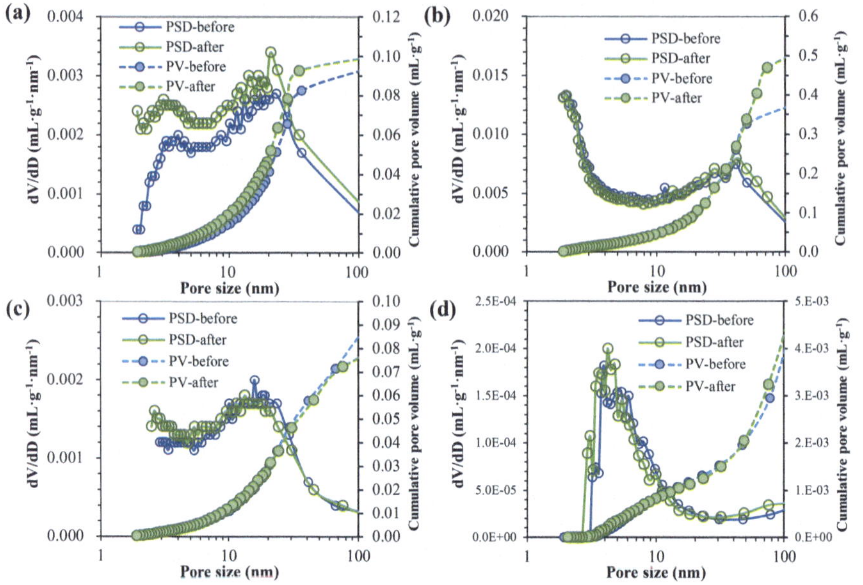

Fig. 7.11 Pore structure evolution under cryogenic attack for CSH **a**, CSH@PAA **b**, CSH@PAAm **c**, and CSH@P(AAm-co-AA) **d**. PSD: pore size distribution; PV: cumulative pore volume. The adsorption isotherm branch was used for the BJH analysis in order to avoid the network-percolation effects. One should note that the scales of the y-axis for **a**–**d** are different

CSH@PAA and CSH@P(AAm-co-AA) samples show excellent pore structure stability upon cryogenic attack. CSH@PAAm is far from stable, even worse than CSH. The CSH@PAAm shows increased macroporosity after the attack, which is not wanted for cement-based materials used in cryogenic engineering [88, 89].

7.4.2 Nanomechanical Properties

After being submitted to the cryogenic attack, the nanoscopic mechanical properties of the composites were assessed again. Their PDDs before and after the attack are compared in Fig. 7.13 (the corresponding Young's moduli contour and statistical analysis are given in Fig. 7.15). The average Young's moduli and the relative decreases are displayed in Fig. 7.14. The cryogenic attack has led to significant variations in Young's modulus of all samples. Except for CSH@PAA, all other samples show a shift toward smaller moduli for the peak of their PDDs. For the reference CSH sample, the distribution is more concentrated in the low-modulus region < 15 GPa, which leads to a decrease of the average Young's modulus to 10.4 GPa with a relative reduction of 34.1%. The poor nanomechanical stability of CSH upon cryogenic attack has also been reported elsewhere [73, 78]. The CSH@PAAm sample exhibits a similar evolution under the attack. Its peak probability is shifted to a lower Young's

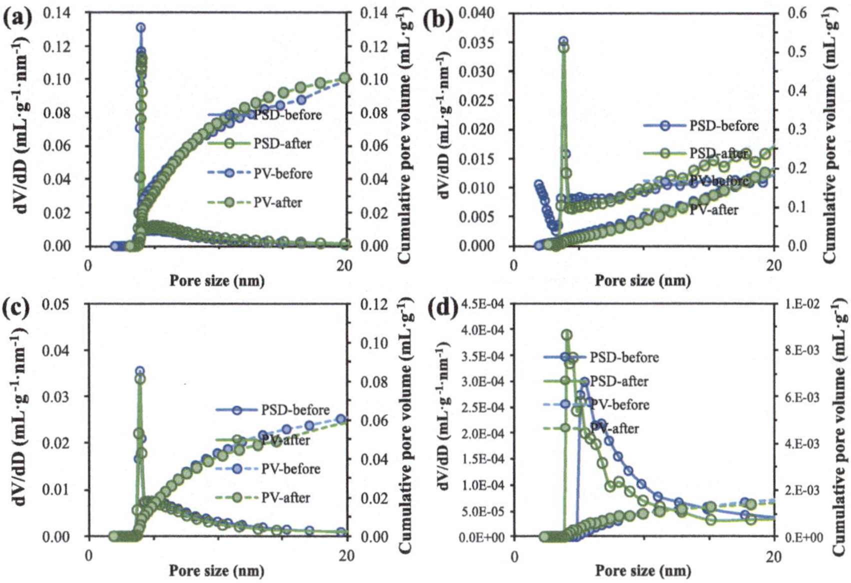

Fig. 7.12 Pore structure evolution under cryogenic attack for CSH **a**, CSH@PAA **b**, CSH@PAAm **c**, and CSH@P(AAm-co-AA) **d**. PSD: pore size distribution; PV: cumulative pore volume. The desorption isotherm branch was used for the BJH analysis. One should note that the scales of the y-axis for **a**–**d** are different

modulus at ~ 20 GPa, with the distribution concentrated within 5 ~ 40 GPa, and the average Young's modulus decreases by 14.2%. For CSH@P(AAm-co-AA), the distribution increases significantly in the domain of low-moduli, mainly around 1 ~ 20 GPa. However, there is also a non-negligible increase in the amount of very stiff regions (>70 GPa). As a result, the decrease in average Young's modulus of CSH@P(AAm-co-AA) is moderate (6.8%) after the attack. CSH@PAA shows a very different evolution among these samples upon the cryogenic attack. Although the distribution decreases in the very stiff domain (>80 GPa), it increases at intermediate stiffnesses within 20 ~ 50 GPa. Therefore, its average Young's modulus remains at the same level as before the attack. Note that the repeatability of our measurements has been validated by comparing the two PDDs of each sample, which was obtained by evenly dividing the measured moduli along the vertical direction in two groups, as shown in Fig. 7.16. Apart from CSH@P(AAm-co-AA) sample, the PDDs of all other seven samples have good repeatability. Even though the surface of CSH@P(AAm-co-AA) is clearly composed of two regions, which indicates an imperfection of the AFM-based indentation measurements, half of the characterized surface of CSH@P(AAm-co-AA) sample exhibits a large stiffness, still revealing a stiffer nature than other CSH@polymer composites. Attention should be given to this imperfection, and further studies are needed to comprehensively identify the nature of this material. One may avoid this imperfection by collecting AFM-based indentation data from more regions.

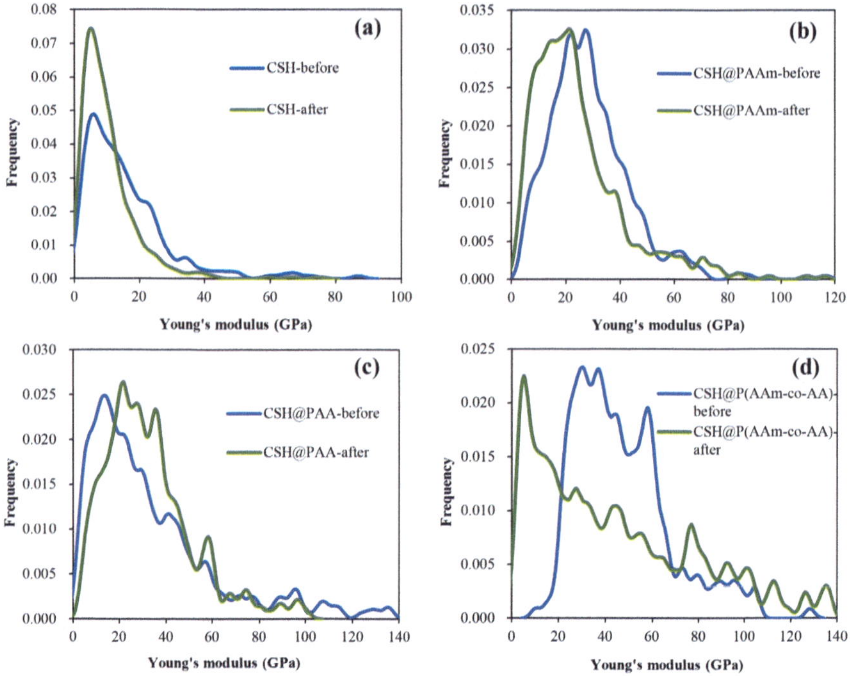

Fig. 7.13 The evolution of Young's modulus after a cryogenic attack for CSH **a**, CSH@PAAm **b**, CSH@PAA **c**, and CSH@P(AAm-co-AA) (**d**). The modulus contours and statistical analysis are given in the Fig. 7.15

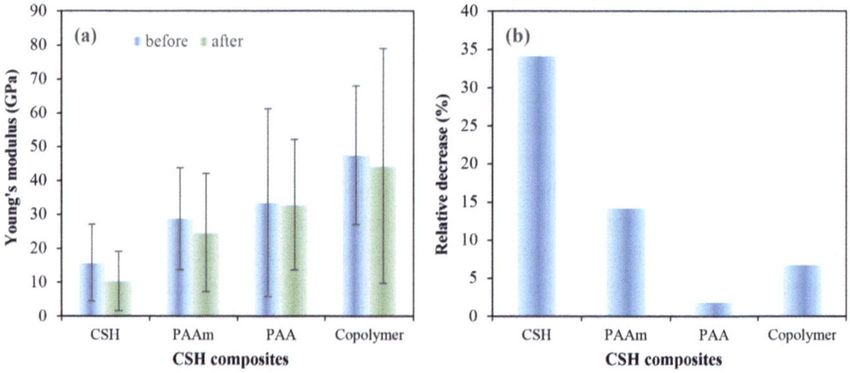

Fig. 7.14 Variations of the average Young's moduli under cryogenic attack. PAAm, PAA, and copolymer represent CSH@PAAm, CSH@PAA, and CSH@P(AAm-co-AA), respectively. **a** The comparison of average Young's moduli; **b** the relative decreases of average Young's moduli

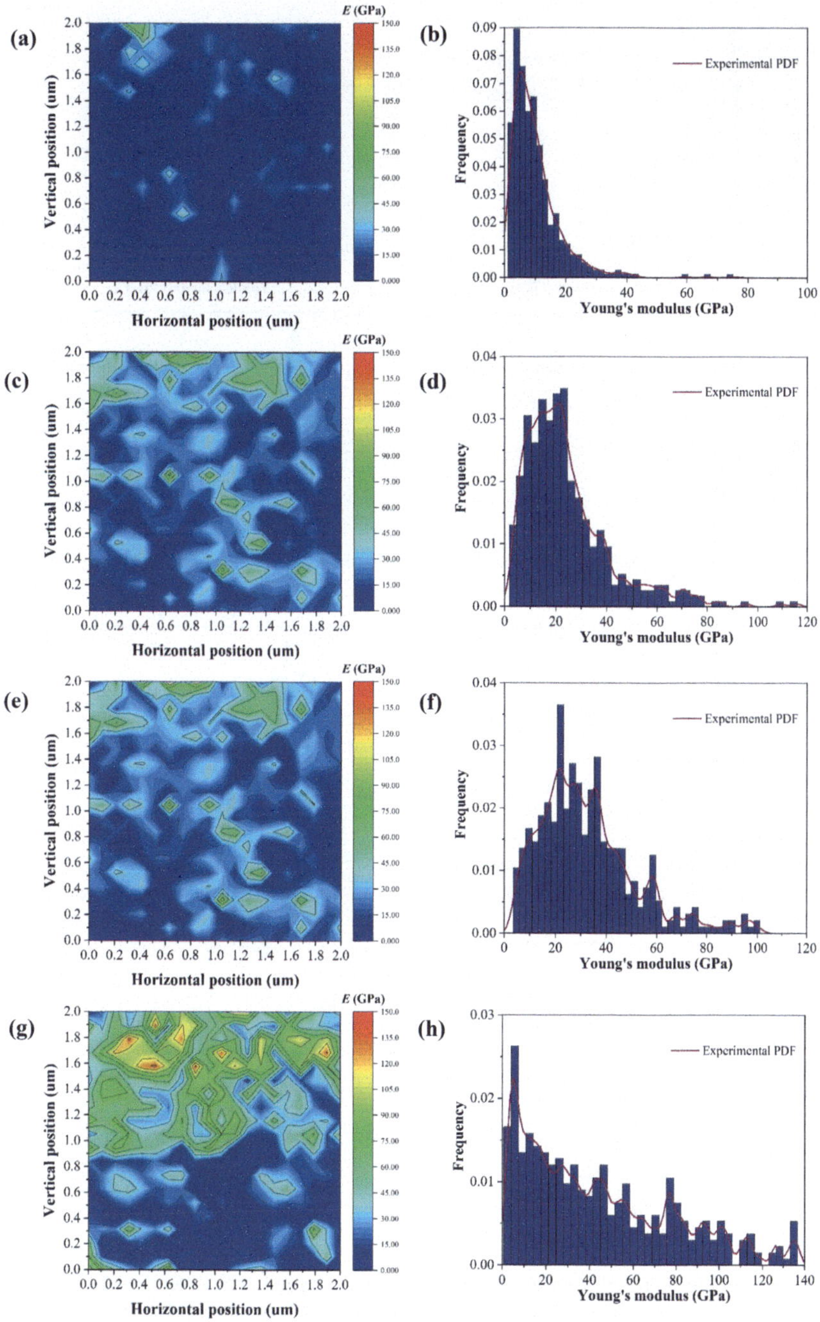

Fig. 7.15 Young's modulus contours and statistical analysis for samples after the cryogenic attack. CSH **a**, **b**, CSH@PAAm **c**, **d**, CSH@PAA **e**, **f**, and CSH@P(AAm-co-AA) (**g**) and (**h**)

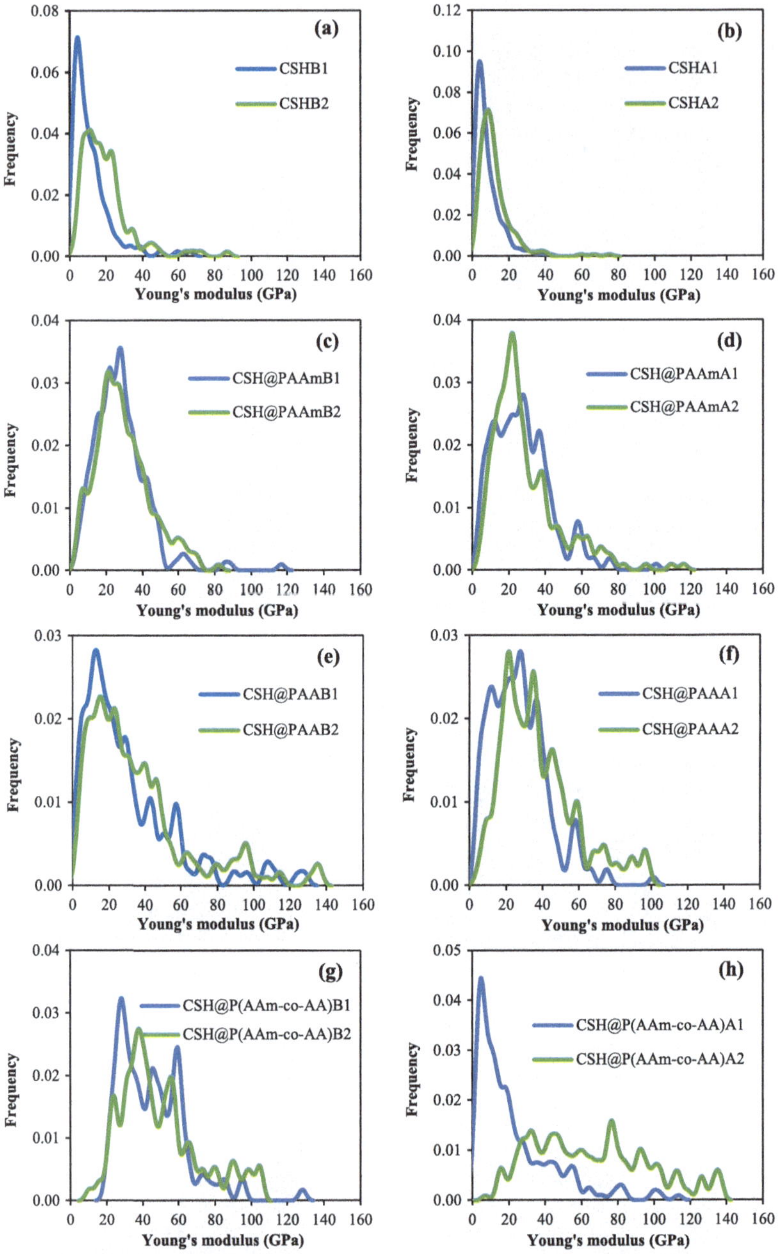

Fig. 7.16 Comparison of statistical analysis for the two groups of each sample (divided horizontally for Young's modulus contours). **a** CSH befroe cryogenic attack; **b** CSH after cryogenic attack; **c** CSH@PAAm before cryogenic attack; **d** CSH@PAAm after cryogenic attack; **e** CSH@PAA before cryogenic attack; **f** CSH@PAA after cryogenic attack; **g** CSH@P(AAm-co-AA) before cryogenic attack; **h** CSH@P(AApm-co-AA) after cryogenic attack. "_B" represents before cryogenic attack, and "_A" represents after cryogenic attack

As already stated before, the packing condition of gel closely corresponds to the nanoscopic mechanical properties. From a conventional point of view, a higher Young's modulus of C-S-H indicates a higher packing density and a more compact gel structure. However, the pore structure evaluations shown in Fig. 7.11 imply that the gel pores have not been apparently densified or loosened and remain very stable for CSH@polymer composites except for CSH, which means that the packing conditions of these samples have not been altered. Therefore, we infer that the decrease in the average Young's modulus observed for our synthetic CSH@polymer composites is derived from the variation in the mechanical properties of CSH@polymer skeleton because of the cryogenic attack. In other words, the modifications of the nanostructure may contribute to the variations of macroscopic stiffness, according to Li et al. [79]. Nonetheless, it is still reasonable to conclude that CSH@PAA and CSH@P(AAm-co-AA) have better nanomechanical properties than CSH and CSH@PAAm as far as the average Young's moduli are considered (Fig. 7.16).

7.4.3 Enhancement Mechanism of C-S-H Superstructure

According to the previous analysis, utilizing the P(AAm-co-AA) organic polymer assembly with C-S-H nanoparticles has successfully led to the creation of a C-S-H superstructure material. This superstructure material possesses extremely low porosity, high rigidity, and lightweight features, almost devoid of micropores, mesopores, or macropores. Under cryogenic attack, both its pore structure and nanomechanical properties remain relatively stable. Considering these characteristics of the C-S-H superstructure material, its mechanism for enhanced cryogenic stability can be analyzed.

(1) The assembly of C-S-H nanoparticles by the P(AAm-co-AA) organic polymer forms an ordered microstructure, preventing the formation of various types of pores. Such ordered structures derived from this organic-inorganic combination exhibit remarkable stability. Even in humid environments, the remarkably low porosity within the C-S-H superstructure significantly reduces the water content, thereby preventing the formation of massive crystalline pressure resulting from the freezing of pore water during cryogenic attack. As shown in Fig. 7.17, ordinary C-S-H structures contain various interconnected pores. Once ice forms in the pores, the pore walls unavoidably bear immense pressure, leading to crack formation and damage. In contrast, the C-S-H superstructure material contains almost no pore water, thus avoiding the need to withstand such pressure.
(2) Due to the penetration of polymer molecules throughout the C-S-H superstructure and their tight integration between C-S-H particles, even if partial defects are inevitably formed within the C-S-H superstructure material due to cold contraction or freezing of pore water during cryogenic attack the polymer interlaced network can still to a certain extent limit the expansion and growth of larger microcracks, as shown in Fig. 7.17.

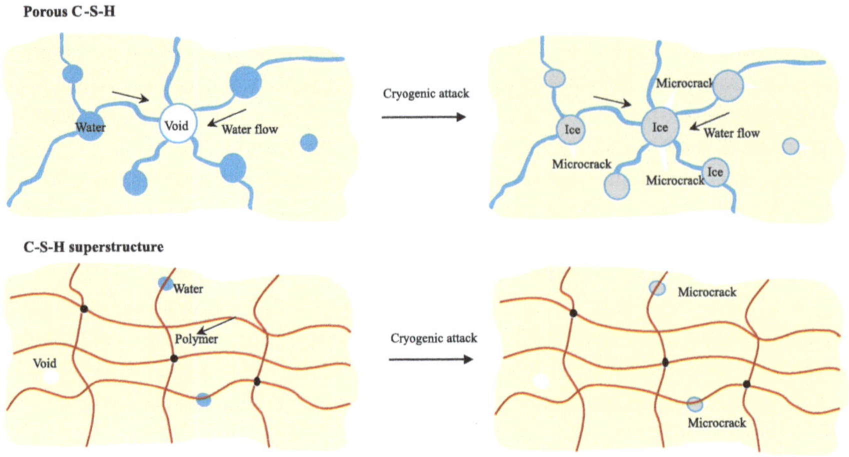

Fig. 7.17 The mechanism of enhanced stability of C-S-H superstructure at ultra-low temperatures

(3) The high rigidity properties enhance the elastic deformation capability of C-S-H. The C-S-H superstructure exhibits an ultra-high Young's modulus, twice as high as that of ordinary C-S-H, providing it with greater resistance against elastic deformation.

7.5 Concluding Remarks

In this chapter, we have shown an innovative molecular editing strategy to enhance the cryogenic stability of C-S-H. The formulation of various CSH@polymer nanocomposites is studied in order to form mesostructured C-S-H superstructure. By incrementally adding polymers with amide, carboxylic groups, and partially anionic carboxylic groups, we emphasize that both hydrophilic residues and anionic groups are indispensable parts of assembling hierarchical C-S-H superstructure with super-low porosity and significantly increased stiffness.

A hierarchical CSH@polymer composite with high Young's modulus and low-porosity microstructure is obtained by incorporating P(AAm-co-AA). The composite shows a hierarchically aligned superstructure with no microporosity. It has a high average Young's modulus at 47.5 GPa, three times higher than the reference synthetic C-S-H. Meanwhile, it has super low porosity and a lightweight structure compared with most usual construction materials. Its specific surface area is two orders of magnitude smaller than that of hydrated cement paste and the reference synthetic C-S-H, comparable to some dense rocks. This composite is quite remarkable in so far as it is at the same time dense, lightweight, and stiff. In a density-Young's modulus diagram, this hierarchical C-S-H superstructure composite is located at the intersection of the domains of ceramics, metals, and polymers and close to carbon fibers. In addition, this composite has satisfying cryogenic stability, showing its potential applicability for cryogenic engineering. This novel material with markedly improved stiffness, low-porosity microstructures, lightweight features, and cryogenic stability could offer a transformative advantage to the construction process. It could empower the cement-based materials in the cold region with higher durability. The reduced weight could enhance transportation and installation efficiency while lowering maintenance costs. Herein, we offer a critical insight into elementary characteristics of the organic groups for the formation of hierarchical C-S-H in an idealistic aqueous solution environment. Given the pore solution of the hydrating cement system is far more complex than in laboratory conditions, the assembly of organic groups with hydration products is more challenging than with C-S-H only. To facilitate the large-scale application of this novel material in the construction field, further studies are required to elucidate the interactions between organic groups, ionic species, and hydration products in the cement hydration process.

References

1. W.T. Li, B.Q. Dong, Z.X. Yang, J. Xu, Q. Chen, H.X. Li, F. Xing, Z.W. Jiang, Recent advances in intrinsic self-healing cementitious materials. Adv. Mater. **30**(17), 1705679 (2018)
2. G. Habert, S.A. Miller, V.M. John, J.L. Provis, A. Favier, A. Horvath, K.L. Scrivener, Environmental impacts and decarbonization strategies in the cement and concrete industries. Nat. Rev. Earth Environ. **1**(11), 559–573 (2020)
3. J. Delatte Norbert, Lessons from Roman cement and concrete. J. Prof. Issues Eng. Educ. Pract. 127(3), 109–115 (2021)
4. C. Hall, On the history of Portland cement after 150 years. J. Chem. Educ. **53**(4), 222 (1976)
5. K.L. Scrivener, V.M. John, E.M. Gartner, Eco-efficient cements: Potential economically viable solutions for a low-CO_2 cement-based materials industry. Cem. Concr. Res. **114**, 2–26 (2018)
6. S.A. Miller, A. Horvath, P.J.M. Monteiro, Impacts of booming concrete production on water resources worldwide. Nat. Sustain. **1**(1), 69–76 (2018)
7. P.J.M. Monteiro, S.A. Miller, A. Horvath, Towards sustainable concrete. Nat. Mater. **16**(7), 698–699 (2017)
8. S.A. Miller, A. Horvath, P.J.M. Monteiro, Readily implementable techniques can cut annual CO_2 emissions from the production of concrete by over 20%. Environ. Res. Lett. **11**(7), 074029 (2016)
9. A. Picker, L. Nicoleau, A. Nonat, C. Labbez, H. Colfen, Identification of binding peptides on calcium silicate hydrate: a novel view on cement additives. Adv. Mater. **26**(7), 1135–1140 (2014)
10. C. Li, J. Li, A. Telesca, D. Marchon, K. Xu, M. Marroccoli, Z. Jiang, P.J.M. Monteiro, Effect of polycarboxylate ether on the expansion of ye'elimite hydration in the presence of anhydrite. Cem. Concr. Res. **140**, 106321 (2021)
11. Y. Zhang, Z. Jiang, Y. Zhu, J. Zhang, Q. Ren, T. Huang, Effects of redispersible polymer powders on the structural build-up of 3D printing cement paste with and without hydroxypropyl methylcellulose. Constr. Build. Mater. **267**, 120551 (2021)
12. R. Sharma, J.G. Jang, P.P. Bansal, A comprehensive review on effects of mineral admixtures and fibers on engineering properties of ultra-high-performance concrete. J. Build. Eng. **45**, 103314 (2022)
13. P. Feng, Y. Wu, T. Liu, Non-uniform fiber-resin distributions of pultruded GFRP profiles. Compos. B Eng. **231**, 109543 (2022)
14. H. Bahmani, D. Mostofinejad, Microstructure of ultra-high-performance concrete (UHPC)—A review study. J. Build. Eng. **50**, 104118 (2022)

15. J. Qin, F. Dai, H. Ma, X. Dai, Z. Li, X. Jia, J. Qian, Development and characterization of magnesium phosphate cement based ultra-high performance concrete. Compos. B Eng. **234**, 109694 (2022)
16. L. Liu, Z. Fang, Z. Huang, Y. Wu, Solving shrinkage problem of ultra-high-performance concrete by a combined use of expansive agent, super absorbent polymer, and shrinkage-reducing agent. Compos. B Eng. **230**, 109503 (2022)
17. A. Moshiri, D. Stefaniuk, S.K. Smith, A. Morshedifard, D.F. Rodrigues, M.J.A. Qomi, K.J. Krakowiak, Structure and morphology of calcium-silicate-hydrates cross-linked with dipodal organosilanes. Cem. Concr. Res. 133 (2020)
18. Y. Zhou, C.A. Orozco, E. Duque-Redondo, H. Manzano, G. Geng, P. Feng, P.J.M. Monteiro, C. Miao, Modification of poly(ethylene glycol) on the microstructure and mechanical properties of calcium silicate hydrates. Cem. Concr. Res. **115**, 20–30 (2019)
19. W. Li, W. Dong, Y. Guo, K. Wang, S.P. Shah, Advances in multifunctional cementitious composites with conductive carbon nanomaterials for smart infrastructure. Cem. Concr. Compos. **128**, 104454 (2022)
20. M. Monasterio, J.J. Gaitero, E. Erkizia, A.M. Guerrero Bustos, L.A. Miccio, J.S. Dolado, S. Cerveny, Effect of addition of silica- and amine functionalized silica-nanoparticles on the microstructure of calcium silicate hydrate (C-S-H) gel. J. Colloid Interface Sci. 450, 109–118 (2015)
21. U. Sharma, L.P. Singh, B. Zhan, C.S. Poon, Effect of particle size of nanosilica on microstructure of C-S-H and its impact on mechanical strength. Cem. Concr. Compos. **97**, 312–321 (2019)
22. M. Oltulu, R. Şahin, Effect of nano-SiO_2, nano-Al_2O_3 and nano-Fe_2O_3 powders on compressive strengths and capillary water absorption of cement mortar containing fly ash: A comparative study. Energy Build. **58**, 292–301 (2013)
23. M.S. Meddah, T.R. Praveenkumar, M.M. Vijayalakshmi, S. Manigandan, R. Arunachalam, Mechanical and microstructural characterization of rice husk ash and Al_2O_3 nanoparticles modified cement concrete. Constr. Build. Mater. **255**, 119358 (2020)
24. S.D. Anitha Selvasofia, E. Sarojini, G. Moulica, S. Thomas, M. Tharani, P.T. Saravanakumar, P. Manoj Kumar, Study on the mechanical properties of the nanoconcrete using nano-TiO_2 and nanoclay. Mater. Today:. Proc. 50, 1319–1325 (2022)
25. H. Madani, A. Bagheri, T. Parhizkar, The pozzolanic reactivity of monodispersed nanosilica hydrosols and their influence on the hydration characteristics of Portland cement. Cem. Concr. Res. **42**(12), 1563–1570 (2012)
26. G. Land, D. Stephan, Controlling cement hydration with nanoparticles. Cem. Concr. Compos. **57**, 64–67 (2015)
27. A.M. Said, M.S. Zeidan, M.T. Bassuoni, Y. Tian, Properties of concrete incorporating nanosilica. Constr. Build. Mater. **36**, 838–844 (2012)
28. S. Ghazizadeh, P. Duffour, N.T. Skipper, Y. Bai, Understanding the behaviour of graphene oxide in Portland cement paste. Cem. Concr. Res. **111**, 169–182 (2018)
29. M.F. Kai, L.W. Zhang, K.M. Liew, Graphene and graphene oxide in calcium silicate hydrates: Chemical reactions, mechanical behavior and interfacial sliding. Carbon **146**, 181–193 (2019)
30. M. Krystek, D. Pakulski, V. Patroniak, M. Gorski, L. Szojda, A. Ciesielski, P. Samori, High-performance graphene-based cementitious composites. Adv. Sci. **6**(9), 1801195 (2019)
31. H. Huang, L. Teng, X. Gao, K.H. Khayat, F. Wang, Z. Liu, Effect of carbon nanotube and graphite nanoplatelet on composition, structure, and nano-mechanical properties of C-S-H in UHPC. Cem. Concr. Res. **154**, 106713 (2022)
32. Y. Oaki, A. Kotachi, T. Miura, H. Imai, Bridged nanocrystals in biominerals and their biomimetics: classical yet modern crystal growth on the nanoscale. Adv. Funct. Mater. **16**(12), 1633–1639 (2006)
33. L. Bergström, E.V. Sturm, G. Salazar-Alvarez, H. Cölfen, Mesocrystals in bominerals and colloidal arrays. Acc. Chem. Res. **48**(5), 1391–1402 (2015)
34. H. Guo, P. Sun, Y. Liang, Y. Ma, Z. Qin, S. Cui, In-situ fabrication of polyelectrolyte-CSH superhydrophilic coatings via layer-by-layer assembly. Chem. Eng. J. **253**, 198–206 (2014)

35. J. Minet, S. Abramson, B. Bresson, A. Franceschini, H. Van Damme, N. Lequeux, Organic calcium silicate hydrate hybrids: a new approach to cement based nanocomposites. J. Mater. Chem. **16**(14), 1379–1383 (2006)
36. Z. Zhu, Z. Wang, L. Xu, Y. Zhou, Y. Chen, K. Wu, G. De Schutter, Synthesis and characterization of an intermediate for C-S-H structure tailoring. Cem. Concr. Res. **160**, 106923 (2022)
37. J. Rieger, M. Kellermeier, L. Nicoleau, Formation of nanoparticles and nanostructures–an industrial perspective on $CaCO_3$, cement, and polymers. Angew. Chem. Int. Ed. Engl. **53**(46), 12380–12396 (2014)
38. H. Matsuyama, J.F. Young, Synthesis of calcium silicate hydrate/polymer complexes: Part II. Cationic polymers and complex formation with different polymers. J. Mater. Res. 14(8), 3389–3396 (1999)
39. M. Kamali, A. Ghahremaninezhad, Effect of biomolecules on the nanostructure and nanomechanical property of calcium-silicate-hydrate. Sci. Rep. **8**(1), 9491 (2018)
40. F. Pelisser, P.J.P. Gleize, A. Mikowski, Structure and micro-nanomechanical characterization of synthetic calcium–silicate–hydrate with Poly(Vinyl Alcohol). Cem. Concr. Compos. **48**, 1–8 (2014)
41. Y. Chen, Y. Zheng, Y. Zhou, W. Zhang, W. Li, W. She, J. Liu, C. Miao, Multi-layered cement-hydrogel composite with high toughness, low thermal conductivity, and self-healing capability. Nat. Commun. **14**(1), 3438 (2023)
42. R. Khoshnazar, J.J. Beaudoin, L. Raki, R. Alizadeh, Interaction of 2-, 3-and 4-nitrobenzoic acid with the structure of calcium-silicate-hydrate. Mater. Struct. **49**(1–2), 499–506 (2016)
43. A. Picker, L. Nicoleau, Z. Burghard, J. Bill, I. Zlotnikov, C. Labbez, A. Nonat, H. Colfen, Mesocrystalline calcium silicate hydrate A bioinspired route toward elastic concrete material. Sci. Adv. 3(11), e1701216 (2017)
44. J. Seto, Y. Ma, S.A. Davis, F. Meldrum, A. Gourrier, Y.Y. Kim, U. Schilde, M. Sztucki, M. Burghammer, S. Maltsev, C. Jager, H. Colfen, Structure-property relationships of a biological mesocrystal in the adult sea urchin spine. Proc. Natl. Acad. Sci. U.S.A. **109**(10), 3699–3704 (2012)
45. M. Wang, P. Zhang, M. Shamsi, J.L. Thelen, W. Qian, V.K. Truong, J. Ma, J. Hu, M.D. Dickey, Tough and stretchable ionogels by in situ phase separation. Nat. Mater. **21**, 359–365 (2022)
46. G. Geng, R.J. Myers, J. Li, R. Maboudian, C. Carraro, D.A. Shapiro, P.J. Monteiro, Aluminum-induced dreierketten chain cross-links increase the mechanical properties of nanocrystalline calcium aluminosilicate hydrate. Sci. Rep. **7**, 44032 (2017)
47. E. Kapeluszna, Ł Kotwica, A. Różycka, Ł Gołek, Incorporation of Al in C-A-S-H gels with various Ca/Si and Al/Si ratio: Microstructural and structural characteristics with DTA/TG, XRD, FTIR and TEM analysis. Constr. Build. Mater. **155**, 643–653 (2017)
48. I.G. Richardson, Model structures for C-(A)-S-H(I), Acta Crystallogr. B: Struct. Sci. Cryst. Eng. Mater. 70(6), 903–923 (2014)
49. B. Walkley, R. San Nicolas, M.A. Sani, G.J. Rees, J.V. Hanna, J.S.J. van Deventer, J.L. Provis, Phase evolution of C-(N)-A-S-H/N-A-S-H gel blends investigated via alkali-activation of synthetic calcium aluminosilicate precursors, Cem. Concr. Res. 89, 120–135 (2016)
50. J. Li, W. Zhang, K. Garbev, G. Beuchle, P.J.M. Monteiro, Influences of cross-linking and Al incorporation on the intrinsic mechanical properties of tobermorite. Cem. Concr. Res. **136**, 106170 (2020)
51. X. Zhu, M. Vandamme, L. Brochard, Z. Zhang, Q. Ren, C. Li, B. He, H. Zhang, Y. Zhang, Q. Chen, Z. Jiang, Nature of aluminates in C-A-S-H: A cryogenic stability insight, an extension of DNA-code rule, and a general structural-chemical formula. Cem. Concr. Res. **167**, 107131 (2023)
52. X. Zhu, Q. Ren, B. He, H. Zhang, Z. Jiang, Upscaling degradation of cementitious calcium (aluminate) silicate hydrate upon ultra-low temperature attack: A multiscale insight and a bottom-up enhancement route. Compos. B Eng. **243**, 110122 (2022)
53. A. Picker, L. Nicoleau, Z. Burghard, J. Bill, I. Zlotnikov, C. Labbez, A. Nonat, H. Colfen, Mesocrystalline calcium silicate hydrate: A bioinspired route toward elastic concrete materials. Sci. Adv. 3(11), e1701216 (2017)

54. A.V. Torres-Figueroa, C.J. Pérez-Martínez, T.D. Castillo-Castro, E. Bolado-Martínez, M.A.G. Corella-Madueño, A.M. García-Alegría, T.E. Lara-Ceniceros, L. Armenta-Villegas, Composite hydrogel of Poly(acrylamide) and starch as potential system for controlled release of amoxicillin and inhibition of bacterial growth. J. Chem. 2020, 5860487 (2020)
55. A. Barth, The infrared absorption of amino acid side chains. Prog. Biophys. Mol. Biol. **74**(3), 141–173 (2000)
56. J.Y. Jiang, Q. Zheng, Y.R. Yan, D. Guo, F.J. Wang, S.P. Wu, W. Sun, Design of a novel nanocomposite with C-S-H@LA for thermal energy storage: A theoretical and experimental study. Appl. Energy **220**, 395–407 (2018)
57. L. Zhang, K. Yamauchi, Z. Li, X. Zhang, H. Ma, S. Ge, Novel understanding of calcium silicate hydrate from dilute hydration. Cem. Concr. Res. **99**, 95–105 (2017)
58. J.J. Thomas., H.M. Jennings., A.J. Allen., The surface of area of cement paste as measured by neutron scattering: Evidence for two C-S-H morphologies, Cem. Concr. Res. 28(6), 897–905 (1998)
59. C.M. Hunt, Nitrogen sorption measurements and surface areas of hardened cement pastes, Symposium on Structure of Portland Cement Paste and Concrete, Highway Research Board Special Report, Washington DC, pp. 112–122 (1966)
60. M.C. Garci, Quantifying microstructural variations in cement pastes: Implications on drying shrinkage, Northwestern University (1999)
61. A. Schmukat, L. Duester, D. Ecker, H. Schmid, C. Heil, P. Heininger, T.A. Ternes, Leaching of metal(loid)s from a construction material: Influence of the particle size, specific surface area and ionic strength. J. Hazard. Mater. **227–228**, 257–264 (2012)
62. X. Zhu, J. Cai, X. Xu, Z. Xie, Discussion on the method for determining BET specific surface area in argillaceous source rocks. Mar. Pet. Geol. **48**, 124–129 (2013)
63. A. Rabbani, S. Jamshidi, S. Salehi, Determination of specific surface of rock grains by 2D imaging. J. Geol. Res. **2014**, 945387 (2014)
64. C. Noiriel, C.I. Steefel, L. Yang, J. Ajo-Franklin, Upscaling calcium carbonate precipitation rates from pore to continuum scale. Chem. Geol. **318–319**, 60–74 (2012)
65. J. Zhu, R. Zhang, Y. Zhang, F. He, The fractal characteristics of pore size distribution in cement-based materials and its effect on gas permeability. Sci. Rep. **9**(1), 17191 (2019)
66. H.M. Jennings, Refinements to colloid model of C-S-H in cement: CM-II. Cem. Concr. Res. **38**(3), 275–289 (2008)
67. H.F.W. Taylor, *Cement chemistry* (Thomas Telford Publishing, London, 1997)
68. V.S. Ramachandran, Differential thermal method of estimating calcium hydroxide in calcium silicate and cement pastes. Cem. Concr. Res. **9**(6), 677–684 (1979)
69. J.H. Schön, Chapter 4–Density, in *Developments in petroleum science*. ed. by J.H. Schön (Elsevier, 2015), pp.109–118
70. P.S. Sharma, Environmental and engineering geophysics, Cambridge University Press (1997)
71. V. Zanjani Zadeh, C.P. Bobko, Nanoscale mechanical properties of concrete containing blast furnace slag and fly ash before and after thermal damage. Cem. Concr. Compos. 37, 215–221 (2013)
72. E.M. Foley, J.J. Kim, M.M. Reda Taha, Synthesis and nano-mechanical characterization of calcium-silicate-hydrate (C-S-H) made with 1.5 CaO/SiO_2 mixture. Cem. Concr. Res. 42(9), 1225–1232 (2012)
73. X. Zhu, Q. Ren, B. He, H. Zhang, Z. Jiang, The nanomechanical nature of cementitious calcium-(alumino)silicate-hydrate under ultra-low temperature attack: An intrinsic transition and the role of aluminum for bottom-up enhancement. Cem. Concr. Compos. **134**, 104811 (2022)
74. M. Vandamme, F.-J. Ulm, P. Fonollosa, Nanogranular packing of C-S-H at substochiometric conditions. Cem. Concr. Res.. Concr. Res. **40**(1), 14–26 (2010)
75. M. Vandamme, F.J. Ulm, Nanogranular origin of concrete creep. Proc. Natl. Acad. Sci. U.S.A. **106**(26), 10552–10557 (2009)
76. G. Constantinides, F.-J. Ulm, The effect of two types of C-S-H on the elasticity of cement-based materials: Results from nanoindentation and micromechanical modeling. Cem. Concr. Res. **34**(1), 67–80 (2004)

77. J.J. Kim, E.M. Foley, M.M. Reda Taha, Nano-mechanical characterization of synthetic calcium–silicate–hydrate (C–S–H) with varying CaO/SiO_2 mixture ratios. Cem. Concr. Compos. 36, 65–70 (2013)
78. X. Zhu, Z. Jiang, B. He, C. Qian, Investigation on the physical stability of calcium-silicate-hydrate with varying CaO/SiO_2 ratios under cryogenic attack. Constr. Build. Mater. **252**, 119103 (2020)
79. J. Li, W. Zhang, P.J.M. Monteiro, Structure and intrinsic mechanical properties of nanocrystalline calcium silicate hydrate. ACS Sustainable Chem. Eng. **8**(33), 12453–12461 (2020)
80. Automation creation Inc., MatWeb, http://asia.matweb.com/
81. J. Rouquerol, F. Rouquerol, P. Llewellyn, G. Maurin, K.S. Sing, Adsorption by powders and porous solids: principles, methodology and applications, Academic press (2013)
82. L. Huang, L. Tang, L. Wadsö, I. Löfgren, N. Olsson, Z. Yang, Using water vapour and N_2 isotherms to unveil effects of SCMs on nanopores and evaluate hydration degree. Cem. Concr. Res. **164**, 107042 (2023)
83. K. Scrivener, R. Snellings, B. Lothenbach, A practical guide to microstructural analysis of cementitious materials, Taylor & Francis (2016)
84. S. Lowell, J.E. Shields, M.A. Thomas, M. Thommes, Characterization of porous solids and powders: surface area, porosity and density, Springer (2004)
85. W. Sekkal, A. Zaoui, M. Benzerzour, N. Abriak, Role of porosity on the stiffness and stability of (001) surface of the nanogranular C-S-H gel. Cem. Concr. Res.. Concr. Res. **87**, 45–52 (2016)
86. Z. Wu, K.H. Khayat, C. Shi, Effect of nano-SiO_2 particles and curing time on development of fiber-matrix bond properties and microstructure of ultra-high strength concrete. Cem. Concr. Res. **95**, 247–256 (2017)
87. H.M. Jennings, J.J. Thomas, J.S. Gevrenov, G. Constantinides, F.-J. Ulm, A multi-technique investigation of the nanoporosity of cement paste. Cem. Concr. Res. **37**(3), 329–336 (2007)
88. H. Zhang, P.K. Sarker, L. Xiao, J. Ai, B. He, Q. Ren, X. Zhu, Y. Zhang, Durability of low-carbon geopolymer mortar: Different responses to cryogenic attack caused by water content and freeze-thaw mediums. Cem. Concr. Compos. **139**, 105065 (2023)
89. X. Zhu, B. He, M. Xie, Y. Zhang, Z. Jiang, A comprehensive nitrogen adsorption measurement on the pore structure of calcium-silicate-hydrate subjected to cryogenic attack. Measurement **184**, 109941 (2021)

Chapter 8
Scaling of Nanoscale Elastic and Failure Properties of Cement-Based Materials

8.1 Introduction

In the previous chapters, we have discussed the stability characteristics of C-S-H under cryogenic attack and proposed two strategies to enhance the cryogenic stability of C-S-H. Here we start to move our sights to the mechanical properties of C-S-H in situ under ultra-low temperature conditions. In laboratory environments, experimental research to explore the mechanical properties of concrete under ultra-low temperatures is challenging for several reasons. Firstly, creating ultra-low temperature conditions is difficult and maintaining a stable ultra-low temperature testing environment for concrete in a steady state is challenging. Secondly, the difficulty in quantitatively controlling the composition and microstructure of concrete materials makes it highly challenging to study the effects of different chemical compositions on the ultra-low temperature mechanical properties of the matrix. Thirdly, the stringent testing conditions lead to high testing costs, limiting large-scale dynamic and static mechanical testing research. According to existing research findings, it is known that even for completely dry concrete, there is an increasing trend in mechanical strength with decreasing temperature [1]. However, the physical mechanisms behind this phenomenon remain unclear. Molecular dynamics simulation research methods may offer new insights into the above-mentioned research challenges.

It is well known that the mechanical properties of disordered cementitious materials are size- and strain rate-dependent, which has been respectively confirmed in macro-, meso-, and microscopic levels by experimental studies [2–7]. The size effect of concrete strength captures the most attention in previously published works. As reported in the literature, the tensile strength of concrete decreases with increasing specimen size. Various theories and models have been proposed to characterize this effect, including Weibull statistical theory [8], Bazant's size effect law [9], multifractality scaling law [10–13], and multiscale model [14]. The Weibull statistical theory [8] adopts the weak chain link model to describe the failure of concrete: it assumes that concrete fails at the weakest link. Its governing equation linearly correlates the

Z. Jiang and X. Zhu, *Cryogenic Stability of Cement-Based Materials at Micro/Nano Scale*, https://doi.org/10.1007/978-981-96-1836-1_8

logarithm of the nominal strength (σ) to the logarithm of the structure length (L). The linear relation between log(σ) and log(L) in Weibull statistical theory tends to underestimate the macroscopic strength. The Bazant's size effect law [9] combines the linear elastic fracture mechanics and limit analysis, yielding another scaling law. This law is only valid for pre-notched specimens and for sizes larger than about 100 nm; for smaller sizes, Bazant's size effect predicts an asymptotical master curve that characterizes the relation between the strength and size of the specimen. In the multifractality scaling law [10–13], the geometrical fractality is considered for the damage domains and correlated to the physico-mechanical quantities, e.g., tensile strength and fracture energy. This scaling law overestimates those macroscopic quantities at large sizes since the master curve of strength versus size becomes a horizontal asymptote. A multiscale model [14] proposed recently combines the Griffith fracture theory and Bazant's size effect law, partly extending the size scaling in a wider range for the abovementioned three laws. Nonetheless, a plateau is observed when the size approaches the nanoscale range [14], like with Bazant's size effect law. In general, these scaling laws and models either underestimate the macroscopic strength or the nanoscopic strength [14].

The strain-rate sensitivity of the strength of concrete is also widely investigated, especially at the large structural size [2, 15–17]. Advanced techniques such as micro/nanoindentation [18], micro-cantilever testing [7], and molecular dynamic simulations [19, 20] allow us to identify this strain-rate sensitivity at the micro- and nanoscale ranges. A clear conclusion is that the strength of cementitious materials increases with the increasing strain rate. Akin to the size effect, several empirical models are proposed to determine the scaling with the strain rate, such as Malvar and Ross equation [21] and Zhou's equation [22]. These models are proposed for structural-size specimens and are purely empirical; thus, they are not adaptive enough to represent the nanoscale strain-rate sensitivity, let alone to elucidate the elementary physics of this time-dependent behavior.

From the kinetic concept theory for the strength of solids, the failure stress is related to temperature-induced fluctuations [23, 24]. Hence temperature is another crucial factor contributing to the mechanical properties of concrete. However, one should note that it is pointless to discuss the strength scaling of concrete at high temperatures (above about 50 °C) from the kinetic concept since high temperatures can lead to significant dehydration of cement paste and severe degradation of concrete microstructure, consequently decreasing the mechanical properties of concrete [25–27]. Instead, the strength scaling with temperature is likely to be captured at low temperatures since compositions and microstructures of the binding phases cannot be catastrophically degraded as much as they are at high temperatures [28–32]. Experimental studies on the cryogenic concrete show that strength increases with a decreasing temperature from ambient temperature to an ultra-low temperature at − 175 °C. This phenomenon persists even when the concrete is dried without the effect of pore water freezing [1]. The increase in the strength of dried concrete has not been well understood so far. To this end, the elementary physics for the mechanical failure of C-S-H requires to be disclosed.

Given the flaws in existing scaling theories of concrete, a general scaling law should be proposed to fit the scaling of mechanical failure down to the nanoscale. Especially the combined temperature-size-strain rate scaling has not attracted much attention on disordered cementitious materials, though some molecular dynamic simulation studies [19, 20, 26, 33–35] have been reported to investigate the effects of strain rate and temperature on C-S-H. In this respect, Brochard et al. [36] formulated a representative temperature-size-rate scaling law for the brittle failure of graphene and an idealized toy model. This scaling law, established from statistical physics, extends the law initially proposed by Zhurkov [24] to size effects. Nevertheless, the C-S-H structure is far more complex than that of graphene and the toy model. And its mechanical behavior differs significantly from that of graphene. Herein, we carry out molecular dynamic simulation studies to investigate the scaling of tensile mechanical failure of various C-S-H in a combination of temperature, strain rate, and system size. The effect of compositional difference on this scaling is also considered.

In this chapter, we will employ molecular dynamics simulation methods to investigate the tensile mechanical properties of C-S-H under ultra-low temperature conditions, aiming to elucidate the physical mechanisms governing the strength development of C-S-H at ultra-low temperatures. Initially, a construction method for a C-S-H model is proposed, further examining the effects of temperature, strain rate, model size, and different chemical compositions on the tensile failure process of C-S-H. The verification of the Zhurkov scaling effect in C-S-H will also be conducted.

8.2 Computational Model for Molecular Dynamics Simulation

8.2.1 Generation of C-S-H Model

In a hydrated cement paste system, the composition of C-S-H is complex and changes dynamically during the maturing and aging periods. The average Ca/Si ratio is around 1.7 for matured cement pastes. To construct molecular structures for molecular dynamic simulation studies, Pellenq et al. [37] proposed a methodology to generate realistic C-S-H atomistic structures with a target Ca/Si ratio by removing a certain amount of bridging and pairing tetrahedra from the Hamid's tobermorite structure [38] which has an initial Ca/Si ratio of 1. In this way, the fractions of bridging and pairing tetrahedra are consistent with ^{29}Si NMR experimental results. This methodology makes it possible to obtain structures compatible with experimental results and is widely accepted in many molecular simulation studies [39–41]. However, generating C-S-H models with various Ca/Si ratios with this methodology is tedious, since one has to perform Grand Canonical Monte Carlo simulations at each target Ca/Si ratio in order to obtain correct compositions.

Abdolhosseini Qomi et al. [42] proposed a combinatorial approach to generate a database of C-S-H atomic structures with various Ca/Si ratios. The topological analysis on the C-S-H database yielded the compositional relations among H_2O/Si, Ca-OH bond/Si, and Ca/Si ratio. Morshedifard et al. [43] adopted these compositional relations and the approach of Ref. [42], to construct C-S-H molecular models. According to its calculation results, the content of water molecules and hydroxyl groups within C-S-H has the following relationship under a given Ca/Si ratio:

$$N(H_2O) = (x - 1) \times N(Si) \tag{8.1}$$

$$N_t(OH) = (1.799x - 1.2539) \times N(Si) \tag{8.2}$$

$$N_{Cw}(OH) = (x - 1) \times N(Si) \tag{8.3}$$

$$N_{Si}(OH) = 2N_t(OH) - 2N(H_2O) - N_{Cw}(OH) \tag{8.4}$$

where $N(H_2O)$, $N(Si)$, $N_t(OH)$, $N_{Cw}(OH)$, and $N_{Si}(OH)$ respectively represent the number of water molecules, silicon atoms, total hydroxyl groups, hydroxyl groups coordinated with interlayer calcium ions, and hydroxyl groups coordinated with silicon atoms.

According to Abdolhosseini Qomi et al. [42], to establish a C-S-H structural model with a given Ca/Si ratio, the Grand Canonical Monte Carlo (GCMC) method is employed to adsorb water molecules, followed by the use of ReaxFF force field simulation to create hydroxyl groups. Based on the quantitative relationship mentioned above, we can directly determine the quantities of water molecules and hydroxyl groups, significantly reducing the computational time required for these steps.

Based on the classic C-S-H model construction method proposed by Pellenq et al. [44] and the quantifiable relationships determined by Abdolhosseini Qomi et al. [42], we propose a simple C-S-H model construction method outlined in Fig. 8.1. Firstly, a supercell structure that meets the requirements is constructed based on the 11 Å tobermorite structure. Next, a certain number of bridging Q^2b units are removed, followed by the removal of a certain number of paired Q^2p units until the Ca/Si requirement is met. Then, the corresponding number of OH groups is added to Si and interlayer Ca according to the relationships given in Eqs. 8.3 and 8.4, respectively. Finally, the appropriate number of H_2O molecules is added between the C-S-H layers as per the relationship defined in Eq. 8.1.

However, in constructing C-S-H as described above, there is still a missing essential condition, namely, the number of deletions corresponding to bridging Q^2b units and paired Q^2p units. To address this, we will provide the corresponding solution in the next section.

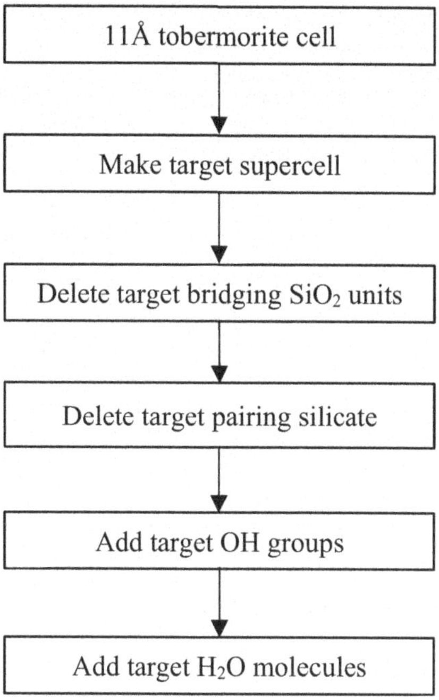

Fig. 8.1 Construction process of C-S-H structure model

8.2.2 Silicate Chains of C-S-H

The approach used by Abdolhosseini Qomi et al. [42] and Morshedifard et al. [43] yields bridging/pairing tetrahedra fractions, which are inconsistent with NMR experiments since bridging tetrahedra are always removed first irrespective of varying Ca/Si ratios. Although this artifact does not impact the mechanical properties [43], one cannot exclude that representative bridging/pairing tetrahedral fractions are important when considering widely different sizes, temperatures, or loading rates as targeted in the present study. Herein, we solve this problem by reconciling the fraction of bridging tetrahedra with ^{29}Si NMR experimental measurements.

To this end, we collect quantitative deconvolution data [45–52] of the bridging tetrahedra fraction present in C-S-H. The relation between bridging tetrahedra fraction (Q^2b) and Ca/Si ratio is obtained by fitting those data points, as shown in Fig. 8.2 and in the following equation:

$$Q^2\mathrm{b} = 14.22(\mathrm{Ca/Si})^{-1.897} \tag{8.5}$$

According to the quantitative relationship provided by Eq. 8.5, we can easily calculate the required number of bridging silicate tetrahedra to be retained in the supercell for any Ca/Si ratio condition.

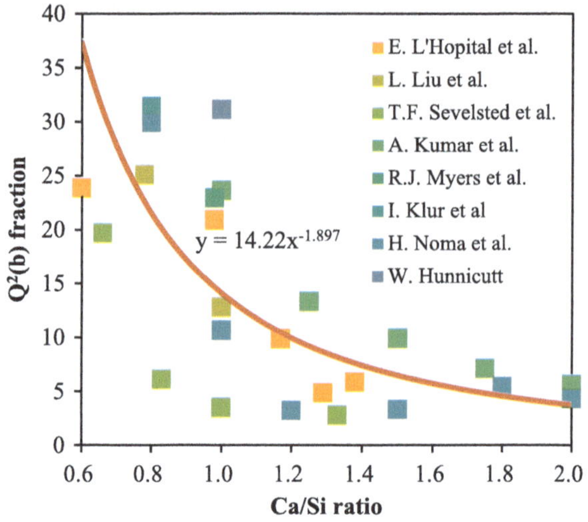

Fig. 8.2 The fraction of bridging silicate tetrahedra of C-S-H at various Ca/Si ratios. The ^{29}Si NMR spectra deconvolution data are collected in published works [45–52]

8.2.3 *Computation Details*

1. C-S-H Model

To investigate the influence of different chemical compositions on the mechanical properties of C-S-H, models were established based on the above-mentioned modeling method. Starting from an initial unit cell of 11 Å tobermorite, supercells of $6a \times 8b \times 1c$ were expanded to create C-S-H computational models with Ca/Si ratios of 1.1, 1.3, 1.5, and 1.7. The initial model dimensions were set as $a = 33.476$ Å, $b = 29.56$ Å, $c = 22.77$ Å. All initial models were supplemented with sufficient hydroxyl groups and water molecules following the relationships provided in Sect. 8.2.1. The resulting initial C-S-H computational models are depicted in Fig. 8.3.

2. Strain rate condition

To investigate the scalings with respect to strain rates, 8 different engineering strain rates ranging several orders of magnitude are used: 10^{-3}, 0.5×10^{-3}, 10^{-4}, 0.5×10^{-4}, 10^{-5}, 0.5×10^{-5}, 10^{-6}, 0.5×10^{-6}, and 10^{-7} ps^{-1}. In addition, the stress–strain curves of C-S-H models in the athermal limit (0 K) are obtained by minimization. In what follows, the reduced mechanical properties presented at each temperature and loading rate, are reduced with respect to the values in the athermal limit. C-S-H structure models with Ca/Si ratios of 1.1 and 1.7 were selected for calculations under three temperature conditions: 100, 200, and 300 K, corresponding to ultra-low temperature and room temperature environments. The simulated strain rates used are listed in Table 8.1.

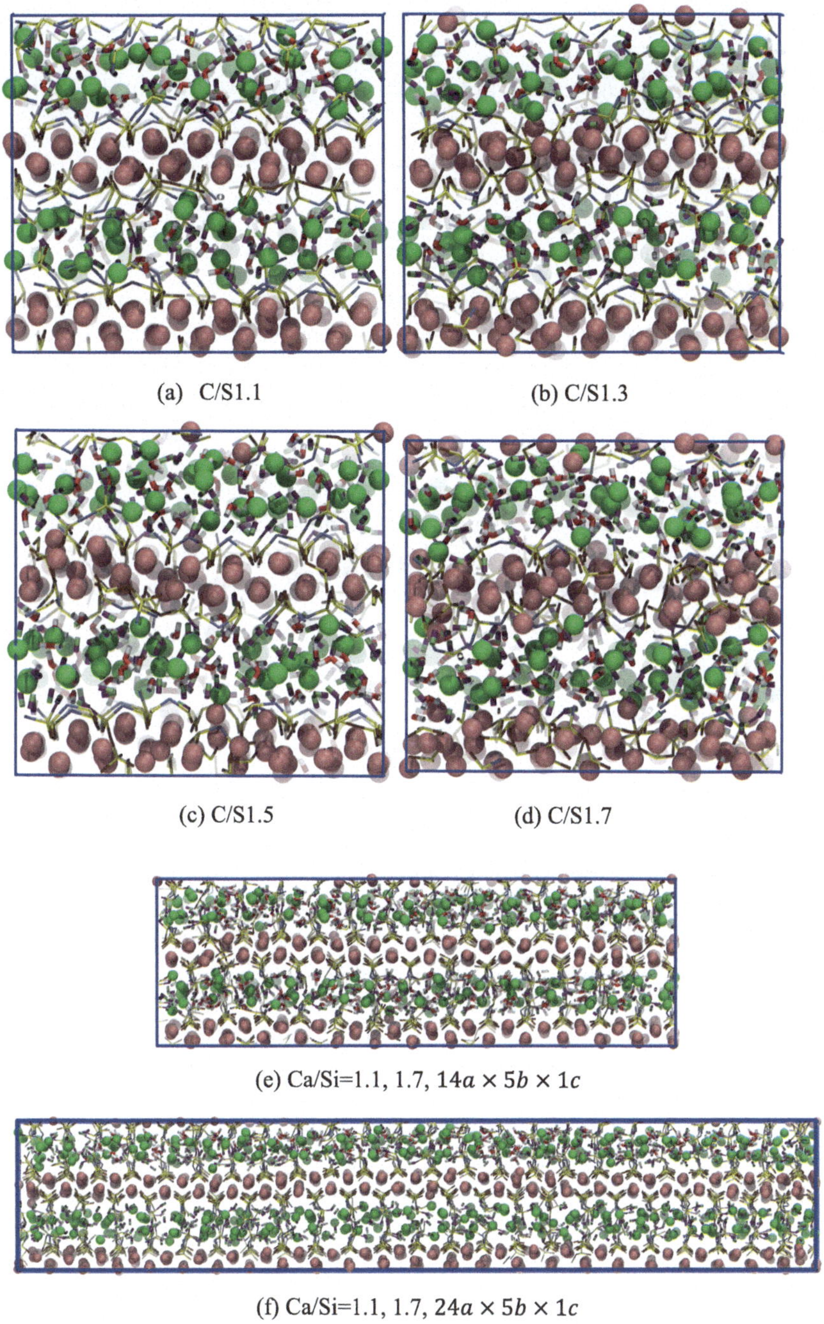

Fig. 8.3 Structure of C-S-H computational models. The green and pink balls are Ca and Cw atoms, respectively. The thin yellow, ice blue, purple, and ochre sticks stand for Si–O bonds. The purple and lime sticks represent hydroxyl groups. The red and white sticks are water molecules

Table 8.1 Strain rate used in the simulation

No	1	2	3	4	5	6	7	8
Strain rate/ps^{-1}	10^{-3}	0.5×10^{-3}	10^{-4}	0.5×10^{-4}	10^{-5}	0.5×10^{-5}	10^{-6}	0.5×10^{-6}
Temperature/K	100, 200, 300							
Model	Ca/Si = 1.1, 1.7, $a = 33.476$ Å, $b = 29.56$ Å, $c = 22.77$ Å							

3. Temperature condition

We investigated the mechanical properties of C-S-H at different temperature environments ranging from 100 to 300 K. Calculations were performed for C-S-H models with Ca/Si ratios of 1.1, 1.3, 1.5, and 1.7, at a fixed strain rate of When investigating the scalings with respect to temperature, the loading rate is fixed at a constant value of 10^{-5} ps^{-1}.

4. Size condition

We also studied the influence of model size on the mechanical properties of C-S-H. Using a C-S-H model with a Ca/Si ratio of 1.7, simulations were conducted with model sizes represented by Fig. 8.3d, e, and f, the loading rate is fixed at a constant value of 10^{-5} ps^{-1}.

5. Force field

CSH-FF is well known for its accuracy in simulating the mechanical properties of silicate minerals in cement paste due to the incorporation of second order properties in the fitting process [53, 54]. In this study, we investigate the elastic and failure properties of C-S-H in the *c* direction, for which failure is expected in the interlayer space, and this force field is adapted to capture the failure in the interlayer. Therefore, the CSH-FF force field is adopted in our simulations. Note that this non-reactive force field is not adapted to study the failure outside the interlayer, and one should consider instead a reactive force field when considering the failure in the *a*- and *b*- directions. The non-bonded interactions among different atom types (Ca, Cw, O, Ob, Oh, Ow, Si, H, Hw, with Ca the intralayer calcium atoms, Cw the interlayer calcium atoms, Ob the bridging oxygen atoms, Oh the hydroxyl oxygen atoms, O the rest oxygen atoms in the silicate chains, Ow the oxygen atoms of water molecules, H the hydrogen atoms of hydroxyl groups, Hw the hydrogen atoms of water molecules, and Si the silicon atoms), bond interactions, angular harmonic contributions, and atom charges are defined in Ref. [42, 53]. A critical detail that deserves special attention is the charges of O, Ob, and Oh atoms, which are not constant at different Ca/Si ratios. After the partial charges of Ca, Cw, Si, Ow, and Hw are explicitly defined as given in Ref. [42], extra charges of the whole system are equally redistributed to all O, Ob, and Oh atoms to ensure electro-neutrality. It is worth noting that CSH-FF is based on SPC water, which is known to inadequately capture the liquid-to-solid phase transition in water. Therefore, using CSH-FF in tandem with SPC water to simulate a gel pore system at cryogenic temperatures would not be physically realistic.

6. Simulation details

All the simulations are performed using LAMMPS package [55]. The initial C-S-H molecular models are relaxed in the NPT ensemble at 1 atm and 300 K for 2 ns, reaching an equilibrium state before further simulations. The integration is done using the Verlet leapfrog integration algorithm with a timestep of 1 fs. Nose–Hoover thermostat and barostat are utilized to control the constant temperature and pressure, respectively. The Ewald summation method is adopted to calculate the long-range interactions, using a cutoff of 10 Å and an accuracy of 10^{-5}. To investigate the effect of temperature, the C-S-H models are further relaxed at varying temperatures from 100 to 300 K with an interval of 20 K for 2 ns before loading. The final relaxed molecular configurations are used for uniaxial tensile loading tests.

It has been extensively proven that the interlayer space is the weak zone of C-S-H [56, 57]. The primary failure mechanism of C-S-H is the breakdown of the interlayer space. The separation of interlayer space can further lead to crack growth and ultimate failure of the materials. In this case, the cohesive pressure within the interlayer is strongly related to the strength, according to Masoumi et al. [58]. Herein, we focus on the uniaxial tensile failure of C-S-H in the interlayer space. In the uniaxial tensile loading simulations, the systems are run in a mixed NPT/NVT ensemble with control of the deformation along the z axis, while setting the pressure in the x and y directions at 1 atm. The normal stress σ_{zz} along the z direction is computed using the virial definition to evaluate the tensile response of the system. All the simulations are performed three times in order to show uncertainty estimates. Distinctive initial velocities are assigned to the three simulations, each utilizing different random distributions to ensure the evolutions follow distinct trajectories. Note that we treat the systems in undrained conditions because the water dynamic is extremely slow at cryogenic temperatures.

8.3 Scaling of C-S-H Tensile Failure Properties

8.3.1 Stress–Strain Curve

In the present work, we focus on the tensile elastic and failure mechanical properties of C-S-H, namely tensile strength f_t, Young's modulus E, fracture energy G, and fracture-process zone length (FPZ). The tensile strength σ is defined as the maximum of the tensile stress–strain curve. Young's modulus E is obtained by fitting the linear elastic portion of the stress–strain curve for strains ε less than 0.04. The fracture energy G is determined by integrating the stress–strain curve until complete failure. Regarding the FPZ length, although various approximation equations have been proposed to estimate this value, all these proposed equations are related to the characteristic length r_{pl} up to a multiplicative factor of order unity [59]. Therefore, we use this characteristic length to represent the FPZ length. Figure 8.4 shows the

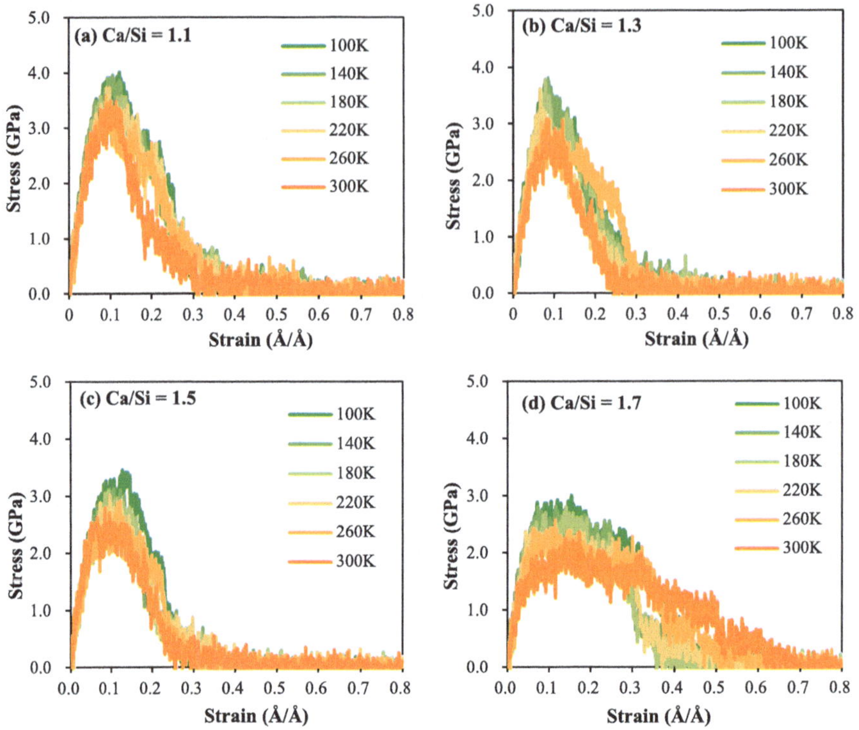

Fig. 8.4 C-S-H tensile stress–strain curves

stress–strain curves from the direct tensile simulations. These mechanical quantities are calculated as follows:

$$f_t = \max(f(x)) \tag{8.6}$$

$$G = \int f(x) \times L_z \tag{8.7}$$

$$r_{pl} = EG/\sigma^2 \tag{8.8}$$

where L_z represents the length in the z-direction of the model.

8.3.2 *Scaling with Temperature*

The scaling of mechanical properties with temperature is shown in Fig. 8.5. Considering the complexity of C-S-H compositions, we analyze four C-S-H models with

Ca/Si = 1.1, 1.3, 1.5, and 1.7. Figure 8.5a displays the tensile strength. The strength decreases linearly with an increasing temperature up to 300 K, irrespective of the Ca/Si ratio. Experimental investigations conducted on dried concrete illustrate a declining trend in macroscopic strength as temperature increases from approximately 100–300 K [1]. In this regard, similarities exist between the effects of cryogenic temperatures on both the macroscopic strength of concrete and the nanoscopic strength of C-S-H. Given that C-S-H serves as the primary binding phase in cement concrete, the reduction in the macroscopic strength of cryogenic concrete may be partly attributed to the decrease in the nanoscopic strength of C-S-H. Of course, C-S-H resistance is not the only factor affecting concrete failure, in particular heterogeneities and microstructure have major impacts, explaining the difference of magnitude between nanoscale resistance and macroscale resistance. Yet, the findings of the present work suggest that at least part of the temperature effect on concrete strength could originate from the C-S-H scaling identified here. The linear dependency of strength on temperature confirms that the temperature effect follows the kinetic concept theory of strength [24], with the reduction of energy barrier that decreases linearly with stress. Note that the lowest Ca/Si ratio at 1.1 yields the highest strength at tensile failure, while the highest Ca/Si ratio at 1.7 corresponds to the lowest strength. The strength of the other two C-S-H models with intermediate Ca/Si ratios of 1.3 and 1.5 are similar and lie between the 1.1 and 1.7 cases. For the tensile response in the c-axis direction, the decline in strength observed with decreasing Ca/Si ratio is likely to originate from the decrease in interlayer charges.

Analogous to tensile strength, a linear trend is also observed for the scaling of Young's modulus with temperature, as shown in Fig. 8.5b, suggesting that C-S-H gels become stiffer in the ***c*** axis direction at low temperatures. The scaling of Young's modulus slightly differs from that of tensile strength regarding the different Ca/Si ratios: while, at 300 K, the Young's modulus of C-S-H decreases with the Ca/Si ratio, this is no more the case at lower temperatures and the trend with Ca/Si ratios becomes nonmonotonous below 270 K. At 100 K, the C-S-H model with Ca/Si = 1.5 has the lowest Young's modulus, while at 300 K the C-S-H model with the lowest Young's modulus is the one with Ca/Si = 1.7. Those changes probably result from the variations of molecular configurations of C-S-H since Young's modulus is highly sensitive to the connectivity and bond energy of a solid material [60]. Overall, our results still suggest that the stiffness of C-S-H generally decreases with the Ca/Si ratio.

Figure 8.5c displays the evolution of the fracture energy with temperature. In comparison with strength and Young's modulus, the estimations of fracture energy show much larger uncertainties. On average, we find a linearly decreasing trend with temperatures from 100 to 300 K, except for the model with Ca/Si = 1.1. But at this lowest Ca/Si ratio, the observed linearly decreasing trend means that fracture failure of the C-S-H models releases more energy at low temperatures than at high ones. Comparing the trends in Fig. 8.5b, c, Young's modulus appears to decrease faster with temperature than fracture energy. As a consequence, toughness, which results from the product of the fracture energy by the Young's Modulus, is estimated to increase at low temperatures (see Fig. 8.6, note that the obtained toughness is merely

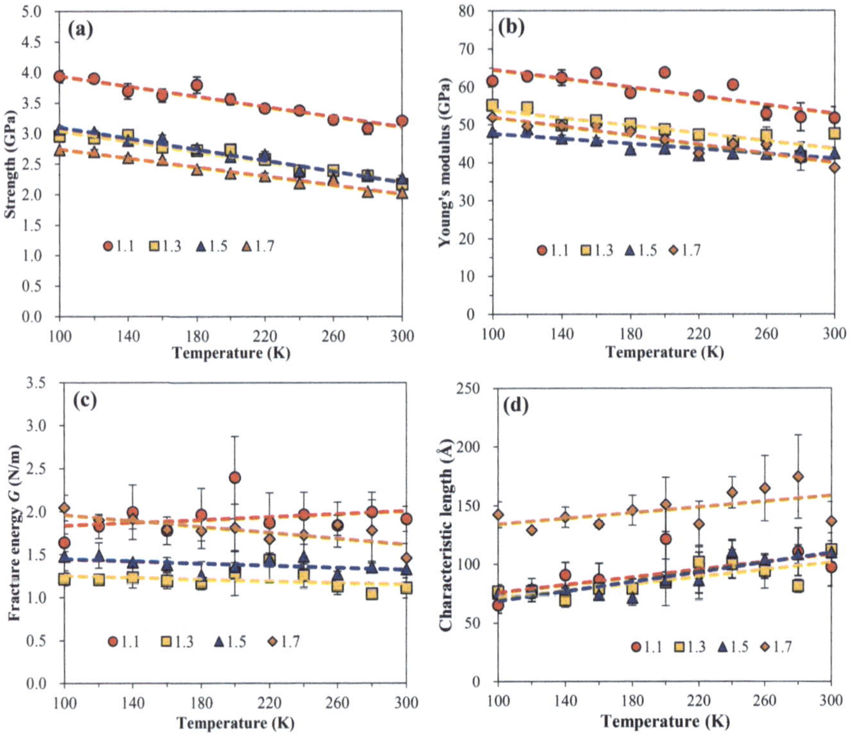

Fig. 8.5 Scaling of mechanical properties of C-S-H with temperature **a** Tensile strength; **b** Young's modulus; **c** fracture energy; **d** characteristic length. Four different Ca/Si ratios, i.e., Ca/Si = 1.1, 1.3, 1.5, and 1.7, are considered

a rough estimation since the Irwin formula is not rigorous for C-S-H at the nanoscale because of the C-S-H heterogeneity and anisotropy). The tensile fracture of C-S-H at the nanoscale shows a ductile behavior rather than a brittle failure, while this is a brittle failure that is usually observed for concrete at the structural size. This ductile behavior is reminiscent of size effects that are ubiquitous in the study of fracture at the nanoscale [61]. Such nanoscopic ductility depends significantly on composition. The fracture energy increases with the Ca/Si ratio.

The fracture energy alone, nonetheless, does not suffice to quantify the ductility. A proper characterization of ductility requires determining the FPZ length, which is given in Fig. 8.5d. FPZ quantifies approximately the size of the domain in which mechanical energy dissipation occurs at the crack tip [59]. This zone is formed spontaneously with the inelastic deformation, therefore reflecting the ductility. As the temperature increases from 100 to 300 K, the FPZ length increases linearly, indicating that C-S-H is more ductile at larger temperatures. To the best of our knowledge, such a linear scaling of C-S-H FPZ length with temperature has never been reported before. This scaling is observed for all C-S-H models irrespective of the composition. However, we note that the estimated FPZ lengths are similar to the

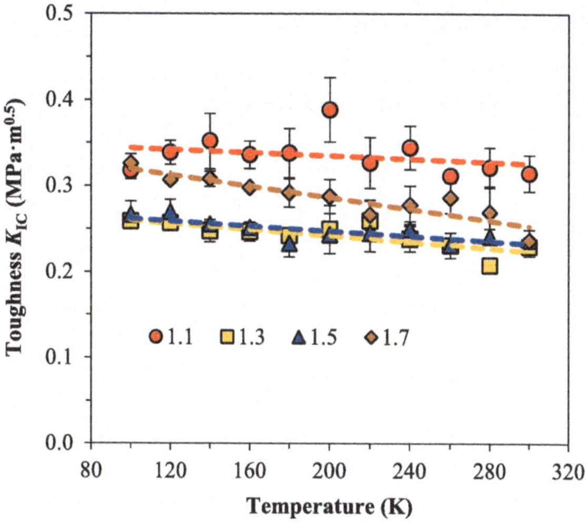

Fig. 8.6 Scaling of toughness with respect to temperature for various C-S-H models

size of the simulated C-S-H models, and even exceed it for Ca/Si = 1.7 for which the process zone length is especially large (see Fig. 8.5d). Accordingly, all the failure responses are quite ductile at the scale of our C-S-H models, and the C-S-H with a Ca/Si of 1.7 exhibits far more ductile response than the other models. According to Bauchy et al. [60, 62], C-S-H has a ductile network when constraint number per atom $n_c < 3$, a brittle network when $n_c > 3$, and an isostatic network when n_c at around 3. Following the topological constraint analysis [60, 62], the isostatic composition of C-S-H is in the domain of 1.3 < Ca/Si < 1.53, for which n_c is around 3: this composition domain is considered as the brittle-to-ductile boundary of C-S-H. In contrast, the composition domains for ductile and brittle C-S-H are in Ca/Si ≤ 1.3 and Ca/Si ≥ 1.53, respectively, with corresponding constraint numbers $n_c > 3$ and $n_c < 3$, respectively. Since C-S-H is expected to reach a local minimum value of brittleness within the isostatic composition window, the domain of 1.3 < Ca/Si < 1.53 is considered the boundary of the brittle-to-ductile transition [60]. The FPZ length results obtained in the present study confirm the existence of the transition boundary. Nevertheless, we further underline here that the brittle-to-ductile boundary of the composition Ca/Si is insignificant in distinguishing the brittle (Ca/Si ≤ 1.3) and isostatic networks, whereas the boundary between isostatic and ductile (Ca/Si ≥ 1.53) networks is marked since the FPZ lengths have similar values when the Ca/Si ≤ 1.5 and a significant increase in the FPZ length of C-S-H occurs only if the Ca/Si ratio is higher than 1.7.

8.3.3 Scaling with Size

8.3.3.1 Nanoscale Size Effect

Herein, we investigate two C-S-H compositions with Ca/Si ratios of 1.7 and 1.1, respectively. Since the general trends of those mechanical quantities are analogous, we present here the results for Ca/Si = 1.7. A few representative stress–strain curves of different system sizes are shown in Fig. 8.7. Given that failures consistently happen within the interlayer and that anisotropic structural effects are absent in the C-S-H systems within these layers, it can be inferred that direct tensile failures would exhibit similarity across all interlayer plane orientations. Consequently, a comparable failure behavior can be anticipated for both the *a* and *b* directions. Therefore, we confine our observations to a through crack in the *b* direction by merely enlarging the dimensions along the *a* direction. Moreover, it appears from Fig. 8.7 that the failure process is mostly localized in the interlayer of the C-S-H, so that most of the mechanical dissipation takes place in the interlayer. Accordingly, there is good reason to believe that increasing the size in the *c* direction would have no effect since the behavior should remain elastic in the supplementary volume. Therefore, in the present study, we limited ourselves to the study of the size effects in the *a* direction. Since the tensile failure is localized in the interlayer, the mechanical properties are highly related to the interparticle cohesive interactions within the interlayer. Indeed, the present results for strength appear consistent with interparticle cohesive pressure estimated by Masoumi et al. [58] (1.9–3.7 GPa). Furthermore, our results for Young's modulus are also consistent with the estimations of Masoumi et al. [58] (52.3–102.43 GPa).

Note that some medium and large systems are not ruptured along a single crack that spans the entire simulation box. Instead, two cracks initiate in a staggered configuration (Fig. 8.7d, e). In such cases, at the final strain considered (0.8), the area of two cracks corresponds approximately to the total horizontal area of a single crack, so that the energy released provides a reasonable estimate of the energy that would have been released by a single crack spanning the entire box. The typical error remains moderate and is of the order of the residual integral at the end of the curve (at a strain around 0.6–0.8). This moderate error is not expected to alter the main trend of the scaling. To show the effect of system size, the four mechanical properties are displayed in Fig. 8.8. One observes that the evolution of the mechanical properties is not monotonous with the system size as generally observed at the microscale and macroscale. The strength and fracture energy evolve significantly with the size; in particular the small system leads to significantly lower estimates of those two quantities. The distinct behaviors observed for the small systems can be attributed to system sizes which are not large enough compared to the process zone to fully capture the failure process. But this could also be due in part to usual artifacts in molecular simulations due to finite size effects and periodic boundary conditions. Consequently, such small systems exhibit premature failure, resulting in the underestimation of both strength and fracture energy. In contrast, system size has no significant impact on Young's modulus and FPZ length. A characteristic length

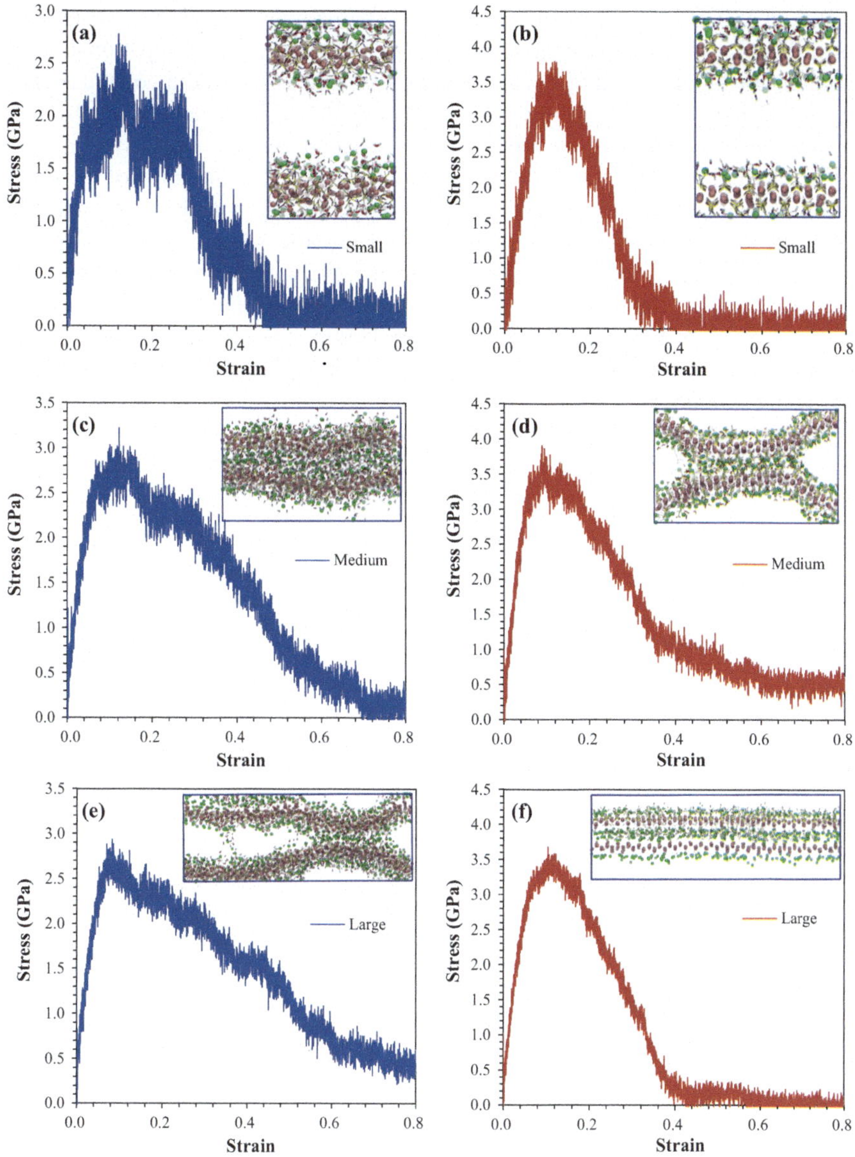

Fig. 8.7 Stress–strain curves of C-S-H systems with small **a**, **b**, medium **c**, **d**, and large **e**, **f** system sizes at 300 K. The blue and red curves are results for C-S-H with a Ca/Si ratio of 1.7 and 1.1, respectively. All stress–strain curves show ductile fracture features, indicating that the system sizes are close to the FPZ length. Under the tension conditions of **d** and **e**, C-S-H systems were not ruptured along a single crack, which leads to misestimating the fracture energy. Nevertheless, the corresponding area in the associated stress–strain curves (around strains of 0.6 to 0.8) remains relatively close to the total area, so that the error in the estimate of fracture energy remains modest

of around 150 Å is obtained for all systems. The periodic C-S-H systems investigated should be at least twice the FPZ length to fully capture the dissipation process arising from the formation of a crack (i.e., two crack tips). Depending on the analytical model, the FPZ is proportional to r_c with a multiplicative factor ranging from 2/3π to 1 [63–68]. Therefore, to fully capture C-S-H failure, the C-S-H system shall not be shorter than 32 Å for Ca/Si = 1.7. The length in the *x*-direction of the small system investigated here is around 33 Å, i.e., approaching or even shorter than the minimum FPZ length. Accordingly, the dissipation processes of C-S-H failure are not fully captured for this small system, which explains that strength and fracture energy are smaller than for larger systems. In other words, the length of the small system is too short to be representative of C-S-H failure. For the medium and large systems, however, one recovers the expected evolution, i.e., a decrease of strength and fracture energy with size.

Note that the size effect on the FPZ length is more pronounced for C-S-H with Ca/Si = 1.1 than for C-S-H at other Ca/Si ratios (see Fig. 8.9d). The FPZ length of the

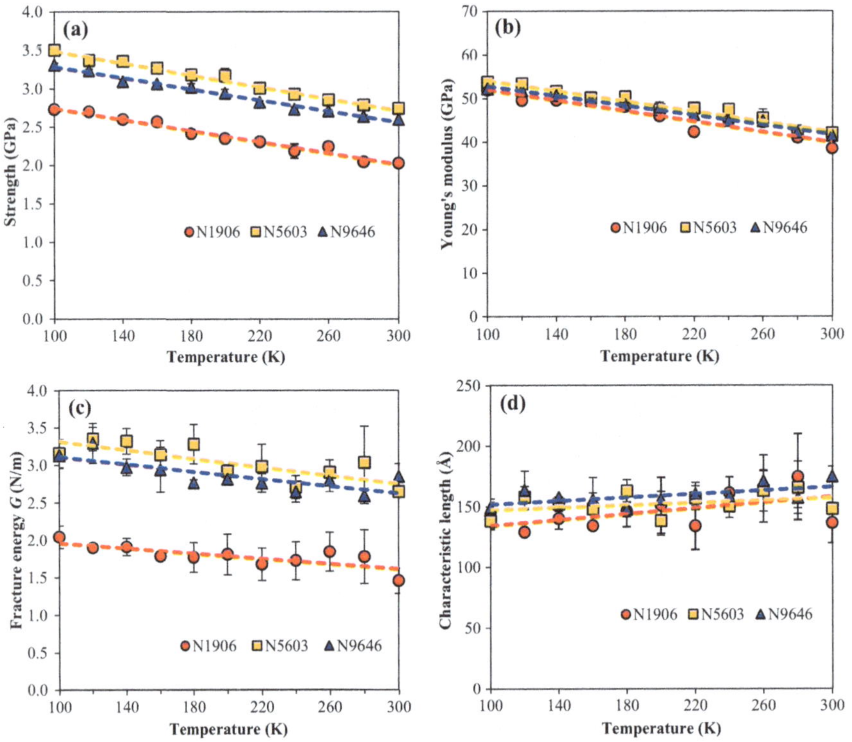

Fig. 8.8 Comparisons of mechanical properties of C-S-H (Ca/Si = 1.7) with different system sizes at varying temperatures. **a** Strength; **b** Young's modulus; **c** fracture energy; **d** characteristic length. N1906, N5603, and N9646 represent the small, medium, and large systems, with 1906, 5603, and 9646 atoms, respectively

small system is significantly shorter than that of the medium and large systems. The characteristic length of the small C-S-H system at this low Ca/Si ratio increases from ~ 60 to ~ 100 Å when the temperature rises from 100 to 300 K. These low values of FPZ length are due to a tensile strength of the small system that approaches that of the medium and large systems (see Fig. 8.9a), whereas the fracture energy is still significantly underestimated in the small system (see Fig. 8.9c). The characteristic lengths obtained for the medium and large systems are around 130 Å, from which we can conclude that the size of the C-S-H system should well exceed 28 Å for Ca/Si = 1.1 to fully capture the failure processes. Accordingly, the small system is again not large enough to be fully representative of the failure processes.

In summary, confronting the small, medium, and large systems, we highlight that the mechanical quantities that we studied do not evolve monotonously with system size at the nanoscale. Once the system size is smaller than its FPZ length, mechanical quantities, i.e., strength, Young's modulus, and fracture energy, are underestimated. This nanoscale size effect differs from the microscale and macroscale size effects observed in previous studies [3, 6].

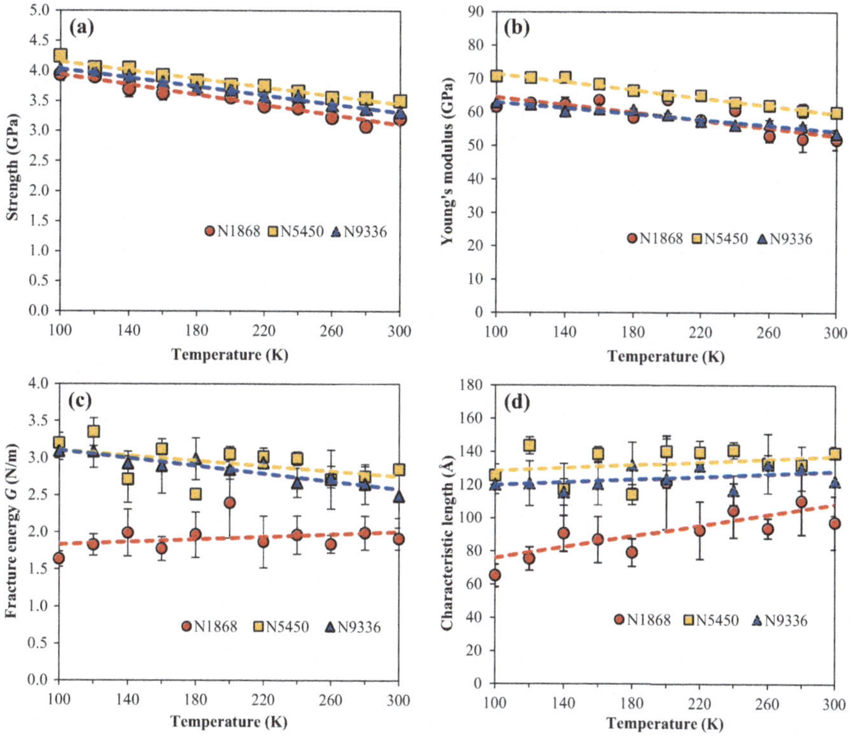

Fig. 8.9 Comparisons of mechanical properties of C-S-H (Ca/Si = 1.1) with different system sizes at varying temperatures **a** Strength; **b** Young's modulus; **c** fracture energy; **d** characteristic length. N1868, N5450, and N9336 represent the small, medium, and large systems, with atom numbers 1868, 5450, and 9336

8.3.3.2 Minimum Cell Size for Fracture Failure

From the observations on C-S-H with a Ca/Si ratio of 1.7 and 1.1, it cannot be overlooked that the interplay between failure properties, and C-S-H system size and composition is critical and calls for a careful analysis. This brings up a crucial issue: the minimum size of a molecular model for capturing the fracture failure of C-S-H with various compositions. At a small scale, the failure of solids exhibits size effects that are reminiscent of the transition from fracture failure to yield failure. According to Brochard et al. [69], fracture failure can be observed only if the system size (L) and FPZ length (l_0) satisfy the inequality $L/(2l_0) > 1.96$. Otherwise, yield failure can only be observed. Following this relation, one can estimate the minimum cell size L_{min} of C-S-H molecular models to capture fracture failure:

$$L_{min} = 1.96 \times 2l_0 = 1.96 \times 2 \times \frac{2}{3\pi} r_c \tag{8.9}$$

where r_c is the characteristic length, $\frac{2}{3\pi}$ is taken from the FPZ analytical model of Hui et al. [63], which has the minimum multiplicative factor among all FPZ models [63–68].

The actual minimum cell size is inferred with the following rationale: If the size of the simulation box is too small to fully capture the fracture energy, then fracture energy increases rapidly with the system size. If the simulation box is large enough instead, one expects a fracture energy that decreases slowly with the logarithm of the system size. Strength, in contrast, exhibits a logarithmic decrease even for the smallest size. Accordingly, the computed length of the FPZ is underestimated as long as it increases with the size of the system, and it is correctly estimated once it stops increasing. By looking for this transition, one can thus infer the actual minimum cell size. Figure 8.10 shows the minimum cell size for fracture failure estimated from Eq. 8.9. It is apparent that the minimum cell size depends on both temperature and composition. It generally increases with increasing temperature, and the system with a high Ca/Si ratio of 1.7 requires a larger cell size than the system with a low Ca/Si ratio of 1.1. At 300 K, the minimum cell size estimated for C-S-H with a Ca/Si ratio of 1.7 and 1.1 are 145.7 ± 6.7 Å and 115.9 ± 1.9 Å, respectively. In the present work, the length of small and medium systems after relaxation are ~ 32 Å and ~ 75 Å, respectively, which are significantly lower than the corresponding minimum cell sizes estimated. The large system after relaxation has a length of ~ 131 Å, which approaches the estimated minimum cell size for Ca/Si = 1.7 and exceeds it for Ca/Si = 1.1, as shown in Fig. 8.10. The stress–strain curves of the large systems remain mostly ductile during the tensile failure, since the system size is still close to the minimum size (see Fig. 8.7). Yet, one can expect that those systems are large enough to correctly capture the full dissipation process at failure. Our results suggest that the minimum cell size for fracture failure of C-S-H is larger at a high Ca/Si ratio (150 Å for Ca/Si = 1.7) than at a low Ca/Si ratio (131 Å for Ca/Si = 1.1). In this respect, the cell size of models used in most relevant published works [19, 20, 42, 56, 70] fail

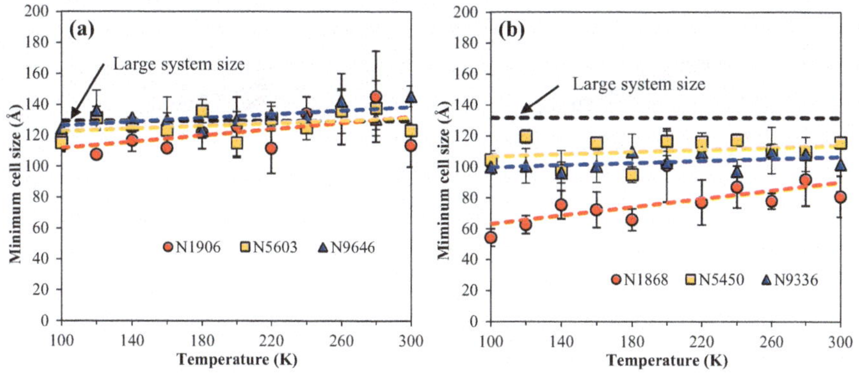

Fig. 8.10 Minimum cell size for fracture failure of C-S-H with Ca/Si ratio of 1.7 (**a**) and 1.1 (**b**). N1906, N5603, and N9646 represent the small, medium, and large systems, with 1906, 5603, and 9646 atoms, respectively

to meet these criteria, therefore leading to underestimations of strength and fracture energy.

8.3.3.3 Scaling Law for Nanoscale Size Effect

In this section, we show the scaling of the mechanical properties of interest with respect to system size. In contrast to what is done to study the size effect at the structural scale, here we use the number N of atoms in the system to characterize the system size instead of the length of the specimen. The number of atoms can be treated as proportional to the number of transition states in a statistical physics view [36]. According to the theory of thermally activated processes, the failure stress is proportional to the logarithmic of the number of transition states. Thus, we plot the mechanical properties of interest versus $T\ln N/N_{cr}$ with N_{cr} the fitting constant in Fig. 8.11, for the C-S-H system with Ca/Si = 1.7.

One expects a single master curve to characterize the temperature-size equivalence, which, however, is not observed for all mechanical properties in Fig. 8.11. The FPZ length is the only property that exhibits a satisfactory scaling (i.e., a scaling that is independent of the system size) with $T\ln N/N_{cr}$ (see Fig. 8.11d). For the other three properties, the results for the small system do not align with those for the medium and large systems. The same is also observed for C-S-H with Ca/Si = 1.1 (see Fig. 8.12). This is expected since the scaling with the size is no longer monotonous at the nanoscale, as discussed before. In the spirit of the theory of thermally activated process, these mechanical quantities are reformulated in a reduced format (see Figs. 8.13 and 8.14): we normalize each property by its value at 0 K. Indeed, according to the theory, all properties should converge to a value independent of system size in the 0 K limit. It is worth noting that the size-dependency of mechanical properties is obtained even at absolute zero temperature, as shown in

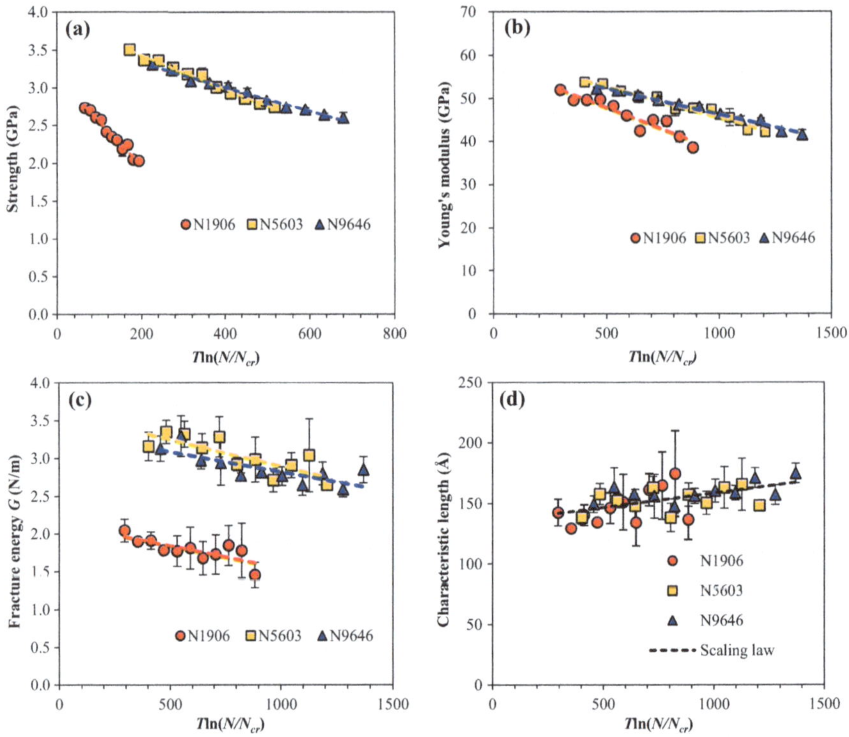

Fig. 8.11 Scaling of mechanical properties of C-S-H (Ca/Si = 1.7) with temperature and size. **a** Strength; **b** Young's modulus; **c** fracture energy; **d** characteristic length. N1906, N5603, and N9646 represent the small, medium, and large systems, with 1906, 5603, and 9646 atoms, respectively

Table 8.2. In particular, small systems tend systematically to underestimate slightly the tensile strength at 0 K. By performing this normalization, the results do indeed align on a single master curve, in particular the relative strength and relative Young's modulus. Therefore, a universal scaling form of mechanical quantities, i.e., tensile strength, Young's modulus, fracture energy, and FPZ length with system size, can be expressed as:

$$\frac{\sigma_{failure}}{\sigma_{failure}(0\text{ K})} \text{ or } \frac{E}{E(0\text{ K})} \text{ or } \frac{G}{G(0\text{ K})} \text{ or } \frac{FPZ}{FPZ(0\text{ K})} \propto T\ln\frac{N}{N_{cr}} \tag{8.10}$$

This unique scaling is valid for the mechanical properties of very small and large systems, even if the size of a system is smaller than the FPZ length. However, it is important to highlight that this scaling holds for the mechanical properties normalized with respect to their value in the 0 K limit.

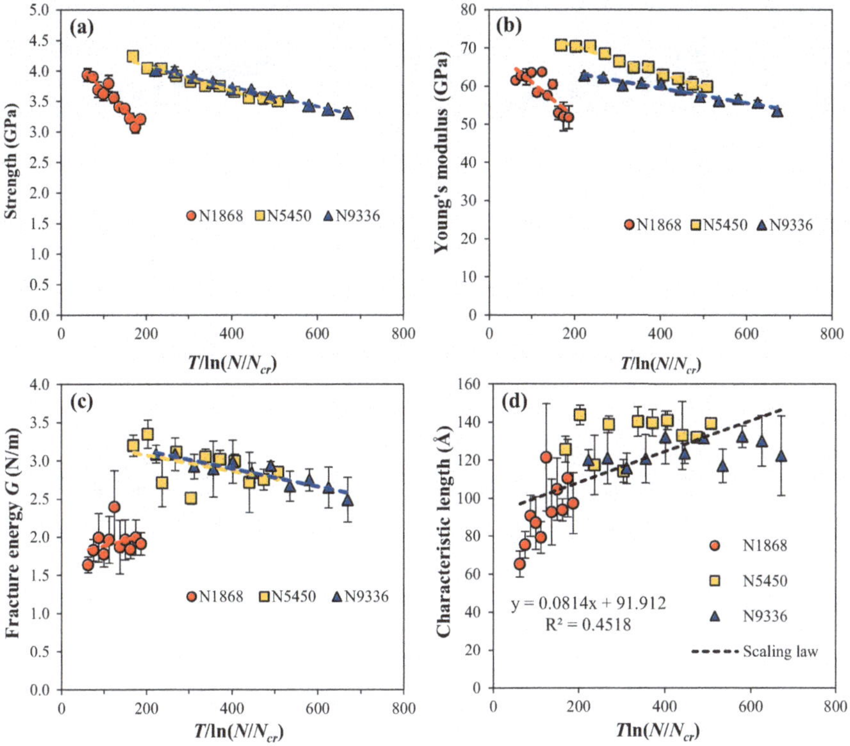

Fig. 8.12 Scaling of mechanical properties of C-S-H (Ca/Si = 1.1) with temperature and size. **a** Strength; **b** Young's modulus; **c** fracture energy; **d** characteristic length

8.3.4 Scaling with Strain Rate

In this section, we focus on the effect of strain rate on the tensile response of C-S-H. Two C-S-H systems with a Ca/Si ratio of 1.7 and 1.1 were stretched at eight different strain rates ranging from 10^{-7} to 10^{-3} ps^{-1}, at three temperatures (i.e., 100 K, 200 K, and 300 K). Following the theory of thermally activated processes, the strength is expected to be linearly proportional to the quantity $\ln(\acute{\varepsilon}_{cr}/\acute{\varepsilon})$, with $\acute{\varepsilon}_{cr}$ a fitting constant with the dimension of a strain rate. Therefore, in Fig. 8.15, we plot the mechanical properties with respect to $\ln(\acute{\varepsilon}_{cr}/\acute{\varepsilon})$ for the C-S-H model with Ca/Si = 1.7. The results for the C-S-H model with Ca/Si = 1.1 are quite similar and are shown in Fig. 8.16.

Both tensile strength and fracture energy increase with the strain rate at all temperatures. As displayed in Fig. 8.15a, all tensile strengths at a specific temperature align on a single master curve, except for those obtained at very high strain rates (> 0.5 × 10^{-3} ps^{-1}). Such a scaling with strain rate and its deviation at large strain rates is also observed for the toy model and graphene of Ref. [36]. The deviation at very high strain rates is attributed to insufficient equilibration of the system, which leads to poor sampling of phase space of atom configurations [36]. Accordingly, those strain

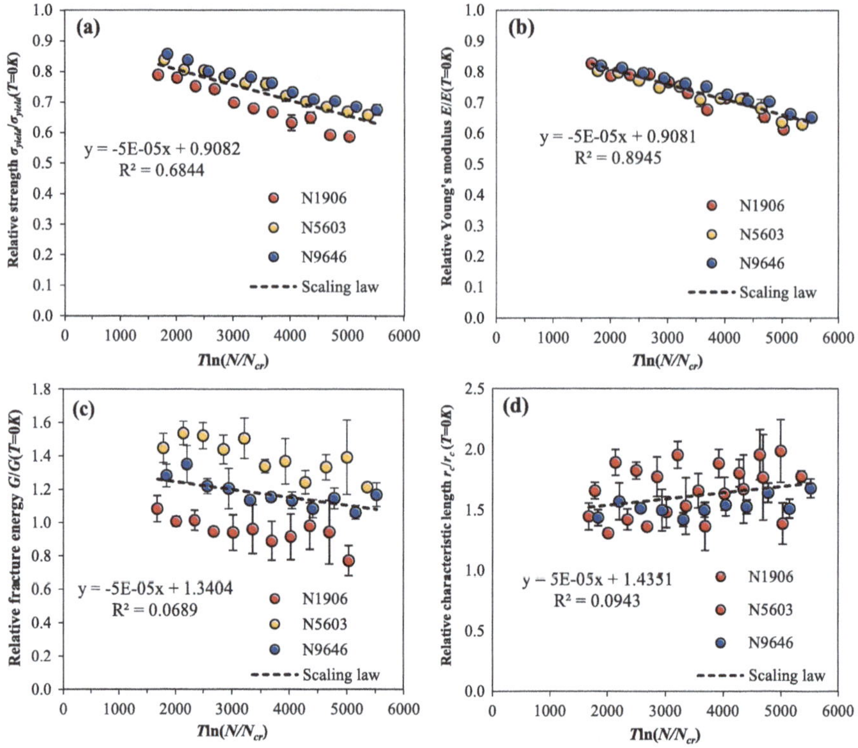

Fig. 8.13 Scaling of reduced mechanical properties of C-S-H (Ca/Si = 1.7) with temperature and size. **a** Strength; **b** Young's modulus; **c** fracture energy; **d** characteristic length. N1906, N5603, and N9646 represent the small, medium, and large systems, with 1906, 5603, and 9646 atoms, respectively

rates are excluded when fitting master curves; the same logic is also adopted when analyzing the scaling of the other properties. The scaling of fracture energy is similar to that of tensile strength (see Fig. 8.15c). The estimated fracture energy aligns well on a master curve at each temperature.

The scaling of Young's modulus with strain rate is slightly different from that of strength and fracture energy (see Fig. 8.15b). The Young's modulus of the systems at each temperature aligns well along a master curve, but the Young's modulus appears almost rate-independent at cryogenic temperatures (100 K, 200 K), whereas it depends significantly on strain rate at ambient temperature (300 K). The increase of Young's modulus with strain rate at ambient temperature is also observed for cement-based materials at the microscale [7] and structural scale [15]. From our results, we highlight that the stiffness of C-S-H is independent of strain rate when the system is at a cryogenic temperature, whereas the sensitivity to strain rate is significant at ambient temperature.

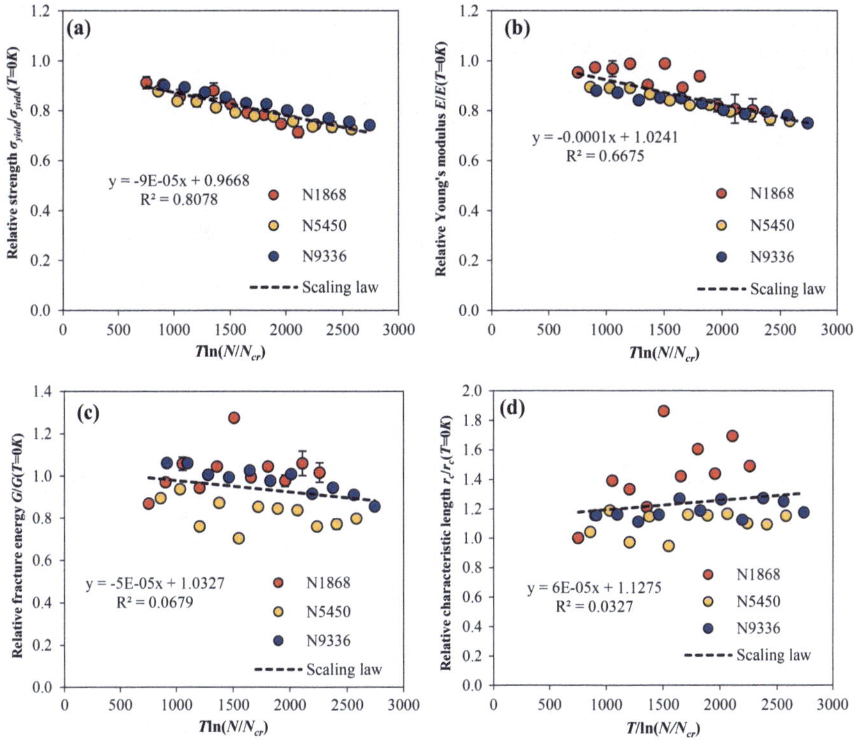

Fig. 8.14 Scaling of reduced mechanical properties of C-S-H (Ca/Si = 1.1) with temperature and size. **a** Strength; **b** Young's modulus; **c** fracture energy; **d** characteristic length

Table 8.2 The mechanical properties of C-S-H at 0 K

Ca/Si ratio	System	σ(0K)	E(0K)	G(0K)	r_c(0K)
1.1	Small	4.31	64.52	1.88	65.34
	Medium	4.83	79.04	3.57	121.04
	Large	4.47	71.41	2.91	104.24
1.7	Small	3.46	62.88	1.88	98.70
	Medium	4.18	66.92	2.18	83.43
	Large	3.86	63.80	2.44	104.16

Regarding the FPZ length, we observe a linear increase with $\ln(\acute{\varepsilon}_{cr}/\acute{\varepsilon})$ (see Fig. 8.15d). To our best knowledge, this scaling is reported for C-S-H for the first time. Among all the mechanical properties, the FPZ length is the only property that decreases with the increasing strain rate, suggesting that C-S-H shows a lower ductility when it is stretched at a higher strain rate. Accordingly, a smaller minimum cell size is required in order to capture fracture failure at a higher strain rate.

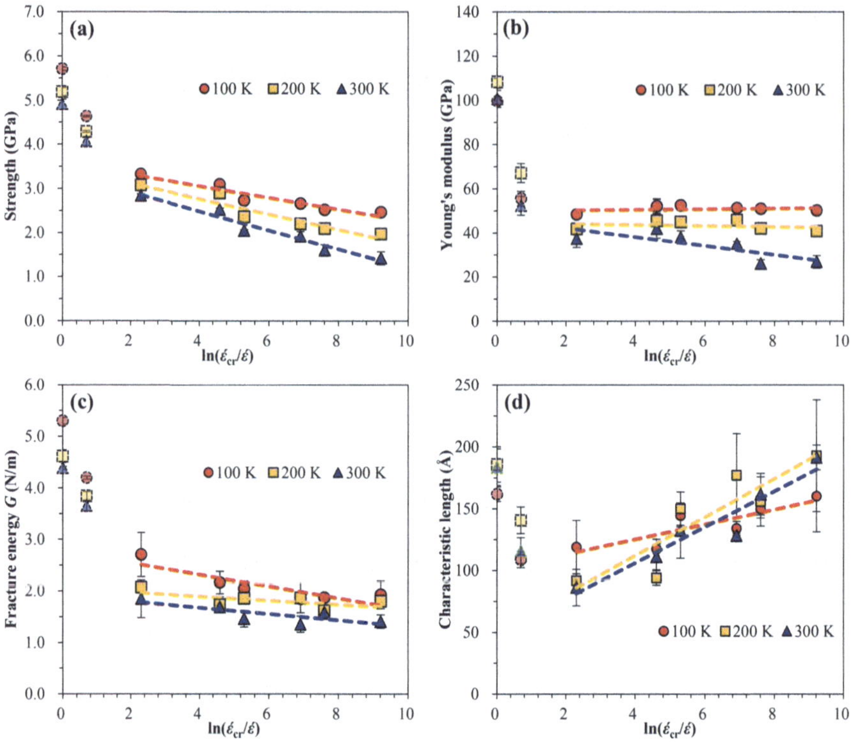

Fig. 8.15 Scaling of mechanical properties of C-S-H (Ca/Si = 1.7) with strain rate at various temperatures. **a** Strength; **b** Young's modulus; **c** fracture energy; **d** characteristic length. The data points filled in light color are excluded from the fitting

Similar scaling behaviors with strain rates are also observed for the C-S-H model with Ca/Si = 1.1 (see Fig. 8.16). All these observations lead to the conclusion that the tensile strength, Young's modulus, fracture energy, and FPZ length of C-S-H do scale linearly with $\ln(\acute{\varepsilon}_{cr}/\acute{\varepsilon})$, irrespective of the composition.

To go further, one can combine the rate effects with the effects of temperature investigated before. Following the theory of thermally activated processes, the general temperature-rate scaling is expected in terms of the quantity $T\ln(\acute{\varepsilon}_{cr}/\acute{\varepsilon})$. Accordingly, we plot in Fig. 8.17 the tensile strength, Young's modulus, fracture energy, and FPZ length in the function of $T\ln(\acute{\varepsilon}_{cr}/\acute{\varepsilon})$ for all the temperatures and strain rates considered. As before, the results obtained at very high strain rates are excluded from the analysis due to the insufficient equilibration of the systems. Doing so, we observe that all the tensile strengths, Young's moduli, and fracture energies are well aligned on their respective master curves. Regarding the FPZ length, the results are rather scattered but with an overall increasing trend (see Fig. 8.17d). Considering the large standard deviations of FPZ length estimates, more data would be needed to confirm the proposed scaling.

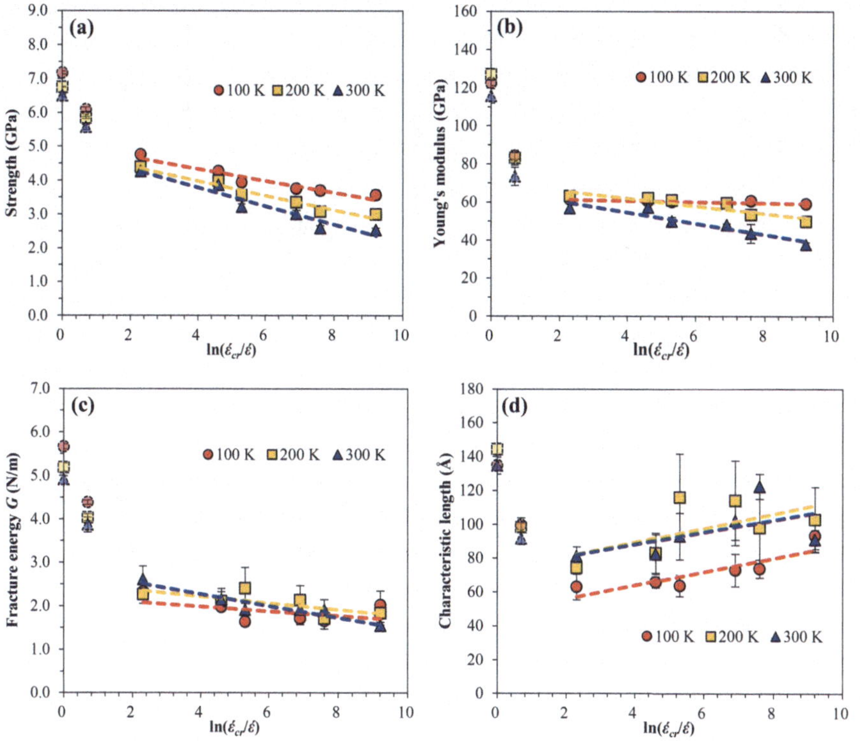

Fig. 8.16 Scaling of mechanical properties of C-S-H (Ca/Si = 1.1) with temperature and strain rate. **a** Strength; **b** Young's modulus; **c** fracture energy; **d** characteristic length

8.3.5 Zhurkov Scaling Law of C-S-H Elastic and Failure Properties

So far, we have investigated the scaling of C-S-H mechanical properties with temperature, system size, and strain rate, respectively. In previous sections, we showed that temperature and rate effects could combine in a single scaling as a function of $T\ln(\acute{\varepsilon}_{cr}/\acute{\varepsilon})$, with $\acute{\varepsilon}_{cr}$ a fitting constant. In Sect. 8.3.3, we also showed that the temperature and size effects could combine in a single scaling as a function of $T\ln(N/N_{cr})$, provided that the mechanical properties are reduced with respect to their 0 K values. A combination of those two laws suggests a general temperature-size-rate scaling in terms of $T\ln\left(\frac{N}{\acute{\varepsilon}}/\left(\frac{N}{\acute{\varepsilon}}\right)_{cr}\right)$. Such expression originates from the statistical physics analysis of failure [36], which leads to a scaling with $\frac{T}{T_{cr}}\ln\left(\frac{N_{TS}}{\acute{\varepsilon}}/\left(\frac{N_{TS}}{\acute{\varepsilon}}\right)_{cr}\right)$ where T_{cr} and $\left(\frac{N_{TS}}{\acute{\varepsilon}}\right)_{cr}$ are two constants with dimensions of temperature and time, and where N_{TS} refers to the number of transition states to failure. Here, one can reasonably consider N_{TS} proportional to the number of atoms in the interlayer because the initial system is intact, and failure can initiate anywhere in the interlayer. And, since the size of the system varies only in the direction of the interlayer, N_{TS} is therefore considered

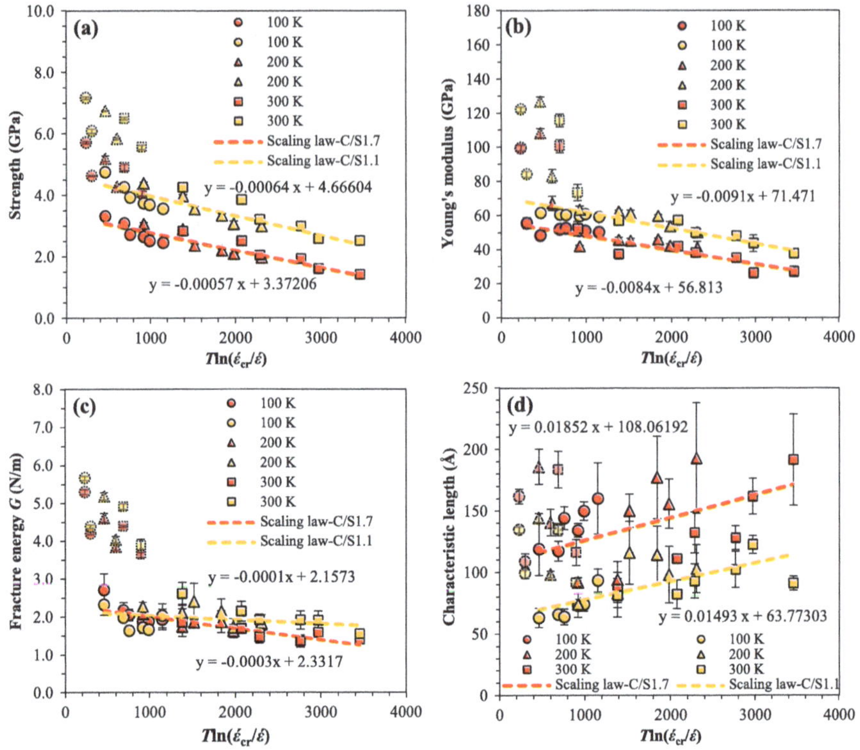

Fig. 8.17 Scaling of mechanical properties of C-S-H with temperature and strain rate. **a** Strength; **b** Young's modulus; **c** fracture energy; **d** characteristic length. The light color-filled data points are excluded from the fitting. C/S1.1 and C/S1.7 stand for Ca/Si = 1.1 and Ca/Si = 1.7, respectively

proportional to the total number of atoms N. Therefore, following the statistical physics analysis, the scaling of the mechanical quantities is expected to follow:

$$\sigma_{failure} \text{ or } E \text{ or } G \text{ or } FPZ \propto \left(\frac{T}{T_{cr}} \ln\left(\frac{N_{TS}}{\dot{\varepsilon}} \Big/ \left(\frac{N_{TS}}{\dot{\varepsilon}}\right)_{cr}\right)\right)_{N_{TS}=constant} \tag{8.11}$$

Note that the law is increasing for the FPZ length scaling whereas it is decreasing for the other properties. Unlike the scalings for the failure properties, which one can relate to transition states to failure, we do not have a clear understanding of the origin of the scaling for Young's modulus. The effects of size and strain rate are actually not very significant for the Young's modulus, and a clear scaling is observed only with temperature, which could be of entropic origin because of the disordered nature of C-S-H and the presence of bound water.

As discussed in Sect. 8.3.3, the nanoscale size effect does not follow Eq. 8.11 (see Figs. 8.11 and 8.12). Instead, the relative mechanical quantities with respect to 0 K scale proportionally with $T\ln(N/N_{cr})$ (see Figs. 8.13 and 8.14). To show the combined temperature-size-strain rate dependency, we, therefore, plot all the relative mechanical quantities as a function of $\frac{T}{T_{cr}}\ln\left(\frac{N_{TS}}{\dot{\varepsilon}}/\left(\frac{N_{TS}}{\dot{\varepsilon}}\right)_{cr}\right)$, as displayed in Fig. 8.18. All the fitting constants and uncertainties are listed in Table 8.3. All the relative tensile strengths of the two C-S-H systems align well along their master curves, with a limit value of 1 at 0 K (see Fig. 8.18a). Accordingly, one can propose the following expression of the temperature-size-strain rate scaling of strength:

$$\frac{\sigma_{failure}}{\sigma_{failure}(0\text{ K})} = 1 - \frac{T}{T_{cr}}\ln\left(\frac{N_{TS}}{\dot{\varepsilon}}/\left(\frac{N_{TS}}{\dot{\varepsilon}}\right)_{cr}\right) \tag{8.12}$$

The same scaling law was found for the failure of graphene by Brochard et al. [36]. In contrast to graphene, however, we highlight here that the failure stress at 0 K is size-dependent when the cell size is close to or smaller than FPZ length. In the case of simple idealistic systems with a crystalline structure and nearest neighbor interactions only, the $k_B T_{cr}$ of strength and toughness can be unequivocally related

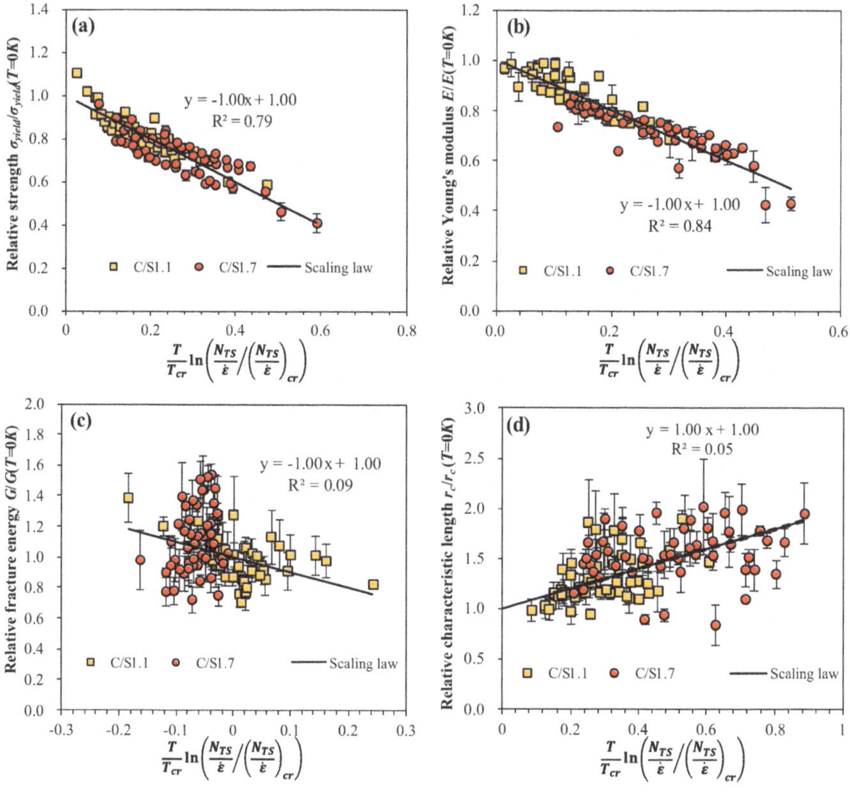

Fig. 8.18 Scaling of mechanical properties of C-S-H with respect to temperature, size, and strain rate. **a** Strength; **b** Young's modulus; **c** fracture energy; **d** characteristic length. C/S1.1 and C/S1.7 stand for Ca/Si = 1.1 and Ca/Si = 1.7, respectively

Table 8.3 Fitting constants of master curves

Property	Ca/Si	T_{cr} (K)	N_{cr}/ε_{cr} (1/ps^{-1})	R^2
σ	1.7	5763.44	219,695	0.725
	1.1	5174.37	5,139,164	0.788
E	1.7	10,625.56	253	0.758
	1.1	7058.73	3,978,321	0.732
G	1.7	15,194.88	72,758,334,914	0.058
	1.1	4845.45	366,862,945	0.217
r_c	1.7	6071.01	88,554	0.390
	1.1	7991.32	1.03	0.226

to the energy to break a bond [36]. But, for C-S-H systems with much more complex structures and interactions, such simple interpretation is no more possible. Although the failure occurs in the interlayer, the interactions within this interlayer are still extremely complex compared to idealistic systems.

Similarly, accurate master curves are also obtained for the relative Young's moduli of the two systems (see Fig. 8.18b). The results of relative fracture energy and relative characteristic length exhibit much more uncertainties and are not very well aligned on their respective master curves, as shown in Fig. 8.18c, d. We can notice that this noise is primarily caused by the small systems that have a length scale approaching or lower than the FPZ length. We propose nevertheless a tentative linear scaling law that will require confirmation through further simulations with larger systems. According to the uncertainties given in Table 8.3, we highlight that the tensile strength, Young's modulus, and fracture energy have the same scaling law in a combination of temperature, system size, and strain rate:

$$\frac{\sigma_{failure}}{\sigma_{failure}(0\ \mathrm{K})} \text{ or } \frac{E}{E(0\ \mathrm{K})} \text{ or } \frac{G}{G(0\ \mathrm{K})} = 1 - \frac{T}{T_{cr}} \ln\left(\frac{N_{TS}}{\dot{\varepsilon}} / \left(\frac{N_{TS}}{\dot{\varepsilon}}\right)_{cr}\right) \tag{8.13}$$

The scaling law of FPZ length is expressed as:

$$\frac{FPZ}{FPZ(0\mathrm{K})} = 1 + \frac{T}{T_{cr}} \ln\left(\frac{N_{TS}}{\dot{\varepsilon}} / \left(\frac{N_{TS}}{\dot{\varepsilon}}\right)_{cr}\right) \tag{8.14}$$

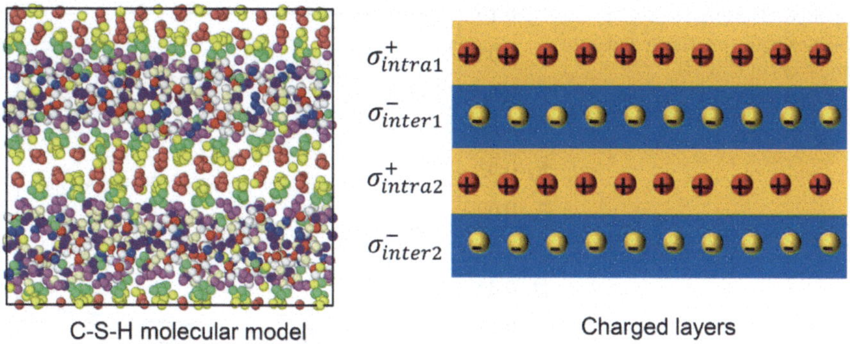

Fig. 8.19 Simplified layered model illustrating the distribution of charges within and between C-S-H layers

8.4 C-S-H Tensile Failure Mechanism

8.4.1 Electrostatic Interactions

According to the analysis results above, it is evident that there is a significant difference in the tensile fracture properties among C-S-H samples with different Ca/Si ratios. The higher the Ca/Si ratio, the lower the tensile strength of C-S-H. The fracture in the z-direction of C-S-H consistently occurs in the interlayer, and its strength is closely related to the Coulomb interaction force between charges. The simplified model of charge distribution in the interlayer and intralayer of C-S-H is depicted in Fig. 8.19. Therefore, we have calculated the charge distribution density in the interlayer and intralayer of C-S-H models with different Ca/Si ratios, and the results are presented in Fig. 8.20.

From Fig. 8.20, it is evident that the single-layer charge density of C-S-H decreases as the Ca/Si ratio increases. Clearly, during the tensile fracture process, the Coulombic forces between C-S-H layers increase with the increase in charge density. Therefore, the Coulombic forces between the layers of low Ca/Si ratio C-S-H are greater than those of high Ca/Si ratio C-S-H, indicating a stronger binding force between the layers of C-S-H with lower Ca/Si ratios. This partly explains the observed phenomenon in the simulation where the tensile strength of C-S-H decreases with an increase in the Ca/Si ratio.

8.4.2 Atom Distribution

The differences in Coulomb forces within C-S-H clearly cannot entirely explain the effect of Ca/Si ratio on the tensile strength of C-S-H. As shown in Fig. 6.5, the trend of decreasing tensile strength of C-S-H with Ca/Si ratio greater than 1.3

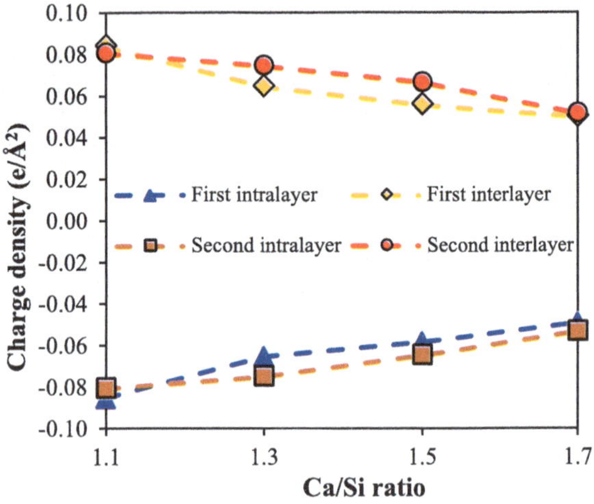

Fig. 8.20 The charge densities in the interlayer and intralayers of C-S-H

is weakened with decreasing Ca/Si. Besides Coulomb forces, the structural status of C S-H, particularly the interlayer atomic distribution, may also be related to its tensile fracture performance. To investigate this, we analyzed the ion distribution in different Ca/Si ratio C-S-H models, and the results are shown in Fig. 8.21.

The distribution of interlayer Ow atoms is shown in Fig. 8.21, where it can be observed that water molecules are mainly concentrated in the middle of the C-S-H interlayer. Additionally, there is a certain amount of water molecules adsorbed near the silicon-oxygen chains on both sides of the interlayer, mainly at defect sites. As the Ca/Si ratio increases, the number of defect sites in the silicon-oxygen chains of C-S-H also increases. Therefore, it is observable that the density of Ow atoms near the silicon-oxygen chains increases with the increase in the Ca/Si ratio. Similarly, the distribution of interlayer Ca^{2+} ions is observed to be adsorbed towards the silicon-oxygen chains, but in a more concentrated manner than water molecules. This is

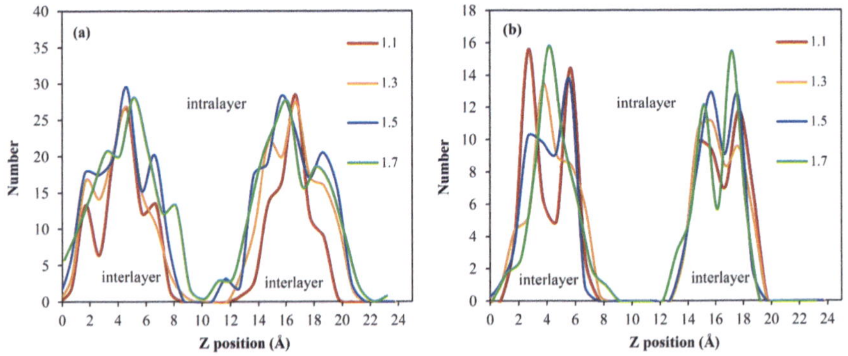

Fig. 8.21 Interlayer atomic distribution of C-S-H with different Ca/Si ratios: **a** Oxygen atoms Ow in water molecules; **b** Interlayer Ca^{2+} ions Cw

primarily due to charge compensation between Ca^{2+} ions and the interlayer hydroxyl groups, oxygen atoms in the bridging silicate tetrahedra of the silicon-oxygen chains, and the counterpart silicon-oxygen tetrahedra at the defect sites of the silicon-oxygen chains.

More importantly, it should be noted that the content of interlayer water molecules in C-S-H tends to increase with an increase in the Ca/Si ratio. The Ow content in low-Ca C-S-H with a Ca/Si ratio of 1.1 is significantly smaller than that in higher Ca/Si ratio systems. The increase in water content simultaneously increases the expansion pressure between the layers of C-S-H, which is also a contributing factor to both the increase and decrease in the tensile strength of C-S-H with the change in the Ca/Si ratio.

8.4.3 The Reduction of Activation Energy Barrier

From the kinetic concept of strength [24], a given solid has to experience three main processes under tension fracture failure. Firstly, the tension stresses applied to the solid excite their bonds; then, the excited bonds are broken due to thermal fluctuations; finally, the large content of broken bonds initiates local instability, leading to the failure of the body. Therefore, the energy barrier is highly correlated to applied stress. According to Brochard et al. [36], the reduction of the activation energy barrier to reaching the transition state of failure can be given as:

$$\Delta E_{TS} = k_B T_{cr}\left(1 - \frac{\sigma_{failure}}{\sigma_{failure}(0\mathrm{K})}\right) \tag{8.15}$$

Combining Eqs. 8.13 and 8.15, we have:

$$\Delta E_{TS} = k_B T_{cr} X \tag{8.16}$$

where $X = \frac{T}{T_{cr}}\ln\left(\frac{N_{TS}}{\dot{\varepsilon}}/\left(\frac{N_{TS}}{\dot{\varepsilon}}\right)_{cr}\right)$ is in the range of (0,1). Since the characteristic temperature T_{cr} of C-S-H has been determined during the fitting of the master curves (see Table 8.3), we can estimate the reduction of activation energy barrier for the tensile failure of C-S-H (see Figs. 8.22 and 8.23).

From our estimation, the maximum reduction of the activation energy barrier for the failure of C-S-H with Ca/Si ratios of 1.7 and 1.1 is 0.497 eV and 0.446 eV, respectively. The reduction of activation energy barrier and failure stress are related by $\Delta E_{TS} = E_0 - \gamma\sigma_{failure}$ with E_0 the interatomic binding energy and γ a constant of the solid. For a given system size with a given strain rate, increasing the temperature increases the reduction of the activation energy barrier, which leads to the damage of the solid before being broken compared with that at the athermal limit. Therefore, C-S-H at cryogenic temperatures has a smaller reduction active energy barrier than at ambient temperature, therefore has a less damaged structure

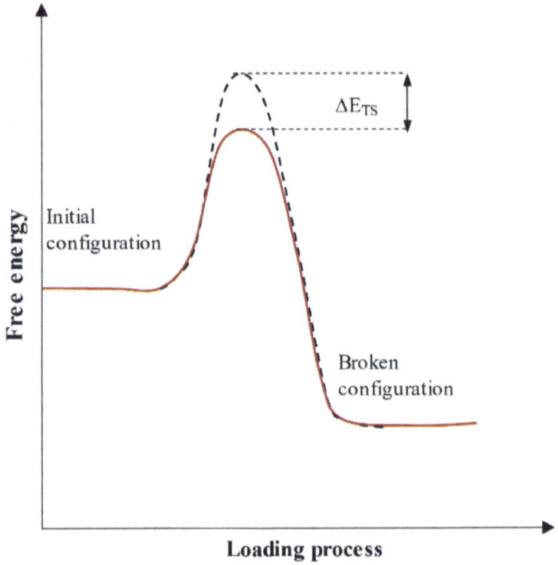

Fig. 8.22 C-S-H tensile failure energy path

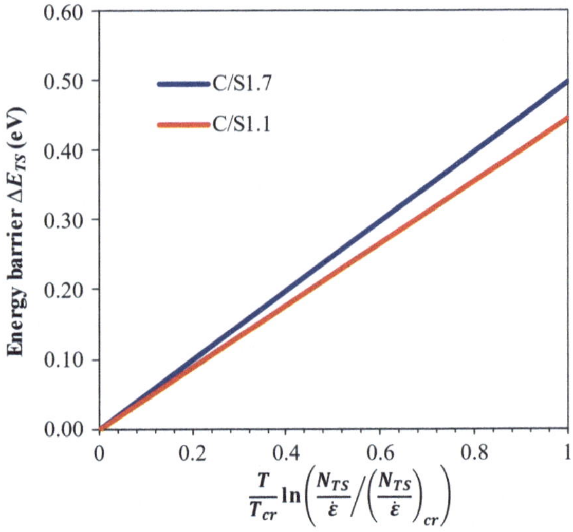

Fig. 8.23 The reduction of activation energy barrier of C-S-H to reach the transition state for tensile failure. C/S1.1 and C/S1.7 stand for Ca/Si = 1.1 and Ca/Si = 1.7, respectively

before the tensile failure, consequently yielding a higher failure stress, which provides an elementary physical explanation for the increase in the strength of oven-dried concrete at cryogenic temperatures observed experimentally [1]. Nevertheless, we note that this is probably not the only reason, given that concrete is heterogeneous and is composed of many components other than C-S-H, e.g., ettringite, portlandite, and unreacted clinker.

8.5 Concluding Remarks

In this chapter, we have shown the nanoscopic tensile elastic and failure properties of C-S-H scale with temperature, system size, and strain rate. Based on molecular simulation results, we demonstrated that Young's modulus, tensile strength, fracture energy, and fracture process zone length of nanoscale C-S-H all follow a Zhurkov-like scaling law. The main conclusions are as follows:

(1) The relative mechanical properties with respect to the 0 K values all scale linearly with $T\ln(N/\dot{\varepsilon})$ (extended Zhurkov law) thus providing a general equivalence between temperature (T), size (N), and loading rate ($\dot{\varepsilon}$). The tensile strength, Young's modulus, and fracture energy of C-S-H all increase with $\ln(N/\dot{\varepsilon})$, while the FPZ length decreases. The constants involved in the scaling law equations have been determined. Following these scaling laws, one can easily predict the nanoscopic mechanical quantities for various temperatures, system sizes, and strain rates.

(2) The decrease of tensile strength, Young's modulus, and fracture energy with temperature suggests a more compliant and less resistant C-S-H as temperature increases. The increase of FPZ length with temperature indicates an increase of C-S-H ductility with temperature. Moreover, since varying temperature in molecular simulation involves a negligible difference in computational cost, in comparison with varying system size or strain rate, the temperature-size-rate equivalence makes it possible to extrapolate properties of large systems at low strain rates by simply considering temperature evolutions of small systems without much computational cost.

(3) The size effect at the nanoscale level is not monotonous when expressed versus $T\ln(N/N_{cr})$. For system sizes smaller than the FPZ length, the tensile strength, and fracture energy of the C-S-H system are underestimated. One can correct this underestimation by considering relative mechanical properties with respect to 0 K, when addressing the combined scaling with temperature and/or strain rate. The minimum system size to fully capture the failure processes of C-S-H depends on the composition. The minimum size is ~ 131 Å for a low Ca/Si ratio (1.1), and ~ 150 Å for a high Ca/Si ratio (1.7).

(4) The tensile strength, Young's modulus, and fracture energy increase with an increasing strain rate, while the FPZ length decreases with the strain rate. However, we note that Young's modulus sensitivity to strain rates almost vanishes at cryogenic temperatures.

(5) The scaling found in this study follows the expression of Zhurkov's law, the origin of which is attributed to the overcome by thermal agitation of an elementary energy barrier to reach a transition path to failure. We give estimates of the reduction of energy barrier and its evolution with temperature, size, and strain rate. Thus, this study provides theoretical ground to explain some experimental observations, such as the increase of strength at cryogenic temperatures. The maximum reduction of activation energy barrier for the failure of C-S-H with a Ca/Si ratio of 1.7 and 1.1 are 0.497 eV and 0.446 eV, respectively.

(6) The influence of C-S-H's chemical composition on its tensile strength is mainly achieved through its internal charge distribution density and atomic distribution status. It is observed that the tensile strength decreases with the increase in the Ca/Si ratio. This is primarily due to two factors: 1) the decrease in the charge distribution density in C-S-H layers with increasing Ca/Si, resulting in a decrease in Coulomb forces; 2) the increase in interlayer water molecule density with the increase in Ca/Si ratio, thereby increasing the interlayer splitting pressure of C-S-H.

References

1. Z. Jiang, B. He, X. Zhu, Q. Ren, Y. Zhang, State-of-the-art review on properties evolution and deterioration mechanism of concrete at cryogenic temperature. Constr. Build. Mater. **257**, 119456 (2020)
2. S. Wu, X. Chen, J. Zhou, Influence of strain rate and water content on mechanical behavior of dam concrete. Constr. Build. Mater. **36**, 448–457 (2012)
3. H. Zhang, B. Šavija, Y. Xu, E. Schlangen, Size effect on splitting strength of hardened cement paste: Experimental and numerical study. Cem. Concr. Compos. **94**, 264–276 (2018)
4. R. Shahrin, C.P. Bobko, Micropillar compression investigation of size effect on microscale strength and failure mechanism of Calcium-Silicate-Hydrates (C-S-H) in cement paste, Cem. Concr. Res. 125 (2019)
5. Z. Wu, J. Zhang, H. Yu, Q. Fang, H. Ma, Specimen size effect on the splitting-tensile behavior of coral aggregate concrete: A 3D mesoscopic study. Eng. Fail. Anal. **127**, 105395 (2021)
6. V. Bindiganavile, N. Banthia, Size effects and the dynamic response of plain concrete. J. Mater. Civ. Eng. **18**(4), 485–491 (2006)
7. Y. Gan, C.R. Rodriguez, E. Schlangen, K. van Breugel, B. Šavija, Assessing strain rate sensitivity of cement paste at the micro-scale through micro-cantilever testing. Cem. Concr. Compos. **121**, 104084 (2021)
8. C. Zweben, B.W. Rosen, A statistical theory of material strength with application to composite materials. J. Mech. Phys. Solids **18**(3), 189–206 (1970)
9. P. Bažant Zdeněk, Size effect in blunt fracture: Concrete, rock, metal, J. Eng. Mech. 110(4), 518–535 (1984)
10. A. Carpinteri, B. Chiaia, G. Ferro, Size effects on nominal tensile strength of concrete structures: multifractality of material ligaments and dimensional transition from order to disorder. Mater. Struct. **28**(6), 311–317 (1995)
11. A. Carpinteri, Scaling laws and renormalization groups for strength and toughness of disordered materials. Int. J. Solids Struct. **31**(3), 291–302 (1994)
12. A. Carpinteri, G. Ferrara, L. Imperato, Scaling laws for strength and toughness of disordered materials: A unified theory based on fractal geometry. Eng. Fract. Mech. **48**(5), 673–689 (1994)
13. A. Carpinteri, P. Cornetti, S. Puzzi, A stereological analysis of aggregate grading and size effect on concrete tensile strength. Int. J. Fract. **128**(1), 233–242 (2004)
14. S. Jin, X. Zhao, Y. Tai, J. Zhou, Multiscale model for the scale effect of tensile strength of hardened cement paste based on pore size distribution. Constr. Build. Mater. **325**, 126775 (2022)
15. X. Gao, L. Zhou, X. Ren, J. Li, Rate effect on the stress–strain behavior of concrete under uniaxial tensile stress. Struct. Concr. **22**(S1), E815–E830 (2021)
16. Q. Fu, Z. Zhang, X. Zhao, M. Hong, B. Guo, Q. Yuan, D. Niu, Water saturation effect on the dynamic mechanical behaviour and scaling law effect on the dynamic strength of coral aggregate concrete. Cem. Concr. Compos. **120**, 104034 (2021)

17. I. Fischer, B. Pichler, E. Lach, C. Terner, E. Barraud, F. Britz, Compressive strength of cement paste as a function of loading rate: Experiments and engineering mechanics analysis. Cem. Concr. Res. **58**, 186–200 (2014)
18. S.M. Liang, Y. Wei, X. Gao, Strain-rate sensitivity of cement paste by microindentation continuous stiffness measurement: Implication to isotache approach for creep modeling. Cem. Concr. Res. **100**, 84–95 (2017)
19. H. Wu, J. Pan, J. Wang, Molecular dynamics simulation study on dynamic mechanical properties of C-S-H with diverse Ca/Si ratios. Mater. Today Commun. **31**, 103755 (2022)
20. D. Hou, Y. Zhu, Y. Lu, Z. Li, Mechanical properties of calcium silicate hydrate (C–S–H) at nano-scale: A molecular dynamics study. Mater. Chem. Phys. **146**(3), 503–511 (2014)
21. L.J. Malvar, C.A.J.M. Ross, Review of strain rate effects for concrete in tension. ACI Mater. J. **95**, 735–739 (1998)
22. X. Zhou, Y. Xie, G. Long, X. Zeng, J. Li, L. Yao, W. Jiang, Z. Pan, DEM analysis of the effect of interface transition zone on dynamic splitting tensile behavior of high-strength concrete based on multi-phase model. Cem. Concr. Res. **149**, 106577 (2021)
23. V.R. Regel, A.I. Slutsker, É.E. Tomashevskiĭ, Reviews of topical problems: The kinetic nature of the strength of solids. Sov. Phys. Usp. **15**, 45–65 (1972)
24. S.N. Zhurkov, Kinetic concept of the strength of solids. Int. J. Fract. **26**(4), 295–307 (1984)
25. S.Y.N. Chan, G.-F. Peng, J.K.W. Chan, Comparison between high strength concrete and normal strength concrete subjected to high temperature. Mater. Struct. **29**(10), 616–619 (1996)
26. Y. Zhang, Q. Chen, J.W. Ju, M. Bauchy, Effects of high temperature on the mechanical behavior of calcium silicate hydrate under uniaxial tension and compression. Int. J. Damage Mech **30**(7), 987–1011 (2021)
27. Y. Zhang, Q. Zhou, J.W. Ju, M. Bauchy, New insights into the mechanism governing the elasticity of calcium silicate hydrate gels exposed to high temperature: A molecular dynamics study. Cem. Concr. Res. **141**, 106333 (2021)
28. X. Zhu, B. He, M. Xie, Y. Zhang, Z. Jiang, A comprehensive nitrogen adsorption measurement on the pore structure of calcium-silicate-hydrate subjected to cryogenic attack. Measurement **184**, 109941 (2021)
29. X. Zhu, C. Qian, B. He, Q. Chen, Z. Jiang, Experimental study on the stability of C-S-H nanostructures with varying bulk CaO/SiO_2 ratios under cryogenic attack. Cem. Concr. Res. **135**, 106114 (2020)
30. E.T. Rodriguez, K. Garbev, D. Merz, L. Black, I.G. Richardson, Thermal stability of C-S-H phases and applicability of Richardson and Groves' and Richardson C-(A)-S-H(I)models to synthetic C-S-H. Cem. Concr. Res. **93**, 45–56 (2017)
31. X. Zhu, Q. Ren, B. He, H. Zhang, Z. Jiang, The nanomechanical nature of cementitious calcium-(alumino)silicate-hydrate under ultra-low temperature attack: An intrinsic transition and the role of aluminum for bottom-up enhancement. Cem. Concr. Compos. **134**, 104811 (2022)
32. B. He, H. Zhang, X. Zhu, Q. Zheng, O. Onuaguluchi, N. Banthia, Z. Jiang, Thermal-dependent brittleness effect of ultra-high performance concrete exposed to cryogenic flexural loads by acoustic emission evaluation. Cem. Concr. Compos. **139**, 105056 (2023)
33. J.K. Zhou, Y.Z. Liang, Reactive molecular dynamics simulation on the structure characteristics and tensile properties of calcium silicate hydrate at various temperatures and strain rates, Mol. Simul. (2020)
34. Y.L. Yaphary, D. Lau, F. Sanchez, C.S. Poon, Effects of sodium/calcium cation exchange on the mechanical properties of calcium silicate hydrate (C-S-H), Constr. Build. Mater. 243 (2020)
35. X.X. Guo, H. Xin, J. Li, Z.H. Wang, Z.Q. Li, Molecular dynamics study on perfect and defective graphene/calcium-silicate-hydrate composites under tensile loading. Mol. Simul. **45**(18), 1481–1487 (2019)
36. L. Brochard, S. Souguir, K. Sab, Scaling of brittle failure: strength versus toughness. Int. J. Fract. **210**(1), 153–166 (2018)
37. R.J.-M. Pellenq., A. Kushim., R. Shahsavari., K.J.V. Vliet., M.J. Buehler., S. Yip., F.-J. Ulm., A realistic molecular model of cement hydrates, Proc. Natl. Acad. Sci. U.S.A. 106(38), 16102–16107 (2009)

38. S.A. Hamid, The crystal structure of the 11 Ä natural tobermorite $Ca_{2.25}[Si_3O_{7.5}(OH)_{1.5}]\cdot 1H_2O$, Z. Kristallogr. Cryst. Mater. 154, 189–198 (1981)
39. Y. Zhou, Y. Chen, M. Jin, J. Liu, C. Miao, Insights at the neutron irradiation-induced structural homogenization effect of calcium silicate hydrates and degradation mechanism of mechanical properties: a molecular dynamics study, J. Sustain. Cem.-Based Mater. 1–13 (2022)
40. Y. Zhou, D.S. Hou, H. Manzano, C.A. Orozco, G.Q. Geng, P.J.M. Monteiro, J.P. Liu, Interfacial connection mechanisms in calcium-silicate-hydrates/polymer nanocomposites: A molecular dynamics study. ACS Appl. Mater. Interfaces **9**(46), 41014–41025 (2017)
41. D.S. Hou, Z.J. Li, T.J. Zhao, P. Zhang, Water transport in the nano-pore of the calcium silicate phase: reactivity, structure and dynamics. Phys. Chem. Chem. Phys. **17**(2), 1411–1423 (2015)
42. M.J. Abdolhosseini Qomi, K.J. Krakowiak, M. Bauchy, K.L. Stewart, R. Shahsavari, D. Jagannathan, D.B. Brommer, A. Baronnet, M.J. Buehler, S. Yip, F.J. Ulm, K.J. Van Vliet, R.J. Pellenq, Combinatorial molecular optimization of cement hydrates, Nat. Commun. 5, 4960 (2014)
43. A. Morshedifard, S. Masoumi, M.J. Abdolhosseini Qomi, Nanoscale origins of creep in calcium silicate hydrates, Nat. Commun. 9(1), 1785 (2018)
44. R.J.M. Pellenq, A. Kushima, R. Shahsavari, K.J. Van Vliet, M.J. Buehler, S. Yip, F.J. Ulm, A realistic molecular model of cement hydrates. Proc. Natl. Acad. Sci. U.S.A. **106**(38), 16102–16107 (2009)
45. E. L'Hopital, B. Lothenbach, D.A. Kulik, K. Scrivener, Influence of calcium to silica ratio on aluminium uptake in calcium silicate hydrate. Cem. Concr. Res. **85**, 111–121 (2016)
46. L. Liu, C. Sun, G.Q. Geng, P. Feng, J.Q. Li, R. Dahn, Influence of decalcification on structural and mechanical properties of synthetic calcium silicate hydrate (C-S-H). Cem. Concr. Res. **123**, 105793 (2019)
47. T.F. Sevelsted, J. Skibsted, Carbonation of C-S-H and C–A–S–H samples studied by ^{13}C, ^{27}Al and ^{29}Si MAS NMR spectroscopy. Cem. Concr. Res. **71**, 56–65 (2015)
48. A. Kunhi Mohamed, P. Moutzouri, P. Berruyer, B.J. Walder, J. Siramanont, M. Harris, M. Negroni, S.C. Galmarini, S.C. Parker, K.L. Scrivener, L. Emsley, P. Bowen, The atomic-level structure of cementitious calcium aluminate silicate hydrate, J. Am. Chem. Soc. 142(25), 11060–11071 (2020)
49. R.J. Myers, E. L'Hôpital, J.L. Provis, B. Lothenbach, Effect of temperature and aluminium on calcium (alumino)silicate hydrate chemistry under equilibrium conditions. Cem. Concr. Res. **68**, 83–93 (2015)
50. I. Klur., B. Pollet., J. Virlet., A. Nonat., C-S-H structure evolution with calcium content by multinuclear NMR, Nuclear Magnetic Resonance Spectroscopy of Cement-Based Materials, (Springer, Verlag Berlin Heidelberg, 1998), pp. 119–141
51. H. Noma., Y. Adachil., M. Takushi., Y. Hideo., Y. Tadashi., N. Yoshihisa., ^{29}Si MAS NMR spectroscopy of poorly-crystalline calcium silicate hydrates (C-S-H), Nuclear Magnetic Resonance Spectroscopy of Cement-Based Materials, (Springer, Berlin, Heidelberg, 1998), pp. 159–168
52. W.A. Hunnicutt, *Characterization of calcium-silicate-hydrate and calcium-alumino-silicate-hydrate* (University of Illinois at Urbana-Champaign, Urbana, Illinois, Civil and Environmental Engineering, 2013)
53. R. Shahsavari, R.J. Pellenq, F.J. Ulm, Empirical force fields for complex hydrated calcio-silicate layered materials. Phys. Chem. Chem. Phys. **13**(3), 1002–1011 (2011)
54. R.K. Mishra, A.K. Mohamed, D. Geissbühler, H. Manzano, T. Jamil, R. Shahsavari, A.G. Kalinichev, S. Galmarini, L. Tao, H. Heinz, R. Pellenq, A.C.T. van Duin, S.C. Parker, R.J. Flatt, P. Bowen, Cemff: A force field database for cementitious materials including validations, applications and opportunities. Cem. Concr. Res. **102**, 68–89 (2017)
55. A.P. Thompson, H.M. Aktulga, R. Berger, D.S. Bolintineanu, W.M. Brown, P.S. Crozier, P.J. in't Veld, A. Kohlmeyer, S.G. Moore, T.D. Nguyen, R. Shan, M.J. Stevens, J. Tranchida, C. Trott, S.J. Plimpton, LAMMPS–a flexible simulation tool for particle-based materials modeling at the atomic, meso, and continuum scales, Comput. Phys. Commun. 271, 108171 (2022)
56. D. Hou, H. Ma, Y. Zhu, Z. Li, Calcium silicate hydrate from dry to saturated state: Structure, dynamics and mechanical properties. Acta Mater. **67**, 81–94 (2014)

57. M. Bauchy, H. Laubie, M.J. Abdolhosseini Qomi, C.G. Hoover, F.J. Ulm, R.J.M. Pellenq, Fracture toughness of calcium–silicate–hydrate from molecular dynamics simulations, J. Non-Cryst. Solids 419, 58–64 (2015)
58. S. Masoumi, S. Zare, H. Valipour, M.J. Abdolhosseini Qomi, Effective interactions between calcium-silicate-hydrate nanolayers, J. Phys. Chem. C 123(8), 4755–4766 (2019)
59. K. Ha, H. Baek, K. Park, Convergence of fracture process zone size in cohesive zone modeling. Appl. Math. Modell. **39**(19), 5828–5836 (2015)
60. M. Bauchy, B. Wang, M. Wang, Y. Yu, M.J. Abdolhosseini Qomi, M.M. Smedskjaer, C. Bichara, F.-J. Ulm, R. Pellenq, Fracture toughness anomalies: Viewpoint of topological constraint theory, Acta Mater. 121, 234–239 (2016)
61. L. Brochard, G. Hantal, H. Laubie, F.-J. Ulm, R.J.M. Pellenq, Capturing material toughness by molecular simulation: Accounting for large yielding effects and limits. Int. J. Fract. **194**(2), 149–167 (2015)
62. M. Bauchy, M.Y. Wang, Y.T. Yu, B. Wang, N.M.A. Krishnan, E. Masoero, F.J. Ulm, R. Pellenq, Topological control on the structural relaxation of atomic networks under stress, Phys. Rev. Lett. 119(3) (2017)
63. C.Y. Hui, J. A, S.J. Bennison, J.D. Londono, Crack blunting and the strength of soft elastic solids, Proc. R. Soc. A: Math. Phys. Eng. Sci. 459(2034), 1489–1516 (2003)
64. G. Irwin, Plastic zone near a crack and fracture toughness, Proceedings of the seventh Sagamore Ordnance materials conference, Syracuse University, (New York, 1960), pp. 63–78
65. A. Hillerborg, M. Modéer, P.E. Petersson, Analysis of crack formation and crack growth in concrete by means of fracture mechanics and finite elements. Cem. Concr. Res. **6**(6), 773–781 (1976)
66. G.I. Barenblatt, The formation of equilibrium cracks during brittle fracture. General ideas and hypotheses. Axially-symmetric cracks, J. Appl. Math. Mech. 23(3), 622–636 (1959)
67. D.S. Dugdale, Yielding of steel sheets containing slits. J. Mech. Phys. Solids **8**(2), 100–104 (1960)
68. Z. Bazant, J. Planas, *Fracture and size effect in concrete and other quasibrittle materials* (CRC Press, Boca Raton, 1998)
69. L. Brochard, I.G. Tejada, K. Sab, From yield to fracture, failure initiation captured by molecular simulation. J. Mech. Phys. Solids **95**, 632–646 (2016)
70. D. Hou, J. Zhang, Z. Li, Y. Zhu, Uniaxial tension study of calcium silicate hydrate (C–S–H): structure, dynamics and mechanical properties. Mater. Struct. **48**(11), 3811–3824 (2014)

Chapter 9
Ice Crystallization and Liquid Transport

9.1 Introduction

In this chapter, we will focus on the freezing liquid in cement pores. Frost damage is a significant concern regarding the long-term durability of porous solids, such as concrete, clay, and rock, particularly in cold regions. This issue is particularly pronounced in partially submerged concrete subjected to cyclic freezing–thawing conditions, resulting in extensive deterioration of concrete structures [1, 2]. Frost-induced damage to concrete is commonly classified as either internal frost attack or surface scaling attack [3]. Internal cracks can rapidly develop as a result of the freezing of liquid (i.e., crystallization of ice) in the pores of the concrete. Surface scaling, on the other hand, is primarily associated with moisture exchange at the concrete surface exposed to saline solutions. Both forms of frost damage are influenced by the phase transition of water from liquid to solid state. The phase transition behavior of liquids within confined pores entails complex physical and mechanical processes. Consequently, frost damage of concrete has attracted significant attention in the past few decades.

Pioneering work dedicated to understanding frost damage was carried out by Powers [4, 5], who developed the hydraulic pressure theory to explain the volume expansion and the internal cracking of concrete during freezing. In this theory, the difference in density between liquid water and ice induces a hydraulic pressure imposed directly on the cement matrix. When the hydraulic pressure exceeds a threshold related to the tensile strength of the cement matrix, internal cracks occur. Although the hydraulic pressure theory accounts well for the volume expansion of concrete during the ice formation, it is inadequate to explain the volume expansion of concrete observed even in the absence of any hydraulic pressure. Indeed, one observes that even though benzene contracts when freezing, a concrete specimen saturated with benzene still swells when the benzene it contains freezes [6]. Hereto, Scherer et al., [7, 8] proposed that the internal frost damage was primarily attributed to the overpressure of the ice crystal with respect to the surrounding liquid still present

Z. Jiang and X. Zhu, *Cryogenic Stability of Cement-Based Materials at Micro/Nano Scale*, https://doi.org/10.1007/978-981-96-1836-1_9

in the unfrozen pores. By taking the thermodynamic stability criteria of a triple-phase condition into account, Setzer [9, 10] derived a micro-ice-lens model that captures the frost contraction and expansion of concrete during the cyclic freezing–thawing conditions. The external water can be sucked into the macroscopic pores due to the micro-ice-lens pump, increasing the saturation degree of concrete. Coussy [11, 12] developed a comprehensive thermo-poromechanical theory that encompasses all physical processes occurring in the freezing porous solids, including the hydraulic pressure, interfacial effects, drainage of the liquid water, the cryo-suction process, and thermomechanical effects. Zhou et al. [13] proposed that freezing–thawing damage can be due to nanofluidic salt trapping.

It is well known that water migrates from the surroundings and flows through the concrete pore networks during the cyclic freezing–thawing. The cement concrete contains pores with widely distributed sizes ranging from 1 nm to 1 μm and may include many larger air voids. The frost damage theories note the critical importance of air voids to increase the freezing–thawing resistance of concrete since the entrained air voids act as expansion reservoirs and cryo-pumps during the ice formation. Mayercsik et al. [14] systematically analyzed the efficiency of those entrained air voids for the freezing–thawing resistance of concrete using a poromechanical framework. The key actions related to the migration of the liquid to the air voids are cryo-suction and drainage of liquid. When the temperature continuously decreases, the freezing process occurs progressively from the large pores to the small pores. In this non-equilibrium process, the formed ice crystals are in local thermodynamic equilibrium with the surrounding unfrozen liquid, which requires the chemical potential of the ice solids and the unfrozen liquid to be equal. Given that the entropy of the ice is lower than that of the liquid water, the difference between the pressure in the ice and the pressure in the remaining liquid water increases when the temperature decreases. To this end, the pore space filled with ice has the potential to suck the liquid from the smaller pores, which is called the cryo-suction process. This cryo-suction process and the thermodynamic mechanisms occurring during freezing and thawing are well-detailed by Coussy [15].

Figure 9.1 shows the multilevel structures in which the cryo-suction process occurs in the cement matrix. For cryo-suction to occur, liquid water needs to be able to flow toward the ice crystal. Consequently, the liquid has to be sucked through the unfrozen premelted film which prevails between the ice and the cement pore wall, as this film is the only flowable path in the already frozen pore space [16]. The underlying significance of this nanometric premelted film has been noted in the frost damage theoretical analysis [9, 16]. In some thermoporometry analyses, the thickness of this unfrozen film was considered constant and equal to 0.8 nm [17, 18]. However, since this nanometric film is difficult to characterize experimentally, the nature of this film and its properties, in particular its transport properties, are still poorly understood. Without knowing the transport properties of the premelted film, the theories that involve cryo-suction lack solid evidence. Complementary to experiments, the molecular dynamics simulations make it possible to model the film and analyze its dynamics and local structures in detail. Liang et al. [19] carried out a molecular dynamics simulation to study the effect of NaCl solution on the freezing

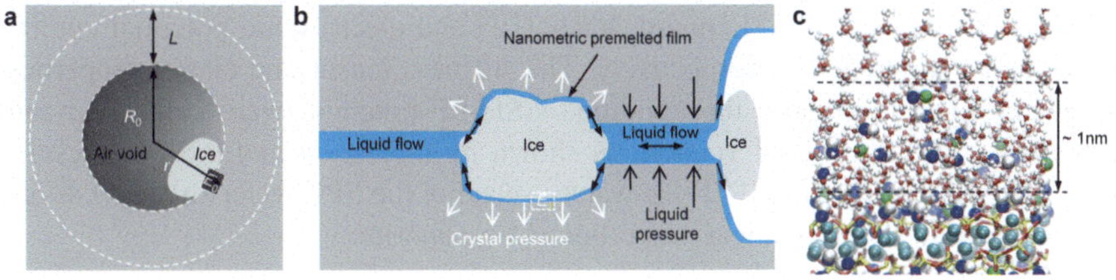

Fig. 9.1 Schematic diagram of cryo-suction in the connected pores of cement concrete. **a** Air void acting as a cryo-pump. **b** Cryo-suction in element E_0. **c** Molecular structure of the premelted film between C-S-H and ice in element E_1. The cyan and green balls stand for Ca and Cw atoms, respectively. Water molecules are presented in ball-stick style (Ow in red and Hw in white). The blue and white balls are the OH groups (Oh, Hh). The yellow–red sticks represent Si–O/Ob bonds

of liquid on a neat tobermorite surface and observed the premelted film. However, the skeleton structure and the surface of the tobermorite are too ordered to represent the cement structure realistically [20–22].

Herein, we tackle two important issues for the freezing liquids in cement pores, i.e., crystallization of ice and transport of liquids (cryo-suction). Specifically, we attempt to present an atomistic view to the two active processes. We will consider a realistic C-S-H pore model and focus mostly on the dynamics and transport properties of the nanometric liquid film between the ice and C-S-H pore wall. Molecular dynamics simulations make it convenient to explore the nature of the premelted film across a wide spectrum of low temperatures, extending down to 220 K and beyond, while exploring the nature of this film in cement-based materials experimentally remains a challenge. Combined with a thermodynamic analysis, we estimate the characteristic time of the cryo-suction process. Based on an estimation of the characteristic time for the migration of liquid through the premelted film, we will show that, at temperatures above 230 K, the cryo-suction process occurs relatively fast (i.e., in about a second at most) in cement-based materials.

9.2 Computational Details

9.2.1 Force Field for C-S-H/Ice/Liquid Mixtures

Herein, we use molecular dynamics simulations to study the crystallization of ice Ih in a cement nanopore. The interactions between the atoms in C-S-H solids are described by the CSH-FF potential [23], which is a re-parameterized Clay-FF potential dedicated to the simulation of C-S-H. In the original CSH-FF potential, SPC water is used to describe the water molecules of C-S-H. However, the SPC water model is not adapted to describe the liquid–solid phase transition. Accordingly, we consider instead the TIP4P/Ice potential, which predicts well the melting line of ice Ih and other ice forms [24]. The parameters of the force field are detailed in Tables 9.1, 9.2 and 9.3. Lennard–Jones (LJ) and electrostatic potential cutoffs are set at 10 Å and 8.5 Å, respectively. The stiffness tensor and elastic properties of tobermorite 11 Å using this modified CSH-FF potential were calculated and compared with the original version, as shown in Tables 9.4 and 9.5. The results show that considering the TIP4P/Ice water instead of the SPC water still reproduces reasonably well the mechanical properties of tobermorite.

Table 9.1 Lennard-Jones potential parameters used in CSH-FF potential combined with TIP4P/Ice

Species i	Species j	ε (kcal/mol)	σ (Å)
Ca	O	0.000869	4.365403719
Ca	Ob	0.000706	5.452300155
Ca	Oh	0.000869	4.365403719
Cw	O	0.001460	4.365403719
Cw	Ob	0.001040	4.445584604
Cw	Oh	0.001040	4.445584604
Si	O	0.000560	3.269598296
Si	Ob	0.000595	3.260689308
Si	Oh	0.000595	3.260689308
O	O	1.240000	2.735059065
Ob	Ob	0.061800	3.447778039
Oh	Oh	0.061800	3.447778039
Ob	O	0.045400	3.625957783
O	Oh	0.045400	3.625957783
Ob	Oh	0.061800	3.447778039
Ca	Ow	0.000876	4.365403719
Cw	Ow	0.000604	4.472311565
Ow	O	0.020000	4.240677898
Ow	Ob	0.872000	2.895420834
Ow	Oh	0.872000	2.895420834
Ow	Ow	0.210840	3.166800000
Si	Ow	0.000530	3.233784167

Table 9.2 Bond interaction constants

Species i	Species j	k_r (kcal/mol)	σ (Å)
H	Oh	500	0.9572

Table 9.3 TIP4P/Ice water model parameters

r_{OH}(Å)	∠HOH (deg.)	r_{OM}(Å)	q_O(e)	q_H(e)	q_M(e)
0.9572	104.52	0.1577	0	0.5897	−1.1794

Table 9.4 Stiffness tensor of tobermorite 11 Å calculated from our re-parameterized CSH-FF potential

Stiffness tensor component	CSH-FF [23] (GPa)	First principle [25] (GPa)	This work (GPa)
C11	151.44	148.25	146.56
C12	54.27	63.25	55.97
C13	25.18	26.75	32.33
C16	1.05	6.63	–0.94
C22	123.16	138.55	133.33
C23	20.90	32.55	32.43
C26	−3.16	1.85	−4.33
C33	61.33	68.40	96.28
C36	−1.99	−1.73	−3.52
C44	37.33	32.75	31.49
C45	−0.15	−1.93	0.81
C55	27.36	25.65	25.42
C66	54.86	53.30	49.35

Table 9.5 Elastic properties of tobermorite 11 Å calculated from our re-parameterized CSH-FF potential

Properties	CSH-FF [23] (GPa)	First principle [25] (GPa)	This work (GPa)
K^{voigt}	59.63	69.03	68.62
G^{voigt}	39.62	37.15	38.28
K^{Reuss}	47.7	48.7	62.72
G^{Reuss}	35.82	33.45	34.97
v	0.23	0.27	0.24
E_x	151	110.31	113.51
E_y	123	106.87	107.22
E_z	61	56.35	85.41
G_{xy}	54.68	52.81	49.12

9.2.2 Preparation of the Initial Configuration

Ice nucleation requires a critical nucleus, whose size has been proven temperature and pressure dependent [26, 27]. However, the formation of ice nuclei involves a high free energy barrier, making it difficult for crystallization to occur in the bulk solution. We build a C-S-H/ice/liquid mixture model by assembling an open-pore C-S-H model, a liquid water box, and an ice Ih block, as shown in Fig. 9.2. We put a piece of ice in the liquid, artificially creating an initial nucleus with two flat crystal surfaces from which ice can grow or melt. To build the mixture model, we generate an ice Ih block with a dimension of $3\boldsymbol{a}_w \times 4\boldsymbol{b}_w \times 6\boldsymbol{c}_w$, containing 576 water molecules replicated from the $\boldsymbol{a}_w \times \boldsymbol{b}_w \times \boldsymbol{c}_w$ ice Ih lattice [28]. The ice block is relaxed at 1 bar and 230 K using the NPT simulation and rotated by 90 degrees along the y axis so that the crystal face orthogonal to the xy plane is the basal face {0001} (face which will be exposed to crystallization ultimately). After rotation, this procedure yields a reference ice block size of $22.086 \times 27.126 \times 31.186$ Å^3. This particular replication of the ice Ih cell is chosen because it almost corresponds to a $4\boldsymbol{a}_{csh} \times 4\boldsymbol{b}_{csh} \times 1\boldsymbol{c}_{csh}$ replicated cell of the C-S-H model in the x and y directions. Of course, the unstressed C-S-H dimensions in the x and y directions are not exactly identical to that of ice Ih and the x and y dimensions are constrained to match the ice Ih block, while the crystallographic structure of ice Ih remains intact to preserve a 1 bar ice configuration since we know that ice pressure impacts the melting behavior. Hence, due to the geometric limitations between C-S-H and ice cells, we consider only crystallization occurring at the basal {0001} face of ice Ih, and do not look at the prism and secondary prism faces. The C-S-H model is built with a Ca/Si ratio of 1.7 and is based on Hamid tobermorite 11 Å unit cell [29]. The approach to generate the C-S-H model follows the procedures adopted by Morshedifard et al. [30]. However, we slightly modified the way neutral SiO_2 units were removed to fit NMR measurements. Precisely, when randomly removing the SiO_2 units, we assume the number N of left bridging silicate tetrahedra is related to the Ca/Si ratio as follows: $N = 14.22(\mathrm{Ca/Si})^{-1.897}$. The number of OH groups and water molecules follow the compositional relations: $Ca_xSiO_{2.75}(OH^{Ca})_{x-1}(OH^{Si})_{0.598x-0.4922} \cdot [x-1]H_2O$ according to Abdolhosseini Qomi et al. [31]. Detailed model construction procedures can be found in our previous work [32]. The original C-S-H dimensions are $22.317 \times 29.560 \times 22.770$ Å^3. Then, the C-S-H model is cleaved in the interlayer region to open a slit nanopore of 70 Å, and this pore is filled with liquid water molecules at a density of ~ 0.7 g/cm^3. After the equilibration at 1 bar and 230 K in the $NP_{xx}P_{yy}P_{zz}T$ ensemble for 1 ns, the dimensions of the C-S-H/liquid mixture model are $21.852 \times 27.277 \times 93.997$ Å^3. The box is then stretched and compressed along the x and y axes, respectively, to the exact size 22.086 Å × 27.126 Å of the relaxed ice block. With respect to the relaxed configuration, the C-S-H model underwent a 1.07% extension along the x-direction and a -0.55% compression along the y-direction. Due to the small magnitudes of these strains, there was minimal impact on the release of ions into the liquid. Analysis of the trajectories reveals that the quantities of free calcium and adsorbed calcium remain the same as for the unstrained model. Effects on surface stress and interfacial energy are also possible, yet the main impact of the deformation is on the stress: C-S-H being quite stiff, the imposed deformation generates significant stresses in the x–y plane (typically a few hundred MPa). Yet, these stresses are in the x and y directions and located within the C-S-H. As an isotropic pressure prevails in the liquid film, only the stress in the z direction is transmitted to the ice, so these deformations of the C-S-H are not expected to affect the crystallization process. The system is then relaxed at 230 K for 1 ns in the $NL_xL_yL_zT$ ensemble, where the box size of the C-S-H/liquid mixture is $22.086 \times 27.126 \times 93.997$ Å^3. Finally, we remove liquid water to create an empty region in the pore (from 42 Å to 75 Å along the z axis). The relaxed ice Ih block is inserted in the center of this empty region, yielding the C-S-H/ice/liquid mixture model. We further relax the mixture model in the $NL_xL_yL_zT$ ensemble at 230 K for 20 ps, and then in the $NL_xL_yP_{zz}T$ ensemble at 230 K and 1 bar for 500 ps, finally obtaining the initial configurations for computations. As presented in Fig. 9.2a, the initial thicknesses of the top and bottom liquid water films are ~ 13 Å and ~ 25 Å, respectively. Similarly, a reference bulk model of ice-liquid mixture with the same x and y dimensions as the C-S-H/ice/liquid model was built, as shown in Fig. 9.2b. Overall, the confined C-S-H/liquid/mixture and bulk ice-liquid models contain 5236 atoms ($Ca_{192}Si_{113}O_{269}{\cdot}OH_{210}{\cdot}1414H_2O$) and 1728 water molecules, respectively.

The initial atom configurations of confined and bulk systems for the crystallization simulations are shown in Fig. 9.2a and b, respectively. The confined model in Fig. 9.2 contains many impurities (i.e., Ca^{2+} and OH^- ions) in the unfrozen water layers. To study the effect of ions on the growth rate of ice in the confined system, we then

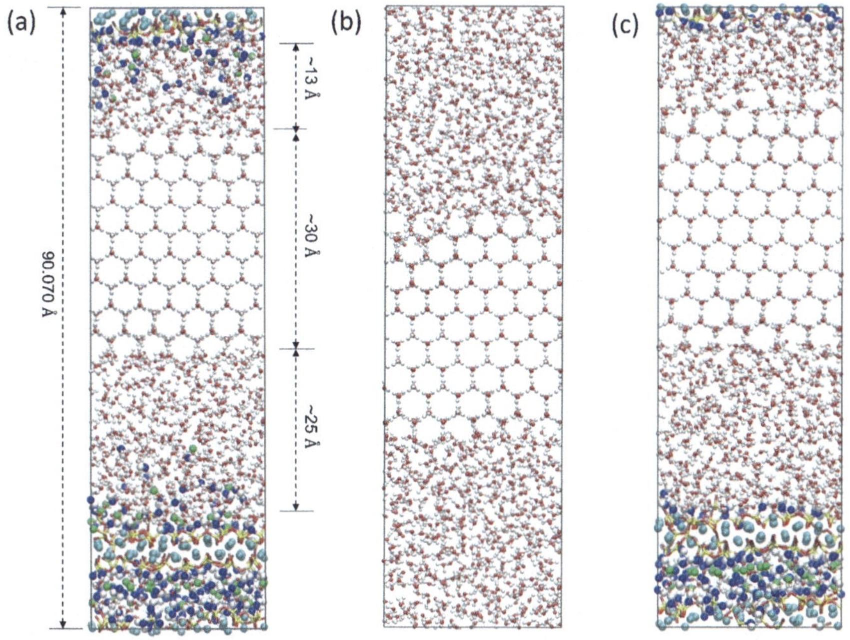

Fig. 9.2 Initial configuration of the C-S-H/ice/liquid systems **a** and bulk systems **b** for ice formation. The confined ice molecular model was built by combining a C-S-H nanopore model with an ice Ih crystal. The cyan and green spheres represent intralayer (Ca) and interlayer calcium (Cw), respectively. The small red and white spheres are the water oxygen (Ow) and hydrogen (Hw), respectively. The large blue and white spheres are the OH groups, with Oh in blue and Hh in white. The yellow–red sticks stand for Si-O/Ob bonds

prepare a confined system without any impurities in the unfrozen water layers. For this purpose, all Ca^{2+} and OH^- ions in the upper and lower unfrozen water layers are removed from the initial atom configurations shown in Fig. 9.2a. Note that we retain all the OH groups bonded to the silicate chains since they are integral components of the C-S-H solids. After the removal of impurities, the system is relaxed in the $NP_{zz}T$ ensemble ($p = 1$ bar, $T = 250$ K) for 1 ns. This relaxation process yields the initial atomic configurations for the crystallization of the confined cases devoid of impurities, as represented in Fig. 9.2c.

9.2.3 Simulation Details

9.2.3.1 Ice Growth

Molecular dynamic simulations are performed using LAMMPS package [33]. To investigate the effect of temperature on the crystallization of ice, simulations are performed at a variety of temperatures: 220 K, 230 K, 240 K, 250 K, and 260 K. We perform NVE simulations for the whole molecular system (i.e., for the system composed of all atoms) while using a Langevin thermostat to control the temperature of the solid parts only. Specifically, the Langevin thermostating is applied to all the atoms of the C-S-H solid and of the initial ice block except for the first layers of atoms (in both the C-S-H solid and in the ice) in contact with the liquid. Doing so, the domains where the process of crystallization occurs follow strictly a time evolution, and the kinetics should not be affected by the Langevin thermostating and can thus be considered representative. The crystallization simulations are conducted over 300 ns, which involves rather long simulations (typically 180 days of 27 CPU). Separately, the crystallization of the reference bulk system is simulated using identical conditions. Since the Langevin thermostat is not applied to all atoms, one may wonder whether the temperature is homogeneous throughout the system. In this respect, some temperature profiles of the confined system are displayed in Fig. 9.3 (case of a thermostating at 260 K): the profiles are reasonably homogeneous, meaning that heat transfers throughout the non thermostated domains are fast enough to ensure a constant temperature in the whole system. We do not observe any temperature gradient that could arise from the latent heat or the selective Langevin thermostating (see Fig. 9.3). Also, we decided to use a Langevin thermostat rather than the Nose–Hoover thermostat because of anomalous results at all temperatures when using the Nose–Hoover thermostat for the confined system: as presented in Fig. 9.4, with a Nose–Hoover thermostat, the potential energy exhibits unphysical evolution after 40 to 50 ns, with no more fluctuations and a steep decrease. The origin of this anomaly is unclear, but visual observation of the atomic configuration suggests that the system may enter in a non-ergodic regime, which is a known drawback of the Nose–Hoover thermostat [34].

Isobaric-isothermal ensemble simulations are performed to determine the melting temperature of nano-confined ice at various imposed pressures along the z axis

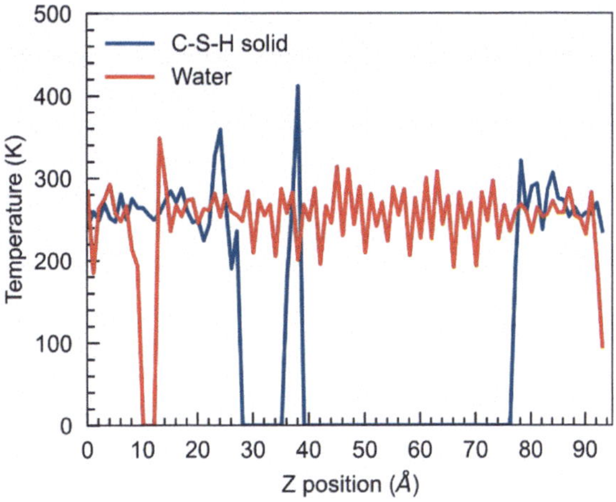

Fig. 9.3 Temperature profiles of C-S-H solid and water molecules along the z position. Case for temperature at 260 K. The zero values are due to the absence of C-S-H solid atoms or water molecules in the regions

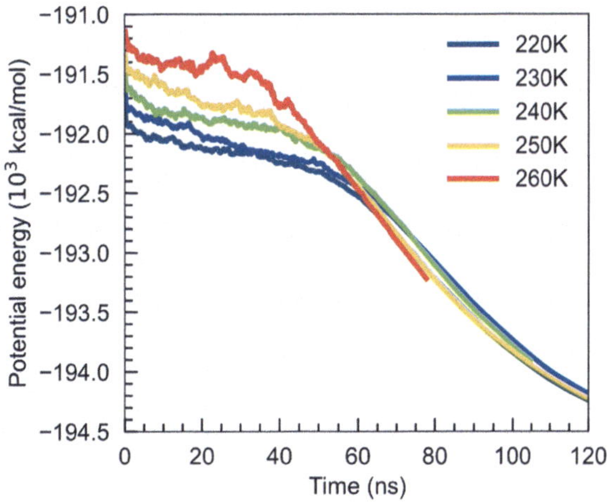

Fig. 9.4 Potential energy evolution of the system during NPT simulations using the Nose–Hoover thermostat

(0.1 MPa, 50 MPa, 100 MPa, 150 MPa, and 200 MPa). In the isobaric-isothermal ensemble simulations, Nose–Hoover thermostat and barostat are employed to control the temperature and pressure, respectively. For each pressure, the confined melting temperature was estimated by considering many temperatures near the melting point of bulk ice, and comparing the evolutions of the potential energy over 20 ns: a decreasing potential energy corresponds to a crystallization, while an increasing energy corresponds to a melting. At the melting temperature of nano-confined ice, the ice block and liquid water coexist with neither crystallization or melting, i.e., with a relatively constant potential energy curve. Note that, for the determination of the melting point, the thermostating/barostating is applied to all atoms and the simulation are much shorter than for the simulation dedicated to the growth kinetics (see previous paragraph). Indeed, here, we are only interested in the direction of evolution (crystallization or melting), and it is not important if this evolution follows the true time. Moreover, the direction of evolution can be clearly established over relatively small times (20 ns), over which the anomaly with the Nose–Hoover thermostat reported in the previous paragraph is not observed.

In all the simulations, time integration follows the Velocity-Verlet algorithm with a time step of 2 fs. The TIP4P/Ice water molecules geometry is constrained using a shake algorithm [35]. The trajectories are visualized with the VMD [36] and Ovito [37] software.

9.2.3.2 Water Transportation in Premelted Film

After ice crystallization was completed, a premelted nanometric water film could be observed between C-S-H and ice (see Fig. 9.5). To study the effect of temperature on the dynamics of this water film, the self-diffusion properties of the premelted water films and calcium cations inside the films were investigated after the molecular systems were fully equilibrated. To this end, we carried out NVE ensemble molecular dynamics simulations for the C-S-H/ice/liquid mixture systems equilibrated at 220, 230, 240, 250, and 260 K. Note that the temperature and potential energy of these equilibrated systems shall be maintained at their equilibrated states during the NVE ensemble simulations in case that the crystallized ice melted, since we aimed to obtain dynamics representative of the system at the target temperatures. For this purpose, the kinetic energy of these systems was imposed manually before the NVE ensemble calculations by setting the initial velocities of all atoms. In practice, since a given equilibrated configuration from a NVT run never corresponds to the exact average energy of the targeted temperature, one must rescale the velocities of the atoms so as to ensure that the NVE runs starting from these rescaled configurations effectively simulate the targeted temperature (220, 230, 240, 250, and 260 K). The time integration was performed using Velocity-Verlet algorithm, with a time step of 1 fs. The simulations were run for 30 ns at least, to make sure that the average displacements of the liquid water molecules exceed the typical size of a molecule.

The interface detection algorithm developed by Rozmanov et al. [38] was adapted to quantify the thickness of the nanometric liquid film. In this algorithm, the water density profile along the z direction is used to detect the ice/liquid–water and liquid–water/solid-C-S-H interfaces. As depicted in Fig. 9.5, ice exhibits a layered structure of water molecules, whereas water molecules in the liquid phase are distributed in a more random fashion. Because of the layered structure of ice, the discrete Fourier transform spectrum $F(z, k_x)$ of the density profiles in x at a given position in z direction can show the presence of periodic motifs of ice layer. In contrast, the random distribution of water molecules in the liquid phase does not yield a periodic order. Using this difference, we slice the simulation box along the z direction into 256 bins and perform fast Fourier transformations on 14 small bins $D(z_a, z_b)$ of the total density profile along the x axis within a z window. We construct a structural order parameter $S(z_i) = F(z_i, k_s)$ as an indicator. From our experience, $k_s = 1$ and 6

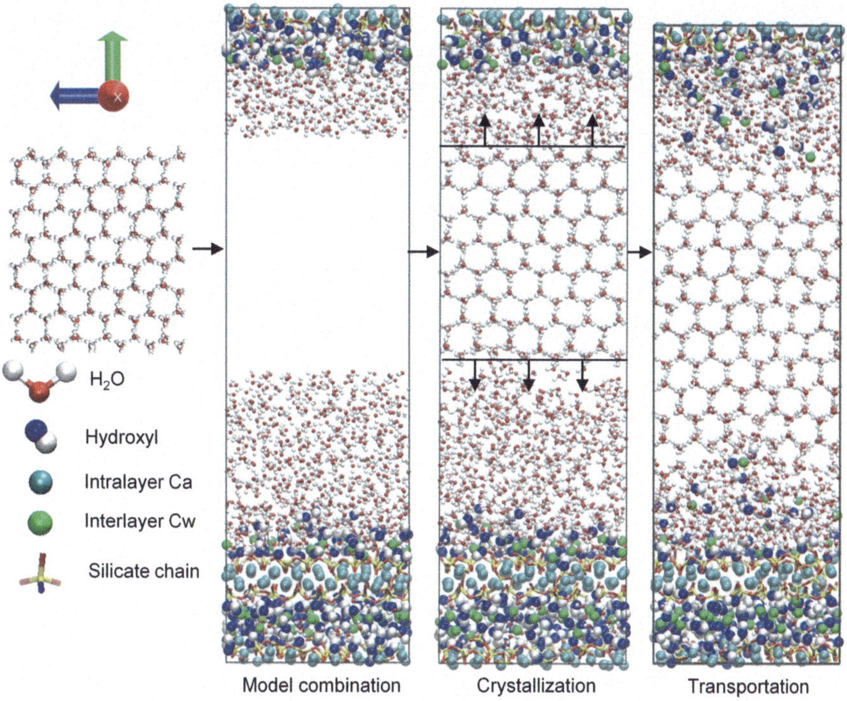

Fig. 9.5 Snapshots showing the elaboration of the molecular systems and their equilibration

are good values to use to identify the liquid–water/solid-C-S-H interface and the ice/liquid–water interface, respectively. The position of the ice/liquid–water and liquid–water/solid-C-S-H interfaces can be identified from the extrema of the derivative dS/dz of the profile of the indicators $S(z_i) = F(z_i, k_s = 1)$ and $S(z_i) = F(z_i, k_s = 6)$, respectively. All the simulations were carried out using the LAMMPS package [33]. The trajectory was visualized using the VMD software [36].

9.2.4 Algorithms for Detecting Ice Crystal and Interfaces

9.2.4.1 Ice Detection Algorithm

For ice crystallization, the number of water molecules in ice and the growth rate is determined by analyzing the trajectories of atom configurations. The number of water molecules in ice can be determined with the CHILL/CHILL + algorithms [39, 40]. In the present work, we slightly modified the CHILL + algorithm [40] to adapt it to our molecular model. In this method, the local order of a water molecule is defined by the local orientational bond parameter, which distinguishes the staggered and eclipsed bonds with respect to the surrounding neighbor water molecules. For each water molecule i, local orientational bond order complex parameters $q_{lm}(i)$ are defined from spherical harmonics $Y_{lm}(\hat{r}_{ij})$ as follows ($l \geq |m|$ integers):

$$q_{lm}(i) = \frac{1}{4}\sum_{j=1}^{4} Y_{lm}(\hat{r}_{ij}) \tag{9.1}$$

where the sum is over the four closest neighbors j of the considered water molecule i, and $\hat{r}_{ij}$ is the vector from i to j. To identify the orientational structure between a pair of water molecules, the correlation of their local order is employed:

$$a(i,j) = \frac{\hat{q}_l(i) \cdot \hat{q}_l(j)}{|\hat{q}_l(i)||\hat{q}_j(i)|} = \frac{\sum_{m=-l}^{l} q_{lm}(i) q_{lm}^*(j)}{\left(\sum_{m=-l}^{l} q_{lm}(i) q_{lm}^*(j)\right)^{1/2}\left(\sum_{m=-l}^{l} q_{lm}(j) q_{lm}^*(i)\right)^{1/2}} \tag{9.2}$$

where q_{lm}^* is the complex conjugate of q_{lm}. One can identify the bond between a pair of water molecules staggered or eclipsed with this correlation. Figure 9.6 illustrates the configurations of the staggered and eclipsed bonds in ice Ih. Specifically, in this study, the staggered bond is in the basal face of ice Ih, and the eclipsed bond is parallel to the secondary prism face of ice Ih. According to Moore et al. [39], the optimum value of l to distinguish between ice and liquid is 3; we adopted this value in the present work. The probability density distributions of $a(i,j)$ for hexagonal ice, liquid water, interfacial water, and interlayer water of C-S-H are presented in Fig. 9.7. Likewise, we defined the staggered and eclipsed bonds based on the $a(i,j)$ value, as shown in Table 9.6. Compared to the CHILL + algorithm, in the present study, we change a bit the limit values for the $a(i,j)$ of staggered bonds: we consider a bond to be staggered when $-0.3 \leq a(i,j) \leq 0.1$. For ice Ih, each water molecule has three staggered bonds and one eclipsed bond with its four water neighbors, while the water molecule of interfacial ice forms at least two staggered bonds, and the liquid and interlayer water exhibit no staggered or eclipsed bond.

9.2.4.2 Ice Interface Detection Algorithm

To determine the growth rate of ice and thickness of water films, one has to detect the positions of the ice front and of the C-S-H/liquid interfaces. To do so, we consider a modified version of the detection algorithm Rozmanov et al. [38] used to identify the ice/liquid and C-S-H/liquid interfaces, which is based on the ice layering structure. Following Rozmanov et al. [38], we constructed a structural profile S based on the density profile of oxygen in water molecules to discriminate between ice, liquid, and C-S-H solids. Rozmanov et al. [38] analyzed this local water density profile along the z axis by implementing a Fourier transformation. Instead, here, we analyze the density profile of water molecules along the x axis in a subvolume around a given z

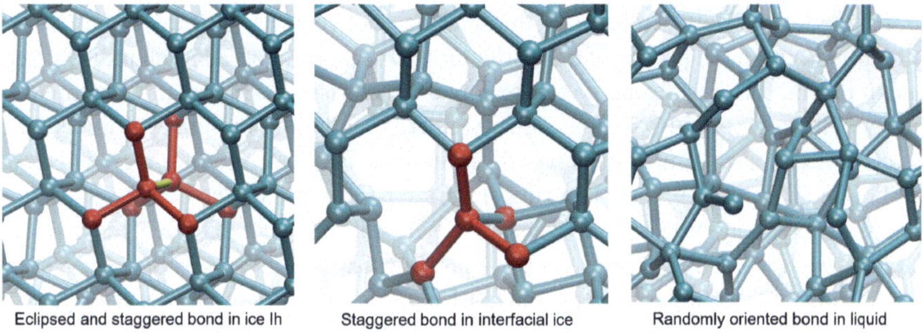

Fig. 9.6 Illustrations for local orientational water-water bonds in ice Ih (left), interfacial ice (center), and liquid (right). In the left-hand snapshot, an eclipsed bond in ice Ih is marked in yellow, and staggered bonds are marked in red. In the center snapshot, three staggered bonds are marked in the interfacial ice. The right-hand snapshot shows the randomly oriented bonds in liquid. Note that we show water oxygen atoms only in order to make the illustration clear

Fig. 9.7 Probability density distribution of the correlation of orientational order

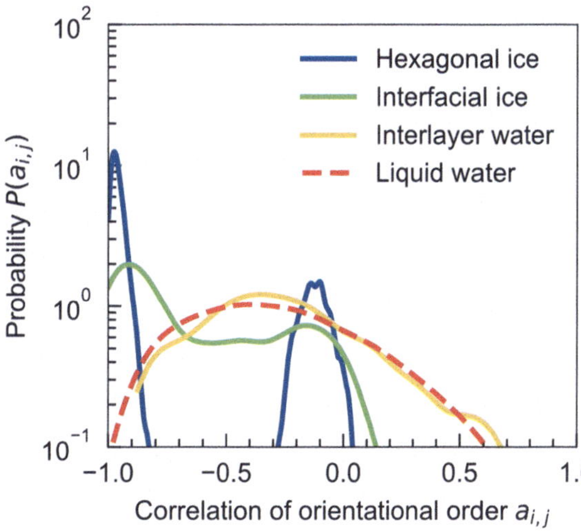

Table 9.6 The value of $a(i, j)$ for staggered and eclipsed bonds

Algorithm	Staggered bond	Eclipsed bond
CHILL [39]	$-0.2 \leq a(i, j) \leq -0.05$	$a(i, j) \leq -0.8$
CHILL + [40]	$-0.35 \leq a(i, j) \leq 0.25$	$a(i, j) \leq -0.8$
This work	$-0.3 \leq a(i, j) \leq 0.1$	$a(i, j) \leq -0.8$

value, from which order parameters can be obtained as indicators of ice/liquid and C-S-H/liquid interfaces.

To measure the growth rate of ice crystals and the thickness of the premelted water films, the positions of the interfaces between C-S-H, ice, and liquid water should be detected accurately. To identify the interface between two different phases, the inflection point in the profile of a specific property can be adopted as an indicator, as shown in Fig. 9.8. The property can be any characteristic feature that discriminates the two phases explicitly, such as density, energy, and atom mobility. In this study, we construct a structural profile based on the layering structure of ice crystals. As can be seen from Fig. 9.9, ice layering structure is captured by the density profile of water oxygen along the z axis. Because of the crystalline structure of ice, the discrete Fourier transform $F(z, k_x)$ of the density profiles in x at a given position in z yields a spectrum that can reflect the presence of periodic motifs such as an ice layer. In contrast, the liquid and interlayer water molecules of C-S-H are positioned rather randomly without clear periodic order. Therefore, we propose hereafter a structural order parameter $S(z)$ based on the discrete Fourier transforms $F(z, k_x)$. To this end, the simulation box is equally sliced into 256 bins along the z axis ($\{z_i\}$ is the set of the center z-position of the bins). Within each bin, the density profile in x is sampled over 10 ps. Then the order parameter $S(z)$ is computed for each bin by considering a particular selection of k-values as follows:

$$S(z_i) = F(z_i, k_s) \tag{9.3}$$

To identify the interface between ice and liquid, Rozmanov et al. [38] performed fast Fourier transformation (FFT) on small subsets $D(z_a, z_b)$ of the total density profile along the z axis. In their algorithm, a window with a fixed width Δz was moved with $D(z_i)$, from which the structural profile $S(z_i)$ was obtained. We note that this detection algorithm can be further improved by performing the FFT along the x axis instead of the z axis, i.e., on the x-axis density profile $D_x(z_a, z_b)$ within a z window. We define a z window:

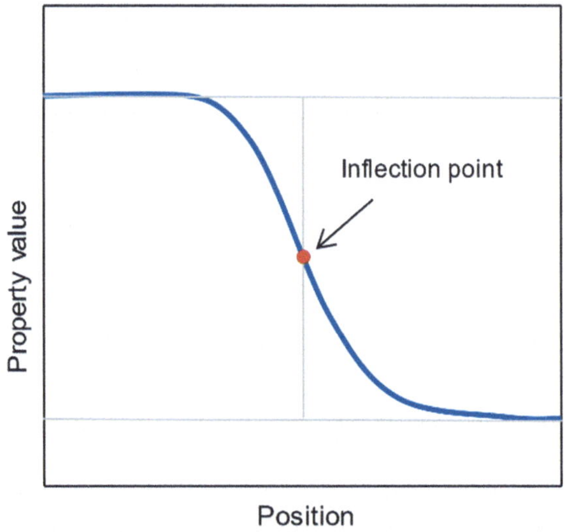

Fig. 9.8 Property profile along a particular direction. The inflection point defines the position of the interface between the two phases

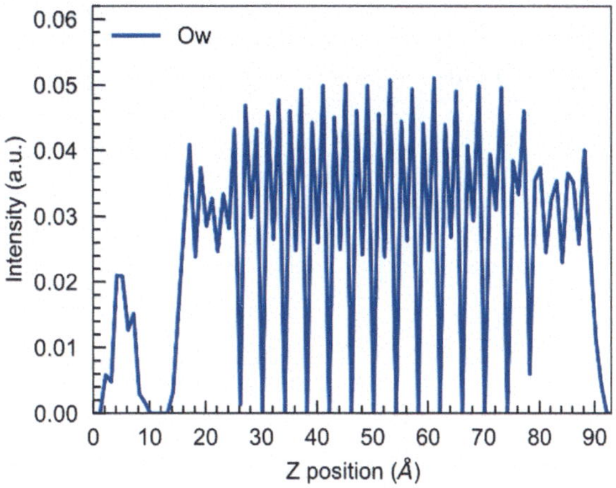

Fig. 9.9 Water oxygen density profile of C-S-H/ice/liquid mixtures

$$\Delta z = z_b - z_a \geq d_l \tag{9.4}$$

where d_l is the distance of the minimum periodic motif, i.e., the length of an ice layer. Herein, the window we used has 14 bins, longer than an ice layer. This ensures that each bin has a non-zero local density profile along the x axis. The local density profile $D_x(z_i)$ used to perform the FFT is also discretized in 256 values along x. After the FFT, we obtained the Fourier spectra for all $\{z_i\}$ positions. Figure 9.10 presents two examples of the Fourier spectra for z positions within the liquid and ice, respectively. From the Fourier spectra, $k_x = 1$ and 6 seems good indicators of the liquid and ice phases, respectively. In contrast, the magnitude of the Fourier spectra terms for the C-S-H solid is always lower than 0.1. Therefore, $k_x = 1$ can be used to discriminate the interface between the C-S-H mineral and the liquid, while $k_x = 6$ can be used to identify the interface between the ice and the liquid, as expressed in Eqs. 9.5 and 9.6:

$$S_{csh/liquid}(z_i) = F(z_i, k_s = 1) \tag{9.5}$$

$$S_{ice/liquid}(z_i) = F(z_i, k_s = 6) \tag{9.6}$$

where $z_i = (z_a + z_b)/2$ with z_a and z_b delimitating the z window in which the density profiles are computed. Examples of the structural profiles $S_{csh/liquid}(z_i)$ and $S_{ice/liquid}(z_i)$ are shown in Fig. 9.11. The inflection points are obtained from the extrema of the profile derivative dS/dz. Note that the structural profiles are smoothed before the derivative in order to reduce the noise. From these inflection points, the thickness of water films and of the ice crystal can be then determined.

With this algorithm, we analyze the movement of the ice fronts and the thicknesses of the water films from the atom configurations. The growth rate of ice is obtained by linearly fitting the curve of film thickness over simulation time.

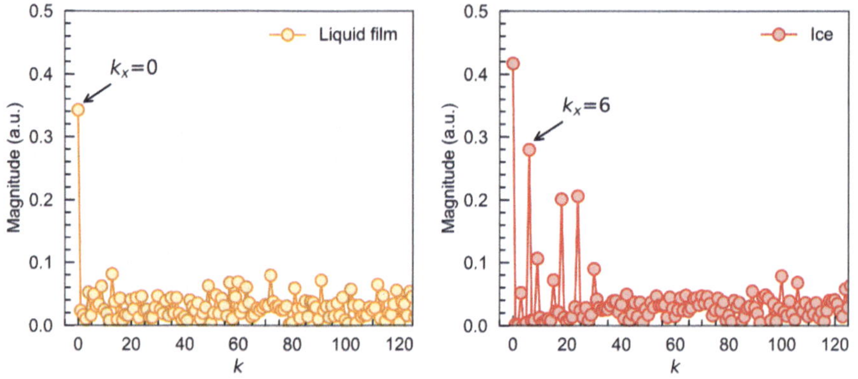

Fig. 9.10 The discrete Fourier spectra of liquid water (left) and ice (right) obtained from the FFT of the local density profile along the x axis

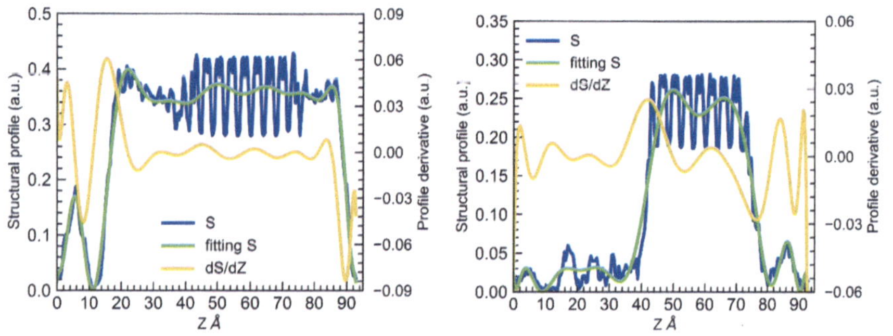

Fig. 9.11 Profiles of $S_{csh/liquid}(z_i)$ (left) and $S_{ice/liquid}(z_i)$ (right). The inflection points are the extrema of the profile derivatives, from which we can determine the positions of C-S-H/liquid and ice/liquid interfaces

9.3 Ice Crystallization in Cement Nanopore

9.3.1 Confinement Effect on Ice Crystallization

In this section, we examine the effect of confinement on the crystallization kinetics of ice. Figure 9.12 shows snapshots of the ice crystallization in bulk and confined systems, respectively. From Fig. 9.12, one clearly observes a layer-by-layer growth of the ice crystal over a few tens of nanoseconds for the bulk system and up to hundreds of nanoseconds for the confined systems. Such a layer-by-layer growth is usually found when the ice growth front is parallel to the basal face {0001} of ice Ih. In contrast, collective volumetric growth behavior is more often observed when the growth front is aligned with the prism face $\left\{11\bar{1}0\right\}$ or with the secondary prism face $\left\{11\bar{2}0\right\}$ of ice Ih [41]. The present comparison shows that the confinement has no impact on the layer-by-layer growth behavior when crystallization occurs on the basal face. Another common feature is that the advance of the ice front is not continuous but rather step-by-step, leading to some asymmetry between the bottom and up faces in both the bulk and confined systems. This reflects the stochastic nature of the ice crystallization process, even in the bulk system whose initial interfaces are very much similar.

Although the bulk and confined systems share some common features regarding the ice growth behavior, some significant discrepancies between the two systems arise. Since the confined system contains free calcium cations and hydroxyl groups in the liquid phase, the ice crystallization is associated with an ion rejection process. As can be seen from Fig. 9.12, both free calcium cations and hydroxyl groups are rejected

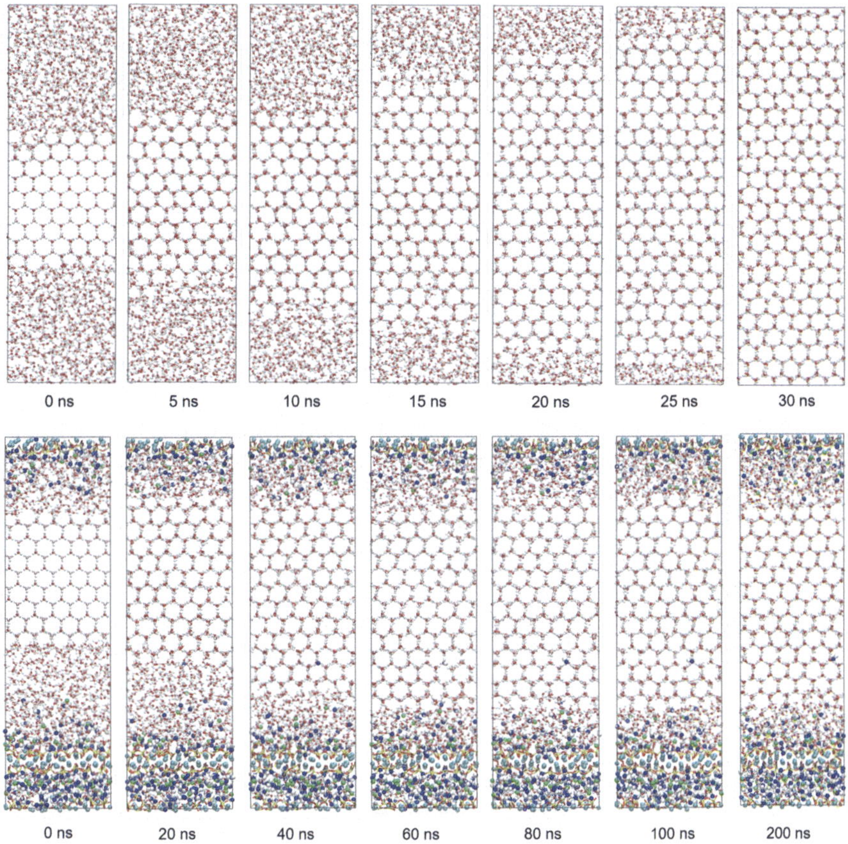

Fig. 9.12 Snapshots of ice crystallization in bulk water at 250 K (top) and in the confined slit pore at 260 K (bottom). The ice in bulk water completed crystallization in 30 ns. The crystallization of ice in the slit pore is mostly completed in 60 ns. Colored balls represent water oxygen (red), hydrogen (white), hydroxyl oxygen (blue), interlayer calcium (green), and intralayer calcium (cyan). Sticks are the skeleton of silicate chains

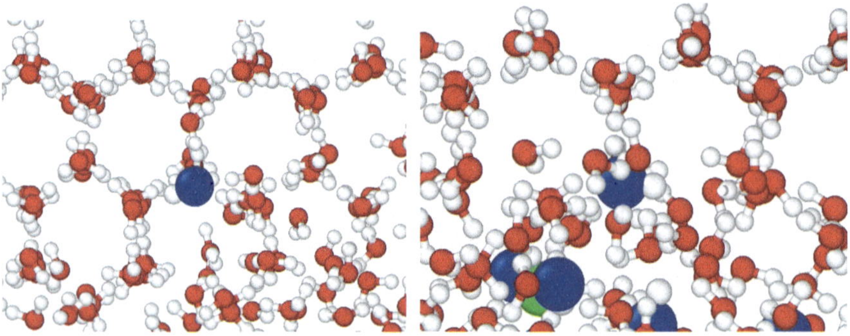

Fig. 9.13 Snapshot of trapped hydroxyl groups inside ice crystal

from the freezing domain into the liquid films during the ice crystallization process. Such ion rejection phenomenon is known for the freezing of bulk brine solutions [42]. Eventually, the ice crystal that forms is almost free of ions. Nevertheless, a few hydroxyl groups can still be trapped inside the ice crystal, as shown in Fig. 9.12. Note that trapped ions are observed only at 260 K, at the temperature of fastest growth, consistent with previous experimental observations [43]. Similar trapped ion cases are reported in previous MD simulation studies on freezing NaCl solutions [42, 44, 45]. After carefully examining snapshots of this study and the previous studies [42, 44, 45], we note that the hydroxyl groups and chlorine are more likely to be trapped than the positive calcium and sodium cations. Concerning the position of the trapped ions, the trapped hydroxyl groups are localized at the lattice site of the ice crystal, replacing water molecules, see Figs. 9.12 and 9.13. In comparison, the trapped Cl^- is found to substitute with two water molecules in the ice lattice [46]. Such a phenomenon can be attributed to the formation of Wokman-Reynolds freezing potential [47, 48].

Let us now look at the the equilibrium reached at the end of the simulation. While the bulk systems show a complete freezing (Fig. 9.12), water in the confined system does not freeze completely and nanometric liquid films can be observed at the lower and upper interfaces of the equilibrated confined system. The liquid of the nanometric film is the so-called unfrozen or premelted water, which is central to explaining the frost damage of porous solids such as concrete and clay [16, 49, 50]. In the confined systems, ice crystals stopped growing once the ice front reached the region of the nanometric film. To exhibit this confinement effect, we intentionally positioned the initial ice nucleus out of the center of the slit pore so that the upper ice front is initially quite close to the C-S-H surface (~13 Å) while the lower ice front is initially further away from the C-S-H surface (~25 Å). From Fig. 9.12, the upper ice front has been hovering in place upon the freezing, while the lower ice front has been progressing downward until the remaining liquid film reached a thickness similar to that of the upper film. The premelted film and adjacent ice layer exhibit sometimes transient non monotonous evolutions with fluctuations, which can be identified by oscillations in the potential energy, the number of water molecules for ice, and the thicknesses of the premelted films over time. Such 'noisy' evolution is reminiscent of the nanometric scale for which thermal agitation becomes ubiquitous. Figure 9.14a, b show the evolution of the potential energy over simulation time and the number of water molecules of ice, respectively. The growth of water molecules in the ice over time is presented in Fig. 9.15. An example of the evolution of ice thickness and film thickness is displayed in Fig. 9.16a (case of 260 K). From these curves, we observed the transient evolution of the potential energy, the number of water molecules of ice, and the thickness of the films upon freezing. While a clear equilibrium is reached at 250–260 K, the simulation time seems insufficient to equilibrate the systems at 220 K and 230 K. The system at 240 K is probably close to equilibration. For the systems at 250–260 K, once equilibrium is reached, the first ice layer at the interface with the liquid film can melt or re-crystallize because of thermal fluctuations (see fluctuations in Fig. 9.15a), but the overall trend remains well stationary.

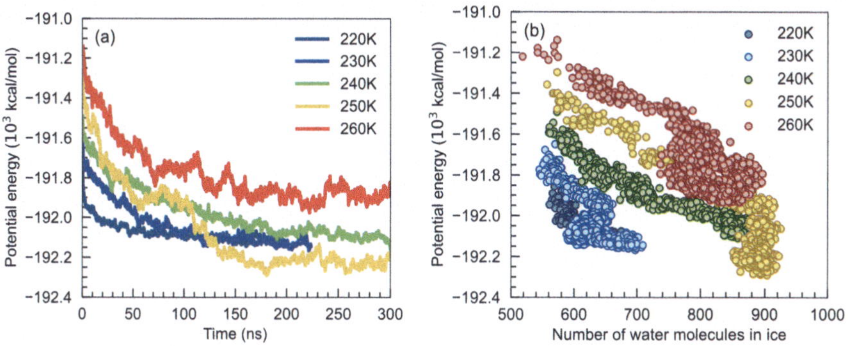

Fig. 9.14 Potential energy evolution of the C-S-H/ice/liquid systems during ice crystallization. Subfigure **a** shows the potential energy as a function of the simulation time. Subfigure **b** shows the potential energy as a function of the number of water molecules in ice

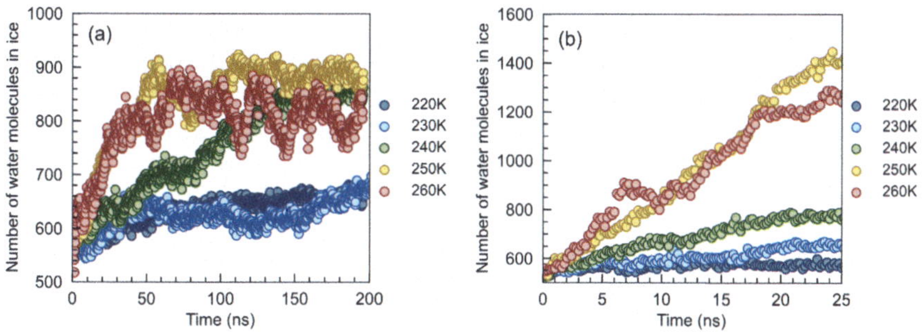

Fig. 9.15 Number of water molecules in ice during the NVE simulation with the Langevin thermostat. **a** Number of water molecules in ice of the confined C-S-H system with impurities; **b** number of water molecules in ice of the bulk system

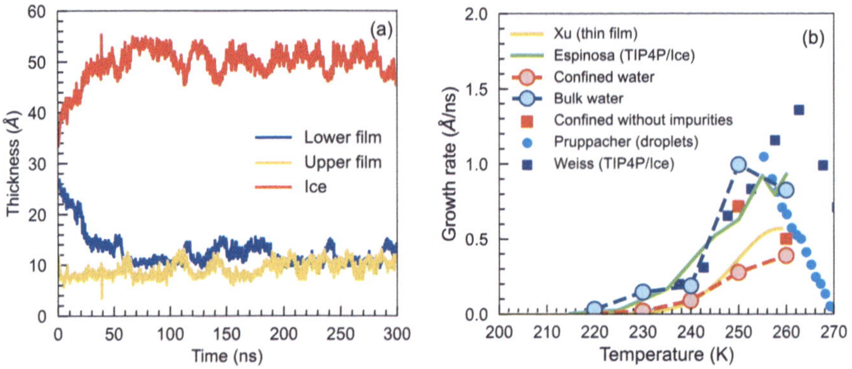

Fig. 9.16 The growth rate of ice in bulk water and in the confined nanopore. **a** An example of confined ice growth at 260 K showing the evolution with time of the thicknesses of the premelted films and of the ice crystal; **b** the growth rates as a function of temperature. The 'Espinosa (TIP4P/Ice)' (green solid line) and 'Weiss (TIP4P/Ice)' (blue squares) show growth rates of ice Ih for bulk water using the TIP4P/Ice water model, collected from Espinosa et al. [51] and Buttersack et al., [52] who obtained the data from Weiss et al. [53] through private communication. The growth rates of thin water film and droplets measured from experiments are collected from Xu et al. [54] (solid yellow line) and Pruppacher [55] (cyan circles), respectively. The 'Confined water' (red dashed line) and 'Bulk water' (dark blue dashed line) are the growth rates of confined and bulk water obtained in this work. The two red square points are the results of confined liquid in C-S-H without any surface impurities. Note that the crystal faces of ice growth in 'Confined water' are the basal face $\{0001\}$ of ice Ih, while the crystal face for 'Espinosa (TIP4P/Ice)' and 'Weiss (TIP4P/Ice)' is the secondary prism face $\{11\bar{2}0\}$ of ice Ih. For the confined system, the growth rate was calculated on the basis of the movement of the lower liquid-ice interface since the upper ice-liquid interface barely moved during the crystallization

Confronting the time scales of crystallization for bulk and confined systems suggests that confinement may significantly hinder ice growth rate. Here we quantitatively estimate the ice growth rates of the two systems at various temperatures and compare them with the reported experiment and simulation results. Since the upper ice front of the confined system barely grows upon the freezing (see Fig. 9.16a), the growth rates u_c of ice in the confined system are estimated from the bottom ice front only (see Eq. 9.7). For the bulk system, since both the upper and lower ice fronts grow significantly over the simulation, the overall growth rates u_b of bulk ice are obtained by averaging the growth rates of the two fronts (see Eq. 9.8):

$$u_c = -\frac{\Delta w_b}{\Delta t} \tag{9.7}$$

$$u_b = \frac{\Delta L}{2\Delta t} \tag{9.8}$$

where u_c and u_b are the growth rate of the confined and bulk systems, respectively, Δw_b and ΔL represent the change of lower film thickness (confined systems) and ice crystal thickness (bulk systems), respectively, and Δt refers to the growth time of ice. The value of Δw_b and ΔL can be determined using our interface detection algorithm.

The ice growth rates are displayed in Fig. 9.16b. For the bulk system, the ice reaches a maximum growth rate at 250 K, consistent with experimental observation. Similar results have also been reported for the secondary prismatic face growth of bulk ice by MD simulations using TIP4P/Ice [56] and TIP4P2005 [41] models. Before Montero de Hijes et al. [56], the existence of a maximum growth rate was challenged by some studies [52, 53], since thermostating could introduce some bias by favoring fast heat dissipation at the ice-liquid interface. Investigating crystallization in the NVE ensemble with no aid of a thermostat, Montero de Hijes et al. [56] observed the same maximum ice growth rate as with the NPT simulations. Therefore, they concluded that thermostating does not affect the growth dynamics of ice. Here, we do not apply any thermostat in the domain of crystallization to avoid perturbating the dynamics, and we observe a similar trend between growth rate and temperature, which further supports the existence of a maximum growth rate of bulk ice. However, the evolution of growth rate with temperature is different for the confined system. As shown in Fig. 9.16b, the ice growth rate of the confined system increases monotonically with the temperature from 220 to 260 K when the temperature is increased by steps of 10 K: the maximum growth rate must occur at a temperature strictly larger than 250 K. In addition, the confinement also led to lower growth rates than in the bulk system, reflecting a significant effect of confinement on the ice growth. Such differences between the bulk and confined systems can be interpreted as a result of the competition between thermodynamic and kinetic effects. According to the Wilson-Frenkel law [57, 58], the growth rate $u(T)$ of crystal can be estimated as follows:

$$u(T) = \frac{D(T)}{\alpha}\left[1 - \exp\left(-\frac{|\Delta\mu(T)|}{k_B T}\right)\right] \tag{9.9}$$

where α represents the size of a molecule or a characteristic length with limited temperature dependence, $\alpha = \frac{\lambda^2}{af}$, where a is the size of the molecule, f is the fraction of the interface growth site, λ represents the liquid-like layer thickness at the ice-water interface, $D(T)$ is the diffusion coefficient reflecting the mobility of molecules, $\Delta\mu(T)$ stands for the chemical potential difference between the liquid and the solid, and k_B is the Boltzmann constant. Following the Wilson-Frenkel law, the growth rate vanishes at the melting point (because of thermodynamic equilibrium, $\Delta\mu(T_m) = 0$) and also in the limit of low temperatures (because of slow diffusion, $D(T) \rightarrow 0$ when $T \rightarrow 0$). The maximum growth rate results from an optimum between the thermodynamic and mobility contributions. This maximum is expected to be quite close to the melting temperature, as the exponential term vanishes in a few tens of K. Based on the Wilson-Frenkel law, a confined growth rate lower than the bulk growth rate could be attributed to lower mobility of the supercooled liquid. Moreover, as we shall see in the next section, the melting point itself is also shifted to lower temperatures because of confinement, which tends to also reduce the thermodynamic contribution to the growth rate. As for the maximum growth rate, it should also exist for the confined system. Our data indicate that the temperature at which the growth rate is maximum must be greater than 250 K, suggesting that it is closer to the melting point than in the bulk. A temperature sensitivity of confined water mobility that would be more pronounced in a confined space than in the bulk could compensate for the decrease in the thermodynamic driving force when approaching 250–260 K, thus leading to a maximum growth rate closer to the melting point. Additional temperatures would be necessary to characterize more precisely the confined growth rate and validate this analysis. However such calculations are particularly expensive computationally, which is why our current results are limited to a few temperatures only. According to Guo et al. [59], rotational dynamics of water molecules also affect the growth rate of ice. The growth of ice involves two kinetic processes, namely the translational diffusion and rotational diffusion (i.e., the re-orientation of oxygen and hydrogen atoms) of water molecules. When considering the effect of rotational dynamics of water molecules, the characteristic length is rewritten as: $\alpha = \frac{af(T)}{\lambda^2}$, *with* $f(T) = \frac{D^R_{interf}}{D^R_{bulk}}$, *where* D^R_{interf} *and* D^R_{bulk} are the rotational diffusion coefficients at the ice-water interface and in the bulk, respectively. Note that the confinement may slow down the rational dynamics of water molecules, which may consequently decrease the ice growth rate [59].

9.3.2 Crystal Pressurization

From our simulation, we observe that the ice growth rate in the confined C-S-H nanopore space is significantly lower than in the bulk. However, the underlying mechanisms of this reduction in growth rate are still unclear. The slowing down of ice growth could be attributed to the effect of pressure, surface impurities, and/or confinement. Given that the water volume expands when it freezes, keeping the simulation box dimensions constant during the simulation could lead to an increase in pressure during the freezing process. To address this point, we record the pressure in the *z*-direction (Fig. 9.17). Although the pressure evolves continuously with the growth of ice, we did not observe any particular associated change in growth rate (Fig. 9.15) for both bulk and confined cases. Accordingly, the slower ice growth in the confined C-S-H nanopore cannot be attributed to the pressure. In a previous study, Montero de Hijes et al. [60] concluded that positive and negative pressures of respectively -100 MPa and 200MPa do not significantly change the growth rate of ice when plotted versus the supercooling, but at high supercooling, at which a small deviation is observed. As illustrated in Fig. 9.17a (bulk case), there is a noticeable rise in pressure resulting from the expansion of ice within the bulk liquid. The pressure increase becomes particularly significant as the liquid approaches complete freezing. From the data in Fig. 9.17b (confined case), the pressure is negative and shows little evolution as ice continues to grow within the confined liquid. We attribute this negative pressure to the initial configuration considered for the confined crystallization simulation. This configuration is the final configuration sampled during the initial

500 ps NPT equilibration at 1 bar. This particular configuration turns out to have a vertical dimension slightly greater than the average vertical dimension of the box during equilibration at 1 bar, leading to the negative pressures observed. The fact that the pressure does not increase during ice growth suggests that the usual molar volume mismatch between liquid water and ice does not hold for the confined liquid. Interestingly, when removing all the surface impurities from the confined liquid, the pressure does increase during freezing like in the bulk case (Fig. 9.17c), which suggests that those surface impurities have a significant impact on the partial molar volume of the confined liquid water.

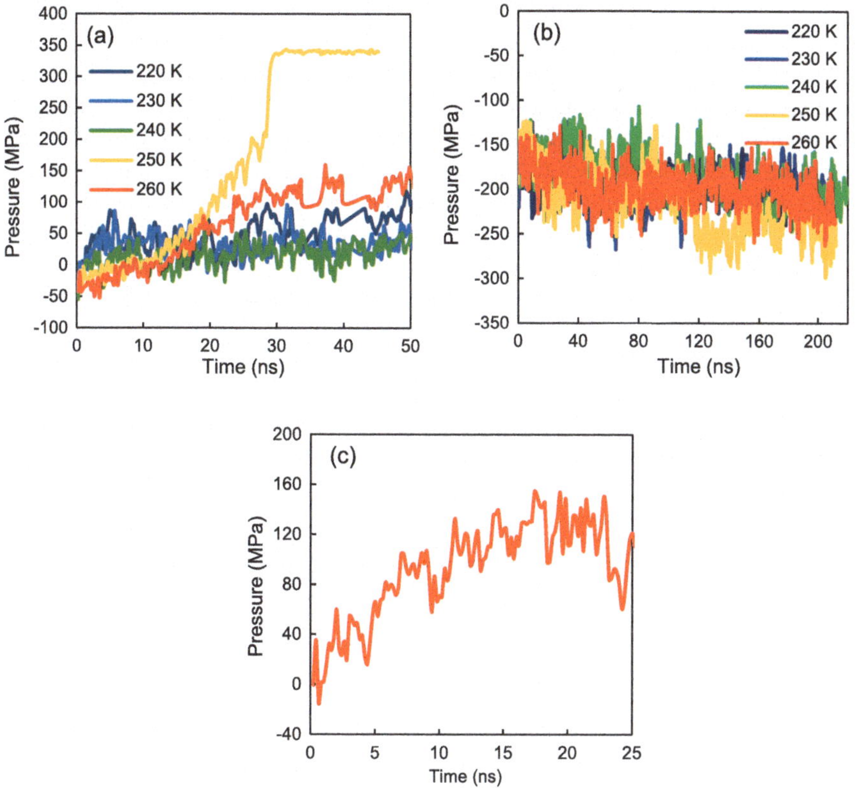

Fig. 9.17 The evolutions of pressure in the z-direction during the ice growth simulations. **a** Pressure evolutions of the bulk cases; **b** pressure evolutions of the confined C-S-H systems with surface impurities; **c** pressure evolution of the confined C-S-H system without surface impurities at 250 K

9.3.3 *Effect of Surface Impurities*

We then focus our attention on the effect of surface impurities and confinement. As revealed by previous studies [61–63], impurities alone can retard the growth of ice in the bulk liquid. In addition, confinement can alter the chemical potential difference between the ice and the liquid, which can also contribute to slowing down the ice growth. To better understand the contribution of those two effects, we investigate a confined system in which all the impurities on the C-S-H nanopore surfaces are removed, and then stimulate the ice growth at 250 K and 260 K. Furthermore, to assess the effect of ionic strength alone on bulk ice growth, we consider three ice-aqueous solutions systems at 250 K, with dissolved $Ca(OH)_2$ in the solution at concentrations of 0.2, 0.4, and 0.6 mol/L, which correspond to concentration in dissolved Ca^{2+} and OH^- of 0.6, 1.2 and 1.8 mol/L, respectively. The estimated growth rates are displayed in Fig. 9.16 (confined without impurities) and Fig. 9.18 (bulk $Ca(OH)_2$ solutions). The results reveal that the confined ice growth rates without impurities at 250 K and 260 K are 0.723 Å/ns and 0.506 Å/ns, respectively, i.e., significantly higher than the results with surface impurities (0.282 Å/ns and 0.394 Å/ns respectively), but remain lower than the corresponding bulk cases (0.998 Å/ns and 0.828 Å/ns respectively). These results show that both confinement and surface impurities contribute to slowing down the growth of ice. Figure 9.18 indicates that the bulk ice growth rate can be slowed down by increasing the ion concentration of the liquid, yet the decrease remains small (i.e., lower than 3%) for $Ca(OH)_2$ concentrations up to 0.6 mol/L. Given the very high concentration of absorbed and free Ca^{2+} and OH^- ions in the confined liquid within our C-S-H pore model ($Ca(OH)_2$ approximately 1.8 mol/L in the lower film before crystallization), a significant impact of on the growth rate seems possible.

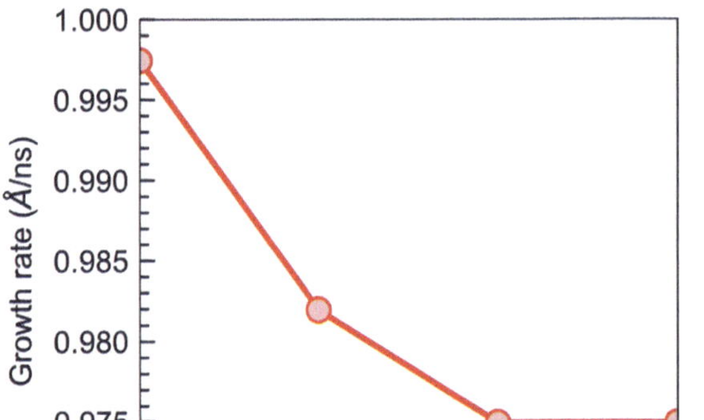

Fig. 9.18 The effect of aqueous $Ca(OH)_2$ concentration on the growth rate of ice in the bulk at 250 K. The concentration in dissolved Ca^{2+} and OH^- is 3 times as large as the concentration in $Ca(OH)_2$

9.3.4 Confinement Effect on the Melting Point

In this section, we show that confinement can also alter the temperature of ice-liquid coexistence in addition to affecting the crystallization kinetics. Hereafter, we start by revisiting the formulation of the coexistence law of unconfined bulk ice and liquid, namely Thomson's law. This reformulation adopts a statistical physics point of view, so that all quantities involved can be evaluated by molecular simulation and thus applied to our study.

For a liquid containing N water molecules at controlled pressure P and controlled temperature T, the thermodynamic potential minimum at equilibrium is the Gibbs free enthalpy $G(N, P, T) = E - S + PV$ with E the internal energy, S the entropy, and V the volume. The associated energy balance is:

$$dG = VdP + \mu dN - SdT \tag{9.10}$$

where μ is the chemical potential. The Gibbs free enthalpy is extensive with respect to the number of water molecules so that $G = \mu N$. Combined with Eq. 9.10, one obtains the Gibbs–Duhem equation:

$$Nd\mu = VdP - SdT \tag{9.11}$$

When the liquid is near its melting point, assuming constant specific volume $v = V/N$ and specific entropy $s = S/N$, thermodynamic integration from the melting point $T_m(P_0)$ at a reference pressure P_0 gives:

$$\mu(T, P) = \mu_m(T_m(P_0), P_0) + v(P - P_0) - s(T - T_m) \tag{9.12}$$

where μ_m is the chemical potential at the melting point.

In the molecular simulation, one does not control T but $\beta = 1/(kT)$ with k the Boltzmann constant. Likewise, one controls βP instead of P and $\beta\mu$ instead of μ. The energy balance of G (Eq. 9.10) can be rewritten in a statistical physics form as follows:

$$d(\beta G) = Vd(\beta P) + (\beta\mu)dN + Ed\beta \tag{9.13}$$

with $E = (G + TS - VP)$ the internal energy. From this reformulated energy balance, one can rewrite the Gibbs–Duhem equation in a form adapted to statistical physics:

$$Nd(\beta\mu) = Vd(\beta P) + Ed\beta \tag{9.14}$$

Assuming a constant specific volume v and a specific internal energy $e = E/N$, integration from the melting point $\beta_m = 1/(kT_m)$ at a reference pressure P_0 gives:

$$\beta\mu(\beta, \beta P) = \beta_m\mu_m(\beta_m, \beta_m P_0) + v(\beta P - \beta_m P_0) + e(\beta - \beta_m) \tag{9.15}$$

Coming back to the usual thermodynamic quantities, Eq. 9.15 gives:

$$\mu(T, P) = \frac{T}{T_m}\mu_m(T_m, P_0) + v\left(P - \frac{T}{T_m}P_0\right) + e\left(1 - \frac{T}{T_m}\right) \tag{9.16}$$

From our simulation of supercooled liquid water (subscript l), considering the specific internal energy e_l constant appears reasonably valid, but the specific volume v_l appears highly pressure-dependent. Applying Eq. 9.16 while taking into account the pressure dependence of the specific volume v_l or not can yield significantly different results, especially at high pressures. Here, considering this dependence but restricting ourselves to a first-order pressure dependence $v_l = A_l(P - P_0) + v_{l0}$, integration yields:

$$\beta\mu_l(\beta, \beta P) = \beta_m\mu_m(\beta_m, \beta_m P_0) + \int_{\beta_m P_0}^{\beta P} v_l(\beta^*, (\beta P)^*)d(\beta P)^* + \int_{\beta_m}^{\beta} e_l(\beta^*, (\beta P)^*)d\beta^* \tag{9.17}$$

After some calculations, we obtain:

$$\beta\mu_l(\beta, \beta P) = \beta_m\mu_m(\beta_m, \beta_m P_0) + v_{l0}(\beta P - \beta_m P_0) + \beta\frac{A_l}{2}(P - P_0)^2 + e_l(\beta - \beta_m) \tag{9.18}$$

Again, returning back to the usual thermodynamic quantities, we have:

$$\mu_l(T, P) = \frac{T}{T_m}\mu_m(T_m, P_0) + v_{l0}\left(P - \frac{T}{T_m}P_0\right) + \frac{A_l}{2}(P - P_0)^2 + e_l\left(1 - \frac{T}{T_m}\right) \tag{9.19}$$

Likewise for an ice crystal (subscript c), assuming $v_c = A_c(P - P_0) + v_{c0}$, the chemical potential at controlled pressure P and controlled temperature T is derived as:

$$\mu_c(T, P) = \frac{T}{T_m}\mu_m(T_m, P_0) + v_{c0}\left(P - \frac{T}{T_m}P_0\right) + \frac{A_c}{2}(P - P_0)^2 + e_c\left(1 - \frac{T}{T_m}\right) \tag{9.20}$$

Equations 9.19 and 9.20 are the chemical potentials of liquid and ice expressed from a statistical physics point of view. Considering now the coexistence of liquid water with an ice crystal, the osmotic and mechanical equilibria require that both phases share the same chemical potential, temperature, and pressure, i.e., $\mu_l = \mu_c =$

μ, $T_l = T_c = T$, and $P_l = P_c = P$. Substituting these equalities in Eqs. 9.19 and 9.20, we obtain:

$$(e_l - e_c)\left(1 - \frac{T}{T_m}\right) = (v_{c0} - v_{l0})\left(P - \frac{T}{T_m}P_0\right) + \frac{A_c - A_l}{2}(P - P_0)^2 \quad (9.21)$$

Equation 9.21 is the statistical physics form of Thomson's law, relating the melting temperature of ice to the pressure. The practical interest of this formulation compared to the usual one (Eq. 9.12) is that all the properties involved (i.e., specific volumes v_{l0}-A_l and v_{c0}-A_c, and specific energies e_l and e_c) can be easily obtained from molecular simulation. To apply Thomson's law, a refence melting point is needed: the melting temperature of ice Ih using TIP4P/Ice model at $P_0 = 1$ bar is $T_m = 272.2$ K [24]. Then, we determine the different liquid and ice properties by carrying out MD simulations of bulk liquid and ice at 1 bar (i.e., 0.1 MPa), 50, 100, 150 and 200 MPa (with 1 bar the reference pressure P_0). As shown in Fig. 9.19, the specific volumes versus pressure data align well on the linear fitting lines. From our simulation, A_l and A_c are 8.29×10^{-15} m^3·mol^{-1}·Pa^{-1} and 9.86×10^{-16} m^3·mol^{-1}·Pa^{-1}, respectively, $v_{c0} - v_{l0}$ and $e_l - e_c$ are 1.37×10^{-6} m^3·mol^{-1} and 4987.65 J·mol^{-1}, respectively. The predicted melting line of ice Ih using Eq. 9.21 is presented in Fig. 9.20, along with the results from MD simulations by Abascal et al. [24]. Thomson's law predictions are well consistent with the MD simulation results. The small shift between the melting line of the predictions/simulations and of the experiments is mainly due to the imperfection of the TIP4P/Ice water potential model. Therefore, one can safely exploit the statistical physics form of Thomson's law of Eq. 9.21 to characterize or predict melting points of ice.

To show the effect of confinement on the liquid-ice coexistence, the melting points of the systems confined in C-S-H at 0.1, 50, 100, 150, and 200 MPa are estimated by performing a series of NPT simulations at various temperatures close to the melting temperature, and monitoring the evolution of the potential energy. An example is given in Fig. 9.21. A decrease in potential energy is the signature

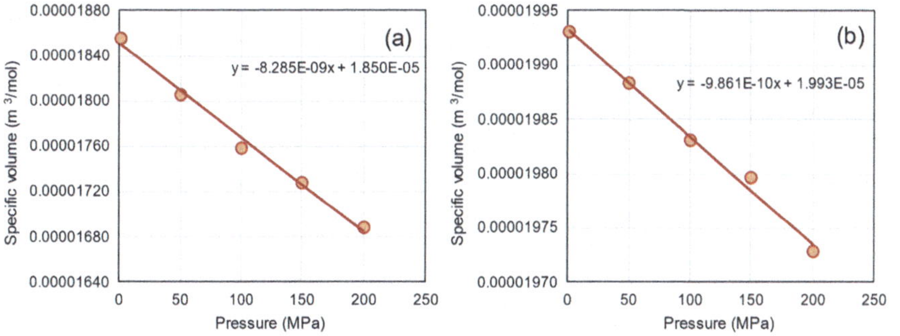

Fig. 9.19 The specific volume of liquid **a** and ice **b** as a function of pressure. The standard deviations of the specific volumes are smaller than 1.6×10^{-8} m^3/mol

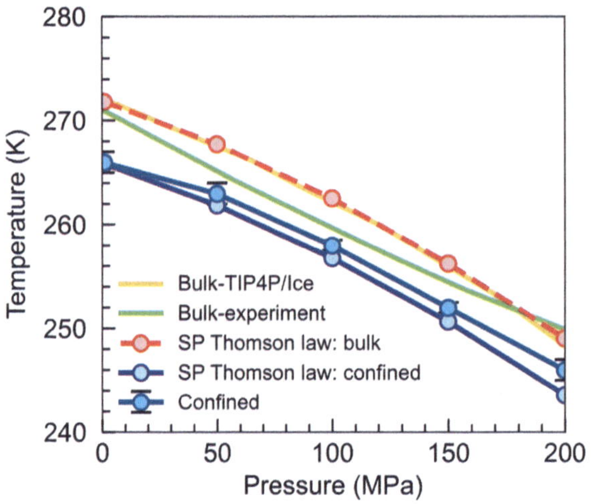

Fig. 9.20 Melting temperatures of ice as a function of pressure. The dark blue line presents the melting temperatures of ice Ih in the current slit nanopore system. The red line exhibits the analytical melting temperatures predicted from the statistical physics (SP) form of Thomson's law (Eq. 9.21). The yellow line shows the melting line of TIP4P/Ice ice Ih, which is obtained from molecular simulation by Abascal et al. [24]. The green line illustrates the melting line of ice determined from experiments by Tulk et al. [71]

of crystallization ($T < T_m$), while an increase corresponds to a melting ($T > T_m$). Thus, observing the evolutions of the potential energy for the different temperatures provides an estimate of the melting temperature T_m (263 ± 1 K in the case illustrated in Fig. 9.21). The estimated melting curve of ice confined in C-S-H is displayed in Fig. 9.20 and compared with the melting curve of bulk ice. We find that the melting temperatures of confined ice Ih are systematically lower than that of the bulk ice Ih, with temperature differences of 4 to 6 K. It is interesting that the trend of the melting line of the confined system is similar to the one of the bulk system. As shown in Fig. 9.20, Eq. 9.21 also well predicts the melting line of the confined case, which indicates that Thomson's law could apply to the melting line of the confined system as well. Confinement-induced reduction in melting temperature has been extensively reported in the literature for many other systems with geometrical constraints [64, 65]. Compared to the bulk phase, the equilibrium point of confined ice-liquid coexistence is highly dependent on the pore size [66, 67]. For a cylindrical or spherical pore system, this shift is usually ascribed to the Gibbs–Thomson effect. In the case of the confined cylindrical or spherical system, the depression of the melting point can be related to the pore width either by balancing the free energies of ice and liquid or by considering the temperature at which the confined ice and the bulk reservoir share the same chemical potential, yielding the following expression for the melting point depression [68]:

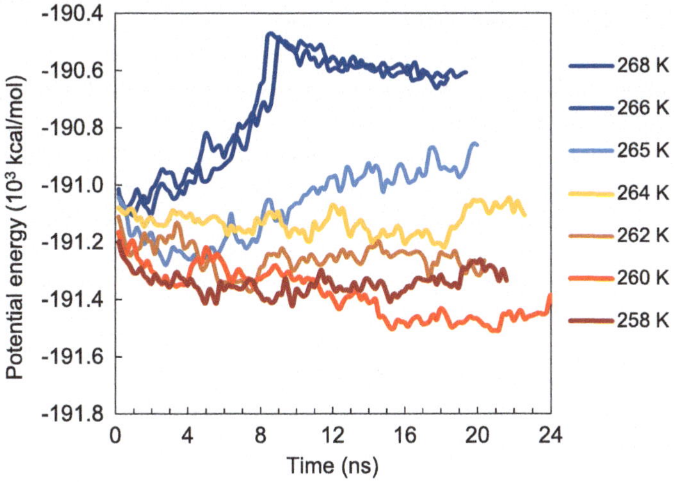

Fig. 9.21 Potential energy curves of the confined systems at 50 MPa and various temperatures. It shows the evolution of the potential energy of the confined system submitted to a pressure of 50 MPa in the z direction. We are interested in the domain of temperatures for which the potential energy neither increases (melting), nor decreases (crystallization), which provides an estimate of the melting temperature. The final melting temperature is determined as the average value of the temperatures satisfying the above conditions. For the example shown below, the melting temperature is estimated at 263 ± 1 K

$$\Delta T_f = T_{f,pore} - T_{f,bulk} = -2\frac{(\gamma_{ws} - \gamma_{wf})\upsilon_l}{H\lambda_{f,bulk}} \tag{9.22}$$

where γ_{ws} and γ_{wf} are the interface energies of wall-ice and wall-liquid, H is the pore width of the confined system, and $\lambda_{f,bulk}$ denotes the melting enthalpy of ice. The difference between the interface energy of wall-ice γ_{ws} and wall-liquid γ_{wf} gives rise to the shift of melting point in the confined system with respect to the bulk phase. For the planar pores which have no effect of ice-liquid interfacial curvature, the interfacial energies still act an essential role in the melting point depression. The impact of interfacial energies on the freezing depression comes from two sources, i.e., the dispersion forces acting between ice/liquid/solid interfaces and the immobile charges (absorbed ions) at the pore wall interface [69]. A strong evidence for the effect of dispersion forces is the melting point depression of the LJ liquid in the planar nanopore [70]. Another reason for the shift of melting point is due to the mobile ions in the liquid. The Ca^{2+} concentration in the premelted film is ~ 5 mol/L at the end of ice growth. According to Zhou et al., [13], increasing the ion concentration of liquid can lead to a substantial melting point depression in the parallel-slit pore.

9.3.5 Premelted Water Film

The configurations of Fig. 9.22b show the presence of premelted water nanofilms between the C-S-H solids and ice, which persist once equilibrium is reached. Likewise, for ice crystallization on a solid surface without the geometrical confinement, experiments and simulations in the literature [72] have captured a quasi-liquid layer between ice and the solid surface (see Fig. 9.22a, b). The thickness of these premelted water nanofilms is of special interest as they are potential paths to transport water from unfrozen pores to freezing pores, which is important for cryo-suction to occur. In particular, we examine here how the film thickness is affected by confinement and the nature of the surface. Using the interface detection algorithm, we calculate the thickness of the premelted films after the confined systems have reached equilibration. The results are also compared with the non-freezable water layer equation proposed by Fagerlund, from water vapor adsorption on cement [73]:

$$t = 19.7 \times 10^{-10} \sqrt[3]{\frac{1}{T_m - T}} \tag{9.23}$$

where t is the thickness of the layer expressed in Å, $T_m - T$ is the freezing depression expressed in K. More precisely, the law proposed by Fagerlund applies to adsorbed water on cement surface at equilibrium with a vapor at the relative humidity for which capillary condensation takes place for the same pore size as melting. Regarding the

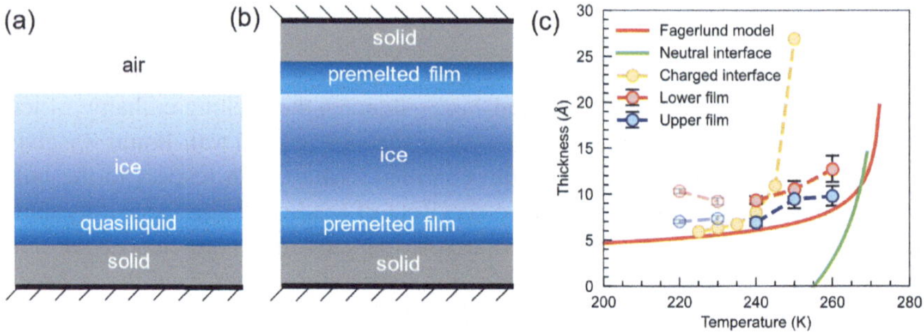

Fig. 9.22 Scenario for interfacial melting without **a** and with **b** confinement, and the thickness of premelted water films at varying temperatures **c**. To obtain the thickness of premelted water nanofilms at 220–230 K, we first ran the simulations at 250 K for 212 ns, then continued to run the simulations at the two very low temperatures for over 280 ns. But the ice growth rates at the two lowest temperatures (220–230 K) are extremely low, and the associated premelted films might not be fully equilibrated after 400 ns, so that the corresponding results should be considered with caution (reduced opacity). The solid red line represents the Fagerlund model [73] proposed to characterize the thickness of vapor adsorption on surfaces of porous materials. The solid green line characterizes the thickness of the premelted film on a neutral SiO_2 surface, measured by Engemann et al. [72]. The dashed yellow line is the thickness of the premelted film on charged tobermorite interface with surface calcium cations obtained with MD simulations by Liang et al. [19]

thickness of the quasi-liquid layer observed for ice on surfaces, Engemann et al. [72] proposed a logarithmic law, well validated for neutral silica surfaces:

$$t(T) = a(0)\ln\frac{T_m - T_0}{T_m - T} \tag{9.24}$$

where a(0) is a constant characterizing the decay length of the non-ordering density of the quasi-liquid layer as measured at the in-plane momentum transfer $q_{||} = 0$, $T_0 = T_m - 17$ K is the temperature below which the quasi-liquid film vanishes, and T_m represents the melting temperature of the bulk ice.

Fig. 9.22c shows the thicknesses of the premelted water nanofilms estimated for our confined systems as a function of the temperature, compared to the predictions using the Fagerlund and Engemann models, and compared also to other MD results for ice at the surface of charged tobermorite [19]. In our confined system, the thicknesses of the upper premelted water nanofilm at 220 K, 230 K, 240 K, 250 K, and 260 K are 7.03 ± 0.22 Å, 7.36 ± 0.33 Å, 6.94 ± 0.39 Å, 9.53 ± 1.06 Å, 10.76 ± 0.96 Å, respectively. As for the lower nanofilms, the thicknesses are slightly thicker than those of the upper nanofilms: 10.35 ± 0.26 Å, 9.24 ± 0.38 Å, 9.37 ± 0.43 Å, 10.60 ± 0.83 Å, and 11.70 ± 1.87 Å, respectively. Since the water mobility at 220–230 K is nearly frozen, the confined ice might not reach complete equilibration despite the simulation being run for ~ 0.5 μs: the results for those two temperatures should be considered with caution (points displayed with reduced opacity in Fig. 9.22c). Nonetheless, one can still confidently conclude from our simulations that the thickness of the premelted water nanofilm increases with the temperature. Interestingly, the Fagerlund model [73] derived for surface adsorption of water vapor on cement captures reasonably well the general trend of the estimated thicknesses with temperature. Such a similarity between the thickness of adsorbed vapor and premelted nanofilm is remarkable, suggesting that the thickness of liquid water films on cement at a given supersaturation is more or less the same irrespective of the nature of the mother phase of water (ice or vapor). We note that our estimated thicknesses are systematically slightly higher than the Fagerlund model prediction. This discrepancy could be due to the presence of impurities, e.g., calcium cations and hydroxyl groups, in the premelted film [19].

As for the model of Engemann et al. [72], the predicted thicknesses are much smaller than our estimated thicknesses (Fig. 9.22c). This model addresses the case of an unconfined quasi-liquid layer formed at the interface of ice and silica. A similarity, though, between our estimates for confined nanofilms and this model for unconfined nanofilm is that both increase sharply as the temperature approaches the melting point. Apart from this aspect, the large discrepancy between our estimates and this model suggests a fundamental difference in the nature of the molecular interactions, composition, and/or structures within the nanofilm. As shown in Fig. 9.22c, our estimated confined thicknesses are greater than 5 Å at all considered temperatures, whereas the premelted film on the SiO_2 neutral interface, observed with a high-energy x-ray transmission-reflection experiment, vanishes as the supercooling exceeds 17 K, i.e., at temperatures lower than ~ 256 K [72]. Other MD simulations for the interfacial melting of ice on the charged tobermorite surface confirm that the premelted film still exists at temperatures down to 225 K [19]. An important difference between silica and charged tobermorite is the presence of calcium ions on the tobermorite surface, which is expected to have a major influence on the structuration of surface water and could well explain the persistence of a premelted film. The thickness of the premelted film on charged tobermorite surface is quite close to our estimation for ice confined in C-S-H at temperatures lower than 245 K. However, a significant difference appears at 250 K, with a much larger thickness of the premelted film on charged tobermorite than in the C-S-H slit pore that we considered.

9.4 Transportation of Liquid in Premelted Film

9.4.1 *Thermodynamic-Based Estimation for Cryo-Suction Process*

In this section, we consider a cylindrical pore of radius r and length L saturated with frozen water and connected to a reservoir of liquid water at the pressure p_∞ at one of its extremities, its other extremity being closed. The liquid water can flow through a premelted liquid water film of thickness w at the interface between ice and the solid. A schematic diagram is illustrated in Fig. 9.23. We consider that the temperature T is sufficiently low for the pore to be filled with ice and that we waited long enough to reach thermodynamic equilibrium, at which point the pressure of the ice crystal is p_I. A small decrease $\Delta T < 0$ of temperature will bring the system out of equilibrium. Cryo-suction of the liquid in the premelted film will make it possible to bring the system back to thermodynamic equilibrium. In this section, we aim to estimate the characteristic time required to come back to thermodynamic equilibrium.

Integrating Gibbs–Duhem equation, the molar chemical potential $\mu_L(p_L, T)$ (in J·mol^{-1}) of the liquid water and $\mu_I(p_I, T)$ (in J·mol^{-1}) of the ice can be expressed as [15]:

$$\mu_L(p_L, T) = \mu_\infty + (p_\mathrm{L} - p_\infty)\overline{V_\mathrm{L}} + \overline{S_\mathrm{L}}(T_{\mathrm{m}\infty} - T) \tag{9.25}$$

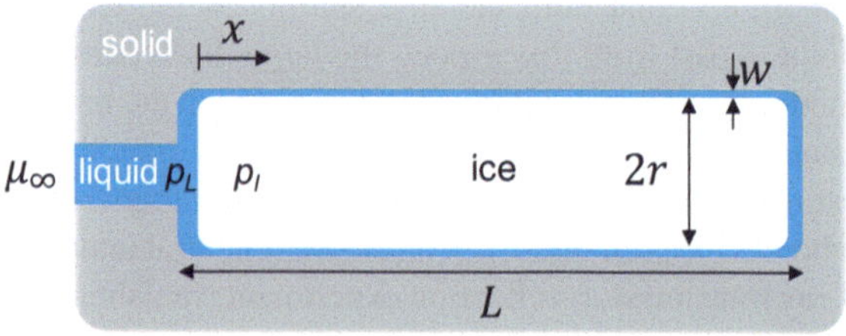

Fig. 9.23 Schematic diagram of the system considered for the thermodynamic-based estimation of characteristic time of cryo-suction process

$$\mu_I(p_I, T) = \mu_\infty + (p_I - p_\infty)\overline{V_I} + \overline{S_I}(T_{m\infty} - T) \tag{9.26}$$

where μ_∞ is the reference chemical potential corresponding to the water melting point at temperature $T_{m\infty}$ and pressure p_∞, $\overline{V_L}$ and $\overline{V_I}$ are the molar volumes (in $m^3 \cdot mol^{-1}$) of liquid water and ice, respectively, and $\overline{S_L}$ and $\overline{S_I}$ are the molar entropies (in $J \cdot K^{-1} \cdot mol^{-1}$) of liquid water and ice, respectively.

We consider a decrease of temperature $T < T_{m\infty}$, while the liquid pressure is unchanged $p_L = p_\infty$. At thermodynamic equilibrium, the chemical potential μ_I of ice must be equal to the chemical potential μ_L of the liquid water in the reservoir, i.e.:

$$\mu_L(p_\infty, T) = \mu_I(p_I, T) \tag{9.27}$$

from which we obtain that the pressure p_I of the ice must increase to satisfy Thomson's equation [15]:

$$p_I - p_\infty = \Delta S(T_{m\infty} - T) \tag{9.28}$$

where $\Delta S = \left(\overline{S_L} - \overline{S_I}\right)/\overline{V_I} > 0$. This equation shows that, at temperatures below the melting temperature, the pressure of the ice crystal at thermodynamic equilibrium must be larger than the pressure of the liquid water surrounding it.

Let us now decrease instantaneously the temperature from the actual temperature T by an amount $\Delta T < 0$, while keeping the pressure p_L of the liquid water equal to p_∞. The chemical potential of the liquid water varies by:

$$\Delta\mu_L = -\overline{S_L}\Delta T \tag{9.29}$$

If during this instantaneous decrease of temperature, the pressure p_I of the ice would remain constant, the chemical potential of the ice would vary by:

$$\Delta\mu_I = -\overline{S_I}\Delta T < \Delta\mu_L \tag{9.30}$$

Since the volumetric thermal expansion of ice (which is on the order of $150 \times 10^{-6}\ K^{-1}$ [74]) is larger than that of C-S-H (which is on the order of $42 \times 10^{-6}\ K^{-1}$ [75]), during the instantaneous decrease of temperature the ice will contract more than the surrounding solid, which will decrease the pressure p_I of the ice and further lower its chemical potential. In any case, the decrease of temperature brings the system out of thermodynamic equilibrium.

For the thermodynamic equilibrium to be reached again, the chemical potential of the ice must increase back to the chemical potential $\mu_L(p_\infty, T + \Delta T)$ imposed by the liquid reservoir, which will occur by an increase in the pressure p_I of the ice. For this pressure to increase, the crystal must grow, i.e., gain in mass, which can only happen by bringing more water molecules to the surface of the crystal along the premelted liquid film, i.e., by cryo-suction.

We associate an x-axis to the pore, such that $x = 0$ stands for the pore extremity in contact with the liquid reservoir and $x = L$ stands for the closed extremity of the pore. We note e the thickness of the layer of unfrozen water at the pore surface and assume that $w \ll r$.

At $t = 0^+$, right after the decrease in temperature, we note μ_0 the chemical potential of the ice, i.e.:

$$\mu_I\left(t = 0^+, x\right) = \mu_0 \tag{9.31}$$

At the pore extremity in contact with the liquid reservoir, we have:

$$\mu_I(t, x = 0) = \mu_0 + \Delta\mu \tag{9.32}$$

where $\Delta\mu > 0$ is the difference of chemical potential between the ice and the liquid water right after the instantaneous decrease of temperature. The imbalance of chemical potential along the x direction causes diffusion of water along the premelted film. Assuming local equilibrium between the film and the ice (i.e., same chemical potential), and assuming that the diffusion of the unfrozen water is governed by Fick's law [15], the flow of water is given by:

$$j = -\frac{D\rho_L}{\overline{M_w}RT}\nabla\mu_I = -\frac{D\rho_L}{\overline{M_w}RT}\frac{\mathrm{d}\mu_I}{\mathrm{d}x} \tag{9.33}$$

where j is the molar flow vector (in mol·m^{-2}·s^{-1}), D is a diffusion coefficient (in m^2·s^{-1}), $\overline{M_w}$ is the molar mass of water (in kg·mol^{-1}), and R is the ideal gas constant (in J·K^{-1}·mol^{-1}). Although the diffusion coefficient D in this equation should be the water collective diffusion coefficient, we will approximate it by the water self-diffusion coefficient (since the collective diffusion is generally faster than the self-diffusion, we provide here a lower estimate of the time of cryo-suction).

We note $M(x)$ the mass of water in the pore between x and $x + dx$. During a small time interval dt, the mass balance can be expressed as:

$$\begin{aligned} dM &= 2\pi\, rw\,(j(x) - j(x + dx))\overline{M_w}\,dt \\ &= -2\pi\, rw\frac{dj}{dx}dx\,\overline{M_w}\,dt = -2\pi\, rw\;div\,(j)\;dx\;\overline{M_w}\,dt \end{aligned} \tag{9.34}$$

If the ice crystal were free to grow radially, a small increase dM of mass would translate into a small increase dr of radius through:

$$dM = 2\pi r\,dx\,dr\,\rho_I \tag{9.35}$$

and therefore:

$$dr = -\frac{w\overline{M_w}}{\rho_I}\;\mathrm{div}\,(j)\;dt \tag{9.36}$$

Since ice is in fact constrained by its surroundings, the mass increase will translate into an increase dp_I of the ice pressure on the order of:

$$dp_I \sim K_{hom}\frac{dr}{r} \tag{9.37}$$

where K_{hom} is a modulus that combines the bulk modulus K_I of the ice and K_s of the C-S-H solids. One can estimate the magnitude of K_{hom} based on the Reuss limit: $K_{hom} \sim \frac{1}{(1/K_I+1/K_s)}$, which is a rough estimate as it does not consider the volume fractions of pores and solids or the pore geometry. This increase in pressure will translate into an increase $d\mu_I$ of chemical potential:

$$d\mu_I = \overline{V_I}dp_I = \overline{V_I}K_{hom}\frac{dr}{r} = -\frac{\overline{V_I}K_{hom}w\overline{M_w}}{\rho_I r}\text{div}(j)dt \tag{9.38}$$

By combining this equation with Fick's law (9.32), we obtain the differential equation that governs the chemical potential $\mu_I(x, t)$ of the ice crystal in the pore, i.e., that governs the cryo-suction process:

$$\frac{\partial \mu_I}{\partial t} = \frac{\overline{V_I}K_{hom}w\overline{M_w}}{\rho_I r}\frac{D\rho_L}{\overline{M_w}RT}\Delta\mu_I = \frac{\overline{V_I}K_{hom}w}{\rho_I r}\frac{D\rho_L}{RT}\frac{\partial \mu_I}{\partial x^2} \tag{9.39}$$

This equation is a diffusion equation of characteristic time τ:

$$\tau \sim \frac{RT}{\overline{V_I}K_{hom}D}\frac{r}{w}L^2 \tag{9.40}$$

where we omitted the liquid and ice densities as they almost cancel out. One can estimate this characteristic time τ of the cryo-suction process for pores with a generic aspect ratio and in particular for pores with an aspect ratio close to 1 by considering $L = r$.

9.4.2 *Diffusivity in the Premelted Water Films*

As one can see from Fig. 9.12, two premelted liquid films at the lower and upper sides between C-S-H wall and ice solids are formed after ice crystallization. From our simulations, the thickness of the premelted water film is around 1 nm and slightly decreases when the temperature decreases from 260 to 220 K (see Fig. 9.22). The surface impurities including Ca^{2+} cations and hydroxyl groups are mostly rejected from the ice to the liquid films. Note that the thicknesses of the lower and upper films are not exactly equal at the same temperature. Given that the number of Ca^{2+} ions and OH^- groups are the same on the two sides of the ice, such a difference is perhaps due to the surface roughness of C-S-H solid walls since the silicate chains are highly defective. The different surface conditions can lead to differences in the hydrogen bonding, dipolar and multipolar interactions, and electrostatic forces felt by the water molecules, which yields a different interfacial energy [76]. Such differences in surface conditions could explain why the thickness of the two premelted films differs.

The premelted liquid film offers a path to transport liquid water toward the ice crystal during the freezing process. The main target of this study is to quantify the transport properties of the premelted liquid films. Herein, the mean square displacement (MSD) was calculated to capture the self-diffusion of water molecules inside the premelted liquid films, which is defined as follows for particle i:

$$MSD_i(t) = \left\langle |r_i(t+\tau) - r_i(\tau)|^2 \right\rangle \tag{9.41}$$

where r_i is the position of the water molecule i or of the cation i and where $f(\tau)$ indicates an average of $f(\tau)$ over multiple times τ.

To illustrate the effect of temperature on the motion of water molecules in the premelted films, Fig. 9.24 exhibits the MSD curves obtained in the NVE ensemble simulation at the various temperatures from 220 to 260 K. One can identify four typical stages [77–79] of diffusion in the premelted film, the time duration of which depends on the temperature. Within the very first 200 femtoseconds, the water molecules move just ballistically (i.e., MSD $\propto t^2$). In contrast, the time duration of the ballistic stage of nanoconfined C-S-H water at 300 K is just within 100 fs [77]. Following the ballistic stage, a plateau associated with the cage stage is seen in the MSD curves. The appearance of the MSD plateau means that water molecules are trapped in cages with limited mobility. The origin of the water cage is ascribed to the neighboring particles of a water molecule, which collide with each other [80]. In the present system, the water molecules, hydroxyl groups, and Ca^{2+} cations in the premelted films can all contribute to the formation of the cages. Note that all particles inside the cages are trapped in the cages as well. As a consequence, the mobility of all particles is limited at this stage; all particles need time to escape from the cages. Also, note that the time duration of this cage stage depends on temperature significantly. At 220 K and 230 K, dozens or even hundreds of picoseconds are required to escape from the cages (see Fig. 9.24b), while the cage stage at 260 K only lasts for a few picoseconds. The stepwise MSD curve of the upper film at 220 K further reflects that the water molecules escaping from the initial sites can be re-trapped in the cages. Typically, the radius of the cage containing the water molecules is around 2 Å, meaning that the water molecules in the upper film need more than 50 ns to escape from their cages when the temperature is 220 K (see Fig. 9.24b). Right after the cage stage, water molecules in the premelted films start to diffuse at a sublinear rate (i.e., MSD $\propto t^\alpha, \alpha < 1$), which is interpreted as the transition stage of motion. Following this transitional state, water molecules begin to diffuse linearly in the premelted films (i.e., MSD $\propto t$). The Ca^{2+} cations in the premelted films are subjected to a caging effect stronger than the water molecules. As illustrated in Fig. 9.25, the MSD curves of Ca^{2+} cations are more stepwise than those of water molecules, exhibiting several plateaus before escaping from the cages. The time duration of the ballistic stage is ~ 2 ps, which is one order of magnitude longer than that of water molecules. Note that the Ca^{2+} cations diffuse in a particular way, with sudden skipping from the cages when they move forward, unlike the progressive motion of water molecules. Such a diffusion behavior of Ca^{2+} cations is due to the hydration shell surrounding them, which limits the motion of ions. In addition, surface Ca^{2+} cations are charge-compensating the C-S-H solids, and are thus strongly attracted by the C-S-H surfaces which limits their mobility. Therefore, the MSDs of Ca^{2+} cations are significantly lower than those of water molecules in the premelted films.

Given that the premelted films are ultra-thin with a thickness of around 1 nm, the water diffusion inside this liquid film can be considered a quasi-two-dimensional motion. The diffusion perpendicular to the film is significantly lower than the diffusion in the plane of the film (i.e., in the x and y directions), as shown in Fig. 9.26. Therefore, we focus on the in-plane MSD, noted MSD_{xy}, which can be calculated using the following equation:

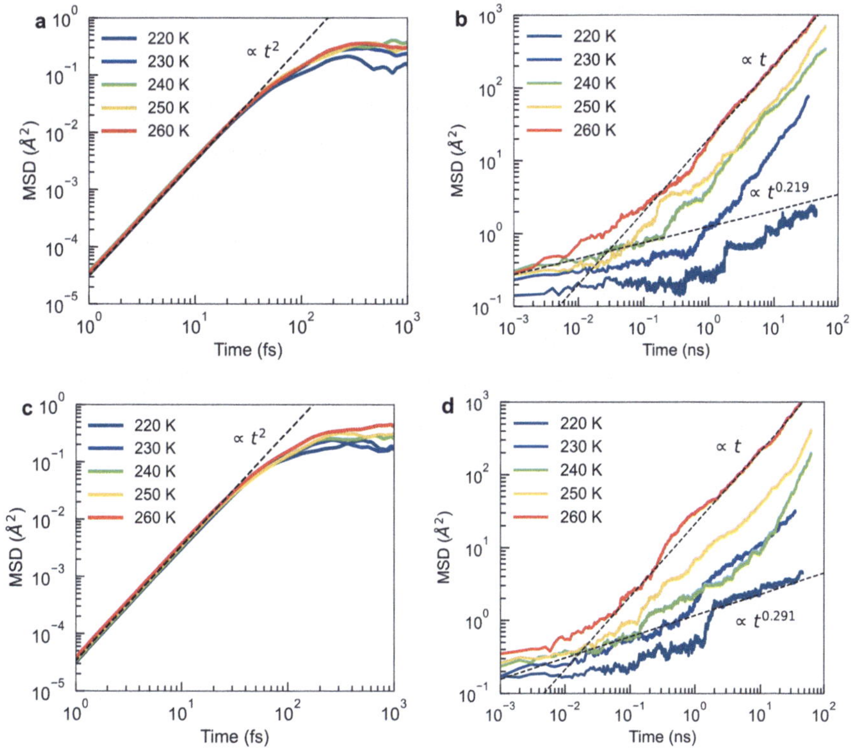

Fig. 9.24 MSD curves of water molecules in the lower **a**, **b** and upper **c**, **d** premelted water films. **a** and **c** show the MSDs within the very first 1000 fs. **b** and **d** plot the MSDs within the full duration. The fitting lines in panels **a** and **c** are for the MSD curves at 220 K. The linear fits (i.e., MSD $\propto t$) in panels **b** and **d** correspond to the MSD curves at 260 K. The sublinear fits (i.e., MSD $\propto t^{\alpha}, \alpha < 1$) in panels **b** and **d** correspond to the curves observed at 240 K

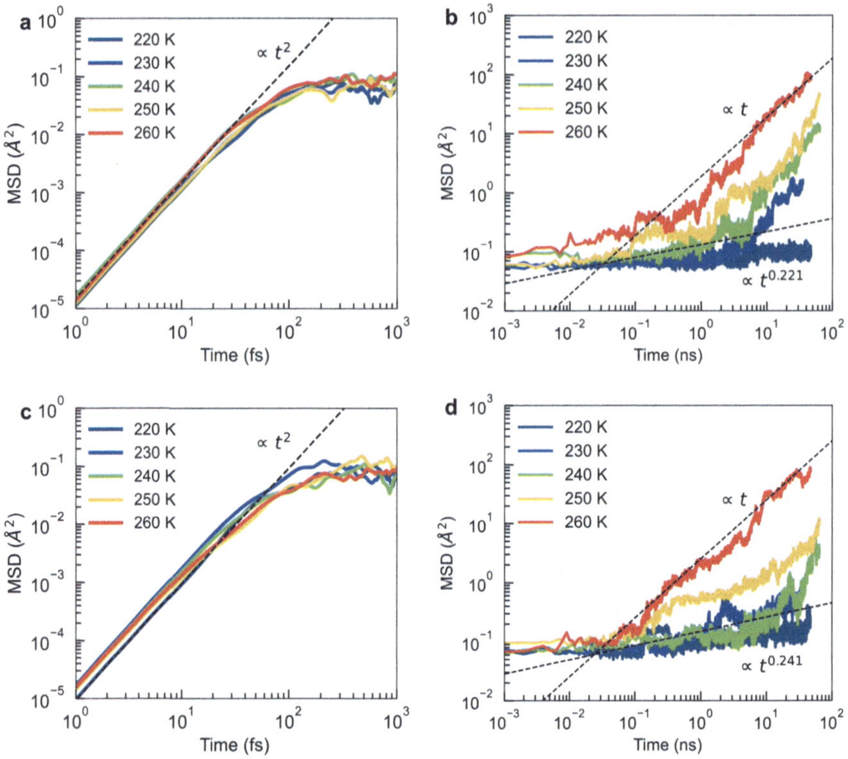

Fig. 9.25 MSD curves of Ca^{2+} cations in the lower **a**, **b** and upper **c**, **d** premelted water films. **a** and **c** show the MSDs within the very first 1000 fs. **b** and **d** plot the MSDs within the full duration. The dashed fitting lines in panels **a** and **c** are for the MSD curves at 220 K (almost valid for all cases). The linear fits in panels **b** and **c** correspond to the MSD curves at 260 K. The dashed-line fits in panels **b** and **c** correspond to the curves observed at 240 K

$$MSD_{xy} = \frac{1}{N}\sum_{i=1}^{N}\left(\Delta x_i^2 + \Delta y_i^2\right) \tag{9.42}$$

where N represents the number of water molecules (or of Ca^{2+} cations), x_i and y_i are the displacements of the i-th water molecule (or of the i-th Ca^{2+} cation) in the x and y directions. With the in-plane displacement, the corresponding 2D self-diffusivity can be estimated using the Einstein relation:

$$D_{xy} = \frac{1}{4}\lim_{t\to\infty}\frac{MSD_{xy}(t)}{t} \tag{9.43}$$

The temperature of the system is examined before analyzing the MSD curves. To avoid perturbating the diffusion process, the MSDs are computed during the NVE simulation, and temperature is not controlled. As shown in Fig. 9.27, some drift in

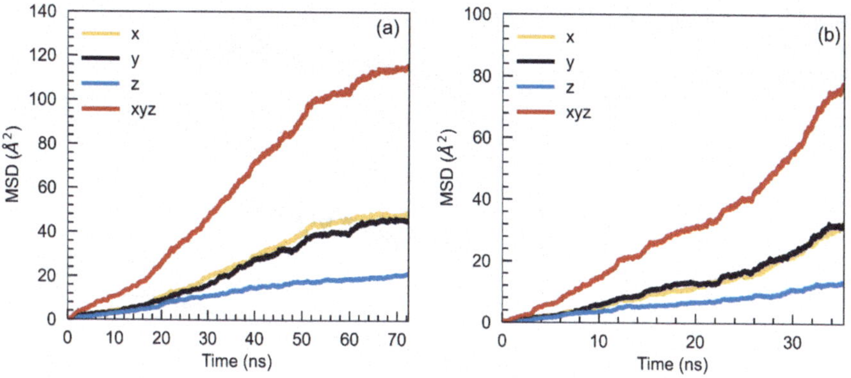

Fig. 9.26 Mean square displacement curves of water molecules in the lower premelted film at 260 K (**a**) and 250 K (**b**)

temperature is observed over time (approximately 5 K over 30 ns), which corresponds to a slight increase of kinetic energy during the NVE runs. It means that the systems still continue to equilibrate. Since self-diffusivity is sensitive to temperature, we limited the evaluation of the MSDs to a time frame of 30 ns so as to avoid excessive drift in temperature. Nevertheless, for the 220 K case, this time frame was enlarged to 45 ns so that the MSD could reach a sufficient distance. The MSD curves for water molecules are divided into three segments (each lasting 10 ns) within the specified time frame. This division allows for the computation of average self-diffusivities, accompanied by their respective standard deviations. In contrast, due to the greater variability in the MSDs of Ca^{2+} cations, average self-diffusivities are determined based on three complete MSD profiles within the specified time frame. Following Eq. 9.43, the self-diffusivities of water molecules and Ca^{2+} cations in the premelted films are shown in Table 9.7 and Fig. 9.28. The self-diffusivities of bulk TIP4P/Ice water reported by Espinosa et al. [51] and Weiss et al. [53] are also displayed for comparison.

From our simulations, the self-diffusivities of water molecules in the premelted films are lower than those of bulk water at the same supercooling by a factor of 1.6 to 5 (Fig. 9.28a). We note that the temperature dependency of the self-diffusivity of water molecules follows an Arrhenius law for temperatures down to 230 K:

$$D = D_0 \exp\left(-\frac{E_a}{RT}\right) \tag{9.44}$$

where D_0 is a pre-exponential factor, where D_0 is a pre-exponential factor, E_a is a diffusion activation energy (i.e., the minimum amount of energy required for a water molecule to diffuse), R is the gas constant, and T is the temperature. Note that the self-diffusivities of bulk water also follow the Arrhenius equation. The fitting parameters are listed in Table 9.7. It is well known that bulk water exhibits a liquid–liquid transition around 230 K, called the fragile-to-strong transition, below which

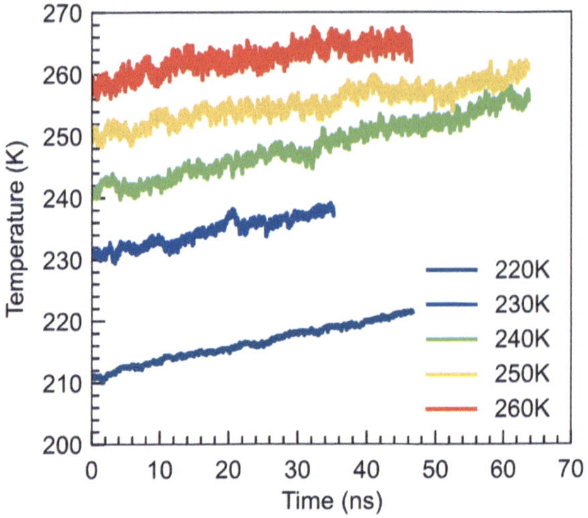

Fig. 9.27 Temperature evolution during NVE ensemble simulation

one observes a significant deceleration of water dynamics [51, 81]. The Arrhenius law effectively describes the dynamics above the transition temperature, and deviations are observed below [51, 82]. In particular, in Fig. 9.28a, the self-diffusivities of the bulk water at 220 K slightly deviate from the fitted Arrhenius line. For the TIP4P/Ice bulk water, the mode coupling theory, expressed as $D = D_0(T - T_{MCT})^{\alpha}$ (with D_0 a pre-exponential factor and T_{MCT} the fragile-to-strong temperature [83]), provides a more precise representation of the dynamics of supercooled bulk liquid below the fragile-to-strong transition temperature. The diffusion within the nanometric premelted film also exhibits a noticeable decrease at 220 K, suggesting a similar liquid–liquid transition takes place in the confined space between C-S-H and ice. Focusing now on the Ca^{2+} cations (Fig. 9.28b), Eq. 9.44 effectively describes their self-diffusivities in the premelted films. Notably, the Ca^{2+} self-diffusivity is about one order of magnitude smaller than that of water at the same temperatures, which is due to the constrained motion of the Ca^{2+} cations (Fig. 9.25). At 220 K, mobility is extremely limited, not really sufficient to properly estimate the diffusivity. Yet, a tentative estimate yields a sudden decrease in diffusivity, as for water at the same temperature.

Apart from the temperature, the thickness of the premelted film is another key factor affecting the diffusion of the molecules. The upper and lower films are not of the exact same thicknesses, and the dynamics of water and Ca^{2+} within those films differ. The self-diffusivities of the atom species in the lower film are often larger than those in the upper film (Table 9.7). The diffusion of confined fluids is known to depend not only on the temperature T but also on the degree of confinement (e.g., thickness H of the premelted film in the present case), while bulk diffusion is recovered for large thicknesses [78, 84]. In the present case, increasing the thickness of the premelted film is expected to promote the diffusion of the fluid. As presented

Table 9.7 Self-diffusivities of water molecules (H_2O) and Ca^{2+} cations in the upper and lower films

T (K)	Lower H_2O (10^{-12} m^2/s)	Upper H_2O (10^{-12} m^2/s)	Lower Ca^{2+} (10^{-13} m^2/s)	Upper Ca^{2+} (10^{-13} m^2/s)	Bulk water [53] (10^{-12} m^2/s)	Bulk water [51] (10^{-12} m^2/s)
220	0.058 ± 0.012	0.122 ± 0.045	0.016 ± 0.002	0.034 ± 0.001	–	1.16
230	4.20 ± 0.74	1.56 ± 0.14	0.86 ± 0.15	0.15 ± 0.05	12	9.76
240	13.16 ± 1.99	4.59 ± 0.89	2.77 ± 0.47	0.52 ± 0.05	30	49.3
250	20.74 ± 3.18	8.00 ± 0.63	5.03 ± 0.42	1.84 ± 0.11	93	136
260	47.34 ± 0.33	57.40 ± 5.68	40.85 ± 4.10	47.19 ± 8.04	215	277
Arrhenius equation description						
D_0 (m^2/s)	0.0025	7.2719	3.2071	4,284,553	4.9805	41.8556
E_a (kJ)	38.4753	56.0808	60.0647	90.9065	51.4283	55.2626

Note The self-diffusivities of TIP4P/Ice bulk water are collected from Espinosa et al. [51] and Weiss et al. [53]

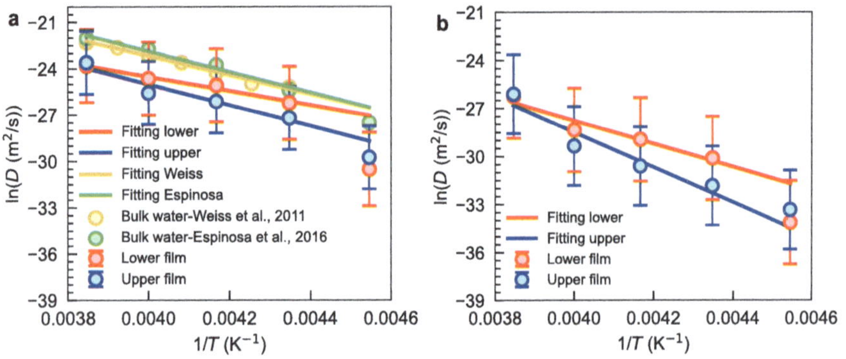

Fig. 9.28 Self-diffusivity of water molecules **a** and Ca^{2+} cations **b** in the premelted water films. For the water molecules, we also display the self-diffusivities of bulk TIP4P/Ice supercooled liquid water from Espinosa et al. [51] and Weiss et al. [53]. Note that the self-diffusivity of bulk water is for 3D diffusion, while the self-diffusivities of water molecules and Ca^{2+} cations in the lower and upper films are for 2D in-plane diffusion. Also, note that only the self-diffusivities above the transition temperature at 230 K are used to fit the Arrhenius equation

in Fig. 9.22, the lower films are slightly thicker than the upper films, which explains why the self-diffusivities are larger in the lower films than in the upper films.

9.4.3 Characteristic Time of Cryo-Suction Process

In the previous section, we quantified the thickness of the premelted films and the self-diffusivity of water in those films at various temperatures. In addition to those properties, Eq. 9.40 shows that we also need the molar volume and bulk modulus of ice and the bulk modulus of C-S-H to assess the characteristic time of cryo-suction. We provide estimates of these temperature-dependent parameters in Table 9.8 together with water self-diffusivity and film thickness. The bulk modulus of TIP4P/Ice ice Ih is collected from Moreira et al. [85], considered a constant at all temperatures. The bulk moduli of C-S-H are estimated using the approximation of the Voigt-Reuss-Hill average [86]: $K_s = \left(K^{Voigt} + K^{Reuss}\right)/2$, where K^{Voigt} and K^{Reuss} are the Voigt upper bound and Reuss lower bound of the bulk modulus, respectively, obtained from the full stiffness $\boldsymbol{C}$ and compliance $\boldsymbol{S}$ tensors of C-S-H. These tensors are computed by finite difference from constant temperature molecular dynamics simulations. The C-S-H model employed in this simulation maintains a constant Ca/Si ratio of 1.7 and is generated from an 11 Å tobermorite supercell with dimensions of $6\boldsymbol{a}_t \times 8\boldsymbol{b}_t \times 1\boldsymbol{c}_t$. Making use of those various parameters, we are now able to evaluate the characteristic time of cryo-suction through the premelted films with Eq. 9.40; the corresponding results are displayed in Fig. 9.29. Note that the diffusion coefficient used to estimate the characteristic time is a self-diffusivity instead of a collective diffusivity. Collective diffusivity is usually larger than self-diffusivity [87], but remains of similar order of magnitude especially in ultra-confined conditions [84, 88, 89].

Figure 9.29 shows that the characteristic time of the cryo-suction process increases when the temperature decreases, which is mostly due to the fact that the water dynamics decreases with a decreasing temperature. The radius of most cement pores falls within the range of 10 nm to approximately 1 μm [90]. According to our calculations, the cryo-suction process at the pore scale occurs within approximately 0.1 s in a pore with a size of 1 μm at a temperature of 260 K, while for pores with a size of 10 nm, the characteristic time is less than 10^{-7} s. As the temperature further decreases to 230 K, the characteristic time ranges from ~ 1 to ~ 10^{-6} s for these pores. Consequently, the calculated characteristic time for cement pores is sufficiently small for water transport within the nanometric premelted films at normal freezing temperatures (i.e., > 230 K) to occur, indicating that the cryo-suction process at the pore scale should be possible on a time scale realistic for freezing–thawing damage (minutes to hours). However, the characteristic time increases significantly when the temperature

Table 9.8 Fitting parameters of characteristic time of cryo-suction process

Temperature (K)	220	230	240	250	260
Average self-diffusivity (10^{-12} m^2/s)	0.076	2.313	6.338	10.374	35.882
Molar volume of ice (10^{-5} m^3/mol)	1.978	1.982	1.984	1.988	1.999
Average thickness (10^{-10} m)	9.271	10.978	11.698	11.569	14.913
C-S-H bulk modulus K_s (GPa)	46.4484	50.8036	45.2523	45.9561	36.8814
Ice bulk modulus K_I [85] (GPa)[a]	18.77	18.77	18.77	18.77	18.77
Bulk modulus K_{hom} (GPa)	13.3680	13.7061	13.2670	13.3269	12.4393

A we assume that the bulk modulus of ice is temperature-independent

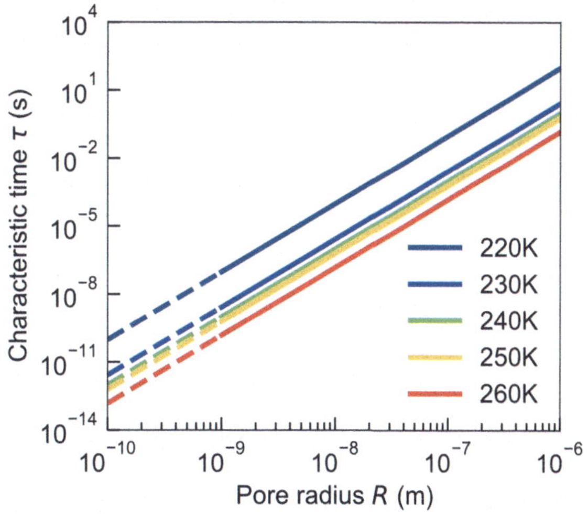

Fig. 9.29 Characteristic time of the cryo-suction process in function of the pore radius. Dashed lines represent extrapolations to pore radii smaller than the simulated premelted films

drops to 220 K, which is below the fragile-to-strong transition point of the premelted film. At this supercooled temperature, the cryo-suction characteristic time is three orders of magnitude longer than that at 260 K, suggesting a much slower process that may be too slow with respect to freezing–thawing time scale. As the temperature continues to decline below 220 K, the dynamical behavior of water molecules further diminishes, resulting in a concomitant reduction in their diffusivity. Consequently, the cryo-suction process is expected to slow down significantly.

Our estimation suggests that cryo-suction can indeed occur in the pores of cement concrete upon freezing. The unfrozen water will continue to migrate through the nanometric premelted films until thermodynamic equilibrium is reached [14]. As the temperature decreases, the unfrozen water in the fine pores migrates toward the pores already filled with ice, so that more ice can be formed, thus increasing the ice pressure until thermodynamic equilibrium with the supercooled liquid is reached. This migration causes the expansion of the cement matrix, which contributes to the cryo-swelling usually observed in concrete upon freezing–thawing conditions [2, 91]. Transport within the pore occurs until local thermodynamic equilibrium is reached, at which point Thomson's equation (Eq. 9.28) is valid. Our results suggest that the cryo-suction can happen within about a second at most at temperatures larger than 230 K, meaning that the hypothesis of local thermodynamic equilibrium and validity of Thomson's equation (Eq. 9.28) used in most poromechanical models of freezing of cement-based materials (see, e.g., Ref. [11, 12, 14, 92, 93], for which the time scales of interest are significantly larger than a second) is reasonable. In partially saturated materials, cryo-suction occurs through the water-filled part of the network, and our conclusion that, for freezing–thawing poromechanical modeling over time scales significantly larger than a second, assuming local thermodynamic equilibrium and the validity of Thomson's equation (Eq. 9.28) is reasonable still holds.

Note that the cryo-swelling observed experimentally mainly happens when the temperature is larger than 220 K [91]. As the temperature decreases further, concrete is observed to contract again. This phenomenon can be interpreted by our results. As analyzed previously, the characteristic time of cryo-suction at 220 K is markedly longer than that at normal freezing temperatures due to the extremely limited water dynamics, which makes it nearly impossible for the very limited amount of unfrozen gel pore water to migrate toward the large air voids. Thus, the results suggest that cryo-suction becomes much slower below the fragile-to-strong transition of the premelted films due to the significant deceleration of water dynamics.

9.4.4 Local Structure in the Premelted Film

It is well known that the properties and behaviors of molecules and ionic species around interfaces differ substantially from bulk ones. In the following, we examine the molecular structure of the confined system, especially the local structure of the premelted water films. Herein, we show a representative case at 260 K only since results at other temperatures exhibit similar trends. The atomic density profiles of the water molecule (Ow), hydroxyl groups (Oh), and interlayer Ca^{2+} cations (Cw) are shown in Fig. 9.30. As can be seen from the Ow profiles, the highly structured ice has periodic water configurations, showing marked differences from the poorly structured premelted films. Some free Ca^{2+} cations and OH^- ions are located within the liquid films, while most Ca^{2+} cations and OH^- ions are adsorbed on the C-S-H surface because of favorable electrostatic attraction with the C-S-H solids. This preferential adsorption explains why the self-diffusivities of Ca^{2+} cations are significantly lower than those of water molecules. We can reasonably infer that more and more Ca^{2+} cations accommodated in the films would absorb on the C-S-H surface as the Ca/Si ratio decreases since it favors charge compensation by the (fixed) silicate skeletons rather than by the (possibly mobile) hydroxyls [94, 95]. In the limit of low Ca/Si (~1), the Ca^{2+} cations in the premelted films could become immobile.

The orientations of water molecules are characterized by the distributions of dipolar angle, H–H angle, and O–H angle with respect to the z-axis, as depicted in Fig. 9.31. Herein, we display the orientation for the lower film at 260 K only and we divide the premelted film into three regions: a bottom layer (layer 1) adjacent to the C-S-H surface (located at 15 Å < $z \leq$ 20.5 Å), a top layer (layer 3) adjacent to the ice slab (located at 26 Å < $z \leq$ 31.5 Å), and a middle layer (layer 2) between layer 1 and layer 3 (located at 20.5 Å < $z \leq$ 26 Å) (see Fig. 9.31a). The adjacent ice layer (located at 31.5 Å < $z \leq$ 45 Å) is also analyzed for comparison. As expected, the

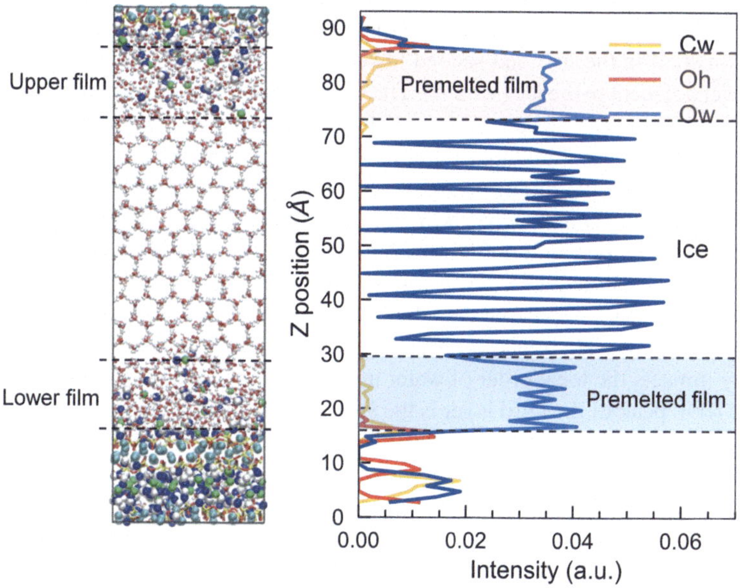

Fig. 9.30 Atomic density profiles of C-S-H/ice/liquid mixtures equilibrated at 260 K. Cw, Oh, and Ow represent interlayer Ca atoms, oxygen atoms of hydroxyl groups, and oxygen atoms of water, respectively. The atomic density profiles of the system at other temperatures exhibit similar distributions

water molecules in the ice layer exhibit significant preferential orientations due to the crystal Ih structure. For the water molecules in the premelted film, all the angle distributions exhibit a peak at $\cos\varphi = 0$, $\cos\tau = 0$, and $\cos\theta = 0$, meaning that the corresponding dipole vector angle, HH vector angle, and OH vector angle are around 90°. Therefore, water molecules accommodated in the liquid film sit preferentially parallelly to the C-S-H surface. As the water molecules are approaching the C-S-H surface, the intensity of the acute dipolar angle ($\cos\varphi > 0$) is systematically higher than that of the obtuse dipolar angle ($\cos\varphi < 0$): the neighboring water oxygen atoms are attracted to orient perpendicularly to the hydrophilic C-S-H surface. The orientation of water molecules in the top layer 3 shares some similarities with that of the water molecules in the adjacent ice layer. In contrast, the preferential orientation of water molecules is less pronounced in the middle layer 2 than in layer 1 and in layer 3.

It is also interesting to examine the relative configurations of the atom species in the premelted film, especially the first and second hydration shells that can impact the mobilities of atoms. The water molecules in the first hydration shell interact strongly with the caged atom through H-bonding, van der Waals forces, and Coulombic interactions. The cage size, coordination numbers of the first hydration shell, and the charge of the solute all play a role in the mobility of the caged atom [96]. Figure 9.32 illustrates the radial distribution functions (RDF) and coordination numbers of Ca^{2+} cations, OH^- groups, and water oxygens in the three layers of the lower premelted film. As shown in Fig. 9.32a, the coordination number of the first hydration shell of water molecules differs significantly among the three layers. The radius of the first hydration shell in the first and second lower film layers is ~ 4 Å, larger than in the third layer adjacent to the ice (~3.2 Å). The Ow-Ow RDF curve in the third premelted liquid layer shows the most similarity with the one in the ice layer and corresponds to two hydration shells around a water molecule, at about 6 Å. In contrast, the first and second premelted film layers are merely observed with a single hydration shell, which also differs from what has been observed on the freezing tobermorite surface [19]. In the premelted film on the tobermorite surface, water molecules uniformly feature two hydration shells, regardless of their location. Since the surface of the tobermorite is far more ordered than the C-S-H surface, this difference suggests that the disordered C-S-H surface with many impurities (Ca^{2+} and OH^- in our case) strongly impacts the local order of water molecules in the premelted film. Note also that the RDF peak of the third layer is the sharpest, implying that this premelted film layer is strongly amorphous. These differences between the various layers show that the water molecules close to the ice surface are more restrained than those further from the ice surface, which can lead to lower mobility. Nevertheless, water molecules in the third layer have one fewer coordination number in the first hydration shell than the ice layer. Consequently, those water molecules in the third layer show less restrain than the ice and diffuse more freely than molecules in the ice.

The RDFs and coordination number of Ca^{2+} cations and OH^- groups with the water molecules are displayed in Fig. 9.32b, c. The Ca^{2+} cations in the first, second, and third premelted film layers are all surrounded by about 7 water molecules in the second hydration shells, but differ in their first hydration shell with ~ 3, ~ 5, and

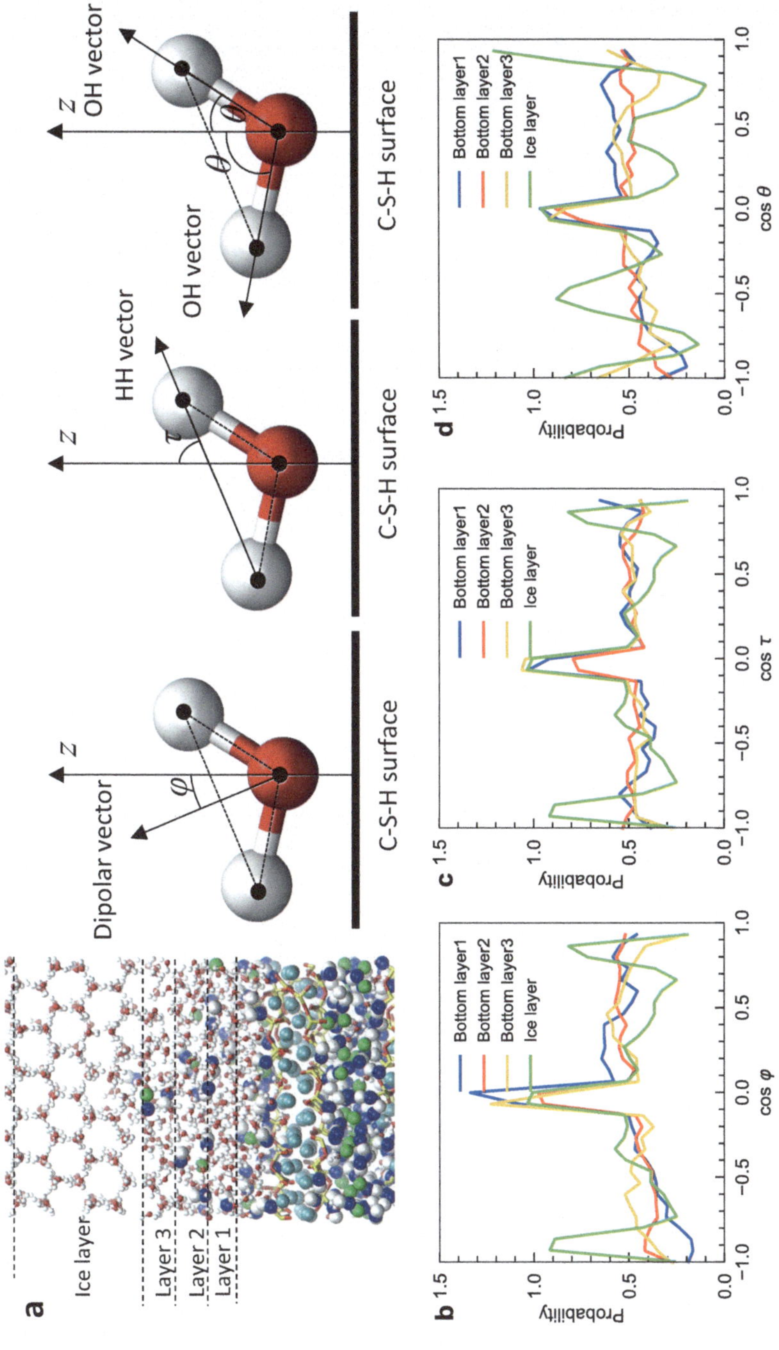

Fig. 9.31 (**b**) The dipolar angle, (**c**) the HH angle, and (**d**) the OH angle distribution of water molecules in the premelted films, as defined in (**a**)

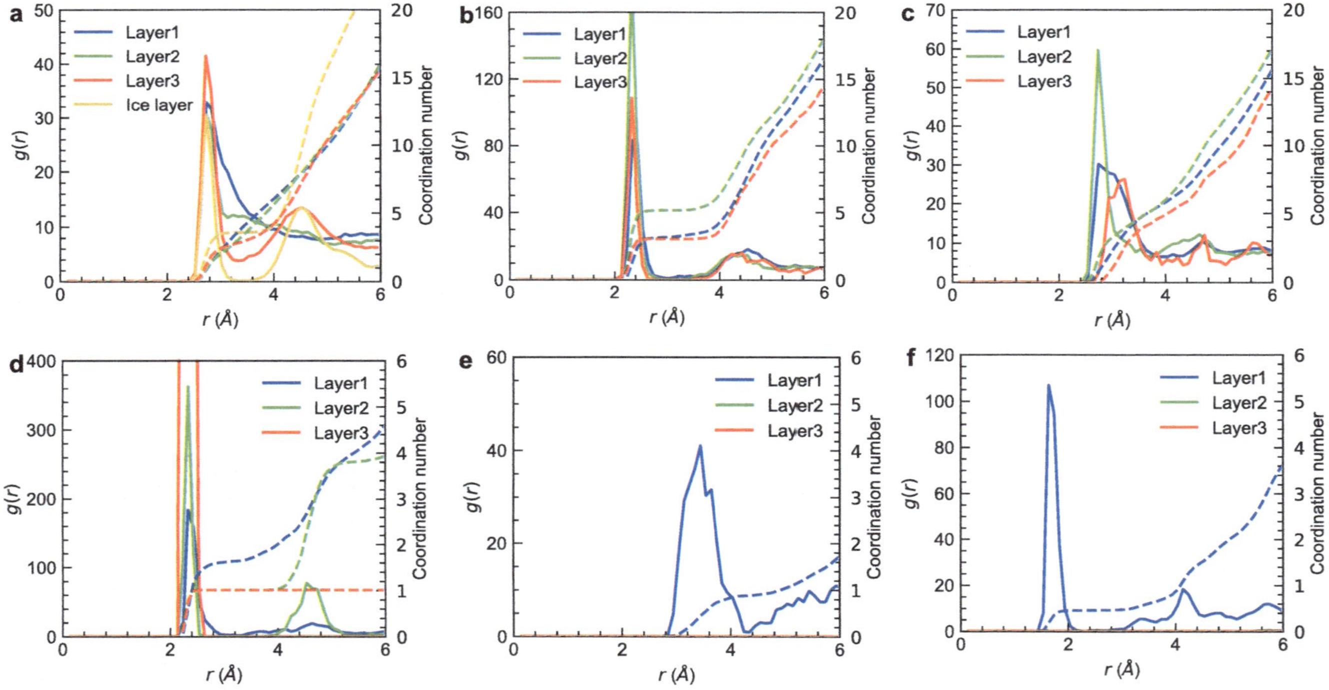

Fig. 9.32 Pair correlation functions of Ow-Ow (**a**), Cw-Ow (**b**), Oh-Ow (**c**), Cw-Oh (**d**), Cw-C-S-H surface (**e**), and Oh-C-S-H surface (**f**) in the layers of the lower premelted film at 260 K. Considering that the C-S-H surface is very disordered, we select Si atoms on the surface to locate the C-S-H surface. Solid lines are probability density functions, while dashed lines are cumulative distribution functions

~ 3 water molecules. Consequently, they form the complexes $[Ca(H_2O)_3]^{2+}\cdot 7H_2O$, $[Ca(H_2O)_5]^{2+}\cdot 7H_2O$, and $[Ca(H_2O)_3]^{2+}\cdot 7H_2O$, respectively. Apparently, the solvation number of Ca^{2+} cations in the premelted film is much lower than in the bulk one (which is around 6 to 8). This difference is mainly because the Ca^{2+} cations in the premelted film are in a massively high concentration and bonded with the C-S-H surface through electrostatic interactions. As shown in Fig. 9.32d, e, the Ca^{2+} cations in the first premelted film layer are more bonded with OH^- groups and C-S-H surface than in the second and third film layers. In this layer close to the C-S-H surface, about one more OH^- group competes against the water molecule to coordinate with the Ca^{2+} cation, compared to the other two layers. These differences in the local structure may explain the increasing restraints of Ca^{2+} cations as they are approaching the C-S-H surface, which explains why the Ca^{2+} cations far from the surface are more mobile than those near the surface. The first hydration shell of the Ca^{2+} cations contains two more water molecules in the second premelted film layer than in the first and third layers. The configuration of the OH^- groups with respect to the water molecules is similar to that of the Ca^{2+} cations. One noticeable difference is that the distance between OH^- groups and water molecules surrounding them in the first hydration shell becomes larger when the OH^- groups are close to the C-S-H surface or the ice surface (see in Fig. 9.32c). This difference is attributed to the attraction of OH^- groups by the C-S-H surface when they are in close proximity to it (as observed in in Fig. 9.32f). When OH^- groups approach the ice interface, there is a likelihood of them becoming trapped at the lattice site.

9.5 Concluding Remarks

In this chapter, we have shown the atomistic views for the two active processes regarding freezing liquids in cement pores, i.e., ice crystallization and liquid transport (cryo-suction). In a pioneering effort, we calculated the characteristic time associated with cryo-suction. Our findings reveal that cryo-suction is an active phenomenon at freezing temperatures typically encountered for cement-based materials. The cryo-suction can happen in about a second at most (for temperatures larger than 230 K), meaning that the hypothesis of local thermodynamic equilibrium used in most poromechanical models of freezing of cement-based materials (see, e.g., Ref. [11, 12, 14, 92, 93]) is reasonable. Conclusions are summarized as follows:

(1) Most free ions in the confined C-S-H slit pore system are rejected into the liquid during ice crystallization. A few hydroxyl groups are trapped in the lattice of ice Ih, replacing some water molecules. Such ion rejection and trapping phenomena are analogous to those observed in bulk water systems in the presence of ions.
(2) The growth rate of ice is significantly reduced by confinement, which can be attributed to a lower mobility of water and surface impurities but, also, to a decrease of the chemical potential difference between ice and liquid since confinement reduces the melting temperature. Unlike bulk ice crystallization,

which has a maximum growth rate at ~ 250 K, the ice growth rate of the confined system increases monotonously up to a temperature strictly larger than 250 K. The maximum growth rate is thus expected to be shifted to a temperature closer to the melting point, which could result from a higher sensitivity of water mobility to temperature.

(3) We revisited the formulation of Thomson's law in the framework of statistical physics, which is more adapted to molecular simulation. Thomson's law predicts accurately the melting line of bulk ice Ih. The confinement systematically lowers the melting temperatures of ice Ih by 4–6 K with respect to the bulk ice. It is the interface energy difference between ice-liquid and liquid–solid that contributes to the melting point depression. The melting point depression is also attributed to the ions in the thin films.

(4) Premelted water nanofilms are observed between the ice and C-S-H solid walls. Confrontations between our estimates and existing models and results in the literature suggests that the presence of ion impurities on the C-S-H surface is essential to explain the existence of the premelted film at the lowest temperatures. Interestingly, the Fagerlund model [73] for the premelted film, which was initially proposed using data from surface adsorption of water vapor on cement, captures quite well our estimations of the evolution of the thickness of the premelted film with temperature.

(5) The premelted films exhibit a fragile-to-strong transition around 230 K, which leads to a significant deceleration of water dynamics. The dynamics of water molecules and ions in the nanometric unfrozen films exhibit marked dependence on temperature and thickness. The self-diffusivity of water molecules in the premelted films is about one order of magnitude lower than the bulk water at the same temperature. Water molecules and Ca^{2+} cations diffuse faster in the thicker premelted film than in the thinner premelted film.

(6) The transport of water through the premelted films is sufficiently fast for cryo-suction to occur at normal freezing temperatures (i.e., at temperatures larger than 230 K). The characteristic time for the cryo-suction is derived from a thermodynamic analysis, expressed in Eq. 9.40. Upon our calculation, above 230 K, the characteristic time of cryo-suction ranges from a few ~ 10^{-7} to ~ 1 s in most cement pores. Consequently, in poromechanical modeling of freezing–thawing of cement-based materials over time scales significantly larger than a second, assuming local thermodynamic equilibrium and the validity of Thomson's equation is reasonable. However, the significantly slowing down of water dynamics at temperatures below the fragile-to-strong transition yields a longer characteristic time for the cryo-suction process, making the cryo-suction significantly slower.

(7) The local structure of water molecules and ions in the premelted films depends significantly on their distances to the C-S-H and ice slab surfaces. Water molecules in the premelted films exhibit a preferential orientation parallel to the C-S-H and ice surfaces. The water film near the ice slab is more amorphous than in other parts of the film. A majority of the Ca^{2+} cations and OH^- ions are adsorbed on the C-S-H surface under the action of the electrostatic interactions.

Fewer water molecules are coordinated with them in the first hydration shell when they are close to the C-S-H or ice surfaces than when they are far from those surfaces.

Under actual conditions, the pore solution in cement-based materials is notably more intricate than the scenarios explored in this study, encompassing a range of ionic species such as Na^+, K^+, Cl^-, SO_4^{2-}. The kinetics of ice growth and the dynamics of water molecules within the premelted films may exhibit substantial deviations in the presence of these ionic species, potentially influencing the cryo-suction behavior. In future research, it is advisable to investigate the influence of these ionic species. Furthermore, the crystallization of ice within the confined spaces of C-S-H structures, maintaining a constant volume, inevitably exerts pressure on the pore wall. This pressure phenomenon is directly related to the freezing–thawing damage observed in cement-based materials. Therefore, it is essential to explore the effects of pressure in more depth in forthcoming studies. Another limitation of this study is that we estimated the characteristic time of the cryo-suction process using a self-diffusivity rather than a collective diffusivity. Although the latter would be more appropriate since the cryo-suction in the premelted film is a collective diffusion behavior, we do not expect that the estimated characteristic times would change drastically.

References

1. Z. Jiang, B. He, X. Zhu, Q. Ren, Y. Zhang, State-of-the-art review on properties evolution and deterioration mechanism of concrete at cryogenic temperature. Constr. Build. Mater. **257**, 119456 (2020)
2. X. Zheng, Y. Wang, S. Zhang, F. Xu, X. Zhu, X. Jiang, L. Zhou, Y. Shen, Q. Chen, Z. Yan, W. Zhao, H. Zhu, Y. Zhang, Research progress of the thermophysical and mechanical properties of concrete subjected to freeze-thaw cycles. Constr. Build. Mater. **330**, 127254 (2022)
3. Q. Zeng, L. Li, X. Pang, Q. Gui, K. Li, Freeze–thaw behavior of air entrained cement paste saturated with 10 wt% NaCl solution. Cold Reg. Sci. Technol. 102, 21–31 (2014)
4. T.C. Powers, A working hypothesis for further studies of frost resistance of concrete. J. Am. Concr. Inst. **16**(4), 245–272 (1945)
5. T.C. Powers, T. Willis, The air requirement of frost-resistant concrete, Highway Research Board Proceedings, Highway Research Board, pp. 184–211 (1949)
6. J.J. Beaudoin, C. MacInnis, The mechanism of frost damage in hardened cement paste. Cem. Concr. Res. **4**(2), 139–147 (1974)
7. G.W. Scherer, Crystallization in pores. Cem. Concr. Res. **29**(8), 1347–1358 (1999)
8. G.W. Scherer, Freezing gels. J. Non-Cryst. Solids **155**(1), 1–25 (1993)
9. M.J. Setzer, Mechanical stability criterion, triple-phase condition, and pressure differences of matter condensed in a porous matrix. J. Colloid Interface Sci. **235**(1), 170–182 (2001)
10. M.J. Setzer, Micro-ice-lens formation in porous solid. J. Colloid Interface Sci. **243**(1), 193–201 (2001)
11. O. Coussy, Poromechanics of freezing materials. J. Mech. Phys. Solids **53**(8), 1689–1718 (2005)
12. O. Coussy, P.J.M. Monteiro, Poroelastic model for concrete exposed to freezing temperatures. Cem. Concr. Res. **38**(1), 40–48 (2008)
13. T. Zhou, M. Mirzadeh, R.J.M. Pellenq, M.Z. Bazant, Freezing point depression and freeze-thaw damage by nanofluidic salt trapping. Phys. Rev. Fluids **5**(12), 124201 (2020)

14. N.P. Mayercsik, M. Vandamme, K.E. Kurtis, Assessing the efficiency of entrained air voids for freeze-thaw durability through modeling. Cem. Concr. Res. **88**, 43–59 (2016)
15. O. Coussy, Mechanics and physics of porous solids, John Wiley & Sons (2011)
16. Z. Sun, G.W. Scherer, Effect of air voids on salt scaling and internal freezing. Cem. Concr. Res. **40**(2), 260–270 (2010)
17. Z. Sun, G.W. Scherer, Pore size and shape in mortar by thermoporometry. Cem. Concr. Res. **40**(5), 740–751 (2010)
18. M. Brun, A. Lallemand, J.-F. Quinson, C. Eyraud, A new method for the simultaneous determination of the size and shape of pores: the thermoporometry. Thermochim. Acta **21**(1), 59–88 (1977)
19. T. Liang, Y. Lai, D. Hou, Q. Yang, Y. Yang, R. Bai, J. Zhang, J. Jiang, Freezing mechanism of NaCl solution ultra-confined on surface of calcium-silicate-hydrate: A molecular dynamics study. Cem. Concr. Res. **154**, 106722 (2022)
20. X. Zhu, M. Vandamme, L. Brochard, Z. Zhang, Q. Ren, C. Li, B. He, H. Zhang, Y. Zhang, Q. Chen, Z. Jiang, Nature of aluminates in C-A-S-H: A cryogenic stability insight, an extension of DNA-code rule, and a general structural-chemical formula. Cem. Concr. Res. **167**, 107131 (2023)
21. X. Zhu, C. Qian, B. He, Q. Chen, Z. Jiang, Experimental study on the stability of C-S-H nanostructures with varying bulk CaO/SiO_2 ratios under cryogenic attack. Cem. Concr. Res. **135**, 106114 (2020)
22. Y. Chen, Y. Zheng, Y. Zhou, W. Zhang, W. Li, W. She, J. Liu, C. Miao, Multi-layered cement-hydrogel composite with high toughness, low thermal conductivity, and self-healing capability. Nat. Commun. **14**(1), 3438 (2023)
23. R. Shahsavari, R.J. Pellenq, F.J. Ulm, Empirical force fields for complex hydrated calcio-silicate layered materials. Phys. Chem. Chem. Phys. **13**(3), 1002–1011 (2011)
24. J.L. Abascal, E. Sanz, R. Garcia Fernandez, C. Vega, A potential model for the study of ices and amorphous water: TIP4P/Ice. J. Chem. Phys. 122(23), 234511 (2005)
25. R.K. Mishra, A.K. Mohamed, D. Geissbuhler, H. Manzano, T. Jamil, R. Shahsavari, A.G. Kalinichev, S. Galmarini, L. Tao, H. Heinz, R. Pellenq, A.C.T. van Duin, S.C. Parker, R.J. Flatt, P. Bowen, Cemff: A force field database for cementitious materials including validations, applications and opportunities. Cem. Concr. Res. **102**, 68–89 (2017)
26. G. Bai, D. Gao, Z. Liu, X. Zhou, J. Wang, Probing the critical nucleus size for ice formation with graphene oxide nanosheets. Nature **576**(7787), 437–441 (2019)
27. V. Bianco, P.M. de Hijes, C.P. Lamas, E. Sanz, C. Vega, Anomalous behavior in the nucleation of ice at negative pressures. Phys. Rev. Lett. **126**(1), 015704 (2021)
28. K. Röttger, A. Endriss, J. Ihringer, S. Doyle, W.F. Kuhs, Lattice constants and thermal expansion of H_2O and D_2O ice Ih between 10 and 265 K, Acta Crystallogr. Sect. B: Struct. Sci 50(6), 644–648 (1994)
29. S.A. Hamid, The crystal structure of the 11 Ä natural tobermorite $Ca_{2.25}[Si_3O_{7.5}(OH)_{1.5}]\cdot 1H_2O$, Z. Kristallogr. Cryst. Mater. 154, 189–198 (1981)
30. A. Morshedifard, S. Masoumi, M.J. Abdolhosseini Qomi, Nanoscale origins of creep in calcium silicate hydrates. Nat. Commun. 9(1), 1785 (2018)
31. M.J. Abdolhosseini Qomi, K.J. Krakowiak, M. Bauchy, K.L. Stewart, R. Shahsavari, D. Jagannathan, D.B. Brommer, A. Baronnet, M.J. Buehler, S. Yip, F.J. Ulm, K.J. Van Vliet, R.J. Pellenq, Combinatorial molecular optimization of cement hydrates, Nat. Commun. 5, 4960 (2014)
32. X. Zhu, L. Brochard, M. Vandamme, Z. Jiang, Scaling of nanoscale elastic and tensile failure properties of cementitious calcium-silicate-hydrate materials at cryogenic temperatures: A molecular simulation study. Cem. Concr. Res. **172**, 107242 (2023)
33. A.P. Thompson, H.M. Aktulga, R. Berger, D.S. Bolintineanu, W.M. Brown, P.S. Crozier, P.J. in 't Veld, A. Kohlmeyer, S.G. Moore, T.D. Nguyen, R. Shan, M.J. Stevens, J. Tranchida, C. Trott, S.J. Plimpton, LAMMPS–A flexible simulation tool for particle-based materials modeling at the atomic, meso, and continuum scales. Comput. Phys. Commun. 271, 108171 (2022)
34. F. Legoll, M. Luskin, R. Moeckel, Non-ergodicity of the Nosé-Hoover thermostatted harmonic oscillator. Arch. Ration. Mech. Anal. **184**(3), 449–463 (2007)

35. J.-P. Ryckaert, G. Ciccotti, H.J.C. Berendsen, Numerical integration of the Cartesian equations of motion of a system with constraints: Molecular dynamics of n-alkanes. J. Comput. Phys. **23**(3), 327–341 (1977)
36. W. Humphrey, A. Dalke, K. Schulten, VMD: Visual molecular dynamics. J. Mol. Graphics **14**(1), 33–38 (1996)
37. A. Stukowski, Visualization and analysis of atomistic simulation data with OVITO–the open visualization tool. Modell. Simul. Mater. Sci. Eng. **18**(1), 015012 (2010)
38. D. Rozmanov, P.G. Kusalik, Temperature dependence of crystal growth of hexagonal ice (Ih). Phys. Chem. Chem. Phys. **13**(34), 15501–15511 (2011)
39. E.B. Moore, E. de la Llave, K. Welke, D.A. Scherlis, V. Molinero, Freezing, melting and structure of ice in a hydrophilic nanopore. Phys. Chem. Chem. Phys. **12**(16), 4124–4134 (2010)
40. A.H. Nguyen, V. Molinero, Identification of cathrate hydrates, hexagonal ice, cubic ice, and liquid water in simulations: the CHILL plus algorithm. J. Phys. Chem. B **119**(29), 9369–9376 (2015)
41. M.S.G. Razul, P.G. Kusalik, Crystal growth investigations of ice/water interfaces from molecular dynamics simulations: Profile functions and average properties. J. Chem. Phys. 134(1) (2011)
42. L. Vrbka, P. Jungwirth, Brine rejection from freezing salt solutions: A molecular dynamics study. Phys. Rev. Lett. 95(14) (2005)
43. U. Shafique, J. Anwar, M. Ali Munawar, W.-u. Zaman, R. Rehman, A. Dar, M. Salman, M. Saleem, N. Shahid, M. Akram, A. Naseer, N. Jamil, Chemistry of ice: Migration of ions and gases by directional freezing of water, Arabian J. Chem. 9, S47-S53 (2016)
44. I. Tsironi, D. Schlesinger, A. Spah, L. Eriksson, M. Segad, F. Perakis, Brine rejection and hydrate formation upon freezing of NaCl aqueous solutions. Phys. Chem. Chem. Phys. **22**(14), 7625–7632 (2020)
45. G.D. Soria, J.R. Espinosa, J. Ramirez, C. Valeriani, C. Vega, E. Sanz, A simulation study of homogeneous ice nucleation in supercooled salty water. J. Chem. Phys. 148(22) (2018)
46. M.M. Conde, M. Rovere, P. Gallo, Spontaneous NaCl-doped ice at seawater conditions: focus on the mechanisms of ion inclusion. Phys. Chem. Chem. Phys. **19**(14), 9566–9574 (2017)
47. Y. Yashima, Y. Okada, M. Harada, T. Okada, Structures of ions accommodated in salty ice Ih crystals. Phys. Chem. Chem. Phys. **23**(33), 17945–17952 (2021)
48. E.J. Workman, S.E. Reynolds, Electrical phenomena occurring during the freezing of dilute aqueous solutions and their possible relationship to thunderstorm electricity. Phys. Rev. **78**(3), 254–259 (1950)
49. G.W. Scherer, Stress from crystallization of salt. Cem. Concr. Res. **34**(9), 1613–1624 (2004)
50. O. Coussy, T. Fen-Chong, Crystallization, pore relaxation and micro-cryosuction in cohesive porous materials, C.R. Mec. 333(6), 507–512 (2005)
51. J.R. Espinosa, C. Navarro, E. Sanz, C. Valeriani, C. Vega, On the time required to freeze water. J. Chem. Phys. 145(21) (2016)
52. T. Buttersack, S. Bauerecker, Critical radius of supercooled water droplets: On the transition toward dendritic freezing. J. Phys. Chem. B **120**(3), 504–512 (2016)
53. V.C. Weiss, M. Rullich, C. Koehler, T. Frauenheim, Kinetic aspects of the thermostatted growth of ice from supercooled water in simulations. J. Chem. Phys. 135(3) (2011)
54. Y. Xu, N.G. Petrik, R.S. Smith, B.D. Kay, G.A. Kimmel, Growth rate of crystalline ice and the diffusivity of supercooled water from 126 to 262 K. Proc. Natl. Acad. Sci. U.S.A. **113**(52), 14921 (2016)
55. H.R. Pruppacher, Interpretation of experimentally determined growth rates of ice crystals in supercooled water. J. Chem. Phys. **47**(5), 1807–1813 (1967)
56. P.M. de Hijes, J.R. Espinosa, C. Vega, E. Sanz, Ice growth rate: Temperature dependence and effect of heat dissipation, J. Chem. Phys. 151(4) (2019)
57. H.W. Wilson, XX. On the velocity of solidification and viscosity of super-cooled liquids, The London, Edinburgh, and Dublin Philosophical Magazine and Journal of Science 50(303), 238–250 (1900)

58. J. Frenkel, Note on a relation between the speed of crystallization and viscosity. J Phisik. Zeit. Sowjetunion **1**, 498–510 (1932)
59. C. Guo, H. Qin, Y. Zhu, M. Chen, Y. Lü, Rotational dynamics of water associated with interfacial dielectric oscillation and its role in crystal growth of ice. J. Chem. Phys. **157**(21), 214703 (2022)
60. P. Montero de Hijes, S. Romano, A. Gorfer, C. Dellago, The kinetics of the ice–water interface from ab initio machine learning simulations. J. Chem. Phys. **158**(20), 204706 (2023)
61. S. Bauerecker, P. Ulbig, V. Buch, L. Vrbka, P. Jungwirth, Monitoring ice nucleation in pure and salty water via high-speed imaging and computer simulations. J. Phys. Chem. C **112**(20), 7631–7636 (2008)
62. M.S. Wahl, A. Aasen, D.R. Hjelme, Ø. Wilhelmsen, Ice formation and growth in supercooled water–alcohol mixtures: Theory and experiments with dual fiber sensors. Fluid Phase Equilib. **522**, 112741 (2020)
63. T. Lin, X. Quan, P. Cheng, J. Li, G. Chen, Interaction between nanoparticles and advancing ice-water interfaces: A molecular dynamics simulation. Int. J. Heat Mass Transfer **163**, 120412 (2020)
64. M.J. Powell-Palm, B. Rubinsky, W. Sun, Freezing water at constant volume and under confinement. Commun. Phys. **3**(1), 39 (2020)
65. E.G. Solveyra, E. de la Llave, D.A. Scherlis, V. Molinero, Melting and crystallization of ice in partially filled nanopores. J. Phys. Chem. B **115**(48), 14196–14204 (2011)
66. A. Consiglio, G. Ukpai, B. Rubinsky, M.J. Powell-Palm, Suppression of cavitation-induced nucleation in systems under isochoric confinement. Phys. Rev. Res. **2**(2), 023350 (2020)
67. K. Morishige, K. Kawano, Freezing and melting of water in a single cylindrical pore: The pore-size dependence of freezing and melting behavior. J. Chem. Phys. **110**(10), 4867–4872 (1999)
68. C. Alba-Simionesco, B. Coasne, G. Dosseh, G. Dudziak, K.E. Gubbins, R. Radhakrishnan, M. Sliwinska-Bartkowiak, Effects of confinement on freezing and melting. J. Condens. Matter Phys. **18**(6), R15–R68 (2006)
69. H. Hansen-Goos, J.S. Wettlaufer, Theory of ice premelting in porous media. Phys. Rev. E Stat. Nonlin. Soft Matter Phys. **81**(3 Pt 1), 031604 (2010)
70. T. Kaneko, Elevation/depression mechanism of freezing points of liquid confined in slit nanopores. Mol. Simul. **43**(13–16), 1364–1369 (2017)
71. C.A. Tulk, J.J. Molaison, A.R. Makhluf, C.E. Manning, D.D. Klug, Absence of amorphous forms when ice is compressed at low temperature. Nature **569**(7757), 542–545 (2019)
72. S. Engemann, H. Reichert, H. Dosch, J. Bilgram, V. Honkimäki, A. Snigirev, Interfacial melting of ice in contact with SiO_2. Phys. Rev. Lett. **92**(20), 205701 (2004)
73. G. Fagerlund, Determination of pore-size distribution from freezing-point depression. Matériaux et Const. **6**(3), 215–225 (1973)
74. D.R. Lide, CRC handbook of chemistry and physics 86th Edition 2005-2006, CRC press, Taylor & Francis, Boca Raton (2005)
75. S. Ghabezloo, Micromechanics analysis of thermal expansion and thermal pressurization of a hardened cement paste. Cem. Concr. Res. **41**(5), 520–532 (2011)
76. M. Maruyama, M. Bienfait, J.G. Dash, G. Coddens, Interfacial melting of ice in graphite and talc powders. J. Cryst. Growth **118**(1), 33–40 (1992)
77. M.J.A. Qomi, M. Bauchy, F.-J. Ulm, R.J.M. Pellenq, Anomalous composition-dependent dynamics of nanoconfined water in the interlayer of disordered calcium-silicates. J. Chem. Phys. **140**(5), 054515 (2014)
78. T. Honorio, H. Carasek, O. Cascudo, Water self-diffusion in C-S-H: Effect of confinement and temperature studied by molecular dynamics. Cem. Concr. Res. **155**, 106775 (2022)
79. Y. Zhou, H. Zheng, W. Li, T. Ma, C. Miao, A deep learning potential applied in tobermorite phases and extended to calcium silicate hydrates. Cem. Concr. Res. **152**, 106685 (2022)
80. B.N. Galimzyanov, V.I. Ladyanov, A.V. Mokshina, Remarkable nuances of crystallization: From ordinary crystal nucleation to rival mechanisms of crystallite coalescence. J. Cryst. Growth 526 (2019)

81. R. Shi, J. Russo, H. Tanaka, Origin of the emergent fragile-to-strong transition in supercooled water. Proc Natl Acad Sci U S A **115**(38), 9444–9449 (2018)
82. P. Gallo, D. Corradini, M. Rovere, Fragile to strong crossover at the Widom line in supercooled aqueous solutions of NaCl. J. Chem. Phys. 139(20) (2013)
83. M. De Marzio, G. Camisasca, M. Rovere, P. Gallo, Mode coupling theory and fragile to strong transition in supercooled TIP4P/2005 water. J. Chem. Phys. 144(7) (2016)
84. P. Simonnin, B.t. Noetinger, C. Nieto-Draghi, V. Marry, B. Rotenberg, Diffusion under confinement: Hydrodynamic finite-size effects in simulation. J. Chem. Theory Comput. 13(6), 2881–2889 (2017)
85. P.A.F.P. Moreira, R.G.D. Veiga, M. de Koning, Elastic constants of ice I-h as described by semi-empirical water models. J. Chem. Phys. 150(4) (2019)
86. R. Hill, The elastic behaviour of a crystalline aggregate. Proc. Phys. Soc. London, Sect. A 65(5), 349 (1952)
87. C. Chmelik, H. Bux, J. Caro, L. Heinke, F. Hibbe, T. Titze, J. Kärger, Mass transfer in a nanoscale material enhanced by an opposing flux. Phys. Rev. Lett. **104**(8), 085902 (2010)
88. C. O'Sullivan, C. Arson, B. Coasne, A perspective on Darcy's law across the scales: From physical foundations to particulate mechanics. J. Eng. Mech. **148**(11), 04022064 (2022)
89. J.-L. Barrat, J.-P. Hansen, *Basic concepts for simple and complex liquids* (Cambridge University Press, Cambridge, 2003)
90. Q. Zeng, S. Xu, Discussion of Numerical simulation of moisture transport in concrete based on a pore size distribution model. Cem. Concr. Res. **73**, 63–66 (2015)
91. J. Zhengwu, Z. Cong, D. Zilong, Z. Xinping, L. Wenting, Thermal strain of cement-based materials under cryogenic temperatures and its freeze-thaw cycles using fibre Bragg grating sensor. Cryogenics **100**, 1–10 (2019)
92. Q. Zeng, T. Fen-Chong, P. Dangla, K. Li, A study of freezing behavior of cementitious materials by poromechanical approach. Int. J. Solids Struct. **48**(22–23), 3267–3273 (2011)
93. D. Eriksson, T. Gasch, R. Malm, A. Ansell, Freezing of partially saturated air-entrained concrete: A multiphase description of the hygro-thermo-mechanical behaviour. Int. J. Solids Struct. **152–153**, 294–304 (2018)
94. E. Duque-Redondo, K. Yamada, H. Manzano, Cs retention and diffusion in C-S-H at different Ca/Si ratio. Cem. Concr. Res. **140**, 106294 (2021)
95. E. Duque-Redondo, P.A. Bonnaud, H. Manzano, A comprehensive review of C-S-H empirical and computational models, their applications, and practical aspects. Cem. Concr. Res. **156**, 106784 (2022)
96. S. Koneshan, J.C. Rasaiah, R.M. Lynden-Bell, S.H. Lee, Solvent structure, dynamics, and ion mobility in aqueous solutions at 25 °C. J. Phys. Chem. B **102**(21), 4193–4204 (1998)

The manufacturer's authorised representative in the EU is Springer Nature Customer Service Centre GmbH, Europaplatz 3, 69115 Heidelberg, Germany. If you have any concerns regarding our products, please contact ProductSafety@springernature.com

Printed and bound by CPI Group (UK) Ltd, Croydon, CR0 4YY
01/10/2026
02225941-0008